T0235159

EXACT SPACE-TIMES
IN EINSTEIN'S
GENERAL RELATIVITY

Einstein's theory of general relativity is a theory of gravity and, as in the earlier Newtonian theory, much can be learnt about the character of gravitation and its effects by investigating particular idealised examples. This book describes the basic solutions of Einstein's equations with a particular emphasis on what they mean, both geometrically and physically.

New concepts, such as big bang and big crunch-types of singularities, different kinds of horizons and gravitational waves are described in the context of the particular space-times in which they naturally arise. These notions are initially introduced using the most simple and symmetric cases. Various important coordinate forms of each solution are presented, thus enabling the global structure of the corresponding space-time and its other properties to be analysed.

The book is an invaluable resource for both graduate students and academic researchers working in gravitational physics.

JERRY GRIFFITHS is an Emeritus Professor at Loughborough University, where he worked since obtaining his Ph.D. from the University of Wales on neutrino fields in general relativity. He has worked on the generation and interpretation of exact solutions of Einstein's equations and published over 80 papers on these topics.

JIŘÍ PODOLSKÝ is a Professor at Charles University in Prague, where he also obtained his Ph.D. and Habilitation in theoretical physics. His research concentrates mainly on the investigation of exact radiative and black hole space-times in general relativity. He has published over 60 papers in these fields.

CAMBRIDGE MONOGRAPHS ON MATHEMATICAL PHYSICS

General Editors: P. V. Landshoff, D. R. Nelson, S Weinberg

Exact Space-Times
in Einstein's
General Relativity

JERRY B. GRIFFITHS

Loughborough University

JIŘÍ PODOLSKÝ

Charles University, Prague

CAMBRIDGE
UNIVERSITY PRESS

CAMBRIDGE UNIVERSITY PRESS
Cambridge, New York, Melbourne, Madrid, Cape Town,
Singapore, São Paulo, Delhi, Mexico City

Cambridge University Press
The Edinburgh Building, Cambridge CB2 8RU, UK

Published in the United States of America by Cambridge University Press, New York

www.cambridge.org
Information on this title: www.cambridge.org/9781107406186

First published 2009
Reprinted with corrections and updates 2012
First paperback edition 2012

A catalogue record for this publication is available from the British Library

ISBN 978-0-521-88927-8 Hardback
ISBN 978-1-107-40618-6 Paperback

Contents

Preface

In the now extensive literature on general relativity and its related subjects, references abound to "known solutions" or even "well-known solutions" of Einstein's field equations. Yet, apart from a few familiar space-times, such as those of Schwarzschild, Kerr and Friedmann, often little more is widely known about such solutions than that they exist and can be expressed in terms of a particular line element using some standard coordinate system.

With the most welcome publication of the second edition of the "exact solutions" book of Stephani *et al.* (2003), an amazing number of solutions, and even families of solutions, have been identified and classified. This is of enormous benefit. However, when it comes to understanding the physical meaning of these solutions, the situation is much less satisfactory – even for some of the most fundamental ones.

Of course, there are now many excellent textbooks on general relativity which present the subject in a coherent way to students with a variety of primary interests. These always describe the basic properties of the Schwarzschild solution and usually a few others as well. Yet, beyond these, when trying to find out what is known about any particular exact solution, there is normally still no alternative to searching through original papers dating back many years and published in journals that are often not available locally or freely available on the internet. Proceeding in this way, it is possible to miss significant contributions or to repeat errors or unhelpful emphases. And, since most original papers deal only with a single topic, it takes a lot of time to develop an appreciation of what is actually already "known". This can frustrate those of us who have been working in the field for many years, and can be daunting for students new to the subject. Moreover, it usually turns out that there are still gaps in the published understanding of even some of the best known solutions.

This situation is not surprising. It appears that it is much easier to find

a new solution of Einstein's equations than it is to understand it. Moreover, the discovery of a solution is not necessarily achieved using the most
appropriate set of coordinates and parameters. For example, the solution
Schwarzschild found in 1915 for a vacuum spherically symmetric gravitational field was initially expressed in terms of coordinates that now look
very strange. And, although he also presented the now familiar coordinates, it still took over 50 years before the correct character of the horizon
was generally appreciated. Similarly, many other long-known solutions have
features and parameters that are still not fully understood. Some have different kinds of singularities, and most are not asymptotically flat and have
unknown sources. Indeed, most solutions of Einstein's equations may have
no satisfactory interpretation as global models of realistic physical situations
at all. On the other hand, locally, many may reasonably represent particular regions of realistic space-times. And some may have applications if
generalised in newer higher-dimensional theories of semi-classical quantum
gravity. It is therefore important that the geometrical and physical properties of solutions of Einstein's equations, even in four dimensions, should be
as widely known as possible.

This book is presented as a contribution towards this end. It seeks to
interpret some of the most basic solutions of standard general relativity.
However, it is not about the applications of the theory. Related topics in
astrophysics and cosmology are hardly mentioned. Our approach is simply
to try to interpret exact solutions analytically and geometrically.

Within the space available, we have decided to concentrate on just a
few solutions that have been known for many decades. The literature on
the interpretation of these particular cases is reviewed. However, there are
remarkably few publications on some of the solutions considered, so our
observations in these cases have been partly based on our own calculations.
In fact, many chapters contain some material that, although not necessarily
new, is not readily available in existing literature. This approach, moreover,
has enabled us to describe a range of solutions in a unified way.

In selecting the particular solutions to be described, we have concentrated
on basic vacuum solutions. Whenever appropriate, we include a cosmological
constant or an electromagnetic field, or sometimes even a pure radiation
field. However, only the most important solutions with a perfect fluid source
are mentioned. We initially describe conformally flat and highly symmetric
space-times. Then we proceed from algebraically special solutions, mostly
of the simplest types D and N, to more general types.

We are only too well aware of the fact that our survey is severely limited
in extent. A glance at the "exact solutions" book of Stephani *et al.* (2003)

will show that we have only begun to scratch the surface of the number of exact solutions that require an interpretation. Moreover, even for the limited number of solutions that we describe here, there are relevant references that we have not quoted. These may have been omitted because we considered them to be less important in terms of direct physical interpretation, or they concentrate on aspects that we do not wish to review, or simply because we were not aware of them at the time of writing. To the authors of such papers, we offer our sincere apologies. In selecting the limited number of papers to be cited, it is our hope but not expectation that we will not have offended almost every researcher in this field!

While working in this subject for many years, we have benefited enormously from discussions with colleagues on most of the topics described here, and we are very grateful to many people for guiding us in our attempts to understand them. In particular, we want to acknowledge the substantial help we have received over the years from Georgii Alekseev, Jiří Bičák, Bill Bonnor and Cornelius Hoenselaers among others. In fact, the influence of Bill Bonnor and Jiří Bičák may be recognised throughout this work, and their comments on an early draft of it were most helpful.

In addition, several other colleagues have helped us significantly in the preparation of this book, either by answering specific questions, or by reading through early drafts of various chapters (or even in some cases the entire book) and giving us some most helpful comments. In this context, we want to explicitly express our thanks to Brian Edgar, Daniel Finley, Valeri Frolov, Cornelius Hoenselaers, Deborah Konkowski, Pavel Krtouš, Tomáš Ledvinka, Jorma Louko, Malcolm MacCallum, Brien Nolan, Marcello Ortaggio, Vojtěch Pravda, Alena Pravdová, Nico Santos, Oldřich Semerák, and Norbert Van den Bergh. All these colleagues have helped us to clarify many misunderstandings and have alerted us to numerous additional references. We are grateful to Pavel Krtouš for providing us with the figures in Chapter 14. Students Liselotte De Groote and Robert Švarc read through a late draft and identified numerous points that required correction. Any remaining errors that have found their way into this book are of course solely our responsibility.

Finally, we are pleased to acknowledge the support we have received from our own institutions that has enabled us to collaborate on a number of projects that have led to the writing of this book. Over the years, our collaboration has also been supported by various grants from the Czech Republic, the Royal Society, the London Mathematical Society and the EPSRC.

Note added in October 2011: For the paperback publication of this book, opportunity has been taken to correct a number of errors that had crept into the first printing. These include our own mistakes as well as a few slips in typing and formatting. Appendix B has also been extended and several further references have been cited. Some of these are of important papers that we had originally overlooked, while others are included that were published since this book was originally submitted but which represent significant additional material that, in our opinion, ought to be noted.

1
Introduction

After Einstein first presented his theory of general relativity in 1915, a few exact solutions of his field equations were found very quickly. All of these assumed a high degree of symmetry. Some could be interpreted as representing physically significant situations such as the exterior field of a spherical star, or a homogeneous and isotropic universe, or plane or cylindrical gravitational waves. Yet it took a long time before some of the more subtle properties of these solutions were widely understood.

In their seminal review of "exact solutions of the gravitational field equations", Ehlers and Kundt (1962) included the following statement. "At present the main problem concerning solutions, in our opinion, is not to construct more but rather to understand more completely the known solutions with respect to their local geometry, symmetries, singularities, sources, extensions, completeness, topology, and stability." Since this was written, considerable progress has been made in the understanding of many exact solutions. However, this development has been very restricted compared to the enormous effort that has been put into the derivation of further "new" solutions. Although significant advance has been achieved in the interpretation of many solutions, it is a fact that some aspects of even the most frequently quoted exact solutions still remain poorly understood. The opinion of Ehlers and Kundt thus still indicates an even more urgent task.

In this work, the very traditional approach will be adopted that an exact solution of Einstein's equations is expressed in terms of a metric in particular coordinates. Specifically, it will be represented in the form of a 3+1-dimensional line element in which the coordinates have certain ranges. Our purpose will be to try to identify its physical interpretation. Does it represent a specific physical situation? Does it include singularities or horizons, and what do these mean? For what range of the coordinates is the solution valid, or can it be extended by the introduction of different

1

coordinates? How does it behave asymptotically? Does it approach other known solutions in specific limits? These and other questions will be used to probe the meaning of the solutions considered.

It must be remembered of course, that an exact solution of Einstein's equations as defined above is initially just a *local* solution of the field equations. Global and topological properties of the associated manifold may be chosen according to our own prejudices. They are not implied by the field equations. Thus it may well occur that some particular exact solution may have a number of very different physical interpretations. For example, part of the well-known Schwarzschild solution may either represent part of the space-time inside the horizon of a black hole, or it could represent the interaction region following the collision of two specific plane gravitational waves.

Such cases in which a particular solution has a number of possible interpretations are, however, unusual. It is far more likely that a solution has no useful physical significance at all. Nevertheless, each space-time may at least be understood in terms of its geometrical properties. And it could well be that realistic physical situations may be approximated by compound space-times formed by patching different local exact solutions in appropriate ways. To construct such space-times, it is necessary to have some understanding of the properties of each component. Thus, although this work will concentrate on the simpler solutions that have clear physical meanings, we will also describe the basic properties of related solutions even when their immediate applicability is unclear.

Einstein's general theory of relativity is a covariant theory. The same physical space-time may be expressed (at least locally) in any number of different coordinate representations. Smooth coordinate transformations can be applied without changing the character of the physical space-time itself. However, for any particular space-time, some coordinate systems are more useful than others. Some may be convenient because they enable the field equations to be expressed in forms that have nice mathematical properties. Others may be more useful for a physical interpretation of the space-time. However, when transforming from one coordinate system to another, the different coordinates may not be in a simple one-to-one correspondence with each other over their natural ranges. In such cases, the different coordinates may span different portions of the complete space-time. Care therefore has to be taken in specifying the ranges of the coordinates employed, and also in identifying whether or not the boundaries of the coordinate patch correspond to the boundaries of the physical space-time being represented.

This principle of general covariance also has significant implications in the derivation of "new" solutions of the field equations. It is generally unclear initially whether a solution that is newly obtained is a known solution in an unfamiliar coordinate system or represents a previously unknown space-time. This is referred to as "the equivalence problem" that has now been widely addressed in the literature (see for example Chapter 9 in Stephani *et al.*, 2003).

In presenting a solution of Einstein's equations, it is now standard practice to identify its local sources (i.e. the structure of the Einstein tensor, and hence of the energy-momentum tensor), the algebraic type of the Weyl tensor, its curvature invariants, and the number and type of its symmetries. These are all essential in assisting to classify the solution, and hence to determine whether or not it is genuinely new rather than a new coordinate representation of some previously known solution. These properties are also important for its physical interpretation.

The solutions described in this work mostly represent vacuum space-times. Often a cosmological constant or an electromagnetic field will be included, or occasionally even a pure radiation field. However, we have severely restricted the number of solutions included that have a perfect fluid source. The reason for this decision is simply that, although such solutions are often interpreted as possible cosmological models or stellar interiors, they are already thoroughly reviewed in published literature. In particular, we would recommend the classic text of Ryan and Shepley (1975) on homogeneous cosmological models and the excellent and complementary book of Krasiński (1997) on inhomogeneous models.

We initially deal with the fundamental Minkowski, de Sitter and anti-de Sitter spaces and the Friedmann–Lemaître–Robertson–Walker universes, which are all conformally flat and highly symmetric. We then address solutions which have Weyl tensors that are of the special algebraic types D and N, which generally represent the simplest non-radiating and radiating solutions. And we finally proceed to address some algebraically more general solutions. As some of the simplest known solutions of Einstein's equations, most of those described here have a high degree of symmetry. However, we will not generally identify all the Killing vectors, and the existence of homothetic and conformal symmetries will be largely ignored.

A basic knowledge of Einstein's theory of relativity is assumed throughout this work. On the other hand, since we will not describe any method by which exact solutions are obtained, it will not be necessary to introduce much unfamiliar notation or advanced techniques. We trust that the notation used will be familiar, modulo certain sign conventions.

For the specific notation employed in this book, we have tried to follow as closely as appropriate that of the "exact solutions" book of Stephani *et al.* (2003). This also includes some very helpful introductory surveys of certain important topics in general relativity and differential geometry that therefore need not be repeated here.

We very willingly acknowledge that there have been many previous reviews of various exact solutions of Einstein's equations which emphasise aspects of their physical interpretation. We are therefore building on a sure foundation. However, the subject has now become so vast that any review must be selective and will inevitably reflect the prejudices of the authors. A balance also has to be struck between pedagogy and a review of current research. Each review has been addressed to a particular need, and we trust that our present contribution to the literature will be sufficiently different as to be considered a welcome addition.

Published work on the particular topics discussed in this book will be cited in the relevant sections. However, some general reviews apply more widely and are more appropriately cited here.

An early review by Ehlers and Kundt (1962) has had a significant impact on all later work. Understanding of the global properties of space-times was greatly advanced by the book of Hawking and Ellis (1973). Another seminal work that has had a major impact on the subject is the "exact solutions" book of Kramer *et al.* (1980). A most welcome second edition of this is now available (Stephani *et al.*, 2003). This provides an exhaustive review of known solutions at that time, but does not usually emphasise their physical interpretation.

Reviews of exact solutions with an emphasis on their interpretation have been given by Bonnor (1982, 1992), Bonnor, Griffiths and MacCallum (1994) and Bičák (2000a). Reviews of general families of radiative space-times have been given by Bičák (1989, 1997, 2000b) and Bičák and Krtouš (2003). For a brief and modern introduction to this subject, see Bičák (2006).

2
Basic tools and concepts

The purpose of this chapter is to define the notation that is used in this book, to introduce some basic tools that are employed, and to make a few initial comments on some of the concepts involved. It is not intended as a review of the topics mentioned, as these are described thoroughly in existing textbooks on general relativity.

2.1 Local geometry

Throughout this book, a solution of Einstein's equations is assumed to be given in terms of a metric, that is expressed in some local coordinate system, and which could represent a particular region of some theoretically possible space-time.

Space-time is assumed to be 3+1-dimensional. Taking the timelike coordinate first, the metric is assumed to have a Lorentzian signature $(-, +, +, +)$ so that timelike vectors have negative magnitude. It is represented (at least locally) by a manifold $\mathcal{M}$ with a symmetric metric $\mathbf{g}$ with coordinate components $g_{\mu\nu}$, where Greek letters span 0,1,2,3. The inverse of $g_{\mu\nu}$ is denoted by $g^{\mu\nu}$.

The speed of light is taken to be unity so that time and distance are measured by the same (unspecified) units, and null cones in space-time diagrams are normally drawn at an angle of 45° to the vertical.

The manifold $\mathcal{M}$ is assumed to be endowed with a linear (metric) connection that can be expressed in a coordinate basis in the form

$$\Gamma^{\lambda}_{\ \mu\nu} = \tfrac{1}{2}\, g^{\lambda\alpha}(g_{\mu\alpha,\nu} + g_{\nu\alpha,\mu} - g_{\mu\nu,\alpha}),$$

where the summation convention is adopted and a comma denotes a partial derivative. A semi-colon is used to denote a covariant derivative, so that the covariant derivative of a vector V^μ is given by

$$V^\mu{}_{;\nu} = V^\mu{}_{,\nu} + \Gamma^\mu{}_{\nu\alpha} V^\alpha.$$

It is frequently important, at any event (or point in space-time), to determine the components of vectors or tensors in particular directions. For this, it is first appropriate to introduce a normalised orthonormal tetrad $\boldsymbol{t}, \boldsymbol{x}, \boldsymbol{y}, \boldsymbol{z}$, composed of a timelike and three spacelike vectors. From these, it is convenient to construct a null tetrad $\boldsymbol{k}, \boldsymbol{l}, \boldsymbol{m}, \bar{\boldsymbol{m}}$, with the two null vectors $\boldsymbol{k} = \frac{1}{\sqrt{2}}(\boldsymbol{t} + \boldsymbol{z})$ and $\boldsymbol{l} = \frac{1}{\sqrt{2}}(\boldsymbol{t} - \boldsymbol{z})$, and the complex vector $\boldsymbol{m} = \frac{1}{\sqrt{2}}(\boldsymbol{x} - \mathrm{i}\,\boldsymbol{y})$ and its conjugate $\bar{\boldsymbol{m}} = \frac{1}{\sqrt{2}}(\boldsymbol{x} + \mathrm{i}\,\boldsymbol{y})$, which span the 2-spaces orthogonal to $\boldsymbol{k}$ and $\boldsymbol{l}$. These null tetrad vectors are mutually orthogonal except that $k_\mu l^\mu = -1$ and $m_\mu \bar{m}^\mu = 1$. With these conditions, the metric tensor can be expressed in terms of its null tetrad components in the form

$$g_{\mu\nu} = -k_\mu \ell_\nu - \ell_\mu k_\nu + m_\mu \bar{m}_\nu + \bar{m}_\mu m_\nu.$$

Such a null tetrad may be transformed in the following ways:

$$\boldsymbol{k}' = \boldsymbol{k}, \qquad \boldsymbol{l}' = \boldsymbol{l} + L\bar{\boldsymbol{m}} + \bar{L}\boldsymbol{m} + L\bar{L}\boldsymbol{k}, \qquad \boldsymbol{m}' = \boldsymbol{m} + L\boldsymbol{k}, \qquad (2.1)$$

$$\boldsymbol{k}' = \boldsymbol{k} + K\bar{\boldsymbol{m}} + \bar{K}\boldsymbol{m} + K\bar{K}\boldsymbol{l}, \qquad \boldsymbol{l}' = \boldsymbol{l}, \qquad \boldsymbol{m}' = \boldsymbol{m} + K\boldsymbol{l}, \qquad (2.2)$$

$$\boldsymbol{k}' = B\boldsymbol{k}, \qquad \boldsymbol{l}' = B^{-1}\boldsymbol{l}, \qquad \boldsymbol{m}' = \mathrm{e}^{\mathrm{i}\Phi}\boldsymbol{m}, \qquad (2.3)$$

where L and K are complex and B and Φ are real parameters. Together, these represent the six-parameter group of Lorentz transformations.

2.1.1 Curvature

Using the above notation, the (Riemann) curvature tensor, defined such that

$$V^\kappa{}_{;\mu\nu} - V^\kappa{}_{;\nu\mu} = -R^\kappa{}_{\lambda\mu\nu} V^\lambda,$$

is given by

$$R^\kappa{}_{\lambda\mu\nu} = \Gamma^\kappa{}_{\lambda\nu,\mu} - \Gamma^\kappa{}_{\lambda\mu,\nu} + \Gamma^\alpha{}_{\lambda\nu}\Gamma^\kappa{}_{\alpha\mu} - \Gamma^\alpha{}_{\lambda\mu}\Gamma^\kappa{}_{\alpha\nu}. \qquad (2.4)$$

This generally has 20 independent components according to the symmetries $R^\kappa{}_{\lambda(\mu\nu)} = 0$, $R^\kappa{}_{[\lambda\mu\nu]} = 0$ and $R_{(\kappa\lambda)\mu\nu} = 0$, where round and square brackets are used to denote the symmetric and antisymmetric parts, respectively. Defining the (symmetric) Ricci tensor and the Ricci scalar by

$$R_{\mu\nu} = R^\alpha{}_{\mu\alpha\nu}, \qquad R = g^{\alpha\beta} R_{\alpha\beta}, \qquad (2.5)$$

the trace-free part of the curvature tensor is given explicitly by

$$C_{\kappa\lambda\mu\nu} = R_{\kappa\lambda\mu\nu} + \tfrac{1}{2}(R_{\lambda\mu}g_{\kappa\nu} + R_{\kappa\nu}g_{\lambda\mu} - R_{\lambda\nu}g_{\kappa\mu} - R_{\kappa\mu}g_{\lambda\nu})$$
$$+ \tfrac{1}{6}R\,(g_{\kappa\mu}g_{\lambda\nu} - g_{\kappa\nu}g_{\lambda\mu}). \qquad (2.6)$$

This is known as the Weyl tensor which, in general, has ten independent components.

The curvature tensor can be expressed in terms of its various tetrad components. In particular, the ten independent components of the Ricci tensor are determined by the scalar quantities defined as[1]

$$\Phi_{00} = \tfrac{1}{2}R_{\mu\nu}\,k^\mu\,k^\nu, \qquad \Phi_{22} = \tfrac{1}{2}R_{\mu\nu}\,l^\mu\,l^\nu,$$
$$\Phi_{01} = \tfrac{1}{2}R_{\mu\nu}\,k^\mu\,m^\nu, \qquad \Phi_{12} = \tfrac{1}{2}R_{\mu\nu}\,l^\mu\,m^\nu, \qquad (2.7)$$
$$\Phi_{02} = \tfrac{1}{2}R_{\mu\nu}\,m^\mu\,m^\nu, \qquad \Phi_{11} = \tfrac{1}{4}R_{\mu\nu}(k^\mu\,l^\nu + m^\mu\,\bar{m}^\nu),$$

in which Φ_{AB} are generally complex but satisfy the constraint $\bar{\Phi}_{AB} = \Phi_{BA}$, and the Ricci scalar R.

The ten independent components of the Weyl tensor are similarly determined by the five complex scalar functions defined as

$$\Psi_0 = C_{\kappa\lambda\mu\nu}\,k^\kappa\,m^\lambda\,k^\mu\,m^\nu,$$
$$\Psi_1 = C_{\kappa\lambda\mu\nu}\,k^\kappa\,l^\lambda\,k^\mu\,m^\nu,$$
$$\Psi_2 = C_{\kappa\lambda\mu\nu}\,k^\kappa\,m^\lambda\,\bar{m}^\mu\,l^\nu, \qquad (2.8)$$
$$\Psi_3 = C_{\kappa\lambda\mu\nu}\,l^\kappa\,k^\lambda\,l^\mu\,\bar{m}^\nu,$$
$$\Psi_4 = C_{\kappa\lambda\mu\nu}\,l^\kappa\,\bar{m}^\lambda\,l^\mu\,\bar{m}^\nu.$$

By considering the equation of geodesic deviation (see below) in a suitably adapted frame, these components (in vacuum space-times) may be shown generally to have the following physical meanings:

Ψ_0 is a transverse component propagating in the l direction,

Ψ_1 is a longitudinal component in the l direction,

Ψ_2 is a Coulomb-like component, $\qquad (2.9)$

Ψ_3 is a longitudinal component in the k direction,

Ψ_4 is a transverse component propagating in the k direction.

According to Einstein's general theory of relativity, the curvature of space-time is related to the distribution of matter. Specifically, components of the

[1] The notation for these components, and those of the Weyl tensor given below, is closely related to that of Newman and Penrose (1962). However, a variant of the Newman–Penrose formalism is required here because a different signature is employed. The notation used here is that given in Stephani *et al.* (2003).

Ricci tensor are directly related to the local energy-momentum tensor $T_{\mu\nu}$ by Einstein's field equations

$$R_{\mu\nu} - \tfrac{1}{2} R g_{\mu\nu} + \Lambda g_{\mu\nu} = 8\pi T_{\mu\nu}, \qquad (2.10)$$

in which units of mass have been chosen so that $G = 1$, and Λ is the *cosmological constant*. This can also be rewritten in terms of the Einstein tensor $G_{\mu\nu} = R_{\mu\nu} - \tfrac{1}{2} R g_{\mu\nu}$.

The trace-free part of the curvature tensor (i.e. the Weyl tensor), however, is determined only indirectly from the field equations. These components may therefore be understood as representing "free components" of the gravitational field that also arise from non-local sources. In seeking to interpret any exact solution physically, these components need to be investigated explicitly.

2.1.2 Algebraic classification

For reasons that will be clarified in Section 2.3.3, a space-time is said to be *conformally flat* if its Weyl tensor vanishes, i.e. if $C_{\kappa\lambda\mu\nu} = 0$. Otherwise, gravitational fields are usually classified according to the Petrov–Penrose classification of their Weyl tensor. This is based on the number of its distinct principal null directions and the number of times these are repeated.

A null vector $\boldsymbol{k}$ is said to be a *principal null direction* of the gravitational field if it satisfies the property

$$k_{[\rho} C_{\kappa]\lambda\mu[\nu} k_{\sigma]} k^\lambda k^\mu = 0. \qquad (2.11)$$

If $\boldsymbol{k}$ is a member of the null tetrad defined above, then the condition (2.11) is equivalent to the statement that $\Psi_0 = 0$. It may then be noted that, under a transformation (2.2) of the tetrad which keeps $\boldsymbol{l}$ fixed, but changes the direction of $\boldsymbol{k}$, the component Ψ_0 of the Weyl tensor transforms as

$$\Psi_0 = \Psi_0{}' - 4K\Psi_1{}' + 6K^2\Psi_2{}' - 4K^3\Psi_3{}' + K^4\Psi_4{}'.$$

The condition for $\boldsymbol{k}$ to be a principal null direction, i.e. that $\Psi_0 = 0$, is thus equivalent to the existence of a root K such that

$$\Psi_0{}' - 4K\Psi_1{}' + 6K^2\Psi_2{}' - 4K^3\Psi_3{}' + K^4\Psi_4{}' = 0. \qquad (2.12)$$

Since this is a quartic expression in K, there are four (complex) roots to this equation, although these do not need to be distinct.

Each root of (2.12) corresponds to a principal null direction which can be constructed using (2.2), and the multiplicity of each principal null direction is the same as the multiplicity of the corresponding root. For a principal null direction $\boldsymbol{k}$ of multiplicity 1, 2, 3 or 4, it can be shown that, respectively

$$k_{[\rho}C_{\kappa]\lambda\mu[\nu}k_{\sigma]}k^\lambda k^\mu = 0 \quad \Leftrightarrow \qquad \Psi_0 = 0, \qquad\qquad \Psi_1 \neq 0,$$
$$C_{\kappa\lambda\mu[\nu}k_{\sigma]}k^\lambda k^\mu = 0 \quad \Leftrightarrow \qquad \Psi_0 = \Psi_1 = 0, \qquad \Psi_2 \neq 0,$$
$$C_{\kappa\lambda\mu[\nu}k_{\sigma]}k^\mu = 0 \quad \Leftrightarrow \qquad \Psi_0 = \Psi_1 = \Psi_2 = 0, \qquad \Psi_3 \neq 0,$$
$$C_{\kappa\lambda\mu\nu}k^\mu = 0 \quad \Leftrightarrow \quad \Psi_0 = \Psi_1 = \Psi_2 = \Psi_3 = 0, \quad \Psi_4 \neq 0.$$

If a space-time admits four distinct principal null directions (pnds), it is said to be *algebraically general*, or of type I, otherwise it is *algebraically special*. The distinct algebraic types can be summarised as follows:

type I : four distinct pnds

type II : one pnd of multiplicity 2, others distinct

type D : two distinct pnds of multiplicity 2

type III : one pnd of multiplicity 3, other distinct

type N : one pnd of multiplicity 4

type O : conformally flat

If either of the basis vectors $\boldsymbol{k}$ or $\boldsymbol{l}$ are aligned with principal null directions, either $\Psi_0 = 0$ or $\Psi_4 = 0$, respectively. If the vector $\boldsymbol{k}$ is aligned with the repeated principal null direction of an algebraically special space-time, then $\Psi_0 = 0 = \Psi_1$. If $\boldsymbol{k}$ and $\boldsymbol{l}$ are both aligned with the two repeated principal null directions of a type D space-time, then the only non-zero component of the Weyl tensor is Ψ_2. For a type N space-time with repeated principal null direction $\boldsymbol{k}$, the only non-zero component of the Weyl tensor is Ψ_4.

Two particularly useful complex scalar polynomial invariants for a vacuum space-time are given in terms of the Weyl tensor components by

$$I = \Psi_0\Psi_4 - 4\,\Psi_1\Psi_3 + 3\,\Psi_2{}^2, \qquad J = \begin{vmatrix} \Psi_0 & \Psi_1 & \Psi_2 \\ \Psi_1 & \Psi_2 & \Psi_3 \\ \Psi_2 & \Psi_3 & \Psi_4 \end{vmatrix}. \qquad (2.13)$$

In fact, the real part of I is $\frac{1}{16}$ times the Kretschmann scalar $R_{\kappa\lambda\mu\nu}R^{\kappa\lambda\mu\nu}$ for the vacuum case.

It may be noted that, for algebraically special space-times, it is necessary that $I^2 = 27J^2$. In this case, if I and J vanish, the space-time must be of types N or III, otherwise is must be of types D or II. Moreover, if $\boldsymbol{k}$ is a repeated principal null direction, a space-time for which I and J are non-zero is of type D if $3\Psi_2\Psi_4 = 2\Psi_3{}^2$, otherwise it is of type II. (For further details see Stephani *et al.*, 2003.) The space-time is also of type D if $\Psi_1 = 0 = \Psi_3$ and $\Psi_0\Psi_4 = 9\Psi_2{}^2$ (see Chandrasekhar and Xanthopoulos, 1986).

In general, including the Ricci scalar R, it is known that there exist 14 real scalar polynomial invariants of the type (2.13), which involve components of both the Ricci and Weyl tensors. However, 14 independent scalars like this are not known explicitly. In practice, it is convenient to define 16 or 17 such scalar quantities and a number of constraints, or syzygies, constraining them. For details of these, their meanings and the relations between them, see Penrose and Rindler (1986), Carminati and McLenaghan (1991), Zakhary and McIntosh (1997) and references contained therein.

2.1.3 Geodesics and geometrical optics

In some local region of space-time, any two events may be joined by a family of curves. Within such a family, the curve which has either maximum or minimum proper length is known as a geodesic. (It is a maximum or minimum according to whether the events have a timelike or spacelike separation, respectively.) For space-times with a metric connection, a geodesic also has the property that its tangent vector is parallelly transported along it (i.e. it is autoparallel). These are two distinct properties of what is intuitively required to generalise the concept of a straight line in flat space to a curved space-time.

Consider a three-parameter family of curves in a region of space-time such that exactly one curve passes through each point. The equations of such a *congruence* can be written in terms of the local coordinates x^μ in the form

$$x^\mu = x^\mu(y^i, s),$$

where y^i, $(i = 1, 2, 3)$ are the parameters identifying particular curves of the congruence and s is a parameter along each. The corresponding vector field $\boldsymbol{v}$ that is tangent to the congruence at all points is given by

$$v^\mu = \frac{\mathrm{d}x^\mu}{\mathrm{d}s}.$$

The congruence is null if $v_\mu v^\mu = 0$, and consists of geodesics if the tangent vectors are parallelly transported along it, i.e. if

$$\frac{D\,v^\mu}{\mathrm{d}s} \equiv v^\mu{}_{;\nu}\, v^\nu = \lambda\, v^\mu,$$

for some $\lambda(y^i, s)$, where $D/\mathrm{d}s$ denotes the derivative along the congruence. The parameter s is called *affine* if $\lambda = 0$, in which case s is defined up to a linear transformation. For such an affine parameter, the above geodesic equation can be written in the alternative forms

$$v^\mu_{;\nu}\, v^\nu = 0, \qquad \text{or} \qquad \frac{\mathrm{d}^2 x^\mu}{\mathrm{d}s^2} + \Gamma^\mu_{\alpha\beta} \frac{\mathrm{d}x^\alpha}{\mathrm{d}s} \frac{\mathrm{d}x^\beta}{\mathrm{d}s} = 0. \qquad (2.14)$$

Small physical bodies of negligible mass which move in some gravitational field can be represented by test particles, which are treated as single points that have no gravitational fields of their own. The motion of such particles in space-time is represented by their worldlines, which are necessarily timelike everywhere. When test particles move only under the action of the gravitational field and have no "internal" structure, their worldlines are timelike geodesics. However, when other forces act on the particle, its worldline will deviate from that of a geodesic.

It is often convenient to analyse the gravitational field in any neighbourhood in terms of the behaviour of some privileged congruence of curves which pass through it: in particular, in terms of its *expansion, twist* (or *rotation*), and *shear*. Such an analysis can appropriately be applied to any particular timelike congruence such as that of some class of privileged observers or test particles, or the worldlines of particles of a fluid. However, for a vacuum field containing gravitational radiation, it is more appropriate to apply it to null rays, particularly if they are geodesic and aligned with a repeated principal null direction of the space-time. Such an approach was developed and promoted by, among others, Jordan, Ehlers and Sachs (1961), Sachs (1961, 1962), Newman and Penrose (1962) and Pirani (1965).

For a null congruence, it is convenient to introduce the Newman–Penrose *spin coefficients*.[2] These are the twelve complex Ricci rotation coefficients associated with the null tetrad $\boldsymbol{k}$, $\boldsymbol{l}$, $\boldsymbol{m}$ and $\bar{\boldsymbol{m}}$ such that

$$
\begin{aligned}
k_{\mu;\nu} =\ & -(\gamma + \bar{\gamma})k_\mu k_\nu - (\epsilon + \bar{\epsilon})k_\mu l_\nu + (\alpha + \bar{\beta})k_\mu m_\nu + (\bar{\alpha} + \beta)k_\mu \bar{m}_\nu \\
& + \bar{\tau}\, m_\mu k_\nu + \bar{\kappa}\, m_\mu l_\nu - \bar{\sigma}\, m_\mu m_\nu - \bar{\rho}\, m_\mu \bar{m}_\nu \\
& + \tau\, \bar{m}_\mu k_\nu + \kappa\, \bar{m}_\mu l_\nu - \rho\, \bar{m}_\mu m_\nu - \sigma\, \bar{m}_\mu \bar{m}_\nu,
\end{aligned}
$$

$$
\begin{aligned}
l_{\mu;\nu} =\ & (\gamma + \bar{\gamma})l_\mu k_\nu + (\epsilon + \bar{\epsilon})l_\mu l_\nu - (\alpha + \bar{\beta})l_\mu m_\nu - (\bar{\alpha} + \beta)l_\mu \bar{m}_\nu \\
& - \nu\, m_\mu k_\nu - \pi\, m_\mu l_\nu + \lambda\, m_\mu m_\nu + \mu\, m_\mu \bar{m}_\nu \qquad (2.15) \\
& - \bar{\nu}\, \bar{m}_\mu k_\nu - \bar{\pi}\, \bar{m}_\mu l_\nu + \bar{\mu}\, \bar{m}_\mu m_\nu + \bar{\lambda}\, \bar{m}_\mu \bar{m}_\nu,
\end{aligned}
$$

$$
\begin{aligned}
m_{\mu;\nu} =\ & -(\gamma - \bar{\gamma})m_\mu k_\nu - (\epsilon - \bar{\epsilon})m_\mu l_\nu + (\alpha - \bar{\beta})m_\mu m_\nu + (\beta - \bar{\alpha})m_\mu \bar{m}_\nu \\
& - \bar{\nu}\, k_\mu k_\nu - \bar{\pi}\, k_\mu l_\nu + \bar{\mu}\, k_\mu m_\nu + \bar{\lambda}\, k_\mu \bar{m}_\nu \\
& + \tau\, l_\mu k_\nu + \kappa\, l_\mu l_\nu - \rho\, l_\mu m_\nu - \sigma\, l_\mu \bar{m}_\nu.
\end{aligned}
$$

---

[2] For details of the following calculations and further applications of the Newman–Penrose formalism, see Newman and Penrose (1962), Alekseev and Khlebnikov (1978), Frolov (1979) and Penrose and Rindler (1986).

It follows immediately that $k^{\mu}{}_{;\nu} k^{\nu} = (\epsilon + \bar{\epsilon})k^{\mu} - \bar{\kappa}\, m^{\mu} - \kappa\, \bar{m}^{\mu}$, which implies that k^{μ} is everywhere tangent to a geodesic null congruence if, and only if, $\kappa = 0$. Moreover, the congruence is affinely parametrised if $\epsilon + \bar{\epsilon} = 0$, and such a parametrisation can always be chosen.

Consider a congruence of null curves and a null tetrad defined at each point such that k^{μ} is everywhere aligned with the congruence. It can immediately be seen from (2.15) that the tetrad is parallelly propagated along the congruence (i.e. $k_{\mu;\nu}k^{\nu} = l_{\mu;\nu}k^{\nu} = m_{\mu;\nu}k^{\nu} = 0$) if, and only if, κ, ϵ and π all vanish. This implies that a parallel propagation of a tetrad along a null congruence is only possible if that congruence is geodesic and affinely parametrised.

Let $\boldsymbol{w}$ be a connecting vector, orthogonal to $\boldsymbol{k}$, between two neighbouring geodesics (rays) at some instant. For this to retain its meaning when dragged along the congruence, it is necessary that $w^{\mu}{}_{;\nu}k^{\nu} = k^{\mu}{}_{;\nu}w^{\nu}$ (i.e. its Lie derivative must vanish). Taking the vector $\boldsymbol{w}$ to be also orthogonal to $\boldsymbol{l}$, it may then be expressed in the form $\boldsymbol{w} = \bar{\zeta}\,\boldsymbol{m} + \zeta\,\bar{\boldsymbol{m}}$, where the complex parameter ζ may be interpreted as the displacement between the two rays in a spacelike plane spanned by $\boldsymbol{m}$ and $\bar{\boldsymbol{m}}$ that is orthogonal to $\boldsymbol{k}$ at any instant, as shown in Figure 2.1. If the tetrad is parallelly propagated along the congruence, it then follows from (2.15) that

$$\frac{D\zeta}{\mathrm{d}s} = -\rho\zeta - \sigma\,\bar{\zeta}, \tag{2.16}$$

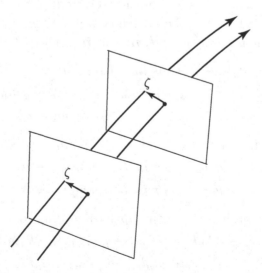

Fig. 2.1 Two neighbouring rays of a null congruence with tangent vector $\boldsymbol{k}$, whose separation is measured by the complex parameter ζ in a spacelike surface, spanned by $\boldsymbol{m}$ and $\bar{\boldsymbol{m}}$, orthogonal to the congruence.

where $D\zeta/\mathrm{d}s = \zeta_{,\nu}k^{\nu}$ is the derivative along the congruence, and s is an affine parameter.

To interpret the coefficients ρ and σ in detail, it is convenient to put $\rho = -\theta - \mathrm{i}\,\omega$ and $\sigma = |\sigma|\,\mathrm{e}^{2\mathrm{i}\psi}$. With this, (2.16) becomes

$$\frac{D\zeta}{\mathrm{d}s} = \theta\,\zeta + \mathrm{i}\,\omega\,\zeta - |\sigma|\,\mathrm{e}^{2\mathrm{i}\psi}\bar{\zeta},$$

in which the three terms on the right-hand side have distinct meanings. It can be seen that $\theta = -\operatorname{Re}\rho$ measures the rate of *expansion* of the congruence. (Alternatively, $\operatorname{Re}\rho$ is its *contraction*, or convergence.) Similarly, $\omega = -\operatorname{Im}\rho$ measures the *twist*, or rotation, of the congruence. The final term indicates a contraction when $\arg\zeta = \psi, \psi + \pi$, but an expansion when $\arg\zeta = \psi \pm \frac{1}{2}\pi$. Thus, σ can be understood as the *shear* of the congruence, with $\arg\sigma = 2\psi$ denoting the orientation of the shear. These properties are illustrated in Figure 2.2. Using (2.15), explicit expressions for the expansion, twist and shear of an affinely parametrised null congruence tangent to $\boldsymbol{k}$ are given respectively by

$$\theta = \tfrac{1}{2}k^{\mu}{}_{;\mu}, \qquad \omega^2 = \tfrac{1}{2}k_{[\mu;\nu]}k^{\mu;\nu}, \qquad \sigma\bar{\sigma} = \tfrac{1}{2}k_{(\mu;\nu)}k^{\mu;\nu} - \tfrac{1}{4}(k^{\mu}{}_{;\mu})^2.$$

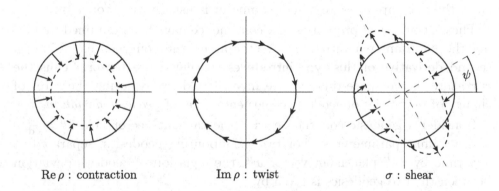

| $\operatorname{Re}\rho$: contraction | $\operatorname{Im}\rho$: twist | σ : shear |

Fig. 2.2 Using a tetrad that is parallelly propagated along an affinely parametrised null geodesic congruence tangent to $\boldsymbol{k}$, the spin coefficients ρ and σ have a geometrical interpretation in terms of the contraction, twist and shear of the congruence in the orthogonal plane spanned by $\boldsymbol{m}$ and $\bar{\boldsymbol{m}}$ as shown.

It can also be shown that the spin coefficient τ measures the change of $\boldsymbol{k}$ in the direction of $\boldsymbol{l}$. However, since it transforms as $\tau' = \tau + L\rho + \bar{L}\sigma + L\bar{L}\kappa$ under the null rotation (2.1) which keeps the direction of $\boldsymbol{k}$ fixed, this interpretation is only well defined for an expansion-free, twist-free and shear-free geodesic null congruence (i.e. for the Kundt space-times that will be described in Chapter 18).

It can in addition be shown that the null tetrad co-vector k_μ is proportional to the gradient of a scalar field if, and only if, it is tangent to a twist-free null geodesic congruence (i.e. $\kappa = 0$ and $\rho - \bar{\rho} = -2i\,\omega = 0$). Such a congruence is said to be *hypersurface orthogonal*. It is actually equal to the gradient of a scalar field when also $\epsilon + \bar{\epsilon} = 0$ and $\bar{\alpha} + \beta = \tau$.

In fact, some geometrical properties of null congruences are related to the algebraic properties of the Weyl tensor. In particular, an important result due to Goldberg and Sachs (1962) states that a vacuum space-time is algebraically special if, and only if, it possesses a shear-free geodesic null congruence (i.e. $\Psi_0 = 0 = \Psi_1 \Leftrightarrow \kappa = 0 = \sigma$). Generalisations of this involving fewer restrictions on the Ricci tensor have been obtained by Kundt and Trümper (1962) and Kundt and Thompson (1962).

It is finally important to note that the spin coefficients ν, μ, λ and π correspond to κ, ρ, σ and τ, respectively, for a congruence tangent to l (i.e. where $\boldsymbol{k}$ and $\boldsymbol{l}$, and $\boldsymbol{m}$ and $\bar{\boldsymbol{m}}$ are both interchanged), where they have the equivalent interpretations.

Similar properties describing the geometrical properties of a timelike congruence can also be obtained, as will be applied to the worldlines of the particles of a perfect fluid in Section 2.2.2. For any such timelike congruence, the most appropriate affine parameter is usually the proper time.

These geometrical properties of a congruence have been obtained by taking the covariant derivative of a vector along the congruence. Taking a second derivative of this type introduces the curvature tensor. Thus the curvature of the space-time can be investigated by exploring the rates of change of properties of geodesic congruences, i.e. of *geodesic deviation*.

Consider a geodesic congruence with the tangent vector field $v^\mu = \frac{dx^\mu}{ds}$, and an affine parameter s. For two neighbouring geodesics, separated at any time by a displacement vector w^μ, the equation of geodesic deviation between the two geodesics is given by

$$\frac{D^2 w^\mu}{ds^2} = -R^\mu{}_{\alpha\beta\gamma} v^\alpha w^\beta v^\gamma, \tag{2.17}$$

where $D^2 w^\mu / ds^2 = (w^\mu{}_{;\alpha} v^\alpha)_{;\beta}\, v^\beta = w^\mu{}_{;\alpha\beta}\, v^\alpha v^\beta$. This describes the relative acceleration of infinitesimally nearby geodesics. It was by using this equation that Szekeres (1965) derived, in a suitable reference frame, the physical meanings of the Weyl tensor components (2.9), (see also Bičák and Podolský, 1999b). An application of this will be used in Section 17.2.

2.1.4 Symmetries

The physical meaning of a space-time does not depend on the particular coordinate system employed in its representation. In general, a transformation of coordinates changes the form of a metric, but not its interpretation. However, a coordinate transformation under which the form of the metric is invariant is known as a *symmetry* or *isometry*. Specifically, if a metric is invariant under the (infinitesimal) transformation

$$x^\mu \to x'^\mu = x^\mu + \varepsilon\, X^\mu,$$

where ε is small and X^μ is a vector field, then it can be shown that

$$X_{\mu;\nu} + X_{\nu;\mu} = 0, \tag{2.18}$$

and vice versa. This is known as Killing's equation[3] and any of its solutions are known as *Killing vectors*.

One of the first steps in seeking to interpret any given space-time is to identify at least some of its symmetries. In fact, it is likely that the solution was actually derived after initially assuming the existence of certain symmetries as this simplifies the field equations sufficiently to enable exact solutions to be obtained. For example, if the metric tensor is independent of one of the coordinates, a symmetry could immediately be observed for which that coordinate is aligned with the corresponding Killing vector field. However, further symmetries may sometimes be more difficult to detect.

If a space-time in some region admits a timelike Killing vector field, that region is said to be *stationary*. Such a vector field may be associated with a time coordinate and the metric, and hence the gravitational field, do not depend on time. However, this does not imply that there are no evolutionary or time-dependent effects. A stronger condition for a space-time to be *static* is that it admits a timelike Killing vector and also that it is invariant under a change in the direction of time.[4] For example, it may be seen that the field of a uniformly rotating body would be stationary, but not static.

It is well known that, in 3+1-dimensions, the maximum number of independent symmetries is 10, in which case the space-time has constant curvature. Most exact solutions admit significantly fewer isometries. In addition, space-times may also admit homotheties, conformal symmetries, Killing tensors, Killing–Yano tensors and other geometrical structures that express "hidden" symmetries. However, these will not generally be referred

---

[3] Killing's equation may alternatively be stated in the coordinate-independent form that the Lie derivative of the metric tensor in the direction of X must vanish.

[4] Geometrically, a stationary space-time is static if, and only if, there exists a spacelike hypersurface which is orthogonal to the orbits of the isometry.

to in this book. For a thorough discussion of the symmetries and related curvature structures that appear in general relativity, see Stephani *et al.* (2003) or Hall (2004).

In fact, in most of the following chapters, the exact solutions described are initially characterised in terms of some of their symmetries. For example, the space-times being considered may be described as static or stationary, or as having plane, axial or spherical symmetry etc. Solutions with the most symmetries are generally covered in the earlier chapters of this book, while those with fewer are treated later. In particular, the space-times described in Chapters 3, 4 and 5 are maximally symmetric, having ten Killing vectors. Chapters 6 and 7 describe space-times with six Killing vectors: they have a maximally symmetric 3-space or two 2-spaces, respectively. The solutions described in Chapters 8 and 9 have four Killing vectors, while those of later chapters, or later sections of those chapters, have fewer symmetries.

2.1.5 Continuity

A matter field contained in a given space-time can be distributed discontinuously as, for example, at the boundary of a physical body. And, since matter density and pressure are related to components of the space-time curvature tensor through Einstein's field equations (2.10) which involve the second derivatives of the metric tensor, this implies that the metric must be C^1 in an appropriate way across such a boundary. Moreover, thin sheets of matter can be conveniently represented in an idealised form using distributions such as Dirac δ-functions. Across such spacelike surfaces, the metric only needs to be C^0.

Pure gravitational fields (represented by components of the Weyl tensor), on the other hand, are normally assumed to be continuous. However, it is possible for shock and even impulsive gravitational waves to exist and, for these, the metric would again be, respectively, C^1 or C^0 across a corresponding null hypersurface. In fact, this would also occur for shock and impulsive matter waves if these propagate at the speed of light.

Detailed conditions across such spacelike hypersurfaces and the matter on them were initially investigated in the classical works of Darmois (1927), O'Brien and Synge (1952), Lichnerowicz (1955), Israel (1966a) and Bonnor and Vickers (1981). The analysis of junction conditions on null hypersurfaces was developed by Stellmacher (1938), Pirani (1965), Penrose (1972), Robson (1973), Taub (1980), Clarke and Dray (1987), Barrabès (1989) and Barrabès and Israel (1991) among others.

For most solutions described here, the metric is locally C^∞ everywhere except at specific singularities. However, compound space-times in which a region containing matter is joined to a vacuum region will sometimes be mentioned. Thin sheets of matter and impulsive or step waves will also be considered when appropriate.

2.2 Matter content

Many of the solutions described in this book represent *vacuum* space-times – or empty regions of space-times exterior to particular material sources. However, the possible presence of a *cosmological constant* is frequently included. Other sources considered in this book correspond to *electromagnetic fields*, *perfect fluids* and *pure radiation*. The notation for such sources is now briefly outlined.

2.2.1 Electromagnetic field

An electromagnetic field is described by an antisymmetric tensor $F_{\mu\nu}$. Introducing also its dual $\widetilde{F}_{\mu\nu} = \frac{1}{2}\varepsilon_{\mu\nu\alpha\beta}F^{\alpha\beta}$, where $\varepsilon_{\mu\nu\alpha\beta}$ is the permutation symbol with $\varepsilon_{0123} = \sqrt{-g}$ and $g = \det(g_{\mu\nu})$, Maxwell's equations can be written as

$$F^{\mu\nu}{}_{;\nu} = 4\pi J^{\mu}, \qquad \widetilde{F}^{\mu\nu}{}_{;\nu} = 0,$$

where J^{μ} is the electric current vector. The electromagnetic field may be determined by a vector potential A^{μ} such that $F_{\mu\nu} = A_{\nu,\mu} - A_{\mu,\nu}$ or, in the notation of differential forms, $F = \mathrm{d}A$. In particular, in a suitable gauge, A^0 may be interpreted as the potential for the electric field and A^i, $i = 1, 2, 3$, as the (vector) potential for the magnetic field.

The energy-momentum tensor for an electromagnetic field is given by

$$4\pi\, T_{\mu\nu} = F_{\mu\alpha}F_{\nu}{}^{\alpha} - \tfrac{1}{4}g_{\mu\nu}F_{\alpha\beta}F^{\alpha\beta}, \qquad (2.19)$$

which is clearly trace-free, i.e. $T^{\mu}{}_{\mu} = 0$. Where an electromagnetic field is included in this book, its source will normally be excluded. Specifically, it will be assumed that $J^{\mu} = 0$. Such a field is referred to as being *electrovacuum*.

Using the null tetrad $\boldsymbol{k}$, $\boldsymbol{l}$, $\boldsymbol{m}$ and $\bar{\boldsymbol{m}}$, the electromagnetic field can be expressed in terms of three complex quantities Φ_A, $A = 0, 1, 2$, defined as

$$\Phi_0 = F_{\mu\nu}k^{\mu}m^{\nu}, \quad \Phi_1 = \tfrac{1}{2}F_{\mu\nu}(k^{\mu}l^{\nu} + \bar{m}^{\mu}m^{\nu}), \quad \Phi_2 = F_{\mu\nu}\bar{m}^{\mu}l^{\nu}. \quad (2.20)$$

For an electrovacuum field, the Ricci tensor components, defined through

Einstein's field equations and (2.7), are given by[5]

$$\Phi_{AB} = 2\Phi_A \bar{\Phi}_B.$$

It may also be noted that a complex invariant of the field is given by

$$\Phi_0\Phi_2 - \Phi_1{}^2 = \tfrac{1}{8}(F_{\mu\nu}F^{\mu\nu} + iF_{\mu\nu}\widetilde{F}^{\mu\nu}).$$

An electromagnetic field is said to be *null* if this invariant vanishes, i.e. if it satisfies the two conditions $F_{\mu\nu}F^{\mu\nu} = 0$ and $F_{\mu\nu}\widetilde{F}^{\mu\nu} = 0$. In this case, the electric and magnetic fields are orthogonal, have equal magnitude, and propagate at the speed of light. Otherwise, the field is described as being *non-null*.

A null vector $\boldsymbol{k}$ is said to be a *principal null direction* of the electromagnetic field if it satisfies the property

$$F_{\mu[\nu}k_{\sigma]}k^\mu = 0.$$

If $\boldsymbol{k}$ is a member of the above null tetrad, then this condition is equivalent to the statement that $\Phi_0 = 0$. In fact, for a principal null direction $\boldsymbol{k}$ of multiplicity 1 or 2, it can be shown that, respectively

$$F_{\mu[\nu}k_{\sigma]}k^\mu = 0 \qquad \Leftrightarrow \qquad \Phi_0 = 0, \qquad \Phi_1 \neq 0,$$
$$F_{\mu\nu}k^\mu = 0 \qquad \Leftrightarrow \qquad \Phi_0 = \Phi_1 = 0, \qquad \Phi_2 \neq 0.$$

In analogy with (2.12), due to the transformation properties of the component Φ_0 under the null rotation (2.2) which keeps $\boldsymbol{l}$ fixed, the condition for $\boldsymbol{k}$ to be a principal null direction of the electromagnetic field is equivalent to the existence of a root K of the quadratic equation

$$\Phi_0{}' - 2K\Phi_1{}' + K^2\Phi_2{}' = 0. \tag{2.21}$$

Obviously, this has just two (complex) roots.

A non-null electromagnetic field has two distinct principal null directions. If the tetrad vectors $\boldsymbol{k}$ and $\boldsymbol{l}$ are each aligned with these, then the only remaining component of the field is Φ_1. On the other hand, a null electromagnetic field has a repeated principal null direction and, if this is aligned with $\boldsymbol{k}$, the only component of the field is Φ_2, and the energy-momentum tensor has the form

$$T_{\mu\nu} = \tfrac{1}{2\pi}\,\Phi_2\bar{\Phi}_2\,k_\mu k_\nu. \tag{2.22}$$

---

[5] In many research papers on electrovacuum fields which use the Newman–Penrose formalism, it is common practice to rescale units so that the gravitational constant is given by $2G = 1$, and then $\Phi_{AB} = \Phi_A\bar{\Phi}_B$.

This section may be concluded by noting an important theorem due to Mariot (1954) and Robinson (1961), which states that the repeated principal null direction of a null electromagnetic field must be aligned with shear-free geodesic null congruence, and is also a repeated principal null direction of an algebraically special gravitational field. Aligning such a direction with the tetrad vector $\boldsymbol{k}$, this theorem states that

$$\Phi_0 = \Phi_1 = 0, \quad \Phi_2 \neq 0 \quad \Rightarrow \quad \kappa = \sigma = 0, \quad \Psi_0 = \Psi_1 = 0.$$

2.2.2 Perfect fluid and pure radiation

In contrast to the above case, a perfect fluid source is not described by quantities satisfying field equations, but is represented phenomenologically by its energy-momentum tensor. This is assumed to have the form

$$T_{\mu\nu} = (\rho + p)u_\mu u_\nu + p\, g_{\mu\nu}, \tag{2.23}$$

where ρ is the energy density, p is the pressure (which is assumed to be isotropic), and u^μ the four-velocity of the fluid. For physical reasons, it should normally be assumed that $\rho \geq p \geq 0$, and that p satisfies a "barotropic" equation of state $p = p(\rho)$.

In particular, it is often appropriate to restrict attention to the case in which the equation of state takes the linear form

$$p = (\gamma - 1)\rho, \tag{2.24}$$

where $1 \leq \gamma \leq 2$ is an appropriately chosen constant. This includes a number of important special cases. That of a pressureless fluid, which is usually referred to as *dust*, occurs when $\gamma = 1$. The case of *incoherent radiation* for which $p = \frac{1}{3}\rho$ is included when $\gamma = 4/3$, and a so-called *stiff fluid* for which $p = \rho$, in which the speed of sound is equal to the speed of light, occurs when $\gamma = 2$.

It may be observed that the special case of (2.24) in which $\gamma = 0$ corresponds to $\rho = -p$. If this is constant, then $8\pi\rho$ is exactly equivalent to the presence of a cosmological constant Λ. In addition, it may be noted that, in a perfect fluid space-time with any equation of state, a cosmological constant can always be included by setting $\rho \to \rho + \Lambda/8\pi$ and $p \to p - \Lambda/8\pi$.

With the four-velocity of the fluid denoted by u^μ, such that $u_\mu u^\mu = -1$, its expansion θ and acceleration a^μ are given by

$$\theta = u^\mu{}_{;\mu}, \qquad a^\mu = u^\mu{}_{;\nu}u^\nu.$$

It is then convenient to introduce a new (projection) tensor

$$h_{\mu\nu} = g_{\mu\nu} + u_\mu u_\nu,$$

which satisfies $h_{\mu\nu}u^\nu = 0$, $h_{\mu\alpha}h^\alpha{}_\nu = h_{\mu\nu}$ and $h^\mu{}_\mu = 3$. A shear tensor $\sigma_{\mu\nu}$ and rotation tensor $\omega_{\mu\nu}$ are then defined by

$$\sigma_{\mu\nu} = \left[\tfrac{1}{2}(u_{\alpha;\beta} + u_{\beta;\alpha}) - \tfrac{1}{3}u^\gamma{}_{;\gamma}h_{\alpha\beta} \right] h^\alpha{}_\mu h^\beta{}_\nu,$$

$$\omega_{\mu\nu} = \tfrac{1}{2}(u_{\alpha;\beta} - u_{\beta;\alpha})h^\alpha{}_\mu h^\beta{}_\nu.$$

Scalar quantities representing the local shear and rotation (or vorticity) of the fluid are then given, respectively, by

$$\sigma = \sqrt{\sigma^{\mu\nu}\sigma_{\mu\nu}}, \qquad \omega = \sqrt{\omega^{\mu\nu}\omega_{\mu\nu}}.$$

We will also occasionally consider a general class of *pure radiation* fields for which the energy-momentum tensor has the form

$$T^{\mu\nu} = \rho\, k^\mu k^\nu, \tag{2.25}$$

where k^μ is a null vector and ρ is its radiation density. This is generally taken to describe some kind of field that propagates at the speed of light. As already noted above, this could represent a null electromagnetic field, see (2.22). It could also represent an incoherent beam of photons or some kinds of idealised (massless) neutrino fields. A source of this type is sometimes referred to as "null dust", since it can be considered to be a limiting case of a pressureless perfect fluid in which the four-velocity becomes null. However, such a source will generally be referred to here as "pure radiation".

2.3 Global structure

The process of interpreting any solution of Einstein's equations requires to set a solution, that has been derived using mainly locally-defined criteria, in a global context and to understand its physical meaning. This involves a number of specific techniques for attempting to identify its properties, particular sources, its effects on test particles and fields, its asymptotic behaviour, and any possible extensions.

In particular, when seeking to interpret an exact solution that has been derived as a local solution of the field equations, some assumptions about its topology are usually made at an early stage. For example, if a metric admits a spacelike symmetry, this could be interpreted either as a translational symmetry associated with a Cartesian-like coordinate $x \in (-\infty, \infty)$, or as a rotational symmetry associated with a periodic angular coordinate relabelled as $\phi \in [0, 2\pi)$, with the events at $\phi = 2\pi$ identified with those at $\phi = 0$. These two natural possibilities could indeed lead to two different physical interpretations of the same metric. The second possibility of rotational symmetry, however, would require further consideration. It could normally only be interpreted in this way if it admits a regular axis. If it does not do so, then it could only be understood as part of some axially symmetric space-time in which it is taken as the exterior solution to some appropriate source that occurs along the axis.

In a flat space-time, the appropriate identification of events with different coordinate values is usually obvious from the desirable topology. Flat space-times with different topologies, such as cylindrical and toroidal, can easily be constructed by making various identifications. However, such procedures are far less straightforward in a curved space-time, and constructions using artificial identifications will generally not be considered here.

As hinted above, it is important to distinguish between a local solution of Einstein's equations and a *complete space-time*. A complete space-time is a maximally extended solution, or a combination of solutions, representing a physical situation over the maximum possible range of space and time. Only for the simplest cases can a complete space-time be represented by the single coordinate patch. For most exact solutions, the limits of the coordinates used do not generally coincide with the boundaries of the physical space-time itself. In such cases, it is appropriate to look for alternative coordinate systems that extend the solution to further regions of the space-time. Alternatively, some exact solutions can be used to describe restricted regions of a *composite space-time* that may represent a particular physical situation (for example, the field exterior to a finite body).

A solution of Einstein's equations, of course, is only valid where the metric is locally Lorentzian. Places where this condition fails to be satisfied are known as *singularities*. These can arise from a wide variety of reasons, both physical and mathematical, and so need to be considered in more detail.

2.3.1 Singularities

The simplest singularities to interpret are those which occur in the metric tensor $g_{\mu\nu}$ and are due purely to the choice of coordinates. For example, the metric of a two-dimensional flat space in polar coordinates $ds^2 = dr^2 + r^2 d\phi^2$ has a coordinate singularity at the origin $r = 0$, but the same space expressed as $ds^2 = dx^2 + dy^2$ obviously has no such singularity. Similarly, spherical polar coordinates are singular on the axis containing the poles of each spherical surface. Such singularities may be referred to as *removable singularities* as it is possible to transform to a different coordinate systems for which the metric is regular at the corresponding points. They have no physical significance and do not appear in the curvature tensor.

In a curved space-time, however, coordinate singularities of $g_{\mu\nu}$ sometimes have a fundamental physical significance. Besides singularities that arise purely as a result of the choice of coordinates, others may arise due to the adaptation of the coordinates to given Killing vectors of the space-time. For example, coordinate singularities can sometimes be interpreted as various kinds of *horizons* (see below).

It may also be expected that there could occur places where the gravitational field becomes infinite. Such locations are familiar in Newtonian theory, for example at the position of a point mass. This particular idealisation may not occur in general relativity, but the gravitational field in Einstein's theory is represented by the curvature of the space-time and there certainly exist solutions for which components of the curvature tensor diverge at particular locations. However, such singularities do not necessarily imply that the space-time curvature there is unbounded. This could still be due to the choice of coordinates. To determine whether or not this is the case, it is necessary to investigate the behaviour of locally invariant scalar quantities. The most convenient of these are the *scalar invariants*, such as I and J defined in (2.13), which are formed as polynomials of the curvature tensor components. Anywhere where any such scalar diverges is referred to as a *scalar polynomial curvature singularity*.

For example, it can be seen from the invariants I and J in (2.13) that a type D solution, expressed in the canonical form in which only the component Ψ_2 is non-zero, has a scalar polynomial curvature singularity whenever this component diverges. However, a similar conclusion cannot be reached for a type N solution which can be expressed in the canonical form in which only Ψ_4 is non-zero. It must be emphasised, however, that the divergence of other types of invariantly defined scalar quantities does not necessarily imply the existence of a curvature singularity (see MacCallum, 2006).

Other (weaker) types of singularities also occur. Points where the metric ceases to be Lorentzian while the curvature tensor remains bounded could simply be places where the coordinates employed become degenerate. If this were the case, a transformation to different coordinates would remove this effect. However, this is not always possible. Such points could alternatively represent some weak kind of singularity that is not surrounded by a diverging gravitational field. In particular, *topological singularities* will often be referred to. These are singularities where the curvature tensor is locally well-behaved, but problems occur in other properties, such as the non-existence of a unique tangent space. The classic example of such a situation is the apex of a cone. Such points are singular for topological reasons rather than because curvature invariants have unbounded limits.

It is also important to emphasise that a curvature singularity is not strictly part of the space-time. Although the solution may be assumed to be valid arbitrarily close to a singularity, this can at most form a boundary of a space-time, which cannot be extended through it. Other parts of the boundary may represent points at infinity, and this structure describes the asymptotic character of the space-time. However, the character of the asymptotic regions needs to be investigated very carefully.

A useful classification scheme for singular boundary points of incomplete space-times has been suggested by Ellis and Schmidt (1977). Basically, such a point is called a curvature singularity if at least one component of the curvature tensor (or its covariant derivatives), evaluated in an orthonormal frame parallelly propagated along a curve approaching the point, does not behave in a continuous way. Otherwise, it is called a quasi-regular singularity. Curvature singularities can further be classified as scalar singularities or non-scalar singularities.

Because of the covariant nature of general relativity, it is very difficult to formulate a completely satisfactory definition of all the different types of singularities that occur. Following the detailed investigation in the 1960s and early 1970s by Penrose, Hawking, Geroch and others of the various singularities that appear in exact solutions of Einstein's equations, an extensive literature now exists on this topic. In fact, the most satisfactory definition may be given in terms of incomplete, inextendible, causal geodesics. Particularly helpful reviews of this topic include those of Geroch (1968), Hawking and Ellis (1973), Tipler, Clarke and Ellis (1980), Clarke (1993) and García-Parrado and Senovilla (2005).

2.3.2 Horizons

The past light cone with vertex at a particular event covers all the points in the space-time that could be seen at that event. Thus, the infinite family of past light cones with vertices on a complete timelike worldline covers all the events that could theoretically be "seen" by an observer with that worldline over its entire history. In a flat space, as will be shown in Chapter 3, this normally spans the complete space-time. However, there are exceptions (such as for uniformly accelerated observers), and even more exceptions occur in a curved space-time. In such cases, events that are not covered by the complete family of past light cones could never be seen by an observer with that worldline. The boundary between events that could be seen and those that cannot forms a kind of horizon for that observer.

In some space-times, it happens that a large family of observers with infinite timelike worldlines could have a common horizon. In such cases, the boundary is appropriately referred to as an *event horizon*.

An analogous horizon also occurs when considering the family of future-oriented light cones. In this case, events that are not covered could never receive light or information from events on past-infinite timelike worldlines.

The presence of these various types of horizon which arise from the global properties of a space-time are frequently, but not necessarily, associated with specific singularities that appear in the metric in some coordinate systems. In such cases, the metric ceases to be Lorentzian on particular hypersurfaces, but the scalar polynomial invariants do not diverge at these locations. The concept of a horizon specifically occurs when such a singularity is found to occur on a null hypersurface. In this case, timelike geodesics representing the motion of test particles could pass through them in one direction only, and without any (locally) detectable effect. The physical interpretation of such singularities depends on an analytic extension through them and on a global analysis of the extended space-time.

The classic example here is part of the coordinate singularity at $r = 2m$ in the Schwarzschild solution (see Chapter 8). This appears as a singularity in the standard coordinates which can be removed by the introduction of at least one null coordinate. It is found that timelike and null geodesics can pass inwards through this hypersurface but not emerge from it. Thus, light emitted from points inside it cannot be seen by observers located outside and the space-time contains an event horizon. The presence of such a horizon thus effectively defines the concept of a *black hole*.

The term "horizon" is in fact widely used to describe a considerable variety of different types of specially identified hypersurfaces which occur in some space-times. Besides event horizons, the following chapters will also refer to *Killing horizons*, which are surfaces where the norm of some Killing vector vanishes. There are also *Cauchy horizons*, which are future boundaries of regions that can be uniquely determined by initial data set on some appropriate spacelike hypersurface. *Cosmological horizons, particle horizons, acceleration horizons, apparent horizons* and other varieties of horizon will also be mentioned below. However, rather than give general and precise definitions of these concepts at this point, we prefer to do so in the context of the particular space-times in which they naturally arise.

For a basic introduction to these concepts see, e.g., Penrose (1964, 1968a), Hawking and Ellis (1973) or Wald (1984).

2.3.3 Causal and conformal relations

Since space-time is locally Lorentzian, any two events in a sufficiently small neighbourhood can be joined by lines that are everywhere either spacelike, timelike or null. In the case in which the events are separated by a timelike line, one occurs "before" the other. Such events are said to be *causally related*. The first event is contained within the past light cone of the second, which it can causally influence. The second event is contained inside the future light cone of the first. Two events that are separated by a spacelike line, on the other hand, are causally unrelated. Information which cannot propagate faster than the speed of light, cannot travel between them.

In a curved space-time, global or large-scale properties can add significantly new features to this simple local causal structure. In a flat space-time, the families of past (or future) light cones with origins on any timelike geodesic span the entire space-time, but this does not necessarily occur in a curved space-time. Indeed, for space-times which include horizons, events occur that can either never causally influence, or never be causally influenced by, any point on families of complete infinite timelike worldlines.

In some space-times, infinite worldlines exist which remain permanently outside each other's light cones. In such cases, observers on these worldlines could never be aware of each other's existence. Distinct regions of space-time that contain families of such worldlines are said to be *causally separated*. When a particular space-time includes such regions, these need to be recognised.

Strictly, nothing is predetermined about the global causal structure of an entire space-time. Extended light cones from given events may be deformed to form caustics. Moreover, some exact solutions of Einstein's equations exist in which particular events are contained within their own future and past light cones. In such cases, extended timelike worldlines could actually intersect themselves. A test particle could thus return to, and hence influence, its own past. Such a possibility would be intolerable philosophically, and would probably not occur in the real world. Thus, the existence of *closed timelike curves* is generally regarded as undesirable in physically significant space-time models. Nevertheless, they are not prohibited by Einstein's field equations, and many interesting exact solutions exist which contain them. Particular examples of such space-times will be described in Subsections 3.4.1 and 11.1.2, Chapters 12 and 13, and Sections 16.3 and 22.6.

A particularly useful way of visualising the global causal structure of a maximally extended space-time is by the use of *conformal diagrams*. As introduced by Penrose (1964, 1965b, 1968a), these represent particular sections of a space-time in which the asymptotic regions ("infinities") have been conformally compactified to construct a finite sized picture. They are achieved by making use of the following properties.

For any space-time with metric $\mathbf{g}$ and manifold $\mathcal{M}$, a related space-time with metric $\tilde{\mathbf{g}}$ and manifold $\tilde{\mathcal{M}}$ is defined by the conformal transformation

$$\tilde{g}_{\mu\nu} = \Omega^2 \, g_{\mu\nu}, \tag{2.26}$$

where the conformal factor Ω can be, in general, an arbitrary function. Tensor indices in the original and conformally related space-times are lowered and raised using the corresponding metric, with $\tilde{g}^{\mu\nu} = \Omega^{-2} \, g^{\mu\nu}$. The significant feature of such transformations is that they preserve the angles between corresponding directions. Hence, the two space-times have the same local light cone structures.

Another remarkable property of the conformal transformation (2.26) is that the Weyl tensor for the original and conformal space-times are identical (i.e. $\tilde{C}^{\kappa}{}_{\lambda\mu\nu} = C^{\kappa}{}_{\lambda\mu\nu}$).[6] Their Ricci tensors, however, are different. Thus, it is generally not possible for both the original and the conformal space-times to be vacuum, or to correspond to the same type of source.

The conformal transformation (2.26) is particularly useful if it maps the asymptotic regions at infinite proper distance in the original space-time to finite regions in the conformal space-time. Taking the conformal factor Ω to be positive in $\mathcal{M}$, the asymptotic boundary of $\mathcal{M}$ maps to the hypersurface in $\tilde{\mathcal{M}}$ on which $\Omega = 0$. This boundary is referred to as *conformal infinity* and denoted as $\mathcal{I}$ (pronounced "scri"). For a review of this topic and its applications, see Frauendiener (2004).

Conformal diagrams, constructed using the relation (2.26), take a particularly simple form when describing spherically symmetric space-times. This enables the space-time to be visualised in a two-dimensional picture in which every point represents a typical point on a 2-sphere at some time. Such diagrams will be given for each of the spherically symmetric space-times described below. For space-times with less symmetry, it is still possible to construct conformal diagrams for specific sections, but to visualise their complete causal structure in a suitable higher-dimensional conformal picture is usually much more difficult.

For further details of the causal properties of space-times, see Carter (1971a), Hawking and Ellis (1973), Penrose and Rindler (1986) or García-Parrado and Senovilla (2005).

---

[6] It is as a consequence of this property that space-times with vanishing Weyl tensor are said to be *conformally flat*. The metrics for such space-times can always be expressed as some conformal multiple of the metric for Minkowski space.

3

Minkowski space-time

The simplest space-time is that which is flat everywhere. Such a space-time contains no matter and no gravitational field. It is known as Minkowski space, and is the space-time of special relativity. It typically occurs as a weak-field limit of many solutions of general relativity, and may also appear as the asymptotic limit of the gravitational field of bounded sources. For all these reasons it is most important that its structure be clearly understood.

As is well known, in a 3+1-dimensional space-time, the maximum number of symmetries is 10. The Minkowski metric has this precise number of isometries, which can be considered to correspond to four translations, three spatial rotations and three special Lorentz boosts. Such a maximally symmetric space-time necessarily has constant curvature, which is zero in this case.

Thus, in any coordinate representation in which it may not be immediately recognisable, Minkowski space (or part of it) can always be uniquely identified by the fact that its curvature tensor vanishes identically.

3.1 Coordinate representations

For this particular space-time, the metric can be expressed in the simple *Cartesian* form

$$ds^2 = -dt^2 + dx^2 + dy^2 + dz^2, \tag{3.1}$$

in which the coordinates t, x, y, z cover their full natural ranges $(-\infty, \infty)$. However, coordinate transformations of this line element represent the same space-time (or part of it) and this may introduce a number of unfortunate features. For example, a transformation to standard *spherical* polar coordinates $x = r \sin\theta \cos\phi$, $y = r \sin\theta \sin\phi$, $z = r \cos\theta$, puts the line element in

27

the form

$$ds^2 = -dt^2 + dr^2 + r^2(d\theta^2 + \sin^2\theta\,d\phi^2), \qquad (3.2)$$

where we may naïvely take $r \in [0,\infty)$, $\theta \in [0,\pi]$, $\phi \in [0,2\pi)$ and $\phi = 2\pi$ identified with $\phi = 0$. In this form, the metric has apparent singularities when $r = 0$, when $\sin\theta = 0$, and when $r \to \infty$. However, it is clear that these are simply coordinate singularities which correspond, respectively, to the origin, the axis of spherical polar coordinates and to spatial infinity. These apparent singularities thus have no physical significance.

In practice, it is often useful to introduce double (retarded and advanced) *null coordinates*. Specifically, putting

$$u = \tfrac{1}{\sqrt{2}}(t - z), \qquad v = \tfrac{1}{\sqrt{2}}(t + z),$$

the line element (3.1) becomes

$$ds^2 = -2\,du\,dv + dx^2 + dy^2. \qquad (3.3)$$

This foliates the space-time into families of *null planes* given by $u = $ const. or $v = $ const., which are orthogonal to the x,y-plane. In fact, it is sometimes convenient to parametrise this plane in terms of the complex coordinate

$$\zeta = \tfrac{1}{\sqrt{2}}(x + i\,y).$$

With this, the metric takes the form

$$ds^2 = -2\,du\,dv + 2\,d\zeta\,d\bar\zeta. \qquad (3.4)$$

With the alternative definitions

$$u = \tfrac{1}{\sqrt{2}}(t - r), \qquad v = \tfrac{1}{\sqrt{2}}(t + r),$$

the line element (3.2) becomes

$$ds^2 = -2\,du\,dv + \tfrac{1}{2}(u - v)^2(d\theta^2 + \sin^2\theta\,d\phi^2). \qquad (3.5)$$

In this case, the hypersurfaces on which u or v are constant are *null cones*, which are, respectively, contracting or expanding spheres whose radius at any value of v and u is given by $r = \tfrac{1}{\sqrt{2}}|v - u|$. These can be represented as past- or future-directed cones on which one dimension is suppressed (e.g. by setting $\theta = \pi/2$). This is illustrated in Figure 3.1, in which (as in many similar subsequent figures) the usual convention is adopted that time is represented in the direction pointing up the page.

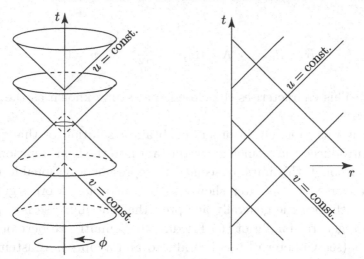

Fig. 3.1 Minkowski space in spherical coordinates (3.2) or (3.5). In the left figure, the θ coordinate is suppressed by setting it to $\pi/2$, so that the contracting and expanding spheres are represented as cones. In the right figure the θ and ϕ coordinates are both suppressed, and each point represents an instantaneous 2-sphere.

3.2 An aside on the Einstein static universe

At this point, it is appropriate to digress for a while to briefly introduce an important exact space-time that Einstein considered in 1917 as an idealised model of a static universe, and that will be described in more detail in Chapter 6. Using the notation introduced below, his initial simplifying assumption that the universe is homogeneous and isotropic leads to the representation of the space-time by the Friedmann–Lemaître–Robertson–Walker metric (6.6), namely

$$ds^2 = -dt^2 + R^2(t)\Big(d\chi^2 + \sin^2\chi\,(d\theta^2 + \sin^2\theta\,d\phi^2)\Big). \qquad (3.6)$$

In this particular case, the spatial sections have constant positive curvature, i.e. they are 3-spheres S^3 (see Appendix B.1).

Back in 1917, it seemed natural to assume that the universe is static, i.e. qualitatively unchanging in its large-scale structure. Einstein observed that an exact solution with this property exists in which the matter in the universe has constant non-zero density ρ and zero pressure p, provided the cosmological constant Λ is strictly positive. This introduces a global repulsion which exactly balances the gravitational attraction between particles of matter in the universe. For such a solution, the field equations (see (6.13) and (6.14)) can indeed be satisfied for a constant radius R of the universe when

$$R = R_0, \qquad \Lambda = 4\pi\rho = \frac{1}{R_0^2}, \qquad p = 0, \tag{3.7}$$

see (6.17). This characterises the closed space-time known as the *Einstein static universe*.

In fact, it was precisely in order to obtain a solution of this type that Einstein introduced the cosmological constant into the field equations (2.10). However, although this initially seemed to represent a reasonable model of a static universe, it was later shown to be unstable. More significantly, it subsequently became generally accepted that the universe is expanding, and a more general family of the Friedmann–Lemaître–Robertson–Walker space-times (see Chapter 6) was introduced and adopted. Einstein's static solution was therefore rejected as a possible cosmological model and the original motivation for the introduction of the cosmological constant ceased to exist. Nevertheless, although the solution (3.6), (3.7) is not a realistic model for our universe, its simple geometry (with seven isometries) is very useful when analysing and describing the global causal structure of other exact solutions.

With the rescaling $t = R_0\,\eta$, the line element (3.6) of the Einstein static universe takes the form

$$ds^2 = R_0^2\Big(- d\eta^2 + d\chi^2 + \sin^2\chi\,(d\theta^2 + \sin^2\theta\,d\phi^2)\Big), \tag{3.8}$$

where all coordinates $\eta \in (-\infty, \infty)$, $\chi \in [0, \pi]$, $\theta \in [0, \pi]$, $\phi \in [0, 2\pi)$ are dimensionless. The spatial sections on which η is constant are 3-spheres, and the Einstein static universe is therefore a manifold with the topology $R^1 \times S^3$. At any time, the universe is spatially closed and without a boundary. As described in detail in Appendix B.1, these spatial sections can be represented schematically as the interior of a single sphere of radius π, spanned by the coordinates χ, θ, ϕ. In this picture, the sections on which χ is constant are represented as 2-spheres of radius χ, even though the actual radii of these 2-spheres in the universe (3.8) is $R_0 \sin\chi$ (their surface area is $4\pi R_0^2 \sin^2\chi$). Thus, as χ increases from zero to π, the actual radii increase to a maximum R_0 at $\chi = \pi/2$ and then decrease again to zero at $\chi = \pi$.

Using this schematic representation of its spatial sections, the particular section of the space-time (3.8) on which η is a constant and $\theta = \pi/2$ is represented as a disc, spanned radially by $\chi \in [0, \pi]$ and with $\phi \in [0, 2\pi)$. By reintroducing the timelike coordinate η, the section of the Einstein static universe with $\theta = \pi/2$ can then be visualised as the interior of an *infinite cylinder* of radius π, as pictured in Figure 3.2 (see Penrose, 1964). Null geodesics with fixed θ, ϕ propagate along the null cones $\eta = \pm\chi + $ const.

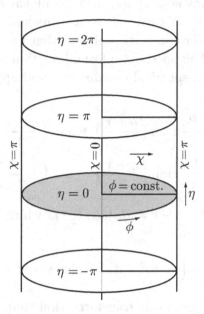

Fig. 3.2 With $\theta = \pi/2$ in the metric (3.8), the Einstein static universe can be represented as the interior of an infinite cylinder of radius π with the radial coordinate $\chi \in [0, \pi]$ and vertical time coordinate $\eta \in (-\infty, \infty)$. With θ reintroduced, each spatial section $\eta = $ const. corresponds to a complete 3-sphere whose actual radius is R_0. While $\chi = 0$ corresponds to the North pole of S^3, its South pole at $\chi = \pi$ is mapped to the complete outer cylindrical surface with ϕ arbitrary.

3.3 Conformal structure of Minkowski space

The global or asymptotic structure of any space-time can often be helpfully analysed by applying a coordinate transformation which maps points at infinity to finite values of some new parameter. This effectively compactifies an infinite space-time to a finite region of some parameter space. In this new "unphysical" space, it is then possible to analyse the properties of the asymptotic regions of the physical space-time. This compactification process is particularly useful when the metric in the new coordinates is *conformally* related (see Section 2.3.3) to some known space-time that has a simple geometrical structure (Penrose, 1963, 1964). The reason for this is that conformal transformations preserve the local causal structure of the space-time. So, if the metric to which it is conformally related is understood, the asymptotic and global properties of the space-time studied can be helpfully pictured.

In the case of flat Minkowski space, such a conformal compactification can be carried out explicitly. It turns out that Minkowski space is conformally related to part of the Einstein static universe that has been described in the previous Section 3.2. This specific conformal relation is obtained by starting with the metric (3.2) in spherical coordinates, and applying the coordinate transformation

$$
\begin{aligned}
t &= \frac{R_0}{2}\left[\tan\left(\frac{\eta+\chi}{2}\right) + \tan\left(\frac{\eta-\chi}{2}\right)\right], \\
r &= \frac{R_0}{2}\left[\tan\left(\frac{\eta+\chi}{2}\right) - \tan\left(\frac{\eta-\chi}{2}\right)\right],
\end{aligned}
\tag{3.9}
$$

i.e. $t + r = R_0 \tan(\frac{\eta+\chi}{2})$, $t - r = R_0 \tan(\frac{\eta-\chi}{2})$, which brings the Minkowski line element into the form

$$
ds^2 = \frac{R_0{}^2}{(\cos\eta + \cos\chi)^2}\left(-d\eta^2 + d\chi^2 + \sin^2\chi\,(d\theta^2 + \sin^2\theta\,d\phi^2)\right). \tag{3.10}
$$

It immediately follows from this transformation that, in sections for which θ and ϕ are constant, the radial null cone structure is preserved, and the infinite ranges of t and r are mapped to finite ranges of η and χ.

The physical metric (3.10) can be seen to be conformal via (2.26) to that of the Einstein static universe as given by (3.8) with the conformal factor

$$
\Omega = \cos\eta + \cos\chi = 2\cos\left(\frac{\eta+\chi}{2}\right)\cos\left(\frac{\eta-\chi}{2}\right).
$$

The boundary of the physical Minkowski space-time is then given by the points where $\Omega = 0$, i.e. by $\eta + \chi = \pi$ and $\eta - \chi = -\pi$ (since $\chi \in [0, \pi]$). These points correspond to infinity in the physical space-time. For a section on which $\theta = \pi/2$, they lie on a pair of null cones within the cylinder that represents the Einstein static universe, as illustrated in Figure 3.2. Such null cones are pictured in Figure 3.3, and Minkowski space can thus be seen to be conformal to the region between them.

This figure uses the schematic representation of the Einstein static universe in which the radial distance is taken to be χ, and in which the sections on which η and χ are constant are 2-spheres, spanned by θ and ϕ, whose radius is maximum at $\chi = \pi/2$ and zero at both $\chi = 0$ and $\chi = \pi$. For the section on which $\theta = \pi/2$, the entire Minkowski space is conformal to the region between the two null cones, as illustrated in Figure 3.3. This immediately illustrates the important fact that the *conformal infinity* of Minkowski

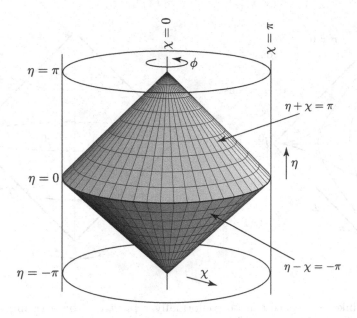

Fig. 3.3 The section $\theta = \pi/2$ of Minkowski space is conformal to the region between the two null cones $\eta + \chi = \pi$ and $\eta - \chi = -\pi$ that is within the timelike cylinder of radius π representing the Einstein static universe, as shown in Figure 3.2.

space is *null*, and denoted as $\mathcal{I}$. Also, since the radius of the 2-spheres vanishes at $\chi = \pi$, it is demonstrated that the "point" at spatial infinity i^0 is indeed just a point.

The space-time (3.10) is spherically symmetric and it is convenient to suppress both the θ and ϕ coordinates. In this case, the space-time is represented by a triangular section $\phi = $ const. through the interior of the double cone that is shown in Figure 3.3. This is illustrated in the left diagram in Figure 3.4. In this section, curves on which the physical coordinates r and t are constant can be drawn explicitly using (3.9).

Figures such as this, in which an infinite space-time is represented by a two-dimensional finite region to which it is conformally related, are known as *Penrose diagrams* (see Penrose, 1964, 1968a). It is sometimes helpful, however, to keep the coordinate ϕ and to represent the space-time as the right picture in Figure 3.4. Even though the θ-coordinate is still suppressed, the null character of future and past conformal infinity $\mathcal{I}$ are illustrated more transparently.

In either of the pictures in Figure 3.4, $\chi = 0$ represents the origin of the spherical coordinates $r = 0$, and $\eta = 0$ corresponds to $t = 0$. Past confor-

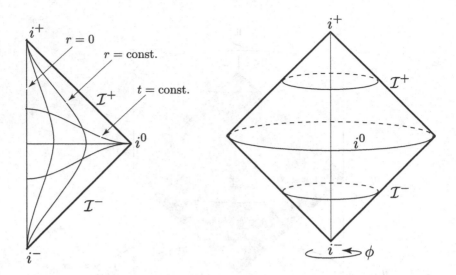

Fig. 3.4 Minkowski space can be conformally represented by a triangular region. The left figure is the Penrose diagram in which the coordinates θ and ϕ are suppressed. In the right figure the ϕ coordinate has been retained, as in Figure 3.3. In either case, the asymptotic and causal structure of Minkowski space is clearly demonstrated.

mal infinity $\eta - \chi = -\pi$ is denoted by $\mathcal{I}^-$, and future conformal infinity $\eta + \chi = \pi$ is denoted by $\mathcal{I}^+$. This is where all complete null geodesics must start and finish, respectively. However, timelike geodesics originate at *past timelike infinity* at the "point" $\chi = 0$, $\eta = -\pi$ which is denoted by i^-, and finish at *future timelike infinity* at the "point" $\chi = 0$, $\eta = \pi$ which is denoted by i^+. Finally, spacelike geodesics both start and finish at *spatial infinity* i^0, which is the "point" at infinity $\chi = \pi$, $\eta = 0$. All points in the two-dimensional Penrose diagram represent 2-spheres, except for i^-, i^+ and i^0, which are single points.

Finally, it may be noted that space-times with the same local structure as Minkowski space, but with different large-scale topological properties, can be constructed by identifying specific points in the manifold. This could possibly introduce closed timelike curves. Generally, space-times that are constructed artificially in this way are of more mathematical than physical interest. The following examples of this procedure, however, lead to space-times which have some possible physical relevance.

3.4 A simple model for a cosmic string

The entire Minkowski space (3.1) can also be expressed in terms of standard cylindrical coordinates using $x = \rho \cos \varphi$, $y = \rho \sin \varphi$, where $\rho \in [0, \infty)$ and $\varphi \in [0, 2\pi)$, giving

$$\mathrm{d}s^2 = -\mathrm{d}t^2 + \mathrm{d}\rho^2 + \rho^2 \,\mathrm{d}\varphi^2 + \mathrm{d}z^2, \qquad (3.11)$$

It is now possible to artificially remove a wedge of angle $2\pi\delta$ from all spatial sections of the space-time by simply taking $\varphi \in [0, (1 - \delta)2\pi)$. Continuity can then be re-established by identifying the two faces of the missing wedge, as illustrated schematically in Figure 3.5. The resulting space-time remains flat everywhere except along the axis $\rho = 0$. The spatial surfaces on which t and z are constant can be represented as cones and, although the curvature tensor still vanishes everywhere off the axis, the axis itself is singular. This kind of topological defect is often referred to as a *conical singularity*.

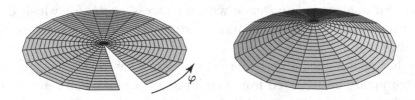

Fig. 3.5 Removing a wedge from a flat disc, as shown on the left, and identifying the new edges produces a cone, as shown on the right. Applying this procedure to all spatial sections of Minkowski space produces a space-time that is flat everywhere except along the axis $\rho = 0$. This may be interpreted as representing a straight cosmic string.

In this case, the space-time representing a conical singularity with deficit angle $2\pi\delta$ may be more conveniently expressed by introducing the rescaling $\varphi = C\phi$, where $\phi \in [0, 2\pi)$ and

$$C = 1 - \delta.$$

With this, the line element takes the form

$$\mathrm{d}s^2 = -\mathrm{d}t^2 + \mathrm{d}\rho^2 + C^2\rho^2 \,\mathrm{d}\phi^2 + \mathrm{d}z^2. \qquad (3.12)$$

In fact, the question of the regularity of the axis is one that must be addressed for all metrics which represent axially symmetric space-times, as considered in Chapters 10–16.

From Figure 3.5, it can be seen that a pair of parallel lines which approach the axis and pass either side of it will then start to converge, and will meet after some finite distance. Thus, although the space-time is locally flat everywhere except along the axis itself so that no tidal gravitational forces occur, the presence of the conical singularity still has a global focusing effect on geodesics which pass either side of it. This mimics the effect of an attractive gravitational source even though the space-time curvature does not diverge near the singularity. In fact, a distributional component occurs in the Ricci tensor and this corresponds, through the field equations, to a non-zero stress component in the energy-momentum tensor (see Israel, 1977). Thus, this space-time can be interpreted as that due to an infinite line source under constant tension.

Such a space-time is often taken as a toy model of a "cosmic string" in which the deficit angle represents the tension in the string. In this case, the mass per unit length of the string, denoted by μ, is given by $\mu = \frac{1}{4}\delta = \frac{1}{4}(1 - C)$. However, it should be stressed that this is not the usual kind of gravitational mass as the surrounding field is locally flat. (And, actually, it is not generally equal to the mass of any interior string solution either.) An analogous line source with an *excess angle* (for which $C > 1$) would equivalently represent an infinite strut (or rod) under compression, but with $\mu < 0$.

Whether or not such objects actually occur in the real world, they have been thoroughly investigated from many points of view. For a discussion of the physics of cosmic strings and their possible significance in cosmology, see in particular the reviews of Vilenkin (1985), Vilenkin and Shellard (1994) and Hindmarsh and Kibble (1995). For a more mathematical discussion, see Anderson (2003). Geodesic motions in this space-time have been analysed in detail by Gal'tsov and Masar (1989). For a rigorous use of distributions and generalised functions to model conical singularities and cosmic strings, see Steinbauer and Vickers (2006).

3.4.1 A spinning cosmic string

The replacement of t by $t + Ca\phi$ in the metric (3.12) leads to the line element

$$\mathrm{d}s^2 = -(\mathrm{d}t + Ca\,\mathrm{d}\phi)^2 + \mathrm{d}\rho^2 + C^2\rho^2\,\mathrm{d}\phi^2 + \mathrm{d}z^2, \qquad (3.13)$$

where $\phi \in [0, 2\pi)$ and $\rho \in [0, \infty)$. This is also clearly flat everywhere except on the axis $\rho = 0$. However, in this case, the hypersurfaces at $\phi = 0$ and $\phi = 2\pi$ should be joined in such a way that points with the same values of ρ, z and $t + Ca\phi$ are identified. This is illustrated in Figure 3.6.

By considering small circles around the axis on the surface on which $t + Ca\phi = $ const. using the above identifications, it may be seen that there still remains a deficit angle $2\pi\delta$ on the axis $\rho = 0$, where $\delta = 1 - C$.

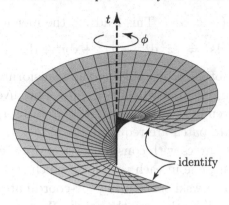

Fig. 3.6 For the surface on which $t + Ca\phi$ is a constant, points at $\phi = 2\pi$ are identified with points at $\phi = 0$ with the same values of ρ and z. This results in the presence of a kind of singularity on the axis which could represent a *spinning* cosmic string. The z-coordinate is suppressed in this figure.

The parameter a in (3.13) represents a kind of rotation. Deser, Jackiw and 't Hooft (1984) and Mazur (1986) have accordingly interpreted the metric (3.13) as describing the external field of a *spinning cosmic string* whose mass and angular momentum per unit length are given, respectively, by $\mu = \frac{1}{4}\delta = \frac{1}{4}(1 - C)$ and $J = \frac{1}{4}(1 - \delta)a$, so that $Ca = 4J$. In the context of a curved space-time that will be described in Section 12.1, Bonnor (2001) has referred to such a conical singularity with a time shift as a "torsion singularity".

In the metric (3.13), the coefficient of $\mathrm{d}\phi^2$ is $C^2(\rho^2 - a^2)$. Thus, when $\rho < |a|$, ϕ is a timelike coordinate, and circles around the axis on which t, ρ and z are constant are closed timelike curves. Although the appearance of such curves is problematic in that they violate all concepts of causality, their presence is a common feature of (analytic extensions of) exact space-times representing rotating systems in general relativity, as will be amply illustrated in following chapters.

Further properties of these space-times have been described by Jensen and Soleng (1992) and Mena, Natário and Tod (2008) (see also the review by Anderson, 2003), who have considered different interior solutions around the axis so as to avoid the occurrence of such closed timelike curves.

3.5 Coordinates adapted to uniform acceleration

Returning to the line element (3.1) for Minkowski space, within the region in which $z > |t|$, consider the transformation $t = \tilde{z}\sinh\tilde{t}$, $z = \tilde{z}\cosh\tilde{t}$, where

$\tilde{z} \in (0, \infty)$ and $\tilde{t} \in (-\infty, \infty)$. This leads to the metric form

$$ds^2 = -\tilde{z}^2 \, d\tilde{t}^2 + dx^2 + dy^2 + d\tilde{z}^2. \qquad (3.14)$$

The same metric is obtained by applying the transformation $t = -\tilde{z} \sinh \tilde{t}$, $z = -\tilde{z} \cosh \tilde{t}$, in the region in which $z < -|t|$. The inverse transformation for both regions is given by $\tilde{z} = \sqrt{z^2 - t^2}$, $\tilde{t} = \operatorname{arctanh}(t/z)$, thus two copies of the same coordinate patch are required.

It can be seen that curves with constant $\tilde{z}$, x and y are two hyperbolae on which $z^2 - t^2 = \text{const.}$, one in each region in which $|z| > |t|$, in the original Cartesian form of Minkowski space using t, z-coordinates, while curves with constant $\tilde{t}$ are straight lines through the origin. These are illustrated in Figure 3.7. (Note that, within the region $z < -|t|$, an increasing $\tilde{t}$ corresponds to a decreasing t.)

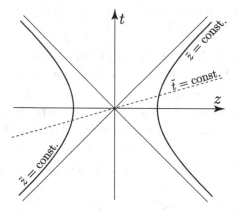

Fig. 3.7 For the metric (3.14), curves with constant $\tilde{z}$ are hyperbolae in the regions $|z| > |t|$ of Minkowski space in the Cartesian coordinates of (3.1). These correspond to worldlines of points with constant uniform acceleration.

Of course, the metric (3.14) is just a different representation of the parts of Minkowski space for which $|z| > |t|$. However, curves on which x, y and $\tilde{z}$ are constants are timelike, and it is appropriate to consider the motion of test particles for which $\tilde{z} = \alpha^{-1}$. Such particles follow trajectories given by

$$t = \pm \frac{1}{\alpha} \sinh \tilde{t}, \qquad z = \pm \frac{1}{\alpha} \cosh \tilde{t},$$

where the signs of both terms are the same. (In fact, these coincide with the orbits of the boost Killing vector $z\partial_t + t\partial_z$.) Introducing the proper time $\tau = \alpha^{-1}\tilde{t}$, the four-velocity of these particles is $u^\mu = (\pm \cosh \tilde{t}, 0, 0, \pm \sinh \tilde{t})$, which satisfies the condition $u_\mu u^\mu = -1$. Similarly, their four-acceleration is $a^\mu = \mathrm{D}u^\mu / \mathrm{d}\tau = (\pm \alpha \sinh \tilde{t}, 0, 0, \pm \alpha \cosh \tilde{t})$, which has constant magnitude α.

Such particles thus have constant acceleration α in the positive or negative $\tilde{z}$ direction, as illustrated in Figure 3.7. The metric (3.14) is therefore referred to as the *uniformly accelerated metric*.[1] However, points with different values of $\tilde{z}$ have a different value for their acceleration.

Clearly the coordinates of the line element (3.14) do not cover the complete space-time. They only cover the two distinct regions for which $|z| > |t|$ that are causally separated. These have null boundaries on which $t = \pm z$. The space-time can obviously be extended through these by reverting to the original coordinates of (3.1).

An equivalent representation of the other regions of the space-time for which $|t| > |z|$ can be obtained from (3.1) by applying the transformation $t = \pm \bar{t} \cosh \bar{z}$, $z = \pm \bar{t} \sinh \bar{z}$, where $\bar{z} \in (-\infty, \infty)$ and $\bar{t} \in (0, \infty)$, and again the choice of signs is the same for the two copies of this region. This takes the metric for these regions to the form

$$\mathrm{d}s^2 = -\mathrm{d}\bar{t}^2 + \mathrm{d}x^2 + \mathrm{d}y^2 + \bar{t}^2 \, \mathrm{d}\bar{z}^2, \tag{3.15}$$

which is a special case of the Kasner metric that will be presented in Section 22.1.

The complete Minkowski space is thus covered by the four above regions that are separated by the null hypersurfaces $t = \pm z$. These can also be interpreted as a kind of event horizon in the sense that any uniformly accelerating observer moving on a worldline on which $\tilde{z}$ is a constant will never become aware of any event on the other side of the corresponding horizon. This type of event horizon is referred to as an *acceleration horizon*. In the coordinates of (3.14), it is given by $\tilde{z} = 0$.

Finally, it must again be emphasised that the metric (3.14) is just a different coordinate representation of part of Minkowski space. Its advantage is that worldlines for which $\tilde{z}$ is constant are uniformly accelerating. Of course, since these are not geodesics, they cannot represent worldlines of particles unless some physical cause for the acceleration is also included.

Minkowski space in these coordinates will again be considered in Section 14.1.4, where the corresponding conformal diagram will be explicitly constructed. This is important in the context of Chapter 14, in which this metric arises as a limit of a solution for an object that is accelerating under the action of a cosmic string.

---

[1] The line element (3.14) is sometimes referred to as the Rindler metric, following Rindler (1966). However, it was widely used to represent accelerated coordinates many years previously (see e.g. Bondi, 1957a).

3.6 Other representations of Minkowski space

Partly because of its importance, either as the simplest background space, or as a limit of many exact solutions of Einstein's equations, Minkowski space will again be referred to in many of the remaining chapters of this book. For example, in the double-null form (3.4), it occurs as a particular case of the direct-product spaces that are described in Section 7.2. However, elsewhere some other important coordinate representations or foliations will be described. Locations where these occur are listed here for convenience.

The next explicit appearance of Minkowski space occurs in the context of cosmological models in Chapter 6. Here, a particular coordinate form for part of Minkowski space is used to represent a flat vacuum model, known as the Milne universe. The metric (6.19) is based on an expanding coordinate frame which originates at a single event. This is described in Section 6.3.1.

When considering static axially symmetric space-times in terms of the Weyl metric (10.2), it is found that Minkowski space (or part of it) is included in three distinct ways. These are presented explicitly in Section 10.1.1.

The flat limit of the stationary axially symmetric Kerr solution will be described in Section 11.1.2. Such a limit is used to partly reveal the ring-like character of the singularity that occurs in this space-time.

Some exact space-times that include radiation are described in Chapters 18 and 19. In their weak-field limits, these Kundt and Robinson–Trautman type N solutions reduce to Minkowski space in forms in which particular coordinate foliations represent corresponding wave surfaces. These are described in detail in Sections 18.4.1 and 19.2.1, respectively.

Impulsive waves in a Minkowski background are described in Chapter 20. For both non-expanding and expanding waves, different representations of Minkowski space may be employed on either side of the impulsive wave surface.

Finally, for colliding plane wave space-times as described in Chapter 21, the initial background region ahead of the approaching waves is taken to be the flat Minkowski space. Moreover, for the collision of impulsive waves as represented by the Khan–Penrose solution, the region behind each wave prior to their collision is also flat. It is this fact that enables certain geometrical properties of the space-time to be determined explicitly as described in Section 21.3.

4

de Sitter space-time

The maximum number of isometries of a four-dimensional space-time is 10. Such maximally symmetric spaces have constant curvature R and are locally characterised by the condition $R_{\alpha\beta\gamma\delta} = \frac{1}{12}R\,(g_{\alpha\gamma}g_{\beta\delta} - g_{\alpha\delta}g_{\beta\gamma})$. The case when R vanishes is the flat Minkowski space that has been described in the previous chapter. The space with $R > 0$ is de Sitter space-time, and the space for $R < 0$ is known as anti-de Sitter space-time. This trio of simplest exact solutions of Einstein's field equations holds a privileged position among the whole class of space-times. It is not surprising that they have been widely used as important models in theoretical physics and cosmology.

It can easily be shown that these three spaces of constant curvature are the only vacuum solutions of (2.10) with a vanishing Weyl tensor ($C_{\alpha\beta\gamma\delta} = 0$). Indeed, contracting Einstein's field equations with $T_{\mu\nu} = 0$ yields $R = 4\Lambda$, and consequently $R_{\alpha\beta} = \Lambda g_{\alpha\beta}$. Using the definition of the Weyl tensor (2.6), it may thus be concluded that Minkowski, de Sitter, and anti-de Sitter space-times are conformally flat vacuum solutions with vanishing, positive, and negative cosmological constant Λ, respectively. In this chapter, we will consider the case in which the cosmological constant is positive.

4.1 Global representation

Soon after the discovery of this solution by de Sitter (1917a), it was realised (see de Sitter, 1917b; Lanczos, 1922) that the de Sitter manifold can be visualised as the hyperboloid

$$-Z_0^2 + Z_1^2 + Z_2^2 + Z_3^2 + Z_4^2 = a^2, \quad \text{where} \quad a = \sqrt{3/\Lambda}, \qquad (4.1)$$

embedded in a flat five-dimensional Minkowski space

$$\mathrm{d}s^2 = -\mathrm{d}Z_0^2 + \mathrm{d}Z_1^2 + \mathrm{d}Z_2^2 + \mathrm{d}Z_3^2 + \mathrm{d}Z_4^2. \qquad (4.2)$$

41

This geometric representation of de Sitter space-time is related to its symmetry structure which is characterised by the ten-parameter group SO(1,4).

The entire hyperboloid is covered most naturally by coordinates (t, χ, θ, ϕ) such that

$$
\begin{aligned}
Z_0 &= a \sinh \frac{t}{a}, \\
Z_1 &= a \cosh \frac{t}{a} \cos \chi, \\
Z_2 &= a \cosh \frac{t}{a} \sin \chi \cos \theta, \\
Z_3 &= a \cosh \frac{t}{a} \sin \chi \sin \theta \cos \phi, \\
Z_4 &= a \cosh \frac{t}{a} \sin \chi \sin \theta \sin \phi.
\end{aligned}
\tag{4.3}
$$

In these coordinates the de Sitter metric takes the Friedmann–Lemaître–Robertson–Walker (FLRW) form with spatial curvature $k = +1$,

$$
\mathrm{d}s^2 = -\mathrm{d}t^2 + a^2 \cosh^2 \frac{t}{a} \left(\mathrm{d}\chi^2 + \sin^2 \chi \left(\mathrm{d}\theta^2 + \sin^2 \theta \, \mathrm{d}\phi^2 \right) \right),
\tag{4.4}
$$

see (6.6). For $t \in (-\infty, +\infty)$, $\chi \in [0, \pi]$, $\theta \in [0, \pi]$, and $\phi \in [0, 2\pi)$ the coordinates cover the whole space, although there are trivial coordinate singularities at $\chi = 0, \pi$, and $\theta = 0, \pi$ which correspond to the familiar location

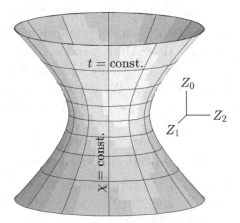

Fig. 4.1 The visualisation of de Sitter space-time as a hyperboloid embedded in a flat five-dimensional space-time in the parametrisation given by global coordinates (t, χ, θ, ϕ). The surface drawn is that for the pole $\theta = 0$ when $Z_2 > 0$ and $\theta = \pi$ when $Z_2 < 0$, so that $Z_3 = 0 = Z_4$. With the range of θ and ϕ reintroduced, each point on this hyperboloid represents a corresponding two-dimensional hemisphere, according as $\theta < \pi/2$ or $\theta > \pi/2$.

of poles in spherical polar coordinates. This natural parametrisation of the de Sitter hyperboloid is illustrated in Figure 4.1.

The spatial sections at a fixed synchronous time t are 3-spheres S^3 of constant positive curvature (see Appendix B) which have radius $a \cosh \frac{t}{a}$. These contract to a minimum size a at $t = 0$, and then re-expand. The de Sitter space-time thus has a natural topology $R^1 \times S^3$.

4.2 Conformal structure

Introducing now a conformal time coordinate η by

$$\sin \eta = \operatorname{sech} \frac{t}{a}, \qquad (4.5)$$

(or equivalently $\tan \frac{\eta}{2} = e^{t/a}$), the metric (4.4) becomes

$$\mathrm{d}s^2 = \frac{a^2}{\sin^2 \eta} \left(- \mathrm{d}\eta^2 + \mathrm{d}\chi^2 + \sin^2 \chi \left(\mathrm{d}\theta^2 + \sin^2 \theta \, \mathrm{d}\phi^2 \right) \right). \qquad (4.6)$$

In view of (3.8), the de Sitter space-time is thus conformal, via (2.26), to part of the Einstein static universe with the conformal factor

$$\Omega = \sin \eta, \qquad (4.7)$$

and identification $R_0 = a$. The boundary of de Sitter space-time, given by $\Omega = 0$, is located at $\eta = 0$ and $\eta = \pi$, which correspond to past and future conformal infinities $\mathcal{I}^-$ and $\mathcal{I}^+$, respectively, as illustrated in Figure 4.2. In contrast to Minkowski space (see Figures 3.3 and 3.4), the infinities $\mathcal{I}^-$ and

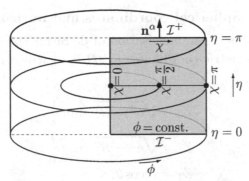

Fig. 4.2 The conformal structure of de Sitter space-time (with $\theta = \frac{\pi}{2}$). The whole de Sitter space is conformal to the region $\eta \in (0, \pi)$ of the Einstein static universe, represented as an embedded solid cylinder whose radius and length are both equal to π. The centre $\chi = 0$ represents the North pole of the 3-sphere S^3, whereas the outer boundary $\chi = \pi$ represents the South pole. The conformal infinities $\mathcal{I}^-$ and $\mathcal{I}^+$ are spacelike.

$\mathcal{I}^+$ now have a *spacelike* character. Indeed, the normal vector defined as $\mathbf{n}^\alpha = \tilde{g}^{\alpha\beta}\,\Omega_{,\beta}$, using (2.26), is obviously timelike.

The two-dimensional Penrose diagram of de Sitter space, corresponding to the shaded section $\theta = \frac{\pi}{2}$, $\phi =$ const. in Figure 4.2, has the form shown in Figure 4.3. For future comparison with alternative sections through the conformal space-time, a complete section (obtained by rotating about $\chi = 0$) has also been indicated by dashed lines.

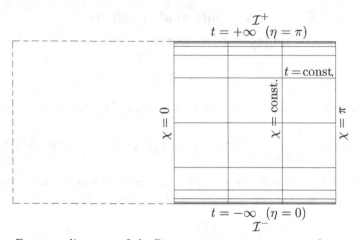

Fig. 4.3 The Penrose diagram of de Sitter space-time corresponds to the shaded region in Figure 4.2 which has a fixed value of $\phi \in [0, 2\pi)$. Taking all possible values of θ and ϕ, each point in the square shown represents a complete 2-sphere of radius $\sin\chi$. For $\chi = 0$ and $\chi = \pi$ these are single points, namely the poles of S^3.

4.3 Spherical coordinates and horizons

Another frequently used parametrisation of de Sitter space-time is given by the spherically symmetric coordinates (T, R, θ, ϕ) which can be introduced by taking

$$
\begin{aligned}
Z_0 &= \sqrt{a^2 - R^2}\,\sinh\frac{T}{a}, \\
Z_1 &= \sqrt{a^2 - R^2}\,\cosh\frac{T}{a}, \\
Z_2 &= R\cos\theta, \\
Z_3 &= R\sin\theta\cos\phi, \\
Z_4 &= R\sin\theta\sin\phi,
\end{aligned}
\tag{4.8}
$$

where $T \in (-\infty, +\infty)$, $R \in [0, a)$, $\theta \in [0, \pi]$ and $\phi \in [0, 2\pi)$. These cover only part of the whole de Sitter space, as illustrated in Figure 4.4. In par-

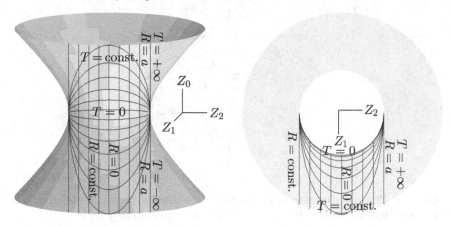

Fig. 4.4 The de Sitter hyperboloid with spherical static coordinates (T, R, θ, ϕ). The points $Z_2 > 0$ correspond to the pole $\theta = 0$ while those with $Z_2 < 0$ represent the opposite pole $\theta = \pi$.

ticular, it may be noticed that the spacelike sections $T = \text{const.}$ are sections such that $Z_0/Z_1 = \tanh(T/a)$, and $R = \text{const.}$ are vertical hyperboloidal lines $Z_1^2 - Z_0^2 = a^2 - R^2$ that degenerate to straight null lines when $R = a$. The corresponding form of the metric is

$$\mathrm{d}s^2 = -\left(1 - \frac{\Lambda}{3}R^2\right)\mathrm{d}T^2 + \left(1 - \frac{\Lambda}{3}R^2\right)^{-1}\mathrm{d}R^2 + R^2(\mathrm{d}\theta^2 + \sin^2\theta\,\mathrm{d}\phi^2). \quad (4.9)$$

Interestingly, this form of the metric also describes the anti-de Sitter spacetime (5.6) when $\Lambda < 0$, and for $\Lambda = 0$ it reduces to the spherically symmetric form (3.2) of Minkowski space. For $R < a = \sqrt{3/\Lambda}$, the metric is obviously static. Performing a simple transformation $R = a \sin \tilde\chi$ in (4.9) gives

$$\mathrm{d}s^2 = -\cos^2\tilde\chi\,\mathrm{d}T^2 + a^2\left(\mathrm{d}\tilde\chi^2 + \sin^2\tilde\chi\left(\mathrm{d}\theta^2 + \sin^2\theta\,\mathrm{d}\phi^2\right)\right). \quad (4.10)$$

It can thus be seen that each section $T = \text{const.}$ is a 3-sphere of constant radius a.

In terms of the (T, R, θ, ϕ) coordinates, the Penrose conformal diagram is drawn in Figure 4.5. In this figure, the conformal space is divided into four regions labelled I–IV. The static coordinates in the range $R \in [0, a)$ cover only the region I which is just a quarter of the complete de Sitter spacetime. However, it is obviously possible to extend the coordinates of (4.9) beyond the limit $R = a$ to two non-static regions II and IV in which $R > a$ becomes a timelike coordinate, while T is a spatial coordinate. To cover the entire de Sitter manifold one needs, in fact, to consider *four* coordinate charts of the type (4.8), each of which has $T \in (-\infty, +\infty)$, with the factor

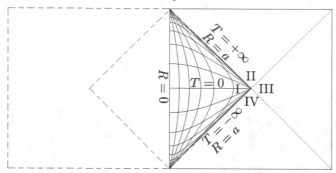

Fig. 4.5 The Penrose diagram of de Sitter space-time with static spherically symmetric coordinates. Each point represents a 2-sphere spanned by θ, ϕ.

$\sqrt{a^2 - R^2}$ replaced by $\pm\sqrt{|a^2 - R^2|}$. The regions I, III and II, IV are covered, respectively, by $R \in [0, a)$ and $R \in (a, \infty)$. These two pairs of regions are distinguished by the choice of the sign $\pm$ and interchanging sinh and cosh in (4.8).

In fact, the null hypersurface $R = a$ forms a *cosmological horizon* of the de Sitter universe, which is an important concept. Geometrically, it is a *Killing horizon* because the norm of the Killing vector ∂_T vanishes there. Its presence is a consequence of the fact that the de Sitter cosmological universe is expanding so fast that there are events which will *never* be seen by an observer at $R = 0$ – they forever lie beyond their horizon.

To define this property more precisely, consider a geodesic timelike observer moving along $R = 0$ (with θ, ϕ fixed) on the de Sitter hyperboloid of Figure 4.4. Due to the homogeneity of the de Sitter space, this in fact represents any natural geodesic observer in this universe. The past null cone with a vertex at $R = 0$ at a given time consists of all events in the space-time that can be seen by the observer at that time. At later times, more events are visible. As can be immediately seen from the conformal diagram shown in Figure 4.5, even after an infinite time the observer will have seen only those events that are located in regions I and IV. All events in regions II and III are outside this limiting past null cone, and thus they are not – and never could be – seen by the observer at $R = 0$. They cannot influence him because they forever lie beyond his horizon. Such a boundary of the past of the geodesic worldline is called the *future event horizon*. For the observer $R = 0$ in de Sitter space-time the future event horizon is located on the null surface $R = a$, $T = +\infty$, as it geometrically separates observable events in the regions I and IV from the unobservable events in regions II and III.

Analogously, it is possible to consider all future null cones "emanating" from the observer's geodesic $R = 0$ in the de Sitter universe. Outside the union of all these cones there are events which will never be influenced by the observer. The corresponding boundary is called the *past event horizon*. Clearly, in the diagram 4.5, such a boundary exists and separates the domains of influence I and II from the two distant domains IV and III. The past event horizon for the geodesic observer $R = 0$ is thus located on the null surface $R = a$, $T = -\infty$.

The existence of the cosmological horizons in the de Sitter universe – namely the future and past event horizons for geodesic observers – is a unique property of this space-time. In fact, it is a direct consequence of the spacelike character of the conformal infinities $\mathcal{I}^+$ and $\mathcal{I}^-$. Such horizons do not occur for *geodesic* observers in Minkowski space-time because its conformal infinity is null (see Figure 3.4). It may appear in flat space only for *accelerating* observers because their trajectories in the conformal diagram terminate at $\mathcal{I}^\pm$ rather than at $i^\pm$. In such a case, the corresponding horizon is called an *acceleration horizon* (see Section 3.5).

It is also important to note that these cosmological horizons are observer dependent. Geodesic observers on the de Sitter hyperboloid with other values of the coordinate χ, as shown in Figure 4.1, would have different horizons. (This is different from the Schwarzschild case that will be discussed later in Chapter 8, in which the horizons are defined relative to a fixed source rather than a specific observer.)

Now, consider natural timelike worldlines in the static coordinates (4.9), namely

$$x^\mu(\tau) = \left(a\tau/\sqrt{a^2 - R_0^2}, R_0, \theta_0, \phi_0 \right), \tag{4.11}$$

where R_0, θ_0, ϕ_0 are constants. The four-velocity is $u^\mu = (a/\sqrt{a^2 - R_0^2}, 0, 0, 0)$ such that $u_\mu u^\mu = -1$, so that τ is the proper time. The four-acceleration is

$$a^\mu \equiv \frac{\mathrm{D}u^\mu}{\mathrm{d}\tau} = u^\mu{}_{;\nu} u^\nu = (0, -R_0/a^2, 0, 0),$$

which is constant with magnitude

$$\alpha \equiv |a^\mu| = \frac{R_0}{a\sqrt{a^2 - R_0^2}}.$$

Since $a^\mu u_\mu = 0$, this constant value of α is identical to the modulus of the 3-acceleration measured in the natural local orthonormal frame of the observer. In addition, a^μ preserves its spatial direction.

The worldlines (4.11) therefore represent the motion of *uniformly accelerating observers* in a de Sitter universe. When $R_0 = 0$ the worldline is a geodesic with zero acceleration. On the other hand, as $R_0 \to a$, the acceleration α becomes unbounded and worldlines approach the null cosmological horizon. In general, these uniformly accelerated trajectories with $R = \mathrm{const.}$ are given by constant values of Z_2, Z_3, Z_4 on the de Sitter hyperboloid, and coincide with the orbits of the isometry generated by the Killing vector ∂_T indicated in Figures 4.4 and 4.5. Coordinates that are naturally adapted to such uniformly accelerated observers in de Sitter space, analogous to (3.14), have been introduced by Podolský and Griffiths (2001c), see Section 14.4.

4.4 Other standard coordinates

Due to the very high degree of symmetry, there exists a large number of different privileged coordinate systems in which the de Sitter metric takes a simple form. Here we briefly describe some of the most important ones. Further details are given, e.g., in Hawking and Ellis (1973), Schmidt (1993), Eriksen and Grøn (1995), and Bičák and Krtouš (2005).

4.4.1 Conformally flat coordinates

Using familiar Cartesian-like coordinates (η_0, x, y, z), the de Sitter hyperboloid is covered by

$$
\begin{aligned}
Z_0 &= \frac{a^2 + s}{2\,\eta_0}, \\[2mm]
Z_1 &= \frac{a^2 - s}{2\,\eta_0}, \\[2mm]
Z_2 &= a\frac{x}{\eta_0}, \\[2mm]
Z_3 &= a\frac{y}{\eta_0}, \\[2mm]
Z_4 &= a\frac{z}{\eta_0},
\end{aligned}
\tag{4.12}
$$

where $s = -\eta_0^2 + x^2 + y^2 + z^2$ and $\eta_0, x, y, z \in (-\infty, +\infty)$ (see Figure 4.6). In the coordinates of (4.12), the de Sitter metric is

$$
\mathrm{d}s^2 = \frac{a^2}{\eta_0^2}\left(-\mathrm{d}\eta_0^2 + \mathrm{d}x^2 + \mathrm{d}y^2 + \mathrm{d}z^2\right).
\tag{4.13}
$$

The metric (4.13) is explicitly conformally flat, in full agreement with the vanishing of the Weyl tensor. Having zero curvature, the spatial sections $\eta_0 = \mathrm{const.}$ are unbounded.

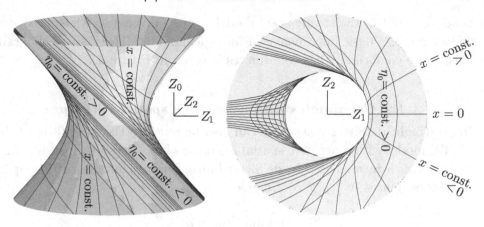

Fig. 4.6 The same de Sitter hyperboloid as in Figure 4.1 but with coordinates η_0, x of (4.12) (and Z_3, Z_4 suppressed by setting $y = 0 = z$), also viewed from above.

The de Sitter universe can subsequently be expressed as an exponentially expanding FLRW metric (6.3) with spatial curvature $k = 0$,

$$\mathrm{d}s^2 = -\mathrm{d}t_0^2 + \exp\left(2\frac{t_0}{a}\right)\left(\mathrm{d}x^2 + \mathrm{d}y^2 + \mathrm{d}z^2\right), \qquad (4.14)$$

by applying the transformation $\eta_0 = a\exp\left(-t_0/a\right)$ onto (4.13). However, for $t_0 \in (-\infty, +\infty)$ the metric (4.14) covers only half of the complete manifold (4.1) on which $Z_0 + Z_1 > 0$ or $\eta_0 > 0$. Another coordinate chart (t_0^*, x, y, z) of the form (4.14) is necessary to cover also $Z_0 + Z_1 < 0$. Such a chart is given by $\eta_0 = -a\exp\left(-t_0^*/a\right)$. The metric (4.14) of the de Sitter space-

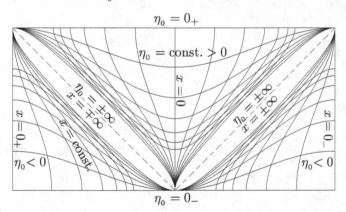

Fig. 4.7 The Penrose diagram of de Sitter space-time with coordinates $(\eta_0, x, y = 0, z = 0)$ in which the metric is explicitly conformally flat. Each point represents a plane spanned by y, z. Unlike the spherically symmetric space shown in Figure 4.3, this is a complete section through the conformal space-time.

time was used in the steady-state model of the universe and has also been considered as a basic inflationary cosmological model. The Penrose diagram with these conformally flat coordinates is given in Figure 4.7.

4.4.2 Coordinates with a negative spatial curvature

Interestingly, de Sitter space-time can also be put into the form (6.7) of the FLRW model with a negative spatial curvature $k = -1$ (see Appendix B.2). In fact, this form was used by de Sitter himself in his landmark 1917 paper. Such coordinates $(t_{-1}, \rho, \theta, \phi)$ are given by

$$
\begin{aligned}
Z_0 &= a \sinh \frac{t_{-1}}{a} \cosh \rho, \\
Z_1 &= a \cosh \frac{t_{-1}}{a}, \\
Z_2 &= a \sinh \frac{t_{-1}}{a} \sinh \rho \cos \theta, \\
Z_3 &= a \sinh \frac{t_{-1}}{a} \sinh \rho \sin \theta \cos \phi, \\
Z_4 &= a \sinh \frac{t_{-1}}{a} \sinh \rho \sin \theta \sin \phi,
\end{aligned}
\tag{4.15}
$$

in which the metric is

$$
\mathrm{d}s^2 = -\mathrm{d}t_{-1}^2 + a^2 \sinh^2 \frac{t_{-1}}{a} \left(\mathrm{d}\rho^2 + \sinh^2 \rho \left(\mathrm{d}\theta^2 + \sin^2 \theta \, \mathrm{d}\phi^2 \right) \right). \tag{4.16}
$$

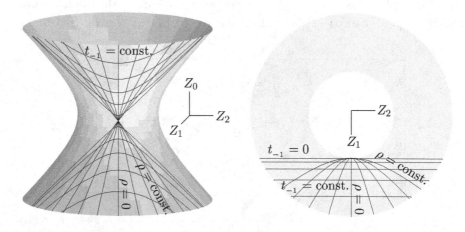

Fig. 4.8 The de Sitter hyperboloid with coordinates $(t_{-1}, \rho, \theta, \phi)$ of (4.16), in which the time slices have negative spatial curvature $k = -1$. As in Figure 4.1, provided $t_{-1} > 0$, the points correspond to the pole $\theta = 0$ when $Z_2 > 0$ and $\theta = \pi$ when $Z_2 < 0$. For $t_{-1} < 0$ the two poles shown here are interchanged.

However, for $t_{-1} \in (-\infty, +\infty)$, $\rho \in [0, +\infty)$, $\theta \in [0, \pi]$, $\phi \in [0, 2\pi)$, these coordinates cover only a small part of the complete manifold, as shown in Figures 4.8 and 4.9. This is not surprising, as the natural spatial sections of the de Sitter space-time are known to have constant positive curvature, while the unbounded spatial sections $t_{-1} = $ const. have constant negative curvature.

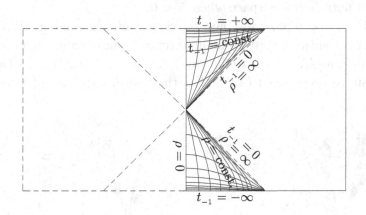

Fig. 4.9 The Penrose diagram of the part of de Sitter space-time that is covered by the coordinates $(t_{-1}, \rho, \theta, \phi)$ of (4.16). Each point represents a 2-sphere spanned by θ, ϕ. Since the space-time has spherical symmetry, the Penrose diagram corresponds to half of the sections through the conformal space-time.

4.4.3 Constant-curvature coordinates

Next consider the parametrisation of the de Sitter space-time given by

$$
Z_0 = \tfrac{1}{\sqrt{2}}(\mathcal{V} + \mathcal{U}) \left[1 - \tfrac{1}{6}\Lambda(\mathcal{U}\mathcal{V} - \xi\bar{\xi})\right]^{-1},
$$

$$
Z_1 = \tfrac{1}{\sqrt{2}}(\mathcal{V} - \mathcal{U}) \left[1 - \tfrac{1}{6}\Lambda(\mathcal{U}\mathcal{V} - \xi\bar{\xi})\right]^{-1},
$$

$$
Z_2 = \tfrac{1}{\sqrt{2}}(\xi + \bar{\xi}) \left[1 - \tfrac{1}{6}\Lambda(\mathcal{U}\mathcal{V} - \xi\bar{\xi})\right]^{-1}, \qquad (4.17)
$$

$$
Z_3 = \tfrac{-i}{\sqrt{2}}(\xi - \bar{\xi}) \left[1 - \tfrac{1}{6}\Lambda(\mathcal{U}\mathcal{V} - \xi\bar{\xi})\right]^{-1},
$$

$$
Z_4 = a \left[1 + \tfrac{1}{6}\Lambda(\mathcal{U}\mathcal{V} - \xi\bar{\xi})\right]\left[1 - \tfrac{1}{6}\Lambda(\mathcal{U}\mathcal{V} - \xi\bar{\xi})\right]^{-1}.
$$

Inversely, this can be expressed in the more simple form

$$
\mathcal{V} = \sqrt{2}\,a\frac{Z_0 + Z_1}{Z_4 + a}, \qquad \mathcal{U} = \sqrt{2}\,a\frac{Z_0 - Z_1}{Z_4 + a}, \qquad \xi = \sqrt{2}\,a\frac{Z_2 + iZ_3}{Z_4 + a}. \quad (4.18)
$$

This gives the alternative conformally flat line element

$$ds^2 = \frac{-2\, d\mathcal{U}\, d\mathcal{V} + 2\, d\xi\, d\bar{\xi}}{[\, 1 - \frac{1}{6}\Lambda(\mathcal{U}\mathcal{V} - \xi\bar{\xi})\,]^2}. \tag{4.19}$$

In fact, this provides an interesting and useful unified form for all spaces of constant curvature – de Sitter space when $\Lambda > 0$, Minkowski space when $\Lambda = 0$, and anti-de Sitter space when $\Lambda < 0$.

For $\mathcal{U}, \mathcal{V} \in (-\infty, +\infty)$ and ξ an arbitrary complex number, these coordinates cover the entire de Sitter manifold (except the events where the coordinate singularities $\mathcal{U}, \mathcal{V} = \infty$ are located), as shown in Figures 4.10 and 4.11. It can easily be seen from (4.18) that the coordinate lines $\mathcal{U} = $ const. and

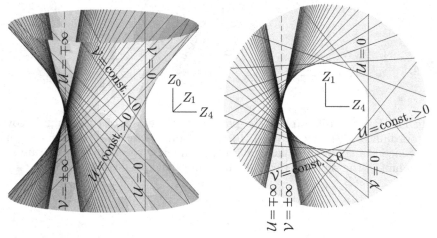

Fig. 4.10 The de Sitter hyperboloid covered by the conformally flat coordinates (4.17), viewed also from above, with $\xi = 0$ so that $Z_2 = 0 = Z_3$.

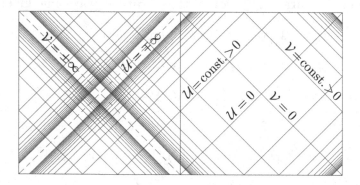

Fig. 4.11 The coordinates $(\mathcal{U}, \mathcal{V})$ with $\xi = 0$ are null straight lines in the Penrose diagram of de Sitter space-time, which is here a complete section through the conformal space-time.

$\mathcal{V} = $ const. are natural straight null lines (hyperplanes) both on the de Sitter hyperboloid and in the conformal Penrose diagram.

4.4.4 Bianchi III and Kantowski–Sachs coordinates

Finally, two other interesting coordinate systems on the de Sitter manifold are sometimes used and should be mentioned. The coordinates (t_B, z_B, θ, ϕ) given by

$$Z_0 = a \sinh \frac{t_B}{a} \cosh \theta,$$

$$Z_1 = a \cosh \frac{t_B}{a} \cos z_B,$$

$$Z_2 = a \cosh \frac{t_B}{a} \sin z_B, \qquad (4.20)$$

$$Z_3 = a \sinh \frac{t_B}{a} \sinh \theta \cos \phi,$$

$$Z_4 = a \sinh \frac{t_B}{a} \sinh \theta \sin \phi,$$

parametrise the Bianchi III homogeneous and anisotropic form (see Section 22.1) of the de Sitter metric

$$ds^2 = -dt_B^2 + a^2 \cosh^2 \frac{t_B}{a} \, dz_B^2 + a^2 \sinh^2 \frac{t_B}{a} (d\theta^2 + \sinh^2 \theta \, d\phi^2), \quad (4.21)$$

while the coordinates (t_K, z_K, θ, ϕ) given by

$$Z_0 = a \sinh \frac{t_K}{a} \cosh z_K,$$

$$Z_1 = a \sinh \frac{t_K}{a} \sinh z_K,$$

$$Z_2 = a \cosh \frac{t_K}{a} \cos \theta, \qquad (4.22)$$

$$Z_3 = a \cosh \frac{t_K}{a} \sin \theta \cos \phi,$$

$$Z_4 = a \cosh \frac{t_K}{a} \sin \theta \sin \phi,$$

represent the Kantowski–Sachs-type metric:

$$ds^2 = -dt_K^2 + a^2 \sinh^2 \frac{t_K}{a} \, dz_K^2 + a^2 \cosh^2 \frac{t_K}{a} (d\theta^2 + \sin^2 \theta \, d\phi^2). \quad (4.23)$$

This is also homogeneous and anisotropic but does not belong to any Bianchi class (see Section 22.2).

4.5 Remarks and references

Of course, by straightforwardly comparing the parametrisations (4.3), (4.8), (4.12), (4.15), (4.17), (4.20) and (4.22), one can easily derive the coordinate transformations connecting the metric forms (4.4), (4.9), (4.13), (4.16), (4.19), (4.21) and (4.23). For example, the transformation from the natural global coordinates (t, χ, θ, ϕ) to the static spherically symmetric coordinates (T, R, θ, ϕ) which relates the metrics (4.4) and (4.9) is

$$\tanh \frac{T}{a} = \tanh \frac{t}{a} \cos^{-1}\chi, \qquad R = a \sin \chi \cosh \frac{t}{a}. \tag{4.24}$$

However, all these relations are not written here explicitly, since there are many such mutual transformations (and even more if the metrics (4.6), (4.10) and (4.14) are taken into account). Such transformations and details of these and other coordinate systems can be found e.g. in Gautreau (1983), Mottola (1985), Schmidt (1993), Eriksen and Grøn (1995), Podolský and Griffiths (1997) and Bičák and Krtouš (2005). Coordinates centred on an accelerating test particle in both de Sitter and anti-de Sitter space-times will be given explicitly in Section 14.4.

Further information about the properties of de Sitter space-time are given in various useful review works, such as Schrödinger (1956), Penrose (1968a), Weinberg (1972), Møller (1972), Hawking and Ellis (1973), Birell and Davies (1982), Bičák (2000a), Grøn and Hervik (2007) and elsewhere.

5

Anti-de Sitter space-time

As explained at the beginning of the previous chapter, anti-de Sitter space-time is maximally symmetric – it is the conformally flat vacuum solution with a negative cosmological constant Λ. It thus has a constant scalar curvature $R = 4\Lambda < 0$, and is the complement of flat Minkowski space ($R = 0$) and de Sitter space ($R > 0$). In this chapter, we will describe the main properties of this constant-curvature exact solution of Einstein's field equations which has recently become important in the context of higher-dimensional theories, in particular due to the conjectured anti-de Sitter space/conformal field theory correspondence.

5.1 Global representation

In full analogy with the de Sitter space-time, the anti-de Sitter space-time can be visualised geometrically as the hyperboloid

$$-Z_0^2 + Z_1^2 + Z_2^2 + Z_3^2 - Z_4^2 = -a^2, \quad \text{where} \quad a = \sqrt{-3/\Lambda}, \quad (5.1)$$

embedded in a flat five-dimensional space

$$\mathrm{d}s^2 = -\mathrm{d}Z_0^2 + \mathrm{d}Z_1^2 + \mathrm{d}Z_2^2 + \mathrm{d}Z_3^2 - \mathrm{d}Z_4^2, \quad (5.2)$$

which has two timelike dimensions Z_0 and Z_4. This representation of anti-de Sitter space as a submanifold is related to its symmetry structure which is characterised by the ten-parameter group of isometries SO(2,3).

Natural coordinates (T, r, θ, ϕ) covering the entire hyperboloid are obtained by putting

$$Z_0 = a \cosh r \sin \frac{T}{a},$$
$$Z_1 = a \sinh r \cos \theta,$$
$$Z_2 = a \sinh r \sin \theta \cos \phi, \quad (5.3)$$

55

$$Z_3 = a \sinh r \sin \theta \sin \phi,$$

$$Z_4 = a \cosh r \cos \frac{T}{a},$$

(see Figure 5.1). In these global static coordinates, in which r is dimensionless, the anti-de Sitter metric reads

$$ds^2 = -\cosh^2 r \, dT^2 + a^2 \Big(dr^2 + \sinh^2 r \, (d\theta^2 + \sin^2 \theta \, d\phi^2) \Big). \qquad (5.4)$$

Any spatial section $T = \text{const.}$ is obviously a 3-space of constant negative curvature spanned by $r \in [0, \infty)$, $\theta \in [0, \pi]$, $\phi \in [0, 2\pi)$, see the metric (B.16) in Appendix B.2. The singularities at $r = 0$, and $\theta = 0, \pi$ are only coordinate singularities because, in fact, the anti-de Sitter space-time is spatially homogeneous and isotropic.

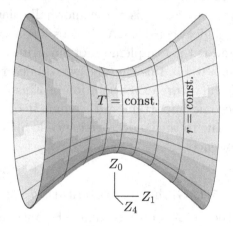

Fig. 5.1 The representation of anti-de Sitter space-time as a hyperboloid embedded in a flat five-dimensional space-time in the parametrisation given by global static coordinates (T, r, θ, ϕ). The surface drawn is that for the pole $\theta = 0$ when $Z_1 > 0$ and $\theta = \pi$ when $Z_1 < 0$, so that $Z_2 = 0 = Z_3$. With the range of θ and ϕ reintroduced, each point on this hyperboloid represents a corresponding two-dimensional hemisphere, with either $\theta < \pi/2$ or $\theta > \pi/2$, respectively.

As can be seen from (5.2), Z_0 and Z_4 are both temporal coordinates in the five-dimensional representation space. These are parametrised using (5.3) by a single time coordinate T that is *periodic*. Indeed, values of T which differ by a multiple of $2\pi a$ represent the same points on the hyperboloid. Thus, anti-de Sitter space-time defined in this way has the topology $S^1 \times R^3$, which contains closed timelike worldlines. However, the periodicity of T is not evident in the four-dimensional metric (5.4), and it is more natural to take $T \in (-\infty, +\infty)$. Such a range of coordinates corresponds to an infinite number of turns around the hyperboloid of Figure 5.1. It is thus appropriate to unwrap the circle S^1 and extend it to the whole R^1 instead, without reference to the parametrisation (5.3). One thus obtains a *universal covering space* of the anti-de Sitter universe with topology R^4 which does not contain any closed timelike curves.

By performing a simple transformation $R = a \sinh r$, (5.3) takes the form

$$Z_0 = \sqrt{a^2 + R^2}\, \sin \frac{T}{a},$$
$$Z_1 = R \cos \theta,$$
$$Z_2 = R \sin \theta \cos \phi, \qquad (5.5)$$
$$Z_3 = R \sin \theta \sin \phi,$$
$$Z_4 = \sqrt{a^2 + R^2}\, \cos \frac{T}{a},$$

and the metric becomes

$$ds^2 = -\left(1 + \frac{R^2}{a^2}\right) dT^2 + \left(1 + \frac{R^2}{a^2}\right)^{-1} dR^2 + R^2(d\theta^2 + \sin^2 \theta\, d\phi^2), \quad (5.6)$$

where $R \in [0, \infty)$. This is another useful static and spherically symmetric coordinate form of the complete anti-de Sitter space-time. In fact, (5.6) has the same form as (4.9) which is valid for any value of the cosmological constant Λ. However, it can be seen from (5.6) that, in contrast to the case $\Lambda > 0$, there are no event horizons in the anti-de Sitter universe because $1 + \frac{R^2}{a^2} \equiv 1 - \frac{\Lambda}{3} R^2 > 1$ everywhere. Note also that the natural worldlines with R, θ, ϕ constant (and hence r constant) represent *uniformly accelerating observers* with acceleration less than $1/a$, analogous to (4.11); these are geodesics for $R = 0$ only. (See Podolský, 2002a, and Krtouš, 2005, for more details.)

The global metric (5.4) can also be rewritten using the transformation $\psi = \tanh(r/2)$, where $\psi \in [0, 1)$, i.e. considering the parametrisation

$$Z_0 = a\, \frac{1 + \psi^2}{1 - \psi^2}\, \sin \frac{T}{a},$$
$$Z_1 = a\, \frac{2\psi}{1 - \psi^2}\, \cos \theta,$$
$$Z_2 = a\, \frac{2\psi}{1 - \psi^2}\, \sin \theta \cos \phi, \qquad (5.7)$$
$$Z_3 = a\, \frac{2\psi}{1 - \psi^2}\, \sin \theta \sin \phi,$$
$$Z_4 = a\, \frac{1 + \psi^2}{1 - \psi^2}\, \cos \frac{T}{a},$$

so that the metric of a complete anti-de Sitter space-time takes the form

$$ds^2 = -\left(\frac{1 + \psi^2}{1 - \psi^2}\right)^2 dT^2 + \frac{4a^2}{(1 - \psi^2)^2}\left(d\psi^2 + \psi^2(d\theta^2 + \sin^2\theta\, d\phi^2)\right). \quad (5.8)$$

All spatial sections $T = $ const. now correspond to the coordinates (B.18) of the *Poincaré ball* $\psi \in [0, 1)$, $\theta \in [0, \pi]$, $\phi \in [0, 2\pi)$ in flat space, which naturally cover the hyperbolic 3-space H^3 via a stereographic-like projection (see Figure B.2). The coordinates of (5.8) were used, for example, to construct specific black holes (with event horizons) from the anti-de Sitter space through a suitable identification of points, see Åminneborg *et al.* (1996).

In fact, the above three metric forms (5.4), (5.6), (5.8) of the complete anti-de Sitter space exactly correspond to its natural foliation $T = $ const. by hyperbolic 3-spaces of constant negative curvature $K = -a^{-2}$, as described by the coordinates (B.16), (B.17), (B.18), respectively, with the identifications $r = \chi$, $R = a\,r$, $\psi = \rho$.

5.2 Conformal structure

To study the global causal structure of anti-de Sitter space, a conformal coordinate χ may be introduced by setting

$$\tan\chi = \sinh r. \quad (5.9)$$

Also, writing $\eta = T/a$, the metric (5.4) then takes the form

$$ds^2 = \frac{a^2}{\cos^2\chi}\left(-d\eta^2 + d\chi^2 + \sin^2\chi\,(d\theta^2 + \sin^2\theta\, d\phi^2)\right). \quad (5.10)$$

It can thus be seen that the whole (universal) anti-de Sitter space is conformal to the region $0 \leq \chi < \frac{\pi}{2}$ of the Einstein static universe (3.8), with the conformal factor

$$\Omega = \cos\chi \quad (5.11)$$

and the identification $R_0 = a$ (see Figure 5.2). The conformal infinity $\mathcal{I}$ of anti-de Sitter space-time given by $\Omega = 0$ is located at the boundary $\chi = \frac{\pi}{2}$ (corresponding to $r = \infty$).

In contrast to Minkowski and de Sitter space (see Figures 3.4 and 4.2), conformal infinity $\mathcal{I}$ for null geodesics (and also spatial infinity i^0) in the anti-de Sitter space-time forms a *timelike surface*. Indeed, the normal vector defined as $\mathbf{n}^\alpha = \tilde{g}^{\alpha\beta}\,\Omega_{,\beta}$, using (2.26), is spacelike on $\chi = \frac{\pi}{2}$. No timelike geodesic can reach this conformal boundary and thus it must always stay within the region $\chi < \frac{\pi}{2}$.

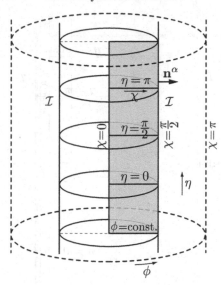

Fig. 5.2 The conformal structure of anti-de Sitter space-time (with $\theta = \frac{\pi}{2}$). The whole anti-de Sitter space is conformal to the region $0 \leq \chi < \frac{\pi}{2}$ of the Einstein static universe represented here (in dashed lines) as a solid cylinder of radius π and infinite length. The boundary $\chi = \frac{\pi}{2}$ of the anti-de Sitter manifold is its conformal infinity $\mathcal{I}$, which has a timelike character and topology $R^1 \times S^2$.

The standard two-dimensional Penrose diagram shown in Figure 5.3, which corresponds to the shaded section indicated in Figure 5.2, is obtained by further fixing the coordinate ϕ. Each point of the Penrose diagram represents a 2-sphere, except at $\chi = 0$ where, at any time, it is a single point.

An important consequence of the timelike character of conformal infinity $\mathcal{I}$ is that there exists *no complete Cauchy surface* whatever in the anti-de Sitter space-time. While the family of sections $\eta = $ const. covers the space-time completely, there are null geodesics which never intersect any particular such section. Given initial data on such a spacelike section, it is not possible to predict the evolution in the region beyond the Cauchy development of this surface. The initial Cauchy data on $\eta = $ const. are not sufficient for a unique determination of the evolution in the complete space-time. The anti-de Sitter space-time lacks global hyperbolicity. It does not have a well-defined Cauchy problem on this background because "boundary conditions on $\mathcal{I}$" must also be supplied (see e.g. Hawking and Ellis (1973), Avis, Isham and Storey (1978), Hawking (1983) or Friedrich (1995) for more details).

Note finally that the conformal structure of the universal covering space for anti-de Sitter space-time shown in Figure 5.2 can be further compactified

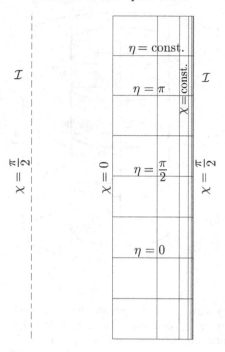

Fig. 5.3 The Penrose diagram of anti-de Sitter space-time corresponds to the shaded region in Figure 5.2. When the coordinates θ, ϕ are reintroduced, each point in the diagram represents a complete 2-sphere of radius $\sin \chi$. For $\chi = 0$ these are single points, while $\chi = \pi/2$ is the conformal infinity $\mathcal{I}$. Since the space-time has spherical symmetry, the Penrose diagram corresponds to half of a section through the conformal space-time. The boundary of the section with $\phi \to \phi + \pi$ is shown by the vertical dashed line on the left.

in the temporal direction to explicitly include timelike future and past infinities i^+ and i^-, as initially given by Tipler (1986). Such a causal representation of anti-de Sitter space-time is obtained by first performing the transformation $\tilde{\eta} = \arctan(\eta/4)$ of the metric (3.8) for the Einstein static universe, to give the form

$$\mathrm{d}s^2 = \frac{16R_0^2}{\cos^4 \tilde{\eta}} \left(- \mathrm{d}\tilde{\eta}^2 + \tfrac{1}{16} \cos^4 \tilde{\eta} \left(\mathrm{d}\chi^2 + \sin^2 \chi \left(\mathrm{d}\theta^2 + \sin^2 \theta \, \mathrm{d}\phi^2 \right) \right) \right), \quad (5.12)$$

where $\tilde{\eta} \in [-\pi/2, +\pi/2]$, if the limiting points $\tilde{\eta} = \pm \pi/2$ corresponding to $\eta = \pm\infty$ are also admitted. In view of the conformal factor, the metric within the large brackets in (5.12) is conformal to the metric of the Einstein static universe. Since their null cones coincide, they both have the same causal structure. It is thus possible to use the metric in large brackets with coordinates $\tilde{\eta}, \chi, \theta, \phi$ to construct a compactified Penrose diagram of the

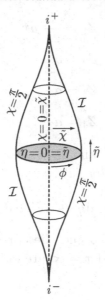

Fig. 5.4 The conformal structure of anti-de Sitter space-time (with $\theta = \pi/2$) is obtained as a further compactification of Figure 5.2, using the radial proper distance $\tilde{\chi}$ and the vertical temporal coordinate $\tilde{\eta} \in [-\pi/2, +\pi/2]$. Timelike future and past infinities i^+ and i^-, which are the endpoints of all timelike geodesics, are located at $\tilde{\chi} = 0$ and $\tilde{\eta} = \pm\pi/2$. The radial proper distance $\tilde{\chi}$ takes the range $\tilde{\chi} \in [0, \frac{1}{8}\pi \cos^2 \tilde{\eta}]$ since the outer boundary, corresponding to $\mathcal{I}$, is located at $\chi = \pi/2$.

anti-de Sitter universe. In this metric, the proper spatial distance between the origin $\chi = 0$ and a comoving observer at $\chi = $ const. (with fixed θ and ϕ) is obviously $\tilde{\chi} = \frac{1}{4}\chi \cos^2 \tilde{\eta}$. Plotting $\tilde{\eta}$ vertically, $\tilde{\chi}$ radially, and keeping the angular coordinate ϕ, gives the Figure 5.4. The vertical dashed line represents the history of the origin $\chi = 0$ while the outer boundary represents conformal infinity $\mathcal{I}$ located at $\chi = \pi/2$, which is separated from $\chi = 0$ by the proper distance $\tilde{\chi} = \frac{1}{8}\pi \cos^2 \tilde{\eta}$. The full range of $\tilde{\eta} \in [-\pi/2, +\pi/2]$ now also explicitly includes the timelike infinities i^- and i^+ (which are located on $\tilde{\chi} = 0$ at $\tilde{\eta} = -\pi/2$ and $\tilde{\eta} = +\pi/2$, respectively) of all timelike geodesic observers.

5.3 Other standard coordinates

Other frequently used coordinates (η_0, x, y, z) covering the hyperboloid (5.1) are given by

$$Z_0 = \frac{1}{2x}(a^2 + s),$$

$$Z_1 = \frac{1}{2x}(a^2 - s),$$

$$Z_2 = a\frac{y}{x},$$

$$Z_3 = a\frac{z}{x}, \qquad (5.13)$$

$$Z_4 = a\frac{\eta_0}{x},$$

where $s = -\eta_0^2 + x^2 + y^2 + z^2$ and $\eta_0, x, y, z \in (-\infty, +\infty)$. The resulting metric is manifestly *conformally flat*,

$$ds^2 = \frac{a^2}{x^2}(-d\eta_0^2 + dx^2 + dy^2 + dz^2). \qquad (5.14)$$

As can be seen from Figure 5.5 and the corresponding Penrose conformal diagram, Figure 5.6, $x = 0$ and $x = \infty$ are only coordinate singularities.

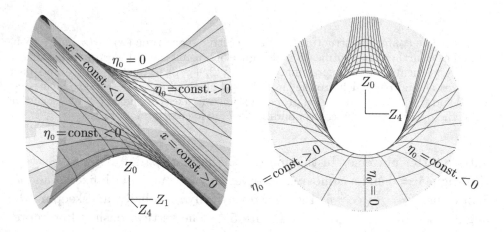

Fig. 5.5 The same anti-de Sitter hyperboloid as in Figure 5.1 but in conformally flat (Poincaré) coordinates η_0, x of (5.14) (Z_2, Z_3 are suppressed by setting $y = 0 = z$). Here, the points on the hyperboloid do not represent 2-spheres.

The so-called Poincaré form (5.14) of the anti-de Sitter space-time is related to the *synchronous* one

$$ds^2 = -dt^2 + a^2 \cos^2\frac{t}{a}\left(d\chi^2 + e^{-2\chi}(dy^2 + dz^2)\right) \qquad (5.15)$$

by the transformation $\eta_0 = e^\chi \tan(t/a), \ x = e^\chi/\cos(t/a)$.

The different transformation $x = \pm a \exp(-\hat{x}/a)$ relates (5.14) to

$$ds^2 = d\hat{x}^2 + \exp\left(2\frac{\hat{x}}{a}\right)(-d\eta_0^2 + dy^2 + dz^2), \qquad (5.16)$$

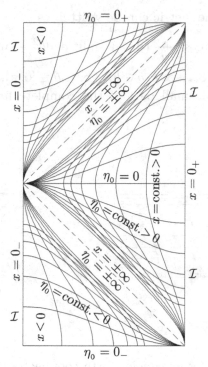

Fig. 5.6 The conformal structure of anti-de Sitter space-time with coordinates $(\eta_0, x, y = 0, z = 0)$ of (5.14). Each point in this Penrose diagram represents a plane spanned by y, z. Unlike in the spherically symmetric cases shown in Figures 5.3 and 5.8, this is a complete section through the space-time.

which is the analogue of (4.14) for de Sitter space.

There also exist standard coordinates $(t_{-1}, \rho, \theta, \phi)$ which represent anti-de Sitter space-time as a particular *FLRW universe* with a negative spatial curvature $k = -1$ (see (6.7) in the following chapter). For these,

$$Z_0 = a \sin \frac{t_{-1}}{a},$$

$$Z_1 = a \cos \frac{t_{-1}}{a} \sinh \rho \cos \theta,$$

$$Z_2 = a \cos \frac{t_{-1}}{a} \sinh \rho \sin \theta \cos \phi, \qquad (5.17)$$

$$Z_3 = a \cos \frac{t_{-1}}{a} \sinh \rho \sin \theta \sin \phi,$$

$$Z_4 = a \cos \frac{t_{-1}}{a} \cosh \rho.$$

In such coordinates the metric can be written as

$$ds^2 = -dt_{-1}^2 + a^2 \cos^2 \frac{t_{-1}}{a} \left(d\rho^2 + \sinh^2 \rho \left(d\theta^2 + \sin^2 \theta \, d\phi^2\right)\right). \qquad (5.18)$$

However, for $t_{-1} \in (-\infty, +\infty)$, $\rho \in [0, +\infty)$, $\theta \in [0, \pi]$, $\phi \in [0, 2\pi)$ these co-ordinates cover only a small part of the complete manifold, as shown in Figures 5.7 and 5.8.

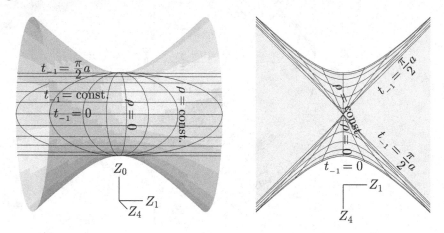

Fig. 5.7 The anti-de Sitter hyperboloid with coordinates $(t_{-1}, \rho, \theta, \phi)$ of (5.18) in which the time slices have negative spatial curvature $k = -1$. As in Figure 5.1, provided $\cos(t_{-1}/a) > 0$, the points correspond to the pole $\theta = 0$ when $Z_1 > 0$ and $\theta = \pi$ when $Z_1 < 0$. For $\cos(t_{-1}/a) < 0$, the two poles shown here are interchanged.

It can be seen from the metric (5.18) that the worldlines with ρ, θ, ϕ constant (which are orthogonal to the surfaces $t_{-1} = $ const.) are *timelike geodesics*. As indicated in Figure 5.7, they all *converge* at specific points separated in t_{-1} by πa. In fact, all the timelike geodesics emanating from *any* point (which may be taken to be, for example, $t_{-1} = -\frac{\pi}{2}a$ since anti-de Sitter space is homogeneous) reconverge to an image point at $t_{-1} = \frac{\pi}{2}a$, diverge again to refocus at $t_{-1} = 3\frac{\pi}{2}a$, and so on. This repeated convergence is related to the occurrence of the coordinate singularities at $\cos(t_{-1}/a) = 0$ in the metric (5.18). (Note that such a periodic effect is more naturally understood in the original anti-de Sitter hyperboloid with closed timelike curves than in its covering space.) Moreover, timelike geodesics (in contrast to the null geodesics) never reach conformal infinity $\mathcal{I}$ located at $\chi = \frac{\pi}{2}$. This results in the existence of regions in the future null cone of any event which cannot be reached from that event by any geodesic. The set of points which can be reached by timelike geodesics forms the interior of the infinite chain of diamond-shaped regions similar to that covered by coordinates $(t_{-1}, \rho, \theta, \phi)$ shown in Figure 5.8. This also demonstrates the non-existence

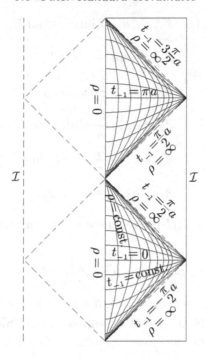

Fig. 5.8 The Penrose diagram of the part of anti-de Sitter space-time covered by the coordinates $(t_{-1}, \rho, \theta, \phi)$. Each point represents a 2-sphere spanned by θ, ϕ.

of a Cauchy surface for the anti-de Sitter space-time. For example, from the surface $t_{-1} = 0$, one can predict only the regions covered by the coordinates $(t_{-1}, \rho, \theta, \phi)$. Any attempt to predict beyond this region is prevented by new additional information, possibly coming from the conformal infinity $\mathcal{I}$. All these effects are related to the fact that a negative cosmological constant Λ has "attractive" properties, in contrast to a positive Λ which causes a repulsion.

Another interesting parametrisation of the anti-de Sitter space-time is obtained by (4.17) which gives the alternative conformally flat line element (4.19), namely

$$ds^2 = \frac{-2\,d\mathcal{U}\,d\mathcal{V} + 2\,d\xi\,d\bar{\xi}}{[1 - \frac{1}{6}\Lambda(\mathcal{U}\mathcal{V} - \xi\bar{\xi})]^2}, \tag{5.19}$$

with $\Lambda < 0$. For $\mathcal{U}, \mathcal{V} \in (-\infty, +\infty)$ and ξ an arbitrary complex number, these coordinates cover the entire anti-de Sitter manifold, except the events where the coordinate singularities $\mathcal{U}, \mathcal{V} = \infty$ are located.

Finally, the coordinates (t_B, z_B, θ, ϕ) introduced by setting

$$Z_0 = a \sin \frac{t_B}{a} \cosh \theta,$$

$$Z_1 = a \cos \frac{t_B}{a} \sinh z_B,$$

$$Z_2 = a \sin \frac{t_B}{a} \sinh \theta \cos \phi, \qquad (5.20)$$

$$Z_3 = a \sin \frac{t_B}{a} \sinh \theta \sin \phi,$$

$$Z_4 = a \cos \frac{t_B}{a} \cosh z_B,$$

parametrise the *Bianchi III* homogeneous but anisotropic form of anti-de Sitter metric

$$ds^2 = -dt_B^2 + a^2 \cos^2 \frac{t_B}{a} \, dz_B^2 + a^2 \sin^2 \frac{t_B}{a} \, (d\theta^2 + \sinh^2 \theta \, d\phi^2). \qquad (5.21)$$

For $t_B \in (-\infty, +\infty)$, $z_B \in (-\infty, +\infty)$, $\theta \in [0, \infty)$, $\phi \in [0, 2\pi)$ they cover only part of the manifold. Note also that, unlike the de Sitter space-time, anti-de Sitter space-time *cannot* have the form of the Kantowski–Sachs metric equivalent to (4.23).

By straightforwardly comparing the parametrisations (5.3), (5.5), (5.7), (5.13), (5.17) and (5.20), one can immediately derive the direct coordinate transformations connecting the anti-de Sitter metric forms (5.4), (5.6), (5.8), (5.14), (5.18) and (5.21), and also the related ones (5.10), (5.15), (5.16), (5.19). However, these are not presented explicitly here.

5.4 Remarks and references

Another important coordinate parametrisation of the anti-de Sitter space-time, adapted to uniformly accelerating test particles, will be given in Section 14.4. More detailed analysis of these coordinates and their explicit visualisations in the global conformal picture have been given recently in Podolský, Ortaggio and Krtouš (2003) and Krtouš (2005).

Further details on anti-de Sitter space are given, e.g., in Hawking and Ellis (1973), Avis, Isham and Storey (1978), Boucher and Gibbons (1984), Siklos (1985), Ozsváth (1987), Polarski (1989), Friedrich (1995) and elsewhere.

6
Friedmann–Lemaître–Robertson–Walker space-times

On a sufficiently large scale, the universe we live in appears to be both spatially homogeneous and isotropic (that is, on an appropriate spatial section its matter content is uniformly distributed on average, and it looks qualitatively the same in all directions). Space-times with these properties were systematically investigated from different points of view in the pioneering work particularly of Friedmann, Lemaître, Robertson and Walker. The solutions they developed underlie the foundation of modern cosmology. They provide a wide range of possible dynamical models of the universe, among which cosmologists can identify that which most closely resembles our own on appropriately large scales. In particular, they have lead to the prediction of an initial cosmological singularity known as the *big bang*.

In this chapter, we will describe such a family of spatially homogeneous and isotropic space-times. These are considered as idealised cosmological models containing a perfect fluid satisfying some equation of state. As such, they represent various possible types of uniformly distributed matter, including the most important special cases of dust and radiation. They also admit a non-trivial cosmological constant. Like the vacuum space-times described in the previous three chapters, they are also conformally flat. The geometrical reason for this is that their natural three-dimensional spatial subspaces have constant curvature. This curvature can be positive, zero or negative, giving rise to different models of closed or open universes whose dynamics are uniquely determined by the specific matter content.

We will briefly summarise the main assumptions, results and properties of this particularly important class of exact solutions. Further information can be found in the very extensive and easily accessible literature on this topic, some of which is cited below.

6.1 Geometry and standard coordinates

The idealising geometric assumption of Friedmann–Lemaître–Robertson–Walker (FLRW) cosmology is that, at "any given time", there are no privileged points or preferred directions in its three-dimensional space. In other words, the space is invariant with respect to spatial "translations" and "rotations". This assumption seems to be in good accord with present-day astronomical observations on a sufficiently large scale.

Technically, the assumed spatial homogeneity and isotropy is expressed by the condition that the space-time everywhere admits a six-parameter group of isometries acting transitively on hypersurfaces Σ which are spacelike 3-spaces. According to a standard theorem of geometry, the maximum number of isometries of a three-dimensional space is 6, and such maximally symmetric spaces have constant curvature. It thus follows that a FLRW space-time manifold must admit a *foliation* by a one-parameter family of *3-spaces* of spacelike character, which all have *constant spatial curvature*.

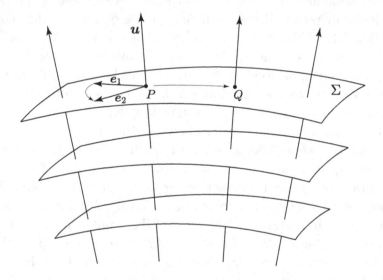

Fig. 6.1 Foliation of a FLRW space-time by a family of spacelike hypersurfaces. These are 3-spaces which all have constant spatial curvature. Each hypersurface Σ is homogeneous, which means that all points on it are equivalent. They are also isotropic, i.e. all spatial directions at any given point are equivalent with respect to a congruence of timelike curves generated by u that is orthogonal to Σ.

Such a foliation is visualised in Figure 6.1, in which Σ denotes a typical 3-space of constant curvature. This schematic picture also enables a more precise meaning to be given to the notion of homogeneity and isotropy. *Spatial homogeneity* occurs when, for any two points $P, Q \in \Sigma$, there exists an isometry of the space-time metric which takes P to Q. To define an isotropy, it is necessary to consider the existence of a congruence of timelike curves in the whole space-time generated by the vector field u. In the case of FLRW space-times, this can be taken to be everywhere orthogonal to Σ. *Spatial isotropy* occurs when, for any two unit spatial vectors e_1, e_2 in the space tangent to Σ at any given point P, there exists an isometry of the space-time metric that leaves both P and u unchanged but which rotates e_1 to e_2.[1] The timelike curves generated by u can be understood as the trajectories of privileged observers for whom the space appears to be isotropic. For all other observers at the same point, the FLRW space-time may look anisotropic. However, such an anisotropy would be considered to be due to the motion of the observers in an isotropic background and this could be evaluated.

It is now possible to proceed with the construction of the most convenient metric forms of the FLRW family of space-times.

These space-times are foliated by three-dimensional hypersurfaces Σ which have constant curvature. As summarised in Appendix B, there are only three types of such 3-spaces. They are distinguished by a parameter k, which is the sign of their curvature. The three possibilities are a three-dimensional *flat space* when $k = 0$, a *3-sphere* S^3 when $k = +1$, and a *hyperbolic 3-space* H^3 when $k = -1$. The metric of all these 3-spaces of constant curvature can be written in a number of useful forms, for example that of (B.3):

$$ds_\Sigma^2 = R^2 \frac{dx^2 + dy^2 + dz^2}{\left(1 + \frac{1}{4}k(x^2 + y^2 + z^2)\right)^2}. \tag{6.1}$$

Here, x, y, z are dimensionless coordinates, while the parameter R (which has dimension of length) determines the magnitude of the curvature $K = k/R^2$. For example, R is the actual radius of the 3-sphere when $k = +1$.

Of course, the value of R on different spatial hypersurfaces Σ may vary, whereas the sign of the curvature k should be the same throughout the whole universe (provided R remains finite). With this restriction, the curvature of the 3-spaces, which is characterised by R, generally depends on time. For the FLRW space-times, these hypersurfaces of homogeneity Σ are orthogonal to the vector field u of isotropic timelike observers. It is thus most convenient to introduce a *global temporal coordinate* t in these manifolds by taking

$$u = \partial_t, \tag{6.2}$$

---

[1] In a general non-homogeneous but isotropic space-time the surfaces Σ need not exist, but it is still required that the vectors e_1, e_2 be orthogonal to u at any P.

synchronised throughout the complete space-time in such a way that the hypersurfaces Σ are given by $t = \text{const.}$ In other words, the one-parameter family of constant-curvature 3-spaces Σ are labelled by the time coordinate t.

With this natural choice, the metric of a general FLRW space-time can be written in the form

$$ds^2 = -dt^2 + R^2(t) \frac{dx^2 + dy^2 + dz^2}{\left(1 + \frac{1}{4}k(x^2 + y^2 + z^2)\right)^2}. \tag{6.3}$$

In this metric, t is a synchronous (proper) time, while x, y, z are dimensionless spatial comoving coordinates. The discrete parameter $k = 0, +1, -1$ distinguishes the three possible cases of 3-spaces given by $t = \text{const.}$ with zero, positive or negative curvature, respectively.

At this point, the only undetermined quantity in the metric is the function $R(t)$. This obviously describes the time evolution of the geometry, specifically the measurement of physical distances. Since it directly determines the scale of the 3-space geometry on the spatial hypersurfaces $t = \text{const.}$, it is sometimes referred to either as the *scale function* or the *expansion factor*.

Other standard metric forms of the FLRW family of space-times can be obtained by using alternative coordinate parametrisations of the 3-spaces of constant curvature given in Appendix B. For example, (B.5) leads to

$$ds^2 = -dt^2 + R^2(t) \left(\frac{dr^2}{1 - k\,r^2} + r^2(d\theta^2 + \sin^2\theta\,d\phi^2) \right), \tag{6.4}$$

while (B.6), (B.7) and (B.8) give, respectively,

$$ds^2 = -dt^2 + R^2(t) \left(d\chi^2 + \chi^2\,(d\theta^2 + \sin^2\theta\,d\phi^2) \right) \qquad \text{for } k = 0, \tag{6.5}$$

$$ds^2 = -dt^2 + R^2(t) \left(d\chi^2 + \sin^2\chi\,(d\theta^2 + \sin^2\theta\,d\phi^2) \right) \qquad \text{for } k = +1, \tag{6.6}$$

$$ds^2 = -dt^2 + R^2(t) \left(d\chi^2 + \sinh^2\chi\,(d\theta^2 + \sin^2\theta\,d\phi^2) \right) \quad \text{for } k = -1. \tag{6.7}$$

In spite of their apparent similarities, these three FLRW metrics represent different types of geometries. The ranges of χ also differ: while $\chi \in [0, \infty)$ for $k = 0, -1$, it is necessary for $k = +1$ to consider only the range $\chi \in [0, \pi]$ since the 3-sphere is compact. However, in all cases $\theta \in [0, \pi]$ and $\phi \in [0, 2\pi)$ span complete 2-spheres.

The above metrics of spatially homogeneous and isotropic space-times were introduced and applied as cosmological models in the 1920s and 1930s, first by Friedmann (1922, 1924) and Lemaître (1927) for the special cases $k = \pm 1$, and then in full generality by Robertson (1929) and Walker (1935).

The previous particular solutions by Einstein (1917) and de Sitter (1917a,b) also belong to this class of space-times. For an account of this early development, see especially the comprehensive review by Robertson (1933).

6.2 Matter content and dynamics

To determine the scale function $R(t)$ and thus the time evolution of the FLRW space-time, it is necessary to apply Einstein's field equations. The dynamics of the universe are thus determined by its matter content.

It can be shown that the assumption of homogeneity and isotropy is only consistent with matter being in the form of a *perfect fluid* for which the total energy-momentum tensor is (2.23), namely

$$T_{\mu\nu} = (\rho + p)\, u_\mu u_\nu + p\, g_{\mu\nu}\,. \tag{6.8}$$

Here the energy density ρ and the pressure p may be functions of time only. For physical reasons, it is usually assumed that $\rho(t) \geq p(t) \geq 0$. Also, $\boldsymbol{u}$ in (6.8) is the four-velocity of the fluid which coincides with the vector field of isotropic timelike observers. In view of (6.2) and the synchronous diagonal form of the FLRW metric (6.3) (or, equivalently, (6.4) or (6.5)–(6.7)), the above energy-momentum tensor only has the temporal component $T_{tt} = \rho$ and spatial components $T_{ij} = p\, g_{ij}$.

The Bianchi identities imply that $T^{\mu\nu}{}_{;\nu} = 0$. Due to the homogeneity and isotropy of the FLRW space-times, the spatial components $\mu = i$ are identically satisfied, while the temporal component $\mu = t$ leads to the energy conservation equation $(\rho R^3)^{\cdot} + p\,(R^3)^{\cdot} = 0$ which can be re-written as

$$\frac{\dot{\rho}}{\rho + p} = -3\frac{\dot{R}}{R}\,. \tag{6.9}$$

Equation (6.9) can be explicitly integrated for physically important barotropic equations of state $p = p(\rho)$. These include the special cases $p = 0$ for *dust* or cold dark matter (that is mutually non-interacting matter with negligible random relative velocities), $p = \frac{1}{3}\rho$ for massless thermal *radiation* (like the cosmic microwave background radiation), or a so-called *stiff fluid* with the equation of state $p = \rho$ in which sound waves propagate with the speed of light. These cases can be treated in a unified way by considering the general linear equation of state (2.24), namely

$$p = (\gamma - 1)\,\rho\,, \tag{6.10}$$

where $1 \leq \gamma \leq 2$ is an appropriate constant. For this, the energy conservation equation (6.9) implies that the energy density is given by

$$\rho = \frac{C}{R^{3\gamma}}, \qquad C = \text{const}. \tag{6.11}$$

For a dust-filled FLRW universe, $\gamma = 1$ and thus $\rho \propto R^{-3}$. In the presence of radiation, $\gamma = 4/3$ and $\rho \propto R^{-4}$. And for stiff fluid, $\gamma = 2$ and $\rho \propto R^{-6}$. For any $\gamma > 0$, the matter density ρ diverges at the curvature singularity where $R = 0$. Notice also that the special case $\gamma = 0$ would correspond to a cosmological constant since (6.11) and (6.10) then formally imply that $\rho = -p = C$. Thus $T_{\mu\nu} = -C\,g_{\mu\nu}$, and $8\pi C$ could effectively be understood as Λ.

Now, for the metric form (6.3) of the FLRW space-times (or, equivalently, (6.4) or (6.5)–(6.7)) the non-vanishing Einstein tensor components are

$$G_{tt} = 3\left(\frac{\dot{R}^2}{R^2} + \frac{k}{R^2}\right), \qquad G_{ij} = -\left(2\frac{\ddot{R}}{R} + \frac{\dot{R}^2}{R^2} + \frac{k}{R^2}\right)g_{ij}. \tag{6.12}$$

Einstein's equations (2.10) with a possible cosmological constant Λ are thus

$$\frac{\dot{R}^2}{R^2} + \frac{k}{R^2} = \frac{\Lambda}{3} + \frac{8\pi}{3}\rho, \tag{6.13}$$

$$2\frac{\ddot{R}}{R} + \frac{\dot{R}^2}{R^2} + \frac{k}{R^2} = \Lambda - 8\pi\,p. \tag{6.14}$$

It can be shown that the field equation (6.14) is identically satisfied as a consequence of (6.13) and (6.9), unless $\dot{R} = 0$, which corresponds to special static FLRW space-times. In view of (6.11), the evolution of the FLRW universe is thus described by the *Friedmann equation*

$$\frac{\dot{R}^2}{R^2} = \frac{\Lambda}{3} - \frac{k}{R^2} + \frac{c}{R^{3\gamma}}, \tag{6.15}$$

where $c = \frac{8}{3}\pi C$. The solution $R(t)$ of this differential equation is determined by the cosmological constant Λ, the spatial curvature parameter $k = 0, \pm 1$, and the parameter γ of the equation of state of the matter content whose density is proportional to $c > 0$. If the FLRW universe is filled with different independent kinds of perfect fluid matter, for example the combination of dust and photon radiation, the last single term in the equation has to be replaced by the sum of the corresponding two terms. (See Stephani *et al.* (2003) for references to explicit solutions of this type.)

A qualitative discussion of all possible solutions to the equation (6.15) can be performed by a "potential method" (see Robertson, 1933). Introducing

$$V(R) \equiv \frac{k}{R^2} - \frac{c}{R^{3\gamma}}, \tag{6.16}$$

it can be seen that solutions of (6.15) are possible only within the range of R for which $V(R) \leq \frac{1}{3}\Lambda$. The scale function $R(t)$ may thus admit only those values for which the curve $V(R)$, for the given choice of the parameters k, γ and c, is *below* the horizontal line corresponding to the constant value of $\frac{1}{3}\Lambda$ (a typical value of this is indicated by the dashed line in Figure 6.2). Turning points in the evolution of $R(t)$ occur when $V(R) = \frac{1}{3}\Lambda$. If $V(R)$ does not intersect $\frac{1}{3}\Lambda$, R will continue to increase or decrease over the complete range $[0, \infty)$, where $R = 0$ corresponds to a curvature singularity.

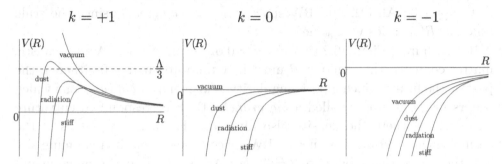

Fig. 6.2 Potentials $V(R)$ of the Friedmann equation (6.15) given by (6.16). The three figures for $k = +1, 0, -1$ correspond to the FLRW space-times in which the 3-space is a sphere, flat space and a hyperbolic space. In each case, the function $V(R)$ is plotted for the vacuum ($c = 0$), dust ($\gamma = 1$), radiation ($\gamma = 4/3$) and stiff fluid ($\gamma = 2$). In all cases, $V \to 0$ as $R \to \infty$, and $V \to -\infty$ as $R \to 0$ (unless $c = 0$, $k = +1, 0$). The solution for the scale function $R(t)$ is possible only in the range of R in which the curve of $V(R)$ is below the horizontal level corresponding to the line of a constant $\frac{\Lambda}{3}$. An intersection of these lines represents a turning point in the evolution of $R(t)$. Static solutions are only possible at stationary points of $V(R)$.

Figure 6.2 contains three parts corresponding to $k = +1, 0, -1$, respectively. In each case, there are typical plots of the potential $V(R)$ for the physically most important possible matter contents of the FLRW universe, namely vacuum ($c = 0$), dust ($\gamma = 1$), radiation ($\gamma = 4/3$) and stiff fluid ($\gamma = 2$). Other possibilities, including $\gamma < 1$, or even $\gamma < 0$, have been thoroughly discussed by Harrison (1967).

Static solutions $R = R_0 = \text{const.}$ of the Friedmann equation and (6.14) correspond to stationary points of the potential function $V(R)$, and occur when $k = +1$ and $\Lambda = \Lambda_0 \equiv 3V(R_0)$. These solutions are given by $R_0 = (4\pi\gamma C)^{1/(3\gamma-2)}$, and are unstable. In particular, for the space-time filled with dust, $\gamma = 1$ and, using (6.11) and (6.10), the parameters take the values

$$R_0 = \frac{3}{2}c, \qquad \Lambda_0 = \frac{1}{R_0^2}, \qquad \rho_0 = \frac{1}{4\pi R_0^2}, \qquad p_0 = 0. \qquad (6.17)$$

This is the *Einstein static universe* (Einstein, 1917), which was initially described in Section 3.2. Analogously, the static closed space-time with radiation ($\gamma = 4/3$) is given by

$$R_0 = \sqrt{2}\,c, \qquad \Lambda_0 = \frac{3}{2R_0^2}, \qquad \rho_0 = \frac{3}{16\pi R_0^2}, \qquad p_0 = \frac{1}{16\pi R_0^2}. \qquad (6.18)$$

Another (trivial) static solution occurs when $k = 0$ and $\Lambda = 0 = c$, with R_0 an arbitrary constant, but this is just the flat Minkowski space described in Chapter 3. All other FLRW space-times are *dynamical* since the scale function $R(t)$ varies with time.

It is seen from Figure 6.2 that, for $k = 0$ or -1 and for any Λ and realistic matter contents, the function R must inevitably reach $R = 0$ either in the past or the future. Such a past singularity is known as a *big bang*, one which occurs in the future is called a *big crunch*. Both correspond to curvature singularities where the density also diverges. For negative values of the cosmological constant, any such FLRW space-time reaches its maximum size and then re-collapses back to $R = 0$. For $\Lambda \geq 0$, the universe may expand indefinitely.

In the case of closed FLRW universes, for which $k = +1$, there exist more possibilities. For non-vacuum space-times of this type there is always one maximum of the effective potential $V(R)$ at R_0, see Figure 6.2. If the value of the cosmological constant is smaller than $3V(R_0)$, the universe may expand from an initial singularity, reach its maximum size and re-collapse to $R = 0$. For $\Lambda > 3V(R_0)$, the universe expands for ever, but its rate of expansion slows down when R is close to R_0.

Interestingly, for $k = +1$, there also exist FRLW *cosmologies without a big bang* singularity. Indeed, if $0 < \Lambda < 3V(R_0)$ and the universe is collapsing from large values of R, it necessarily stops its collapse at some value $R_{\min} > R_0$, which is given by the intersection of the potential curve $V(R)$ with the horizontal (dashed) line of constant $\frac{\Lambda}{3}$. After this "bounce", it re-expands to an infinite value of R. Such a scenario is possible for any perfect fluid matter (dust, radiation etc.), and also for vacuum FLRW space-times with $k = +1$ and any $\Lambda > 0$.

6.3 Explicit solutions

The most important FLRW universes will now be described explicitly. Vacuum cases will be dealt with first, then universes with a zero cosmological constant, and finally cases which contain both matter and Λ.

6.3.1 Vacuum FLRW space-times

It is natural to start with vacuum FLRW space-times. These have $c = 0$, and the Friedmann equation (6.15) can easily be solved.

For $\Lambda = 0$, it is necessary (apart from the static case with $k = 0$ described above) to take $k = -1$, which leads to the metric with $R(t) = t$, namely

$$ds^2 = -dt^2 + t^2 \left(d\chi^2 + \sinh^2 \chi \left(d\theta^2 + \sin^2 \theta \, d\phi^2\right)\right). \qquad (6.19)$$

This is known as the Milne universe, see Milne (1933), Kermack and McCrea (1933) or Robertson and Noonan (1968), and is shown in Figure 6.3. In fact, this is just (part of) flat Minkowski space (3.2) expressed in comoving coordinates t, χ such that $\bar{r} = t \sinh \chi$, $\bar{t} = t \cosh \chi$, in which the world-lines of isotropy all start from a common point at $t = 0$, and the surfaces of homogeneity have negative curvature. This simple vacuum space-time can be used to approximate low-density cosmological models for large R, but it is not appropriate to epochs when the matter content of the universe dominates the dynamics.

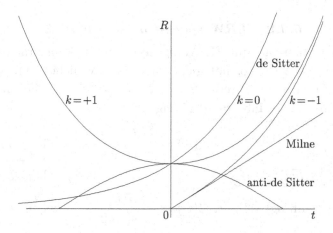

Fig. 6.3 The scale functions $R(t)$ for vacuum FLRW space-times, namely the Milne (Minkowski), de Sitter and anti-de Sitter universes which correspond to a vanishing, positive and negative cosmological constant Λ. The de Sitter universe admits different foliations in which the homogeneous 3-space has the curvatures $k = +1, 0, -1$.

For $\Lambda > 0$, the equation (6.15) admits the following three solutions:

$$R(t) = a \cosh(t/a) \qquad \text{when} \quad k = +1,$$
$$R(t) = a \exp(t/a) \qquad \text{when} \quad k = 0,$$
$$R(t) = a \sinh(t/a) \qquad \text{when} \quad k = -1,$$

where $a = \sqrt{3/\Lambda}$. With the FLRW metric forms (6.6), (6.5) and (6.7), these are exactly the three standard coordinate representations of the de Sitter universe described in Chapter 4, namely the line elements (4.4), (4.14) and (4.16), respectively. The de Sitter universe thus admits different foliations in which the surfaces of homogeneity have positive, zero and negative constant spatial curvature. Such metric forms were first introduced by de Sitter (1917a,b), Lanczos (1922) and Lemaître (1925) (see Robertson, 1933). In all cases the scale functions are asymptotically exponential for large t.

Finally, for $\Lambda < 0$ the only non-trivial solution is

$$R(t) = a \cos(t/a) \qquad \text{when} \quad k = -1 \,,$$

where $a = \sqrt{3/|\Lambda|}$. Putting this into (6.7), the standard comoving form (5.18) on the anti-de Sitter universe is recovered, see Chapter 5.

To summarise, it has thus been found that the only possible vacuum FLRW space-times are Minkowski, de Sitter and anti-de Sitter universes which all have a constant space-time curvature. Of course, this is fully consistent with the fact that the FLRW space-times are conformally flat.

6.3.2 FLRW space-times without Λ

Turning now to non-vacuum FLRW space-times, it is natural to first consider the cases when the cosmological constant Λ vanishes. The Friedmann equation (6.15) with $\Lambda = 0$ reduces to $\dot{R}^2 = c\,R^{2-3\gamma} - k$. Following Harrison (1967), this can be integrated giving

$$R(\psi) = (c \sin^2 \psi)^{1/(3\gamma-2)}, \quad t(\psi) = \frac{2 \int R(\psi)\,\mathrm{d}\psi}{3\gamma - 2} \qquad \text{when} \quad k = +1 \,,$$

$$R(t) = (\tfrac{3}{2}\gamma\sqrt{c}\,t)^{2/3\gamma} \qquad\qquad\qquad\qquad\qquad \text{when} \quad k = 0 \,,$$

$$R(\psi) = (c \sinh^2 \psi)^{1/(3\gamma-2)}, \quad t(\psi) = \frac{2 \int R(\psi)\,\mathrm{d}\psi}{3\gamma - 2} \qquad \text{when} \quad k = -1 \,,$$

in which ψ is defined implicitly, see also Tauber (1967) and Vajk (1969).

Explicit solutions can be obtained for the case $k = +1$, in which the space is a 3-sphere and the FLRW metric takes the form (6.6). The first Friedmann (1922) model with dust has $\gamma = 1$, and thus $R = c \sin^2 \psi$, and $t = c\,(\psi - \sin\psi\cos\psi)$. In this case, $R(t)$ represents a cycloid, and the universe expands from a big bang to a maximum size and then re-collapses to $R = 0$ in a finite time, see Figure 6.4. Similarly, for the universe filled with radiation, $\gamma = 4/3$, which implies $R = \sqrt{c}\,\sin\psi$, $t = \sqrt{c}\,(1 - \cos\psi)$, so that $R(t) = \sqrt{t\,(2\sqrt{c} - t)}$. This is known as the Tolman (1931) model.

Particular cases of spatially flat models (6.5) with $k = 0$ are the Einstein–de Sitter (1932) universe with dust ($\gamma = 1$) for which $R \propto t^{2/3}$, the Tolman

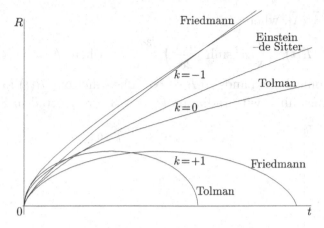

Fig. 6.4 The scale functions $R(t)$ of the FLRW space-times with a vanishing cosmological constant $\Lambda = 0$. For all three values of k, the dust solutions with $\gamma = 1$, namely the Friedmann ($k = \pm 1$) and Einstein–de Sitter ($k = 0$) universes, and the Tolman solutions for radiation with $\gamma = 4/3$ are drawn.

(1934b) space-time with radiation ($\gamma = 4/3$) for which $R \propto \sqrt{t}$, or stiff fluid matter ($\gamma = 2$) in which case $R \propto t^{1/3}$. They all start at a big bang and expand indefinitely.

When $k = -1$, the FLRW space-times with $\Lambda = 0$ can be written in the metric form (6.7). For the Friedmann (1924) second model with dust the scale function $R(t)$ takes the form $R = c \sinh^2 \psi$, $t = c \left(\sinh \psi \cosh \psi - \psi \right)$. If the universe contains radiation then $R = \sqrt{c} \sinh \psi$, $t = \sqrt{c} \left(\cosh \psi - 1 \right)$, i.e. $R(t) = \sqrt{t \left(2\sqrt{c} + t \right)}$. These space-times expand for ever. For large times $R \approx t$, so that they approach the Milne universe (6.19) asymptotically.

6.3.3 FLRW space-times with Λ

If the cosmological constant Λ is non-vanishing, there are a number of non-vacuum FLRW space-times which are solutions of the complete Friedmann equation (6.15) (see e.g. Heckmann, 1932). As can be seen from Figure 6.2, for $\Lambda < 0$, all the models have finite duration. They start from the big bang singularity at $t = 0$ but their expansion necessarily stops, and they collapse back to $R = 0$ after a finite time. For $\Lambda > 0$, the corresponding space-times may have either finite or infinite temporal duration.

Explicit expanding and collapsing solutions of this type for $k = 0$ are

$$R(t) = \left(ca^2 \sin^2 \frac{3\gamma t}{2a} \right)^{1/3\gamma} \qquad \text{when } \Lambda < 0,$$

where $a = \sqrt{3/|\Lambda|}$, whereas

$$R(t) = \left(ca^2 \sinh^2 \frac{3\gamma t}{2a} \right)^{1/3\gamma} \qquad \text{when} \quad \Lambda > 0,$$

which monotonically expands to $R \to \infty$. The functions $R(t)$ for the particular space-times filled with radiation ($\gamma = 4/3$) are plotted in Figure 6.5.

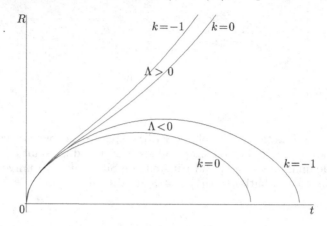

Fig. 6.5 The FLRW space-times with radiation, $\Lambda \neq 0$ and $k = 0, -1$.

When $k = \pm 1$, solutions of the Friedmann equation are generally given in terms of elliptic functions (see e.g. Lemaître, 1933, and Edwards, 1972, for the dust and radiation models), or elementary functions in special cases.

Interestingly, the FLRW $k = +1$ space-times (6.6) with radiation ($\gamma = 4/3$) and a cosmological constant $\Lambda \neq 0$ can be written explicitly as

$$R(t) = \frac{a}{\sqrt{2}} \left(1 - \cosh \frac{2t}{a} + \frac{a_0}{a} \sinh \frac{2t}{a} \right)^{1/2} \qquad \text{when} \quad \Lambda > \Lambda_0,$$

$$R(t) = R_0 \left(1 + \exp \frac{2t}{a_0} \right)^{1/2} \qquad \text{when} \quad \Lambda = \Lambda_0, \quad R > R_0,$$

$$R(t) = R_0 \left(1 - \exp \left(-\frac{2t}{a_0} \right) \right)^{1/2} \qquad \text{when} \quad \Lambda = \Lambda_0, \quad R < R_0,$$

$$R(t) = \frac{a}{\sqrt{2}} \left(1 + \sqrt{1 - a_0^2/a^2} \cosh \frac{2t}{a} \right)^{1/2} \qquad \text{when} \quad 0 < \Lambda < \Lambda_0, \quad R \text{ large},$$

$$R(t) = \frac{a}{\sqrt{2}} \left(1 - \cosh \frac{2t}{a} + \frac{a_0}{a} \sinh \frac{2t}{a} \right)^{1/2} \qquad \text{when} \quad 0 < \Lambda < \Lambda_0, \quad R \text{ small},$$

where $a_0 = \sqrt{3/\Lambda_0} = \sqrt{2} R_0 = 2\sqrt{c}$. In fact, such (necessarily positive) values of Λ_0, R_0 and c correspond to the static Einstein-type solution (6.18) which is unstable. The plots of these functions are given in Figure 6.6. Similar results are obtained for dust solutions.

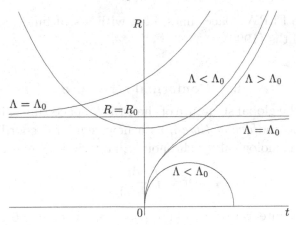

Fig. 6.6 Possible FLRW space-times with radiation, $\Lambda \neq 0$ and $k = +1$.

The solutions with $\Lambda > \Lambda_0$ start at $R(0) = 0$ and expand to $R = \infty$ as $t \to \infty$. However, for R close to R_0, their expansion slows down and the universe subsequently accelerates. For this reason it is sometimes called the Lemaître "hesitation" universe. The two dynamical solutions with $\Lambda = \Lambda_0$ asymptotically recede from or approach the static value $R = R_0$ as $t \to -\infty$ or $+\infty$.[2] When $0 < \Lambda < \Lambda_0$, there are two solutions which start either at $R = \infty$ or at $R = 0$. The first is a non-singular solution which collapses to a minimum size at $t = 0$ and then re-expands. The other expands from a big bang to a maximum size after which it re-contracts back to $R = 0$. (A qualitatively similar solution occurs when $\Lambda < 0$.) See, e.g., Bondi (1961) for a discussion of these possibilities.

The analogous radiation FLRW space-times (6.7) with $k = -1$ have the scale function

$$R(t) = \frac{a}{\sqrt{2}} \left(\cosh \frac{2t}{a} - 1 + \frac{a_0}{a} \sinh \frac{2t}{a} \right)^{1/2} \quad \text{when } \Lambda > 0,$$

$$R(t) = \frac{a}{\sqrt{2}} \left(1 - \cos \frac{2t}{a} + \frac{a_0}{a} \sin \frac{2t}{a} \right)^{1/2} \quad \text{when } \Lambda < 0,$$

where $a_0 = 2\sqrt{c}$. These are shown in Figure 6.5. The former solution describes open universes which expand from the big bang to an infinite size. The latter stop their expansion and re-collapse to the singularity $R = 0$.

Special explicit solutions for the more extreme cases such as $\gamma < 1$, which corresponds to matter with negative pressure, can be found in Harrison

---

[2] For dust these are the Eddington–Lemaître and the Eddington–Lemaître–Bondi models.

(1967). For the FLRW space-times filled with a stiff fluid ($\gamma = 2$) see, e.g., Bičák and Griffiths (1996).

6.4 Conformal structure

To investigate the global structure of the FLRW space-times, it is convenient to introduce the *conformal time* η as a new temporal coordinate. This is related to the cosmological synchronous time t via the relation

$$\eta = \int \frac{dt}{R(t)}. \tag{6.20}$$

The constant of integration can be chosen such that $\eta = 0$ corresponds to $t = 0$. With this, the three standard metric forms (6.5)–(6.7) become

$$ds^2 = R^2(\eta) \left(-d\eta^2 + d\chi^2 + \sin^2 \chi \, (d\theta^2 + \sin^2 \theta \, d\phi^2) \right) \quad \text{for } k = +1,$$

$$ds^2 = R^2(\eta) \left(-d\eta^2 + d\chi^2 + \chi^2 \, (d\theta^2 + \sin^2 \theta \, d\phi^2) \right) \qquad \text{for } k = 0, \tag{6.21}$$

$$ds^2 = R^2(\eta) \left(-d\eta^2 + d\chi^2 + \sinh^2 \chi \, (d\theta^2 + \sin^2 \theta \, d\phi^2) \right) \quad \text{for } k = -1,$$

where $R(\eta) \equiv R(t(\eta))$. Possible non-vacuum space-times will now be discussed. Vacuum FLRW models are Minkowski, de Sitter and anti-de Sitter universes whose global structures are described in Chapters 3–5.

6.4.1 FLRW space-times without Λ

Non-vacuum FLRW space-times with vanishing cosmological constant are explicitly given by

$$R(\eta) = R_c \left(\sin \left(\tfrac{3\gamma - 2}{2} \, \eta \right) \right)^{\frac{2}{3\gamma - 2}}, \qquad \text{when } k = +1,$$

$$R(\eta) = R_c \, \eta^{\frac{2}{3\gamma - 2}}, \qquad \text{when } k = 0, \tag{6.22}$$

$$R(\eta) = R_c \left(\sinh \left(\tfrac{3\gamma - 2}{2} \, \eta \right) \right)^{\frac{2}{3\gamma - 2}}, \qquad \text{when } k = -1,$$

where R_c is a constant length which determines the scale of the universe (when $k = +1$ this is its maximum size). Notice that the power coefficient for the space-times filled with dust is $\frac{2}{3\gamma - 2} = 2$, while $\frac{2}{3\gamma - 2} = 1$ for the case of radiation. Using the conformal time η, the corresponding metrics of the Friedmann, Tolman and Einstein–de Sitter universes (see Figure 6.4) are thus very simple.

Since the FLRW space-times are spherically symmetric around any point, it is possible to restrict the study of their properties to any two-dimensional section spanned by η, χ. On this section the metric is the same for any value of k, namely $R^2(\eta)\left(-\mathrm{d}\eta^2 + \mathrm{d}\chi^2\right)$, and each point represents a 2-sphere spanned by θ and ϕ whose radius is $R(\eta)\sin\chi$, $R(\eta)\chi$ or $R(\eta)\sinh\chi$ when $k = +1, 0$, or -1, respectively.

To properly capture the causal structure of the complete four-dimensional FLRW universe, however, it is most convenient to follow the Penrose (1964, 1968a) approach (see Penrose and Rindler, 1986), which relates a given space-time to part of the Einstein static universe

$$\mathrm{d}s^2 = R_0^2 \left(-\mathrm{d}\bar{\eta}^2 + \mathrm{d}\bar{\chi}^2 + \sin^2\bar{\chi}\,(\mathrm{d}\theta^2 + \sin^2\theta\,\mathrm{d}\phi^2)\right). \qquad (6.23)$$

This does not vary in time and its spatial sections are 3-spheres (see Section 3.2). Its section through the equatorial plane $\theta = \frac{\pi}{2}$ can be pictured as the interior of an infinite solid cylinder of radius $\bar{\chi} = \pi$, where $\bar{\eta} \in (-\infty, +\infty)$ is plotted vertically as in Figure 3.2.[3] The global structure of a FLRW space-time can then be visualised by means of the specific conformal transformation, which depends on the scale function $R(\eta)$, and also on the allowed ranges of the dimensionless coordinates η and χ.

The simplest situation occurs for closed FLRW space-times with $k = +1$, in which the homogeneous 3-space is a 3-sphere S^3. The metric (6.21) is obviously conformal to (6.23) via (2.26). Taking $\eta = \bar{\eta}$ and $\chi = \bar{\chi}$, the conformal factor is

$$\Omega = \frac{R_0}{R(\eta)}. \qquad (6.24)$$

Infinity $\mathcal{I}$ of the space-time corresponds to the points where $\Omega = 0$, that is $R(\eta) = \infty$. Any $k = +1$ FLRW universe is thus conformal to the part of the cylinder (see Figure 3.2) representing the Einstein static universe that is bounded by two sections $\eta = \text{const}$. Such boundaries represent either the *singularity* when $R(\eta) = 0$ or *conformal infinity* $\mathcal{I}$ when $R(\eta) = \infty$.

The corresponding Penrose diagram is readily obtained by fixing both the angular coordinates θ, ϕ. The range of χ (plotted in the horizontal direction) is $[0, \pi]$. Since the universe given by (6.22) with $k = +1$ expands from the big bang to a maximum size R_c and then re-collapses back to the final singularity $R = 0$, the Penrose diagram is finite also in the temporal direction η (plotted vertically), as shown in Figure 6.7.

The conformal diagram on the left corresponds to the closed FLRW space-time with $\Lambda = 0$ filled with dust ($\gamma = 1$), i.e. the Friedmann universe for which $R(\eta) = R_c \sin^2(\eta/2)$, the diagram on the right represents the Tolman radiation universe ($\gamma = 4/3$) with $R(\eta) = R_c \sin\eta$. Both Penrose diagrams

---

[3] Note that the outer boundary $\bar{\chi} = \pi$ corresponds to a single point (the South pole of S^3) as described in Appendix B, in particular Figure B.1.

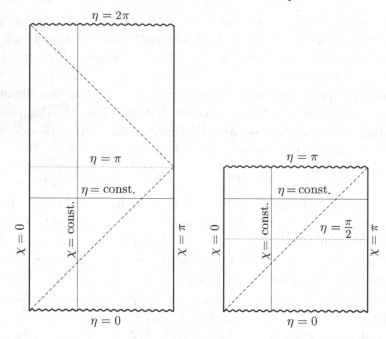

Fig. 6.7 Penrose diagrams for the $k = +1$ FLRW space-times with $\Lambda = 0$, which contain dust (left) or radiation (right). The universe expands from the big bang singularity $R = 0$ at $\eta = 0$, reaches its maximum size $R = R_c$ (indicated by the horizontal dotted line) and than re-collapses to the final singularity. The dashed lines denote the trajectory of a photon emitted from $\chi = 0$ (the North pole of S^3) soon after the big bang. In the universe filled with radiation such a photon reaches only as far as $\chi = \pi$ (the South pole). The dust universe lasts longer so that the photon encircles the whole S^3 and returns to the initial point (as it passes the South pole at $\eta = \chi$, its value of the angle χ starts to decrease).

are bounded by the initial big bang singularity located at $\eta = 0$, and the final singularity at $\eta = 2\pi$ or $\eta = \pi$, respectively. The universe containing radiation thus collapses sooner than that with dust. Any null particle that starts to propagate in the Tolman universe from some point (the North pole $\chi = 0$, say) soon after the big bang can only reach the antipodal point (the South pole at $\chi = \pi$) by the time of the final singularity. By contrast, in the Friedmann dust universe, such a particle has enough time to encircle the whole 3-sphere, returning to its initial point (the North pole) at $\eta = 2\pi$. The Penrose diagram for a general FLRW space-time with $k = +1$ and $\Lambda = 0$ is very similar. The only difference is that the final singularity occurs at $\eta = 2\pi/(3\gamma - 2)$, where γ determines the specific equation of state (6.10).

In order to construct Penrose diagrams for the FLRW space-times (6.21) in the spatially flat case when $k = 0$, it is necessary to perform a compactification by applying the specific conformal transformation

$$\eta = \tan\left(\frac{\bar{\eta}+\bar{\chi}}{2}\right) + \tan\left(\frac{\bar{\eta}-\bar{\chi}}{2}\right) \quad \equiv \frac{2\sin\bar{\eta}}{\cos\bar{\eta}+\cos\bar{\chi}},$$
$$\chi = \tan\left(\frac{\bar{\eta}+\bar{\chi}}{2}\right) - \tan\left(\frac{\bar{\eta}-\bar{\chi}}{2}\right) \quad \equiv \frac{2\sin\bar{\chi}}{\cos\bar{\eta}+\cos\bar{\chi}}, \tag{6.25}$$

that is, $\frac{\eta+\chi}{2} = \tan\left(\frac{\bar{\eta}+\bar{\chi}}{2}\right)$, $\frac{\eta-\chi}{2} = \tan\left(\frac{\bar{\eta}-\bar{\chi}}{2}\right)$. Thus, infinite ranges of η, χ are mapped to finite ranges of $\bar{\eta}, \bar{\chi}$, and the FLRW metric

$$ds^2 = \Omega^{-2}\left(-d\bar{\eta}^2 + d\bar{\chi}^2 + \sin^2\bar{\chi}\,(d\theta^2 + \sin^2\theta\,d\phi^2)\right) \tag{6.26}$$

becomes conformal to (6.23) with the factor

$$\Omega = \frac{R_0}{R(\eta(\bar{\eta},\bar{\chi}))}\cos\left(\frac{\bar{\eta}+\bar{\chi}}{2}\right)\cos\left(\frac{\bar{\eta}-\bar{\chi}}{2}\right). \tag{6.27}$$

In the $\bar{\eta}, \bar{\chi}$-space, null geodesics thus still propagate along the lines with $45°$ inclination. The *conformal infinity* $\mathcal{I}$ of the space-time consists of those points where $\Omega = 0$. This explicitly occurs at $\bar{\eta} + \bar{\chi} = \pm\pi$ and $\bar{\eta} - \bar{\chi} = \pm\pi$, or $R = \infty$. The initial *singularity* at $\eta = 0$ corresponds to $\bar{\eta} = 0$, and similarly $\chi = 0$ corresponds to $\bar{\chi} = 0$, while $\eta = \infty$ and also $\chi = \infty$ represent conformal infinity $\mathcal{I}$ given by $\Omega = 0$ (unless $\bar{\chi} = 0, \pm\pi$ and $\bar{\eta} = \pm\pi, 0$).

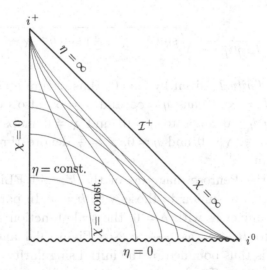

Fig. 6.8 Penrose diagram for the FLRW space-times with $k = 0$ and $\Lambda = 0$ in the compactified coordinates $\bar{\eta}$ (vertical) and $\bar{\chi}$ (horizontal). The universe is bounded by the initial big bang singularity at $\bar{\eta} = 0$ and the future conformal infinity $\mathcal{I}^+$ located at $\bar{\eta} + \bar{\chi} = \pi$. The asymptotic "points" i^+ and i^0 given by $\bar{\eta} = \pi, \bar{\chi} = 0$ and $\bar{\eta} = 0, \bar{\chi} = \pi$ denote the infinities of timelike and spacelike geodesics, respectively. Notice that the coordinate lines $\eta = $ const. are all tangent to $\eta = 0$ near i^0.

Of course, in constructing the Penrose diagram, it is necessary to consider the specific behaviour of the scale function $R(\eta)$. For the FLRW $k = 0$ space-times without a cosmological constant Λ (e.g. the Tolman and Einstein–de Sitter universes), the function $R(\eta)$ grows monotonically from the physical singularity $R = 0$ at $\eta = 0$ to $R \to \infty$ as $\eta \to \infty$, see Figure 6.4 and (6.22). In terms of the coordinates $\bar{\eta}$, $\bar{\chi}$ introduced by (6.25), the Penrose diagram of all such space-times is thus bounded by the initial singularity at $\bar{\eta} = 0$ and the future conformal infinity $\mathcal{I}^+$, given by $\bar{\eta} + \bar{\chi} = \pi$ corresponding to $\eta = \infty, \chi = \infty$. This is shown in Figure 6.8.

Finally, the Penrose diagrams of the FLRW space-times (6.21), in the case when $k = -1$, are obtained by performing the transformation

$$\eta = \operatorname{arctanh}\left[\tan\left(\frac{\bar{\eta}+\bar{\chi}}{2}\right)\right] + \operatorname{arctanh}\left[\tan\left(\frac{\bar{\eta}-\bar{\chi}}{2}\right)\right] \equiv \operatorname{arctanh}\left(\frac{\sin\bar{\eta}}{\cos\bar{\chi}}\right),$$

$$\chi = \operatorname{arctanh}\left[\tan\left(\frac{\bar{\eta}+\bar{\chi}}{2}\right)\right] - \operatorname{arctanh}\left[\tan\left(\frac{\bar{\eta}-\bar{\chi}}{2}\right)\right] \equiv \operatorname{arctanh}\left(\frac{\sin\bar{\chi}}{\cos\bar{\chi}}\right),$$

$$(6.28)$$

i.e. $\tanh\left(\frac{\eta+\chi}{2}\right) = \tan\left(\frac{\bar{\eta}+\bar{\chi}}{2}\right)$, $\tanh\left(\frac{\eta-\chi}{2}\right) = \tan\left(\frac{\bar{\eta}-\bar{\chi}}{2}\right)$. With this, the metric takes the form (6.26) which is conformal to the Einstein static universe (6.23) with the conformal factor

$$\Omega = \frac{R_0}{R(\eta(\bar{\eta},\bar{\chi}))}\sqrt{\sin(\bar{\eta}+\bar{\chi}-\tfrac{\pi}{2})\sin(\bar{\eta}-\bar{\chi}-\tfrac{\pi}{2})}. \qquad (6.29)$$

The *conformal infinity* $\mathcal{I}$, given by $\Omega = 0$, thus occurs at $\bar{\eta} + \bar{\chi} = \pm\frac{\pi}{2}$ and $\bar{\eta} - \bar{\chi} = \pm\frac{\pi}{2}$, or $R = \infty$. Thus $\eta = \infty$ and $\chi = \infty$ also occur on $\mathcal{I}$. The initial *singularity* $\eta = 0$ maps to $\bar{\eta} = 0$, and $\chi = 0$ maps to $\bar{\chi} = 0$. The special points $\bar{\eta} = \frac{\pi}{2}$, $\bar{\chi} = 0$ and $\bar{\eta} = 0$, $\bar{\chi} = \frac{\pi}{2}$ denote timelike infinity i^+ and spacelike infinity i^0.

Consequently, the Penrose diagrams of the $k = -1$ FLRW space-times, given in Figure 6.9, are similar to those with $k = 0$. In particular, for both these Friedmann universes with $\Lambda = 0$, the scale function $R(\eta)$ grows from $R = 0$ at $\eta = 0$ to $R \to \infty$ as $\eta \to \infty$, see Figure 6.4 and (6.22). Their Penrose diagram is thus bounded by the initial singularity at $\bar{\eta} = 0$ and the future conformal infinity $\mathcal{I}^+$ given by $\bar{\eta} + \bar{\chi} = \frac{\pi}{2}$. The main difference is that the range of the $\bar{\chi}$ is now $[0, \frac{\pi}{2}]$, as opposed to $[0, \pi]$. Figure 6.9 resembles Figure 6.8, but i^+ is now located at $\bar{\eta} = \frac{\pi}{2}$, and i^0 is located at $\bar{\chi} = \frac{\pi}{2}$. Thus, in this case, i^0 represents a 2-sphere of radius $\sin\frac{\pi}{2} = 1$ spanned by θ and ϕ (contrary to the case $k = 0$, for which i^0 corresponds to a single point because it is given by $\bar{\chi} = \pi$, i.e. the South pole).

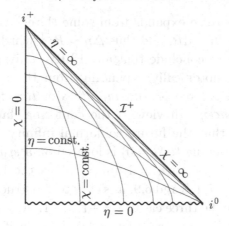

Fig. 6.9 Penrose diagram for the FLRW space-times with $k = -1$ and $\Lambda = 0$ in the coordinates $\bar\eta$ (vertical) and $\bar\chi$ (horizontal). The universe is bounded by the big bang singularity at $\bar\eta = 0$ and the future conformal infinity $\mathcal{I}^+$ located at $\bar\eta + \bar\chi = \frac{\pi}{2}$. The "points" i^+ and i^0 given by $\bar\eta = \frac{\pi}{2}, \bar\chi = 0$ and $\bar\eta = 0, \bar\chi = \frac{\pi}{2}$ denote the timelike and spacelike infinities, respectively. In contrast to Figure 6.8, the coordinate lines $\eta = $ const. are not tangent to $\eta = 0$ near i^0 in this case.

6.4.2 FLRW space-times with Λ

In order to investigate the global structure of the FLRW space-times with a *non-vanishing cosmological constant* Λ, it is necessary to determine the specific dependence of the scale function $R(\eta)$ on conformal time. Instead of performing the integration (6.20), inverting it and substituting into $R(t)$, it is more convenient to re-write the Friedmann equation (6.15) directly as $(\mathrm{d}R/\mathrm{d}\eta)^2 = \frac{\Lambda}{3}R^4 - k\,R^2 + c\,R^{4-3\gamma}$. This equation for $R(\eta)$ can be written as

$$\int \frac{\mathrm{d}\,R}{\sqrt{\frac{1}{3}\Lambda R^4 - k\,R^2 + c\,R^{4-3\gamma}}} = \eta, \qquad (6.30)$$

which leads to complicated functions (in the case of dust they are elliptic, see Edwards, 1972). However, for realistic equations of state (6.10) such that $\gamma \in [1, 2]$ some general conclusions can immediately be drawn.

Close to the initial singularity $R = 0$, the two terms in (6.30) with coefficients Λ and k are negligible, so that $R(\eta) \approx \eta^{\frac{2}{3\gamma-2}}$. This behaviour (which is fully consistent with the solutions (6.22) when $\Lambda = 0$) implies that any universe expands from $R(0) = 0$ to a small but finite R in a *finite* conformal time η.

Similarly, when $\Lambda > 0$ and $R(\eta)$ is very large and growing, the two terms in (6.30) that are proportional to k and c can be neglected. The interval

$\Delta\eta$, in which the universe expands from some (large) value R_L to $R = \infty$, is approximately $\int_{R_L}^{\infty} R^{-2}\,dR$, and thus $\Delta\eta \sim R_L^{-1}$ which is *also finite*.

It is thus possible to conclude that any FLRW universe with $\Lambda > 0$ and $\gamma \in [1,2]$, that is monotonically expanding from the big bang singularity $R(0) = 0$ to $R(\eta) \to \infty$, reaches its *infinite size $R = \infty$ in a finite value of the conformal time η.* In view of the conformal factors (6.24), (6.27) or (6.29), it follows that the future conformal infinity $\mathcal{I}^+$ of these FLRW space-times (corresponding to $\Omega = 0$) is located on a *spacelike surface* given by $\eta = $ const., that is positive and finite. Since the lines $\eta = $ const. are indicated on Figures 6.7, 6.8 and 6.9, it is easy to construct the corresponding Penrose diagrams for the three cases $k = +1, 0, -1$. As a particular example, in Figure 6.10, these diagrams are plotted for the case of radiation ($\gamma = 4/3$) universes whose scale functions $R(t)$ are shown in Figures 6.5 and 6.6. In the case $k = +1$, to describe this Lemaître-type "hesitation" universe it is necessary to assume that $\Lambda > \Lambda_0$ (see Figure 6.2).

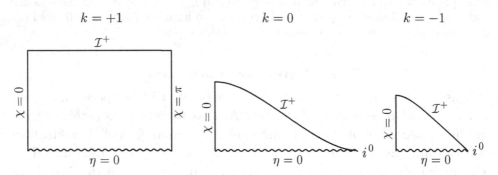

Fig. 6.10 Typical Penrose diagrams of the FLRW space-times with $\Lambda > 0$ for the three possible spatial geometries $k = +1, 0, -1$ (in the case $k = +1$, it is assumed that $\Lambda > \Lambda_0$). All the space-times begin at the initial big bang singularity at $\eta = 0$ and expand to an infinite size $R = \infty$ which corresponds to $\Omega = 0$. Due to the presence of a positive cosmological constant, such an expansion is very rapid, so that the future conformal infinity $\mathcal{I}^+$ is reached in a finite conformal time η. Consequently, $\mathcal{I}^+$ has a spacelike character, as in the case of the de Sitter universe.

Essentially the same diagrams are obtained for other $\Lambda > 0$ FLRW space-times with different matter content. The only difference is the precise location of the future conformal infinity $\mathcal{I}^+$ which corresponds to a different value $\eta = $ const. For example, in the case of dust, such a value is greater than in the case of radiation.

When the cosmological constant is negative ($\Lambda < 0$), it follows from the potentials $V(R)$ shown in Figure 6.2 that any FLRW universe expands from the big bang singularity to a maximum size R_c and then re-collapses back to $R = 0$. (This also applies to $k = +1$ space-times with a big bang when $0 \leq \Lambda < \Lambda_0$.) In all these cases, the temporal duration of the universe is finite. It is finite also in the conformal time η since it has been demonstrated above that, for any reasonable equation of state, the integral (6.30) is finite close to $R = 0$. The Penrose conformal diagram of any such FLRW space-time is thus *bounded by the initial big bang singularity* at $\eta = 0$ and the *final singularity* at $\eta = \text{const.} > 0$ that is finite. The three possibilities, according to the sign of the spatial curvature k, are shown in Figure 6.11. There are no conformal infinities $\mathcal{I}$ since $\Omega = 0$ cannot be reached in these cases.

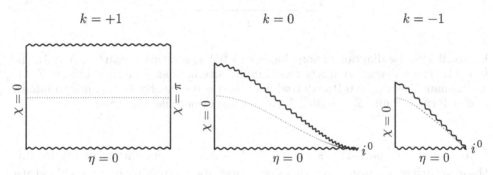

Fig. 6.11 Penrose diagrams of the FLRW space-times with $\Lambda < 0$ for the three spatial geometries $k = +1, 0, -1$. The case $k = +1$ also applies to $0 \leq \Lambda < \Lambda_0$. Any such space-time originates at the big bang singularity at $\eta = 0$, expands to a maximum size R_c (indicated by the dotted line), and then collapses to the final singularity $R = 0$ at a finite value of the conformal time η. Both the initial and the final singularities have spacelike character.

Note that the diagrams in Figure 6.11 apply to *non-vacuum* space-times. The vacuum universe with $\Lambda < 0$ (that is the anti-de Sitter space) is exceptional. It can only be written in the FLRW form (6.7), see (5.18), or equivalently (6.21) with $k = -1$ and $R(\eta) = R_0 / \cosh \eta$, where $\eta \in (-\infty, +\infty)$. For this particular dependence and range, the factor (6.29) takes the special form $\Omega = \cos \bar{\chi}$, which is everywhere regular. In this case, conformal infinity $\mathcal{I}$ given by $\Omega = 0$ is located at $\bar{\chi} = \frac{\pi}{2}$ and has a timelike character in full agreement with (5.11).

Another possible FLRW universe is a non-singular space-time which may arise when $0 < \Lambda < \Lambda_0$ in the case $k = +1$ (see Figure 6.6). Its Penrose diagram is shown in Figure 6.12. In the temporal direction this is bounded by the past conformal infinity $\mathcal{I}^-$ and the future conformal infinity $\mathcal{I}^+$, since $R = \infty$ occurs in a finite value of η both in the past and in the future. Such a universe contracts to a minimum size (at $\eta = 0$) and then re-expands.

It only remains to investigate the global structure of the two exceptional

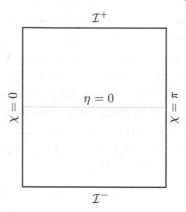

Fig. 6.12 Penrose diagram of non-singular FLRW space-times with $0 < \Lambda < \Lambda_0$ and $k = +1$. The universe contracts from infinite size in past conformal infinity $\mathcal{I}^-$ to a minimum size at $\eta = 0$ (the dotted line), and subsequently re-expands to infinite size at future infinity $\mathcal{I}^+$. Both $\mathcal{I}^+$ and $\mathcal{I}^-$ have spacelike character.

FLRW $k = +1$ space-times for which $\Lambda = \Lambda_0 > 0$. As shown in Figure 6.6, these solutions diverge from or asymptotically approach the unstable static universe characterised by $R = R_0$ which, in the case of dust ($\gamma = 1$), is the Einstein static universe (6.17) described in Section 3.2, and in the case of radiation ($\gamma = 4/3$) is analogously given by (6.18). The first of these solutions is an Eddington–Lemaître-type model. It diverges from the static space-time at $t = -\infty$, and approaches the de Sitter universe as $t \to +\infty$. In terms of the conformal time, $R = R_0$ at $\eta = -\infty$, while

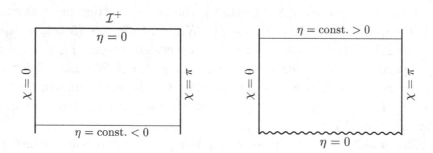

Fig. 6.13 Penrose diagrams of the FLRW space-times with $\Lambda = \Lambda_0 > 0$ and $k = +1$. The Eddington–Lemaître-type model (left) diverges from the static space-time of radius R_0 in the past and approaches the de Sitter universe with a spacelike future conformal infinity $\mathcal{I}^+$ at $\eta = 0$. The complementary Eddington–Lemaître–Bondi-type model (right) expands from the big bang singularity $R = 0$ at $\eta = 0$ and asymptotically approaches the static space-time.

$R = \infty$ at $\eta = 0$ (which, in view of (6.24), corresponds to future conformal infinity $\mathcal{I}^+$). The Penrose diagram is shown in the left part of Figure 6.13 for $\gamma = 4/3$, in which case $R(\eta) = R_0 \coth(-\eta/\sqrt{2})$. The second is an Eddington–Lemaître–Bondi-type model that expands from the big bang singularity $R = 0$ at $t = 0 = \eta$ and approaches the static universe as $t \to \infty$ ($\eta \to \infty$). For $\gamma = 4/3$, $R(\eta) = R_0 \tanh(\eta/\sqrt{2})$, and the corresponding Penrose diagram is plotted in the right part of Figure 6.13.

Note finally that FLRW space-times with matter that obeys other (less physical) equations of state have, in general, different causal structures. See, for example, Senovilla (1998) or García-Parrado and Senovilla (2003) for the FLRW models with $k = 0$ and any value of γ in (6.10).

6.5 Some principal properties and references

Because the FLRW space-times are dynamical, a non-trivial question arises: how much of the complete universe can, in principle, be observed from a given event P, or by a timelike observer throughout its entire history? The boundary between the events (or particles) that could theoretically be seen and those that cannot is naturally called a *horizon*. In fact, as shown by Rindler (1956) and Penrose (1964, 1968a), there may be different kinds of horizons in these cosmological models, including an event horizon and a particle horizon.

An *event horizon* occurs for a timelike observer whenever either the future conformal infinity $\mathcal{I}^+$, or the final singularity $R = 0$, is spacelike. As shown in Figure 6.14, as an observer approaches such a future boundary, there must

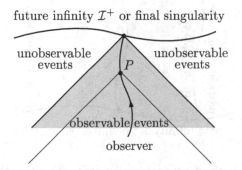

future infinity $\mathcal{I}^+$ or final singularity

unobservable events

unobservable events

P

observable events

observer

Fig. 6.14 An event horizon is the null boundary between the events that could be seen by a given timelike observer during his or her complete history (grey region) and those which will never be observable (white region). It occurs in the FLRW space-times whose future infinity $\mathcal{I}^+$ or final singularity have a spacelike character.

always remain some parts of the space-time that could never have been seen by this observer. The event horizon for a given timelike observer is defined as the unique boundary between the observable and ever unobservable events of the space-time. Formally, it is the boundary of the union of all past light-cones with vertices on all points P of the observer's trajectory (indicated in grey). From the Penrose diagrams given in Section 6.4 it is obvious that an event horizon is present in all FLRW space-times with $\Lambda \neq 0$ (see Figures 6.10, 6.11, 6.12, and the left part of 6.13), and also in the case $k = +1$ when $\Lambda = 0$ (see Figure 6.7). On the other hand, there are no event horizons in the $\Lambda = 0$ FLRW universes with $k = 0$ or $k = -1$ because their future conformal infinity $\mathcal{I}^+$ is null (see Figures 6.8 and 6.9). Also, the Einstein static universe and the Eddington–Lemaître–Bondi-type models do not contain event horizons since the past light-cone of their future timelike infinity i^+ contains the whole universe (see the right part of Figure 6.13).

Analogously, the existence of a *particle horizon* in FLRW space-times is related to the spacelike character of a past conformal infinity $\mathcal{I}^-$ or an initial big bang singularity (see Figure 6.15). An observer at an event P could only have seen a small part of the whole universe during its previous history. Particles located outside its past light-cone are not yet visible. Of course, at a later time, the same observer will have seen more particles, but there still remain others which are beyond its (now larger) horizon. Technically, a particle horizon at P in a FLRW model is the boundary between the timelike

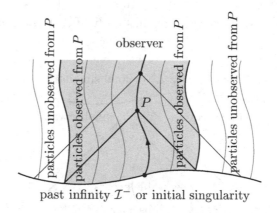

past infinity $\mathcal{I}^-$ or initial singularity

Fig. 6.15 A particle horizon separates all test particles in the universe into the subset that are observable from a given event P on the observer's timelike trajectory (grey region), and the complementary subset of particles which are not yet seen from P. A particle horizon occurs if the past infinity $\mathcal{I}^-$ or the big bang singularity has a spacelike character.

world-lines of fundamental (isotropic) geodesics that could be seen from P, and those that could not. It is a history of particles that are located at the furthest possible limit of the observer's vision at P.

Almost all FLRW models have particle horizons for all timelike curves because the initial singularity at $\eta = 0$ has a spacelike character (see the conformal Penrose diagrams in Figures 6.7, 6.8, 6.9, 6.10, 6.11, and the right part of 6.13). Similarly, there exists a particle horizon in the non-singular FLRW universe shown in 6.12 since its past conformal infinity $\mathcal{I}^-$ is also spacelike. The only exceptions are the Einstein static universe and the Eddington–Lemaître-type models, since all timelike particles in these universes emanate from past timelike infinity i^-, as shown in the left part of 6.13.

It should also be emphasised that, during the evolution of an observer in a given universe, its particle horizon may disappear. For example, in the FLRW $k = +1$ model without Λ which contains dust (shown in the left part of Figure 6.7), the particle horizon ceases to exist at the moment of maximum expansion $\eta = \pi$. After this time, any observer could receive signals from any other fundamental observer. This is evident because light signals emitted at the big bang reach their antipodal points within this closed universe when $\eta = \pi$. This cannot occur in the analogous radiation universe (right part of the same figure) in which particle horizons remain present when the final singularity occurs.

It is now appropriate to briefly consider some other key properties of FLRW cosmologies. It is common to introduce the *Hubble parameter H*, which measures the rate of expansion of the universe,[4] and the dimensionless *deceleration parameter q* which measures the rate at which the expansion is slowing down. These functions of t are defined by

$$H \equiv \frac{\dot{R}}{R}, \qquad q \equiv -\frac{R\ddot{R}}{\dot{R}^2}. \tag{6.31}$$

In terms of these quantities, the field equations (6.13), (6.14) can be re-written as

$$H^2 = \frac{\Lambda}{3} + \frac{8\pi}{3}\rho - \frac{k}{R^2}, \qquad (1 - 2q)H^2 = \Lambda - 8\pi p - \frac{k}{R^2}. \tag{6.32}$$

It is also convenient to introduce the dimensionless *density parameters*

$$\Omega_\Lambda \equiv \frac{\Lambda}{3H^2}, \qquad \Omega_m \equiv \frac{8\pi\rho}{3H^2}, \qquad \Omega_k \equiv -\frac{k}{H^2R^2}, \qquad \Omega_p \equiv \frac{8\pi p}{H^2}. \tag{6.33}$$

---

[4] A dimensionless Hubble parameter h is also sometimes introduced, which is defined as $h = H/(100\,\text{km}\,s^{-1}\text{Mpc}^{-1})$.

These epoch-dependent functions represent the actual energy density corresponding to the cosmological constant Λ (vacuum energy or dark energy), ordinary matter density ρ, the spatial curvature parameter k, and ordinary matter pressure p, respectively. The Einstein equations (6.32) thus take the very simple form

$$\Omega_\Lambda + \Omega_m + \Omega_k = 1, \qquad 3\Omega_\Lambda - \Omega_p + \Omega_k = 1 - 2q. \qquad (6.34)$$

In particular, for the simple FLRW dust models without a cosmological constant, $\Omega_m + \Omega_k = 1$ and $\Omega_k = 1 - 2q$. For such a spatially flat space-time (the Einstein–de Sitter universe), which has $k = 0$, these equations imply that $q = \frac{1}{2}$ and $\Omega_m = 1$, and the matter density ρ must therefore have the special "critical" value

$$\rho_c \equiv \frac{3H^2}{8\pi}. \qquad (6.35)$$

The closed Friedmann universe with $k = +1$ is characterised by $q > \frac{1}{2}$ and $\rho > \rho_c$, while the spatially open Friedmann universe ($k = -1$) occurs for $q < \frac{1}{2}$ and $\rho < \rho_c$.

Note also that, using ρ_c, the parameters (6.33) can be re-written as

$$\Omega_\Lambda \equiv \frac{\Lambda}{8\pi\rho_c}, \qquad \Omega_m \equiv \frac{\rho}{\rho_c}, \qquad \Omega_k \equiv -\frac{3k}{8\pi\rho_c R^2}, \qquad \Omega_p \equiv \frac{3p}{\rho_c}. \qquad (6.36)$$

In general, it follows from (6.34) that $q = \frac{1}{2}(\Omega_m + \Omega_p) - \Omega_\Lambda$. It was traditionally believed that the cosmological constant Λ is either vanishing or negligible in the real universe. In such a case $q > 0$, so that the expansion of the universe always slows down (see (6.31)). However, if the vacuum energy density Ω_Λ associated with the cosmological term (dark energy) is dominant, $q < 0$ and the universe accelerates its expansion. This scenario now seems to be supported by observations of the cosmic microwave background radiation and distant supernovae.

Another important aspect of the FLRW models is the presence of a *cosmological redshift*. This is a direct consequence of the expansion of these universes. Consider a family of natural isotropic timelike observers whose velocity is everywhere given by (6.2), that is $\boldsymbol{u} = \partial_t$. In the metrics (6.5)–(6.7) or (6.21), these stay at constant values of the spatial coordinates χ_0, θ_0, ϕ_0. Light emitted or detected by any such observer propagates along null geodesics. In particular, when θ and ϕ do not change, these null geodesics are simply given by $\Phi \equiv \pm\chi - \eta = \text{const.}$ (see the metrics (6.21)) and the corresponding future-oriented tangent (co-)vector is $k_\alpha = \Phi_{,\alpha}$. In view of (6.20), the frequency of light with respect to the given observer is

$$\omega = -k_\alpha u^\alpha = R^{-1}(t)\,. \qquad (6.37)$$

This depends on the size of the universe, which is characterised by the actual value of the scale function R at a given time t. In an expanding universe the frequency ω decreases. Denoting the frequency of light *emitted* at t_e by ω_e, and the frequency of the same light *observed* at t_o by ω_o, it follows that $\omega_e/\omega_o = R(t_o)/R(t_e)$. In terms of the wavelength of the light signal $\lambda = 2\pi/\omega$, the *redshift factor* is

$$z \equiv \frac{\lambda_o - \lambda_e}{\lambda_e} = \frac{\omega_e}{\omega_o} - 1 = \frac{R(t_o)}{R(t_e)} - 1\,. \qquad (6.38)$$

The redshift factor z is thus a direct measure of the expansion of the universe. For example, if $z = 1$ then cosmic distances double during the time the light travels from the source to the observer. In a contracting universe z is negative and a blueshift would be observed.

Using a Taylor expansion of the scale function $R(t)$ it follows that

$$z = H_o(t_o - t_e)\left[1 + (1 + \tfrac{1}{2}q_0)H_o(t_o - t_e) + \ldots\right], \qquad (6.39)$$

where $H_o \equiv H(t_o)$ and $q_o \equiv q(t_o)$ are the values of the Hubble and deceleration parameters at the observation time t_o.

It follows from (6.5)–(6.7) that the proper distance between the source of the emitted light and the observer at time t is $l(t) = R(t)|\chi_o - \chi_e|$. Consequently, their relative velocity is $v(t) = \dot{R}(t)|\chi_o - \chi_e|$, that is

$$v(t) = H(t)\,l(t)\,. \qquad (6.40)$$

This relation is well-known as *Hubble's law*, according to which the recession velocity of the source is proportional to its distance from the observer. (For large l, this can even exceed 1, which is the speed of light.)

It should however be emphasised that this simple relation evolves with time. At the observation time t_o, $v_o = H_o l_o$, but l_o is not simply related to the observed redshift factor z, interpreted directly as the Doppler effect due to v_o, unless z is very small. The exact relation is obtained from the expression $|\chi_o - \chi_e| = \eta_o - \eta_e = \int_{t_e}^{t_o} R^{-1}(t)\,dt$, which is valid for the null geodesic connecting the emission and observation events. The Taylor expansion of the integrand yields $|\chi_o - \chi_e| = R_o^{-1}(t_o - t_e)[1 + \tfrac{1}{2}H_o(t_o - t_e) + \ldots]$ so that, using (6.38), the proper distance l_o at t_o is related to the redshift z via

$$l_o = H_o^{-1} z \left[1 - \tfrac{1}{2}(1 + q_0) z + \ldots\right]\,. \qquad (6.41)$$

The Hubble law (6.40) thus reads $v_o = z\,[1 - \frac{1}{2}(1 + q_o)\,z + \ldots]$ which, for small redshifts, indeed reduces to $v_o \approx z$.

Other properties of the FLRW space-times, both theoretical and observational, are described in the very extensive literature that is particularly dedicated to this important family of exact cosmological solutions of Einstein's equations. There are many fundamental textbooks, monographs and reviews which cover different aspects of this topic. See, for example, Weinberg (1972), Misner, Thorne and Wheeler (1973), Hawking and Ellis (1973), Wald (1984), Felten and Isaacman (1986), Kolb and Turner (1990), Carroll, Press and Turner (1992), Peebles (1993), Krasiński (1997), Peacock (1999), Landau and Lifshitz (2000), Islam (2002), Bonometto, Gorini and Moschella (2002), Grøn and Hervik (2007) and Weinberg (2008). These also contain extensive lists of further references.

7

Electrovacuum and related background space-times

The Minkowski, de Sitter and anti-de Sitter space-times described in Chapters 3–5 are the only conformally flat solutions of Einstein's vacuum field equations with a possibly non-zero cosmological constant. However, there also exist non-vacuum conformally flat space-times. The most important of these are the perfect fluid FLRW cosmologies that were reviewed in the previous chapter. Conformally flat radiative space-times with pure radiation will be discussed in Chapters 17 and 18. Another important conformally flat space-time is the Bertotti–Robinson universe which contains a uniform non-null electromagnetic field.

Interestingly, the Minkowski, de Sitter and anti-de Sitter solutions are constant-curvature 4-spaces (with ten isometries), the FLRW cosmologies (having six isometries) are foliated by constant-curvature 3-spaces, while the Bertotti–Robinson space-time is a direct product of two 2-spaces of constant curvature (and so also has six isometries). In fact, it belongs to a larger family of geometries of this type, which also include the Nariai, anti-Nariai and Plebański–Hacyan direct-product space-times. In general, these are electrovacuum solutions, which apart from the Bertotti–Robinson solution are of algebraic type D. These will all be described in this chapter.

It is also natural here to include a short description of the Melvin universe, which is another interesting electrovacuum type D solution.

7.1 The Bertotti–Robinson solution

The conformally flat solution of the Einstein–Maxwell equations for a non-null electromagnetic field was obtained independently by Bertotti (1959) and Robinson (1959), see also Levi-Civita (1917). It is commonly expressed as the diagonal static metric

$$ds^2 = \frac{e^2}{r^2}\left(-dt^2 + dr^2 + r^2(d\theta^2 + \sin^2\theta\,d\phi^2)\right), \tag{7.1}$$

where $t \in (-\infty, +\infty)$, $r \in [0, \infty)$, $\theta \in [0, \pi]$, $\phi \in [0, 2\pi)$. This form is explicitly conformally flat with the conformal factor $\Omega = r/e$, so that $r = 0$ corresponds to infinity. In fact, it is a non-singular spherically symmetric and homogeneous space. The Maxwell field associated with the line element (7.1) is, up to a constant duality rotation,

$$F = \frac{e}{r^2}\,dr \wedge dt. \tag{7.2}$$

This is a non-null electromagnetic field since $F_{\mu\nu}F^{\mu\nu} = -2/e^2 \neq 0$. In a null tetrad with vectors $\boldsymbol{k} = \frac{1}{\sqrt{2}}(r/e)(\partial_t + \partial_r)$, $\boldsymbol{l} = \frac{1}{\sqrt{2}}(r/e)(\partial_t - \partial_r)$ and $\boldsymbol{m}$ in the transverse space spanned by the coordinates θ and ϕ, the only non-vanishing component (2.20) of the Maxwell field is $\Phi_1 = (2e)^{-1}$. Since this is constant, the non-null electromagnetic field is uniform.

According to a theorem of Tariq and Tupper (1974), the Bertotti–Robinson space-time is the only conformally flat solution of the Einstein–Maxwell equations for a *non-null* source-free electromagnetic field. See also the works of Cahen and Leroy (1966), Singh and Roy (1966), Stephani (1967) and McLenaghan, Tariq and Tupper (1975).[1] Also, it was proved by Khlebnikov and Shelkovenko (1976) and Podolský and Ortaggio (2003) that there are no conformally flat solutions of the Einstein–Maxwell equations with a non-zero cosmological constant.

Other forms of the Bertotti–Robinson universe are frequently used in the literature. For example, with the transformation

$$t = \frac{e\sqrt{e^2 + R^2}\,\sin(T/e)}{R + \sqrt{e^2 + R^2}\,\cos(T/e)},$$

$$r = \frac{e^2}{R + \sqrt{e^2 + R^2}\,\cos(T/e)}, \tag{7.3}$$

the metric (7.1) becomes

$$ds^2 = -\left(1 + \frac{R^2}{e^2}\right)dT^2 + \left(1 + \frac{R^2}{e^2}\right)^{-1}dR^2 + e^2(d\theta^2 + \sin^2\theta\,d\phi^2). \tag{7.4}$$

It is obvious that the first part of the Bertotti–Robinson metric covered by

---

[1] For a *null* field the only possible such solution is represented by the special plane wave of Baldwin and Jeffery (1926) and Brdička (1951) given by the metric (17.7) with $A = 0$ and $B(u) > 0$. This is summarised in Theorems 37.18 and 37.19 of (Stephani *et al.*, 2003).

the (T, R) coordinates represents exactly a complete two-dimensional anti-de Sitter space-time of radius e, as can be seen by comparing it with the metric (5.6). The second part spanned by the coordinates (θ, ϕ) is a 2-sphere (A.3) of constant radius e. Therefore, the Bertotti–Robinson universe has the topology of a direct product $\mathrm{AdS}_2 \times S^2$.

An alternative transformation

$$t = e\left(\frac{e+v}{e-v} - \frac{2e-u}{2e+u}\right),$$

$$r = e\left(\frac{e+v}{e-v} + \frac{2e-u}{2e+u}\right), \tag{7.5}$$

accompanied by the standard relation $\zeta = \sqrt{2}\, e\, \tan(\theta/2)\mathrm{e}^{\mathrm{i}\phi}$ (see Appendix A), brings the metric of the Bertotti–Robinson universe (7.1) to the form

$$\mathrm{d}s^2 = -\frac{2\,\mathrm{d}u\,\mathrm{d}v}{(1 + \frac{1}{2}e^{-2}\,uv)^2} + \frac{2\,\mathrm{d}\zeta\,\mathrm{d}\bar\zeta}{(1 + \frac{1}{2}e^{-2}\,\zeta\bar\zeta)^2}. \tag{7.6}$$

This nicely exhibits the direct-product character of the space-time geometry in an explicit and symmetric form. Note also that the conformal infinity located at $r = 0$ is now given by two hyperbolae $uv = -2e^2$ in the double-null coordinates u, v.

With the subsequent transformation $u = -e^2/\tilde u$, $v = 2(\tilde u - 2e^2/\tilde v)$, (7.6) becomes

$$\mathrm{d}s^2 = -2\,\mathrm{d}\tilde u\,\mathrm{d}\tilde v - \frac{\tilde v^2}{e^2}\,\mathrm{d}\tilde u^2 + \frac{2\,\mathrm{d}\zeta\,\mathrm{d}\bar\zeta}{\left(1 + \frac{1}{2}e^{-2}\,\zeta\bar\zeta\right)^2}, \tag{7.7}$$

which is a particular subcase of the Kundt family of nonexpanding space-times that will be described in Chapter 18. (See (18.49) with $\Lambda = 0$.)

Another useful form of the Bertotti–Robinson metric is

$$\mathrm{d}s^2 = -\mathrm{d}\tau^2 + \cos^2(\tau/e)\,\mathrm{d}x^2 + \cos^2(z/e)\,\mathrm{d}y^2 + \mathrm{d}z^2, \tag{7.8}$$

which is obtained from (7.1) by the transformation

$$t = \frac{e\cos(\tau/e)\cosh(x/e)}{\sin(\tau/e) - \cos(\tau/e)\sinh(x/e)},$$

$$r = \frac{e}{\sin(\tau/e) - \cos(\tau/e)\sinh(x/e)}, \tag{7.9}$$

$$\theta = \frac{z}{e} + \frac{\pi}{2}, \qquad \phi = \frac{y}{e}.$$

The Bertotti–Robinson universe is important because it appears in the context of various other space-times. For example, it characterises the region near the degenerate horizon of an extreme limit of the Reissner–Nordström black holes (see end of Section 9.2.3), or it could represent the interaction region of the Bell–Szekeres space-time which describes the collision of shock electromagnetic plane waves (see Equation (21.17) in Section 21.6).

7.2 Other direct-product space-times

It has been demonstrated explicitly by the metric forms (7.4) and (7.6) that the Bertotti–Robinson universe is a space-time which is geometrically a direct product of two spaces of constant curvature, namely a two-dimensional anti-de Sitter space covered by the coordinates T, R (or u, v), and a 2-sphere spanned by the spatial coordinates θ, ϕ (or ζ). In this section we will present a complete family of such space-times with the other possible choices of two-dimensional (pseudo-)Riemannian spaces of constant curvature that also include the Minkowski, Nariai, anti-Nariai and Plebański–Hacyan space-times.

The most natural coordinates to parametrise the metric of all direct-product space-times are

$$ds^2 = -\frac{2\,du dv}{(1 - \frac{1}{2}\epsilon_1 a^{-2} uv)^2} + \frac{2\,d\zeta d\bar{\zeta}}{(1 + \frac{1}{2}\epsilon_2 b^{-2} \zeta\bar{\zeta})^2}, \tag{7.10}$$

where a and b are positive constants which determine the curvature, whereas $\epsilon_1, \epsilon_2 = 0, +1, -1$ (see Appendix A). Using the null tetrad $\boldsymbol{k} = \Omega\,\partial_v, \boldsymbol{l} = \Omega\,\partial_u,$ $\boldsymbol{m} = \Sigma\,\partial_{\bar{\zeta}}$, where $\Omega = 1 - \frac{1}{2}\epsilon_1 a^{-2} uv$ and $\Sigma = 1 + \frac{1}{2}\epsilon_2 b^{-2}\zeta\bar{\zeta}$, the only non-vanishing Weyl and Ricci scalars are

$$\Psi_2 = -\frac{1}{6}\left(\frac{\epsilon_1}{a^2} + \frac{\epsilon_2}{b^2}\right), \quad \Phi_{11} = \frac{1}{4}\left(-\frac{\epsilon_1}{a^2} + \frac{\epsilon_2}{b^2}\right), \quad R = 2\left(\frac{\epsilon_1}{a^2} + \frac{\epsilon_2}{b^2}\right). \tag{7.11}$$

Various possibilities now exist according to the Gaussian curvatures of the two 2-spaces, $K_1 = \epsilon_1 a^{-2}$ and $K_2 = \epsilon_2 b^{-2}$. Mathematically, there are nine distinct cases, given by the choice of ϵ_1 and ϵ_2. However, only six are physically reasonable since the condition that the energy density Φ_{11} must be non-negative eliminates three.

Up to this point, a and b are arbitrary parameters which determine the magnitude of the Gaussian curvature of each of the 2-spaces. Generally the metric (7.10) describes type D electrovacuum space-times with a cosmological constant $\Lambda = \frac{1}{4}R$. However, particularly interesting cases occur when $a = b$. With this choice, these space-times are vacuum when $\epsilon_1 = \epsilon_2$, or conformally flat with $\Lambda = 0$ when $\epsilon_1 = -\epsilon_2$. They are summarised in Table 7.1 and illustrated schematically in Figure 7.1.

When both the coefficients ϵ_1 and ϵ_2 vanish, the metric (7.10) obviously describes flat Minkowski space-time in the form (3.4). The case $\epsilon_1 = -1$, $\epsilon_2 = +1$ is the Bertotti–Robinson universe (7.6) with $a = b = e$ that was

ϵ_1	ϵ_2	geometry	universe	Φ_{11}	Λ
0	0	$M_2 \times E^2$	Minkowski	$= 0$	$= 0$
+1	+1	$dS_2 \times S^2$	Nariai	$= 0$	> 0
−1	−1	$AdS_2 \times H^2$	anti-Nariai	$= 0$	< 0
−1	+1	$AdS_2 \times S^2$	Bertotti–Robinson	> 0	$= 0$
0	+1	$M_2 \times S^2$	Plebański–Hacyan	> 0	> 0
−1	0	$AdS_2 \times E^2$	Plebański–Hacyan	> 0	< 0

Table 7.1 Possible physically relevant space-times which are the direct product of two constant-curvature (pseudo-)Riemannian 2-spaces. The identifying names, and the ranges shown for Φ_{11} and Λ, correspond to the choice $a = b$.

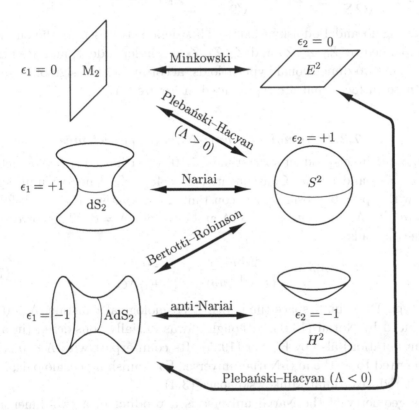

Fig. 7.1 The possible space-times of type (7.10). The parameter ϵ_1 determines the conformal structure, whereas ϵ_2 gives the geometry of the transverse 2-space. The identifying names correspond to the choice $a = b$.

discussed in Section 7.1. The remaining possibilities will be briefly described in Subsections 7.2.1 and 7.2.2 below.

It may finally be remarked here that all these direct-product space-times can be obtained geometrically if the six-dimensional flat space

$$ds^2 = -dZ_0{}^2 + dZ_1{}^2 + \epsilon_1 dZ_2{}^2 + dZ_3{}^2 + dZ_4{}^2 + \epsilon_2 dZ_5{}^2, \qquad (7.12)$$

is constrained to the four-dimensional submanifold for which

$$\epsilon_1(-Z_0{}^2 + Z_1{}^2) + Z_2{}^2 = a^2, \qquad \epsilon_2(Z_3{}^2 + Z_4{}^2) + Z_5{}^2 = b^2. \qquad (7.13)$$

The coordinates of the metric (7.10) are introduced by the parametrisation

$$Z_0 = \frac{v+u}{\sqrt{2}\,\Omega}, \qquad Z_1 = \frac{v-u}{\sqrt{2}\,\Omega}, \qquad Z_2 = a\frac{1 + \frac{1}{2}\epsilon_1 a^{-2}uv}{\Omega},$$

$$Z_3 = \frac{\zeta + \bar\zeta}{\sqrt{2}\,\Sigma}, \qquad Z_4 = -\mathrm{i}\frac{\zeta - \bar\zeta}{\sqrt{2}\,\Sigma}, \qquad Z_5 = b\frac{1 - \frac{1}{2}\epsilon_2 b^{-2}\zeta\bar\zeta}{\Sigma}, \qquad (7.14)$$

(see Ortaggio and Podolský, 2002). This demonstrates that the submanifolds spanned by Z_0, Z_1, Z_2 and Z_3, Z_4, Z_5 are independent, and are planes, spheres, or two-dimensional hyperboloids, according to the sign of ϵ_2 and ϵ_1. It is these surfaces that are represented in Figure 7.1.

7.2.1 Nariai and anti-Nariai space-times

All *vacuum* direct-product space-times (7.10) must have $\Phi_{11} = 0$, which implies $\epsilon_1 = \epsilon_2$ and $a = b$. Consequently, $R = 4\epsilon_1 a^{-2}$. Since vacuum space-times with a possible cosmological constant Λ must satisfy $R = 4\Lambda$, it follows that $\epsilon_1 a^{-2} = \Lambda$, and thus $\Psi_2 = -\frac{1}{3}\Lambda$ and $\epsilon_1 = \operatorname{sign}\Lambda = \epsilon_2$. The corresponding metric reads

$$ds^2 = -\frac{2\,du\,dv}{(1 - \frac{1}{2}\Lambda uv)^2} + \frac{2\,d\zeta\,d\bar\zeta}{(1 + \frac{1}{2}\Lambda\zeta\bar\zeta)^2}. \qquad (7.15)$$

This type D vacuum space-time with a cosmological constant $\Lambda > 0$ was introduced by Nariai (1951), although it was initially considered (in a Euclidean notation) also by Kasner (1925). Its counterpart with $\Lambda < 0$ is usually referred to as the anti-Nariai universe. For vanishing cosmological constant it reduces to flat Minkowski space (3.4).

The geometry of the Nariai universe is a product of a two-dimensional de Sitter space-time dS_2 of radius $a = 1/\sqrt{\Lambda}$, and a 2-sphere S^2 of the same radius, i.e. $R^1 \times S^1 \times S^2$. Its metric can be written in the alternative Kantowski–Sachs form

$$ds^2 = -dt^2 + a^2 \cosh^2 \frac{t}{a}\, d\chi^2 + a^2(d\theta^2 + \sin^2\theta\, d\phi^2), \qquad (7.16)$$

with $t \in (-\infty, +\infty)$, $\chi \in [0, 2\pi]$, $\theta \in [0, \pi]$, $\phi \in [0, 2\pi)$. The R^1 factor thus corresponds to the temporal coordinate t while the S^1 factor describes a spatial circle, covered by the coordinate χ, which contracts to a minimum radius a at $t = 0$ and then re-expands. In contrast to the four-dimensional de Sitter universe in the form (4.4), for which the complete 3-sphere S^3 contracts and re-expands isotropically, the sphere S^2 in the Nariai universe spanned by θ, ϕ has a constant radius a at any time. The Nariai universe is non-singular, spherically symmetric and homogeneous. Its conformal structure for any fixed θ and ϕ is identical to that of the de Sitter universe indicated by the Penrose diagram in Figure 4.3. However, each point now represents a complete 2-sphere of the *same* constant radius a. The Nariai universe admits de Sitter-like cosmological horizons, as discussed e.g. in Ortaggio (2002).

The complementary anti-Nariai metric (7.15) for $\Lambda < 0$, i.e. $\epsilon_1 = \epsilon_2 = -1$, is a product of a two-dimensional anti-de Sitter space-time AdS$_2$ of radius $a = 1/\sqrt{-\Lambda}$, and a two-dimensional pseudo-sphere, i.e. hyperbolic space H^2 with negative Gaussian curvature $-1/a^2$. It can also be written in the form

$$ds^2 = -\left(1 + \frac{R^2}{a^2}\right) dT^2 + \left(1 + \frac{R^2}{a^2}\right)^{-1} dR^2 + a^2(d\theta^2 + \sinh^2\theta \, d\phi^2). \quad (7.17)$$

Its Penrose conformal diagram resembles that of the anti-de Sitter space-time shown in Figure 5.3. The difference is that each point in the diagram now represents a complete two-dimensional pseudo-sphere of the same curvature.

The Nariai and anti-Nariai space-times have recently attracted some attention as specific extremal limits of Schwarzschild–de Sitter or topological black holes, see for example Ginsparg and Perry (1983) or Bousso and Hawking (1996).

7.2.2 Plebański–Hacyan space-times

The remaining direct-product space-times are such that one, and only one, of the 2-spaces has zero curvature. There are two possible cases, namely $\epsilon_1 = 0$, $\epsilon_2 = +1$ and $\epsilon_1 = -1$, $\epsilon_2 = 0$. It follows from (7.11) that

$$\frac{\epsilon_1}{a^2} + \frac{\epsilon_2}{b^2} = 2\Lambda, \quad (7.18)$$

and, since here either ϵ_1 or ϵ_2 vanishes, $\Phi_{11} = \pm\frac{1}{2}\Lambda > 0$ and $\Psi_2 = -\frac{1}{3}\Lambda$, and thus the curvature scalars satisfy the condition $2\Phi_{11} \pm 3\Psi_2 = 0$. Hence, these space-times are two of the "exceptional electrovacuum type D metrics with cosmological constant" investigated by Plebański and Hacyan (1979).[2]

---

[2] All type D solutions of the Einstein–Maxwell equations with a non-zero cosmological constant were investigated by Plebański (1979) under the assumptions that both repeated principal null

For $\epsilon_1 = 0$, $\epsilon_2 = +1$, it follows from (7.18) that $\frac{1}{2}b^{-2} = \Lambda$, which requires $\Lambda > 0$. The metric (7.10) reduces to $ds^2 = -2du\,dv + 2(1 + \Lambda\zeta\bar\zeta)^{-2}d\zeta d\bar\zeta$, which can be rewritten as

$$ds^2 = -dt^2 + dx^2 + b^2(d\theta^2 + \sin^2\theta\,d\phi^2), \tag{7.19}$$

so that this Plebański–Hacyan universe is the direct product $M_2 \times S^2$ of a two-dimensional Minkowski space with a 2-sphere, thus admitting a six-dimensional group of isometries $ISO(1,1) \times SO(3)$.

For $\epsilon_1 = -1$, $\epsilon_2 = 0$, the relation (7.18) implies that $\frac{1}{2}a^{-2} = -\Lambda$, which requires $\Lambda < 0$, so that (7.10) reduces to $ds^2 = -2(1 - \Lambda uv)^{-2}du\,dv + 2d\zeta d\bar\zeta$. This is equivalent to the metric

$$ds^2 = -\left(1 + \frac{R^2}{a^2}\right)dT^2 + \left(1 + \frac{R^2}{a^2}\right)^{-1}dR^2 + dy^2 + dz^2. \tag{7.20}$$

which is the direct product $AdS_2 \times E^2$ with isometries $SO(2,1) \times E(2)$. The original coordinate form given in Plebański and Hacyan (1979) is recovered after the transformation $w = v(1 - \Lambda uv)^{-1}$.

7.3 The Melvin solution

Another interesting one-parameter family of solutions of the coupled Einstein–Maxwell equations is given by the metric

$$ds^2 = \left(1 + \tfrac{1}{4}B^2\rho^2\right)^2\left(-dt^2 + d\rho^2 + dz^2\right) + \left(1 + \tfrac{1}{4}B^2\rho^2\right)^{-2}\rho^2d\phi^2, \tag{7.21}$$

with $t, z \in (-\infty, +\infty)$, $\rho \in [0, \infty)$, $\phi \in [0, 2\pi)$, and with a non-null electromagnetic field expressed by the complex self-dual Maxwell tensor

$$F + i\tilde F = e^{-i\psi} B\left(dz \wedge dt + i\left(1 + \tfrac{1}{4}B^2\rho^2\right)^{-2}\rho\,d\rho \wedge d\phi\right), \tag{7.22}$$

where ψ is an arbitrary real constant which parametrises the freedom in performing a duality rotation. In particular, for $\psi = 0$, the Maxwell tensor is $F = B\,dz \wedge dt$ which describes an electric field pointing along the z-direction, whereas for $\psi = \frac{\pi}{2}$ one obtains $F = B\left(1 + \tfrac{1}{4}B^2\rho^2\right)^{-2}\rho\,d\rho \wedge d\phi$, which represents a purely magnetic field oriented along the z-direction.

This solution was originally obtained by Bonnor (1954) and subsequently rediscovered and studied by Melvin (1964). It is now usually referred to

directions are non-expanding, non-twisting and aligned with the principal null directions of the (non-null) electromagnetic field. When $2\Phi_{11} \pm 3\Psi_2 \neq 0$, the Bianchi identities imply that the principal null directions are also geodesic and shear-free, so that the space-times belong to Kundt's class (see Sections 16.4 and 18.6). The exceptional cases $2\Phi_{11} \pm 3\Psi_2 = 0$ were analysed in detail in Plebański and Hacyan (1979).

as "the Melvin universe". It is a static nonsingular cylindrically symmetric space-time in which there exists an axial electric and/or magnetic field aligned with the z-axis, whose magnitude is determined by the parameter B. It is often described as representing a universe which contains a parallel bundle of electromagnetic flux held together by its own gravitational field.

The metric (7.21) is a member of the Weyl family of static axially symmetric space-times that will be described in Chapter 10. This particular case has the metric (10.2) with $2U = \gamma = \log(1 + \frac{1}{4}B^2\rho^2)^2$. For vanishing B, it exactly reduces to Minkowski space-time in cylindrical coordinates (3.11). When $B \neq 0$, the space-time is not asymptotically flat because of the presence of the electromagnetic field at any z. For large values of the radial coordinate ρ, the metric (7.21) approaches the Levi-Civita (1919) metric (10.9) for the case $\sigma = 1$, which is isometric to the so-called $BIII$-metric (10.14) whose interpretation is unclear.

The intrinsic geometry of the Melvin solution is not trivial. For example, a circle with constant ρ (and t and z fixed) has radius $\rho(1 + \frac{1}{12}B^2\rho^2)$ and circumference $2\pi\rho/(1 + \frac{1}{4}B^2\rho^2)$. Close to the axis of symmetry, the circumference is approximately 2π times the radius, but its growth then becomes slower and reaches a maximum value $2\pi/B$ at $\rho = 2/B$. For larger values of ρ, the circumference decreases and approaches zero as $\rho \to \infty$. The geometry of this transverse plane was represented by Thorne (1965b) in a suitable embedding diagram which resembles a tall narrow-necked vase.

In addition, Thorne, and also Melvin and Wallingford (1966), studied geodesics in the Melvin universe. They have shown that no motion can get too far from the axis of symmetry $\rho = 0$. In particular, for large values of ρ, there are no circular geodesics in the plane $z = \text{const.}$, and all timelike radial geodesics oscillate through the symmetry axis and cannot escape to radial infinity. The reason for this is that the gravitational attraction toward the central axis due to the electromagnetic field is so great that it prevents any object escaping to infinity in the radial direction. This is analogous to the attractive effect of a negative cosmological constant in the anti-de Sitter universe (see Section 5.3).

With the tetrad $\boldsymbol{k} = \frac{1}{\sqrt{2}}(\boldsymbol{e}_t + \boldsymbol{e}_z)$, $\boldsymbol{l} = \frac{1}{\sqrt{2}}(\boldsymbol{e}_t - \boldsymbol{e}_z)$, $\boldsymbol{m} = \frac{1}{\sqrt{2}}(\boldsymbol{e}_\rho - \mathrm{i}\boldsymbol{e}_\phi)$, constructed from the natural orthonormal frame $\boldsymbol{e}_t = D^{-1}\partial_t$, $\boldsymbol{e}_z = D^{-1}\partial_z$, $\boldsymbol{e}_\rho = D^{-1}\partial_\rho$, $\boldsymbol{e}_\phi = D\rho^{-1}\partial_\phi$, where $D = 1 + \frac{1}{4}B^2\rho^2$, the only non-vanishing components of the Weyl and Ricci tensors are

$$\Psi_2 = \tfrac{1}{2}B^2 D^{-4}\left(-1 + \tfrac{1}{4}B^2\rho^2\right), \qquad \Phi_{11} = \tfrac{1}{2}B^2 D^{-4}. \qquad (7.23)$$

The Melvin universe is thus of algebraic type D. The tetrad component of the Maxwell field is $\Phi_1 = \frac{1}{2}B(1 + \frac{1}{4}B^2\rho^2)^{-2} e^{i\psi}$, which corresponds to the physical component measured by static observers in the above frame,

$$F = \frac{B}{(1 + \frac{1}{4}B^2\rho^2)^2} \left(\cos\psi\, \mathbf{e}^z \wedge \mathbf{e}^t + \sin\psi\, \mathbf{e}^\rho \wedge \mathbf{e}^\phi \right). \qquad (7.24)$$

It is obvious that the magnitude of the electromagnetic field oriented along the z-axis is independent of the coordinates z and ϕ (it is cylindrically symmetric), and in this sense it may be considered as "homogeneous". However, it depends on the transverse radial direction ρ from the axis of symmetry. The field has the maximum value B at $\rho = 0$, but decreases monotonically with increasing distance from this axis, and disappears as $\rho \to \infty$.

The null congruences generated by the doubly degenerate principal null directions $\mathbf{k}$ and $\mathbf{l}$ are non-twisting and non-expanding (as well as being shear-free and geodesic). The Melvin space-time thus belongs to the family of type D electrovacuum solutions which was thoroughly analysed by Plebański (1979). In particular, it belongs to the more general Kundt class of space-times (see Chapter 18). The explicit transformation to the canonical Kundt form of the metric (18.1) with $H = 0$,

$$ds^2 = -2du(dr + Wd\zeta + \bar{W}d\bar{\zeta}) + 2P^{-2}d\zeta d\bar{\zeta}, \qquad (7.25)$$

is achieved by putting

$$u = \tfrac{1}{\sqrt{2}}(t - z), \qquad r = \tfrac{1}{\sqrt{2}}(t + z)\,\tilde{z}^2, \qquad \zeta = \tfrac{1}{\sqrt{2}}(x + iy), \qquad (7.26)$$

where

$$\tilde{z} = 1 + \tfrac{1}{4}B^2\rho^2, \qquad x = \int \frac{\tilde{z}^2}{B^2(\tilde{z} - 1)}\, d\tilde{z}, \qquad y = \frac{2}{B^2}\,\phi. \qquad (7.27)$$

The metric functions in (7.25) are given by

$$W = -\frac{\sqrt{2}\,r}{\tilde{z}\,P^2}, \qquad P^2 = \frac{\tilde{z}^2}{B^2(\tilde{z} - 1)}. \qquad (7.28)$$

This is a particular subcase of the Kundt electrovacuum space-times (18.43) which will be discussed in Chapter 18 for the choice $e^2 + g^2 = 2n = B^2$, with γ, ϵ_0, ϵ_2 and Λ vanishing (see also Equation (31.58) in Stephani *et al.*, 2003). Note also that the Plebański (1979) form of this metric is

$$ds^2 = 2\tilde{z}^2(-dt^2 + dz^2) + P^2 d\tilde{z}^2 + P^{-2}dy^2. \qquad (7.29)$$

In addition to obvious symmetries generated by the static, translational and rotational Killing vectors $\partial_t, \partial_z, \partial_\phi$, the Melvin space-time (7.21) also admits the boost Killing vector $z\partial_t + t\partial_z$. This property has recently been used by Ortaggio (2004) to construct explicit impulsive gravitation waves in the Melvin universe by boosting the Schwarzschild–Melvin black hole to the speed of light (see the metric (9.25) in Section 9.3). Note that, in this case, the alternative "spherical" form of the Melvin universe has been employed, namely

$$ds^2 = D^2(-dt^2 + dr^2 + r^2 d\theta^2) + D^{-2} r^2 \sin^2 \theta \, d\phi^2, \tag{7.30}$$

where $D = 1 + \frac{1}{4} B^2 r^2 \sin^2 \theta$, which is obtained from (7.21) by a simple transformation $\rho = r \sin \theta$, $z = r \cos \theta$. Also, it has been shown by Havrdová and Krtouš (2007) that the Melvin universe can be obtained as a specific limit of a charged C-metric (see Section 14.2), which is an important boost-rotation symmetric space-time that describes a pair of uniformly accelerated charged black holes.

The above Melvin magnetic solution has been considered as a useful model in studies of astrophysical processes, gravitational collapse, quantum black hole pair creation and other aspects. Its importance derives also from the fact that it appears as a limit in more complicated solutions and is therefore considered as a background for a number of interesting solutions. It was shown already by Melvin (1965) and Thorne (1965b) that the space-time is, somewhat surprisingly, stable against small radial perturbations, as well as arbitrarily large perturbations which are confined to a finite region about the axis of symmetry. The asymmetries are radiated away in gravitational and electromagnetic waves, see also Garfinkle and Melvin (1992) and Ortaggio (2004).

Recently, generalisations of the Melvin universe to nonlinear electrodynamics theories were also considered by Gibbons and Herdeiro (2001).

8

Schwarzschild space-time

The Schwarzschild solution is undoubtedly the best known nontrivial exact solution of Einstein's equations. It was found only a few months after Einstein published his field equations. And, not only is it one of the simplest exact vacuum solutions, but it is also the most physically significant. It is widely applied both in astrophysics and in considerations of orbital motions about the Sun or the Earth. Until recently, it was only on the assumption of the applicability of this space-time that general relativity had been demonstrated to be a superior theory to the classical gravitational theory of Newton, in a quantitatively precise manner. It predicts the tiny departures from Newtonian theory that are observed in orbital motions in the solar system, in the deflection of light by the Sun, in the gravitational redshift of light and in time-delay effects. In addition, it provides a model for a theory of strong gravitational fields that is widely applied in astrophysics in the final stages of stellar evolution and the formation of black holes.

For all these reasons, the properties of the Schwarzschild solution are explained even in the most introductory texts on general relativity. Nevertheless, it is still useful to describe this space-time here as some important concepts, such as black hole horizons and analytic extensions, are best introduced in this context. These concepts and some associated techniques, which arise naturally in the Schwarzschild space-time, will be developed further and applied in the more complicated solutions that will be described in following chapters.

In this chapter, we present the familiar interpretation of the Schwarzschild space-time that is based on the assumption of global spherical symmetry. A very different interpretation of part of this space-time will be presented later, in Chapter 21. These different possibilities arise through different *global* representations of the space-time symmetries.

8.1 Schwarzschild coordinates

According to an important result known as *Birkhoff's theorem*,[1] when the cosmological constant is zero, the unique vacuum spherically symmetric space-time is the asymptotically flat solution of Schwarzschild (1916a), which can be expressed most easily in the form

$$ds^2 = -\left(1 - \frac{2m}{r}\right)dt^2 + \left(1 - \frac{2m}{r}\right)^{-1}dr^2 + r^2\left(d\theta^2 + \sin^2\theta\, d\phi^2\right), \quad (8.1)$$

where m is an arbitrary parameter whose interpretation, and coordinate ranges, will be clarified below.

It may first be observed that the metric (8.1) reduces to Minkowski space (3.2) in spherical polar coordinates either when $m = 0$ or as $r \to \infty$. To analyse the curvature of this space-time, it is appropriate to introduce the null tetrad

$$\begin{aligned}
k &= \tfrac{1}{\sqrt{2}}\left(\left(1 - \tfrac{2m}{r}\right)^{-1/2}\partial_t + \left(1 - \tfrac{2m}{r}\right)^{1/2}\partial_r\right), \\
l &= \tfrac{1}{\sqrt{2}}\left(\left(1 - \tfrac{2m}{r}\right)^{-1/2}\partial_t - \left(1 - \tfrac{2m}{r}\right)^{1/2}\partial_r\right), \qquad (8.2)\\
m &= \tfrac{1}{\sqrt{2}}\left(-\tfrac{1}{r}\partial_\theta + \tfrac{i}{r}\operatorname{cosec}\theta\,\partial_\phi\right).
\end{aligned}$$

With this, the only non-zero component of the Weyl tensor is

$$\Psi_2 = -\frac{m}{r^3},$$

which shows that the space-time is of algebraic type D, and that it becomes flat as $r \to \infty$. It can then be seen from (2.13) that, when $m \neq 0$, a (scalar polynomial) curvature singularity occurs at $r = 0$. And, since r cannot be extended through such a singularity, it may be assumed that $r > 0$.

By comparing (8.1) with (3.2), it may be inferred that, at least when m is small, r may be interpreted as a kind of radial coordinate. However, it may be seen that the metric (8.1) degenerates when $r = 2m$. Moreover, r is a timelike and t a spacelike coordinate when $0 < r < 2m$ as the factor $(1 - \frac{2m}{r})$ there becomes negative. Thus, r is *not* the distance from the coordinate origin $r = 0$. Nevertheless, with $\theta \in [0, \pi]$ and $\phi \in [0, 2\pi)$, the area of the surface on which r and t are both constants is given by $4\pi r^2$, which is the familiar expression in Euclidean space for the area of a spherical surface of radius r. The Schwarzschild coordinate r is therefore referred to as the *areal radius* as it provides the familiar measure of the area of the

---

[1] Although the result now known as Birkhoff's theorem was published in his textbook of 1923, it had actually been published by a number of people prior to that date. For a discussion of these historical points, see Johansen and Ravndal (2006).

surfaces spanned by θ and ϕ, even though it is not the distance from the centre in the curved space-time represented by the metric (8.1).

Clearly, the metric (8.1) is not well behaved at $r = 2m$, although this does not correspond to a singularity in the curvature tensor. The precise nature of this coordinate singularity will be clarified in the following sections. However, it may be immediately noted that for a radially infalling photon $dt = (1 - \frac{2m}{r})^{-1}dr$. This implies that the coordinate t diverges as $r \to 2m$, even though the proper time on this worldline remains finite. In order to analyse the character of the space-time near $r = 2m$, it is therefore necessary to introduce a different time parameter.

Actually, the Schwarzschild metric in the form (8.1) with $r \gg 2m$ is very well suited to the study of the gravitational field of isolated stars, planets and most spherically symmetric physical bodies. In such cases, the value $2m$ would be very significantly less than the radius r_0 of the object. For example: for the Sun $2m/r_0 \sim 10^{-6}$, while for the Earth $2m/r_0 \sim 10^{-9}$. Thus, the singularity does not occur in the relevant range of coordinates, and the above metric for $r > r_0$ can readily be applied to any vacuum exterior region around such objects. In these circumstances, the coordinate r is almost exactly the distance from the centre of the body.

In fact, it is found that timelike geodesics of the metric (8.1) are almost identical to the trajectories of particles in orbit around a spherical body of mass m according to Newtonian theory. This supports the argument that the Schwarzschild metric can be interpreted as representing the gravitational field of a spherical object in which the parameter m is the (gravitational) mass of the object. Moreover, where there are very small differences between the geodesics of (8.1) and the orbits of particles in Newtonian theory, it is found that the former are in closer agreement with the orbital motions of real physical bodies than those evaluated in the corresponding Newtonian theory. This application of geodesics in the Schwarzschild metric indicates that Einstein's theory of general relativity is an even more accurate theory of gravitation than the highly successful theory of Newton.

All this is described in detail in introductory textbooks on general relativity and need not be repeated here. It may simply be noted that the equations for geodesics of the metric (8.1) in the plane $\theta = \pi/2$ are given by

$$\dot{r}^2 + \left(1 - \frac{2m}{r}\right)\left(\epsilon_0 + \frac{L^2}{r^2}\right) = E^2, \qquad r^2\dot{\phi} = L, \qquad (8.3)$$

where E and L are constants of the motion (related to the existence of the Killing vectors ∂_t and ∂_ϕ, respectively), and $\epsilon_0 = 1$ for a timelike geodesic or $\epsilon_0 = 0$ for a null geodesic. It may be recalled that the equivalent equations for timelike trajectories in Newtonian theory are

$$\dot{r}^2 + \left(1 - \frac{2m}{r} + \frac{L^2}{r^2}\right) = E^2, \qquad r^2\dot{\phi} = L.$$

These equations are very similar, with L being interpreted classically as the angular momentum per unit mass of the test particle, and $\frac{1}{2}(E^2 - 1)$ as its energy (per unit mass). The differences between the two theories are therefore significant only when r is small.

Based on the comparison of orbits, it may be concluded that the Schwarzschild solution (8.1) can be interpreted as representing the external gravitational field of a spherical object of gravitational mass m. Moreover, just as in Newtonian theory, the mass of the source of a stationary asymptotically flat gravitational field can be determined by evaluating an integral over a closed surface near spatial infinity. Such an integral was introduced in a covariant way by Komar (1959). Arnowitt, Deser and Misner (1962) have introduced a similar general concept of energy in their Hamiltonian-type formulation of general relativity. For the Schwarzschild space-time, both the so-called Komar integral and the ADM mass give precisely the parameter m. All these approaches are consistent so that, at least for $r > r_0 > 2m$, the parameter m can be interpreted as the total (spherically distributed) mass inside $r = r_0$.

In the real world, the mass of all particles is necessarily positive. Thus, it is generally assumed that $m > 0$. However, as an exact mathematical solution of Einstein's equations, (8.1) is also valid when m is negative. In such a case, the apparent singularity at $r = 2m$ would not appear, and the corresponding space-time would not possess a similar interpretation.

For the Schwarzschild space-time, the equations (8.3) with $\epsilon_0 = 0$ imply that null geodesics also slightly bend around a central source (assumed to have radius $r_0 > 2m$). This leads to the prediction that rays of light from distant objects would be deflected if they pass near the Sun by a precise amount that depends on the parameter m and the minimum value of r. Such a deflection was in fact observed during a solar eclipse in 1919, providing significant early evidence for Einstein's general theory of relativity. Subsequent observations have confirmed this evidence to a much greater degree of accuracy (see e.g. Will, 1993, 2006, and references therein). Moreover, this phenomenon is also observed in the deflection of light around other stars and galaxies which therefore appear to act like gravitational lenses – a topic that is at the forefront of modern research. For a thorough analysis of this topic see Schneider, Ehlers and Falco (1999) and Schneider, Kochanek and Wambsgass (2006).

8.2 Eddington–Finkelstein coordinates and black holes

To clarify the meaning of the singularity that occurs at $r = 2m$ in the Schwarzschild metric (8.1), it is convenient to first introduce the coordinate $r_\star$ (which is sometimes referred to as a tortoise coordinate) such that

$$r_\star = \int \left(1 - \frac{2m}{r}\right)^{-1} \mathrm{d}r = r + 2m \log \left|\frac{r}{2m} - 1\right|. \qquad (8.4)$$

(The constant of integration has been chosen for simplicity.) With this, the line element becomes

$$\mathrm{d}s^2 = \left(1 - \frac{2m}{r}\right)\left(-\mathrm{d}t^2 + \mathrm{d}r_\star^2\right) + r^2 \left(\mathrm{d}\theta^2 + \sin^2\theta\,\mathrm{d}\phi^2\right), \qquad (8.5)$$

which enables appropriate radial null coordinates to be identified. In particular, the *advanced null coordinate* v defined by

$$v = t + r_\star,$$

can be introduced to eliminate the problematic coordinate t. The line element for the Schwarzschild space-time can thus be expressed in the form

$$\mathrm{d}s^2 = 2\,\mathrm{d}v\,\mathrm{d}r - \left(1 - \frac{2m}{r}\right)\mathrm{d}v^2 + r^2 \left(\mathrm{d}\theta^2 + \sin^2\theta\,\mathrm{d}\phi^2\right), \qquad (8.6)$$

which is clearly nonsingular at $r = 2m$. The metric was first put in this form by Eddington (1924) and later reconstructed by Finkelstein (1958). This explicitly demonstrates that $r = 2m$ is just a removable singularity of the Schwarzschild coordinates.

From the line element in the form (8.6), it can immediately be seen that radial null geodesics with θ and ϕ constant, are given by

$$v = \text{const.}, \qquad \text{or} \qquad \left(1 - \frac{2m}{r}\right)\mathrm{d}v - 2\,\mathrm{d}r = 0. \qquad (8.7)$$

In order to picture these in a space-time diagram, it is convenient to introduce a new coordinate

$$\bar{t} = v - r = t + 2m \log \left|\frac{r}{2m} - 1\right|,$$

which (since $g^{\mu\nu}\bar{t}_{,\mu}\bar{t}_{,\nu} = -1 - \frac{2m}{r} < 0$) is timelike for all $r > 0$, and may be taken to be future directed. The first of the null geodesics given by (8.7) on which $\bar{t} + r = \text{const.}$ are everywhere ingoing, while the second are "outgoing" null geodesics for which

$$\mathrm{d}\bar{t} = \left(\frac{r + 2m}{r - 2m}\right)\mathrm{d}r.$$

From this, it can be seen that $\mathrm{d}r > 0$ when $r > 2m$. However, when $r < 2m$,

$dr < 0$, and thus the coordinate r decreases as $\bar{t}$ increases on all null geodesics in this region. This is illustrated in Figure 8.1. In particular, it shows that light cones, which have their usual structure asymptotically, are deformed so that, for $r < 2m$, all null geodesics (and hence all timelike geodesics) remain within this region and in fact terminate at the singularity at $r = 0$. The coordinates $(\bar{t}, r, \theta, \phi)$ are known as the Eddington–Finkelstein coordinates and Figure 8.1 is referred to as the Eddington–Finkelstein diagram.

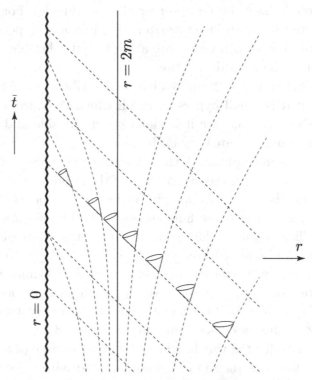

Fig. 8.1 Radial null geodesics in Eddington–Finkelstein coordinates with θ and ϕ constant. Ingoing null geodesics are represented by lines on which $\bar{t} + r = $ const., while null geodesics propagating in the opposite direction have increasing values of r for $r > 2m$, but decreasing values for $r < 2m$. Some future light cones are also indicated.

Real bodies must follow worldlines that remain inside the corresponding null cones at all points. It can thus be seen from the light-cone structure illustrated in Figure 8.1 that, if a body at a point outside $r = 2m$ emits light, some of this will propagate inwards to $r = 0$ and some also propagates out to infinity. On the other hand, if the body were inside $r = 2m$, its own trajectory would inevitably lead into the curvature singularity at $r = 0$. Any light emitted towards the singularity would reach it before the body itself. However, light emitted in the opposite direction would also fall into the singularity. Thus, even "outgoing" null geodesics in this sense propagate inwards towards the singularity in this region.

It is important to stress that nothing significant happens locally at $r = 2m$. This is a regular part of the space-time covered by the metric (8.6). Particles (timelike or null trajectories) can pass in through $r = 2m$, but they cannot come out. The space-time is therefore called a *black hole* and the hypersurface $r = 2m$ is known as a *horizon*. No event that occurs inside the horizon can ever be seen by an observer that is outside. For this reason, $r = 2m$ is referred to as an *event horizon*.[2] It is a null hypersurface: null particles such as photons can propagate along it, but other photons can only pass through it in the inward direction.

The event horizon at $r = 2m$ may also be described as a *Killing horizon* in the sense that it is a null hypersurface on which the norm of the Killing vector ∂_t vanishes. In this case it is timelike on one side and spacelike on the other. It has constant area[3] given by $\mathcal{A} = 16\pi m^2$.

As has already been emphasised, the Schwarzschild space-time outside the horizon provides an accurate representation of the gravitational field exterior to a large spherically symmetric massive object such as a star. However, it is now known that, towards the final stages in the life of a star, its density may become sufficiently large that its outer surface may approach the critical value of $r = 2m$. In its final state, a star may even collapse to form a black hole. Once all possible nuclear reactions have ceased, if its mass is sufficiently large, no interactions between the residual particles may be sufficient to prevent its collapse through its own event horizon to a curvature singularity. Such a collapse is illustrated schematically in Figure 8.2.

Figure 8.2 illustrates the way in which the outer surface of a massive star in its final state may contract under its dominant gravitational field. All the residual material of the star would contract through $r = 2m$ and end in a singularity. As the star collapses, light emitted from its surface is radiated in the usual way and may be seen by a distant observer. However, once the surface is inside the horizon, the light emitted also remains inside the horizon and terminates at the singularity. Thus the final state of the star could never be observed. Radially outgoing light trajectories are indicated in

---

[2] The surface $r = 2m$ in the Schwarzschild space-time is an absolute event horizon in the sense that no events that occur inside it could, even in principle, ever be seen by any (external) observer whose worldline propagates from past timelike infinity to future timelike infinity. It could be defined as the boundary of the region containing events that could be seen from future null infinity.

[3] Although the area of the horizon is uniquely determined, there is no unique measure of the volume contained within this surface. For the vacuum solution, the three-dimensional spatial volume depends on the choice of time, and can be time-dependent or even zero. See DiNunno and Matzner (2010).

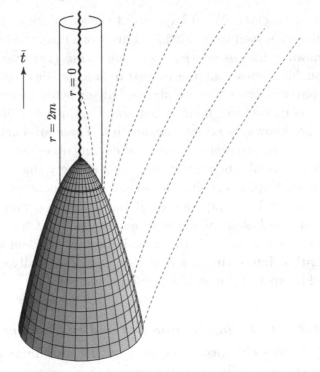

Fig. 8.2 The final stages of the remnant of a massive star are illustrated in terms of Eddington–Finkelstein coordinates $(\bar{t}, r, \phi, \theta = \frac{\pi}{2})$ as it collapses through its own event horizon and forms a black hole. Some paths of light emitted from the surface at different stages are also indicated as broken lines.

Figure 8.2. These confirm that an external observer could never in principle see beyond the point where the star passes through the horizon. In practice, however, to the observer, the brightness of the star would appear to rapidly decay and it would effectively disappear from sight.

It should also be noted that the Schwarzschild metric, and the process of gravitational collapse, are stable with respect to small non-spherical perturbations. These important results have been established by Regge and Wheeler (1957), Vishveshwara (1970), Zerilli (1970a,b), Price (1972) and Wald (1979). These studies show that small perturbations are radiated away leaving the static spherically symmetric Schwarzschild space-time. (For a recent review and further details and applications, see Nagar and Rezzolla, 2005, and references therein.)

The properties described above have some most important physical implications. As the final remnants of a large star collapse to form a black hole, the outer surface passes through the event horizon and can never again be seen by a distant observer. What happens at the singularity, or near it, can never be actually observed from outside. This is one illustration of what has come to be known as the *cosmic censorship hypothesis* (Penrose, 1969). This hypothesis roughly states that the curvature singularities that are caused by plausible physical processes are always hidden behind event horizons.[4] Contrary to this conjecture, many exact solutions of Einstein's equations contain what are known as *naked singularities*. These are curvature singularities that could, in principle, be seen by distant observers. However, in these cases, it can usually be argued fairly convincingly that such solutions do not correspond to physical situations that could be observed in the real world (except for the big bang – a past global spacelike curvature singularity that occurs in models of the universe, as discussed in Chapter 6). When interpreting exact solutions, it is therefore important to identify cases with naked singularities. Interestingly, a naked singularity actually also occurs in the Schwarzschild space-time as will now be demonstrated.

8.2.1 *Outgoing coordinates and white holes*

The metric (8.6) was obtained from (8.1) by the introduction of the advanced null coordinate v. However, the form of (8.5) suggests an alternative possibility of introducing a *retarded null coordinate* defined by

$$u = t - r_\star.$$

With this, the Schwarzschild metric (8.1) takes the alternative form

$$ds^2 = -2\,du\,dr - \left(1 - \frac{2m}{r}\right)du^2 + r^2\left(d\theta^2 + \sin^2\theta\,d\phi^2\right). \tag{8.8}$$

In this case, radial null geodesics with θ and ϕ constant, are given by

$$u = \text{const.,} \qquad \text{or} \qquad \left(1 - \frac{2m}{r}\right)du + 2\,dr = 0.$$

In order to picture this in a space-time diagram, it is convenient to introduce another timelike coordinate

$$\tilde{t} = u + r = t - 2m\log\left|\frac{r}{2m} - 1\right|.$$

With this, the above null geodesics are given by $\tilde{t} - r = \text{const.}$ and

$$d\tilde{t} = -\left(\frac{r + 2m}{r - 2m}\right)dr,$$

---

[4] The cosmic censorship hypothesis, and counter-examples in which naked singularities possibly occur, will be further discussed in Sections 9.2.2, 9.5.2 and 22.7. For a technical review of attempts to resolve this conjecture, see e.g. Clarke (1994).

for which r increases as $\tilde{t}$ increases when $r < 2m$, but r decreases as $\tilde{t}$ increases when $r > 2m$. These represent both outgoing and "ingoing" radial null geodesics, respectively. They are illustrated in Figure 8.3 in which the outgoing null geodesics are drawn at 45°.

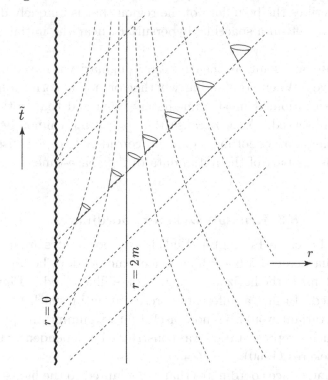

Fig. 8.3 Radial null geodesics in Eddington–Finkelstein-like coordinates with θ and ϕ constant for the Schwarzschild space-time interpreted as a white hole. Some outgoing and ingoing null geodesics and some future light cones are indicated.

It may immediately be observed that Figure 8.3 is actually the time-reverse of Figure 8.1. This is not unexpected. Einstein's field equations have no preferred time direction, so it is not surprising that a time-reversed solution also occurs. This kind of singularity can be interpreted as representing a *white hole* (a time-reversed black hole) from which light can emerge but which no light can reach. It is a white hole because the singularity could be seen – it is naked! Moreover, nothing from outside $r = 2m$ can cross this surface in an inward direction. In addition, nothing can reach the singularity.

In this case, the white hole horizon at $r = 2m$ could also be regarded as an event horizon in the sense that events that occur outside it could never be seen from inside. However, this is not the normal use of the term "event horizon", which is usually considered with respect to null infinity. Clearly, like the black hole horizon, it is a Killing horizon. Strictly, however, as will become clear in the following sections, this particular null hypersurface is more properly referred to as a *Cauchy horizon*. This term is employed because it describes the boundary of the region that is uniquely determined by Cauchy data set on a spacelike hypersurface near the initial singularity at $r = 0$.

For this white hole solution, the curvature singularity is naked in the sense described above. When considering whether or not it corresponds to any real physical situation, it must immediately be stated that nothing like it has yet been observed. Also, because the space-time cannot be extended through the singularity, nothing could be considered to give rise to it. It could only exist as part of the initial data of the space-time.

8.3 Kruskal–Szekeres coordinates

As described above, the Eddington–Finkelstein coordinates are useful for describing the character of a black hole space-time outside the horizon. However, near and inside the horizon, the light-cones illustrated in Figure 8.1 are highly deformed. In fact a different representation in which the light-cone structure is invariant would be more useful for examining the properties of this region. Such a representation was constructed independently by Kruskal (1960) and Szekeres (1960).

In order to introduce coordinates that are adapted to the light-cone structure, it is appropriate to employ *simultaneously* both advanced and retarded null coordinates, $v = t + r_\star$ and $u = t - r_\star$. In terms of these coordinates, the Schwarzschild metric becomes

$$\mathrm{d}s^2 = -\left(1 - \frac{2m}{r}\right)\mathrm{d}u\,\mathrm{d}v + r^2\left(\mathrm{d}\theta^2 + \sin^2\theta\,\mathrm{d}\phi^2\right), \qquad (8.9)$$

which follows immediately from (8.5), and where $r(u, v)$ is defined implicitly by the equation

$$r + 2m\log\left|\frac{r}{2m} - 1\right| = r_\star = \tfrac{1}{2}(v - u).$$

Unfortunately, the metric (8.9) in double-null coordinates is also badly behaved at $r = 2m$. This problem can be overcome, however, by suitably reparametrising the null coordinates by putting

$$U = -4m\,\mathrm{e}^{-u/4m}, \qquad V = 4m\,\mathrm{e}^{v/4m}.$$

With this, the line element takes the form

$$ds^2 = -\frac{2m}{r}\, e^{-r/2m}\, dU\, dV + r^2 \left(d\theta^2 + \sin^2\theta\, d\phi^2\right), \tag{8.10}$$

where r is now defined implicitly by the equation

$$16m^2\left(\frac{r}{2m} - 1\right) e^{r/2m} = -U\,V.$$

The metric (8.10) is clearly continuous, and in fact also regular, at the horizon at $r = 2m$ corresponding to $UV = 0$. It can also be seen that the curvature singularity at $r = 0$ now occurs when $UV = 16m^2$.

To interpret this space-time, it is convenient to reintroduce timelike and spacelike coordinates T and R defined simply by

$$\begin{aligned}
T = \tfrac{1}{2}(V + U), \qquad R = \tfrac{1}{2}(V - U), \qquad \text{for} \quad r > 2m, \\
T = \tfrac{1}{2}(V - U), \qquad R = \tfrac{1}{2}(V + U), \qquad \text{for} \quad r < 2m.
\end{aligned} \tag{8.11}$$

With these, the metric is expressed in what are known as *Kruskal–Szekeres coordinates* in the form

$$ds^2 = \frac{2m}{r}\, e^{-r/2m} \left(-dT^2 + dR^2\right) + r^2 \left(d\theta^2 + \sin^2\theta\, d\phi^2\right). \tag{8.12}$$

Combining the above transformations, these coordinates are related to the Schwarzschild coordinates of (8.1) by

$$\begin{cases}
T = 4m\sqrt{\dfrac{r}{2m} - 1}\; e^{r/4m} \sinh\left(\dfrac{t}{4m}\right), \\[2mm]
R = 4m\sqrt{\dfrac{r}{2m} - 1}\; e^{r/4m} \cosh\left(\dfrac{t}{4m}\right),
\end{cases} \quad \text{for} \quad r > 2m,$$

$$\begin{cases}
T = 4m\sqrt{1 - \dfrac{r}{2m}}\; e^{r/4m} \cosh\left(\dfrac{t}{4m}\right), \\[2mm]
R = 4m\sqrt{1 - \dfrac{r}{2m}}\; e^{r/4m} \sinh\left(\dfrac{t}{4m}\right),
\end{cases} \quad \text{for} \quad r < 2m, \tag{8.13}$$

for which the inverse transformation is given implicitly by

$$16m^2\left(\frac{r}{2m} - 1\right) e^{r/2m} = R^2 - T^2, \qquad \begin{cases} t = 4m\,\text{arctanh}\,(T/R), & \text{for } r > 2m, \\[1mm] t = 4m\,\text{arctanh}\,(R/T), & \text{for } r < 2m. \end{cases}$$

It immediately follows from this that lines with $r = $ const. correspond to hyperbolae in the R, T-plane. Correspondingly, lines with $t = $ const. are straight lines through the origin.

It may be noted that the transformation (8.13) is not well behaved at the horizon $r = 2m$. However, this is necessary to remove the pathological behaviour of the Schwarzschild coordinates at this point. The resulting space-time metric (8.12) is well behaved everywhere except, of course, at the curvature singularities.

The transformation (8.13) is given only for the regions in which $V > 0$, i.e. $T + R > 0$. However, the coordinates used in the metric (8.12) can easily be extended over their complete natural ranges up to the singularity at $r = 0$. The fully extended space-time is represented in Figure 8.4. Moreover, since this construction is based on double-null coordinates, incoming and outgoing null rays are both pictured in this figure at 45° to the vertical T-direction.

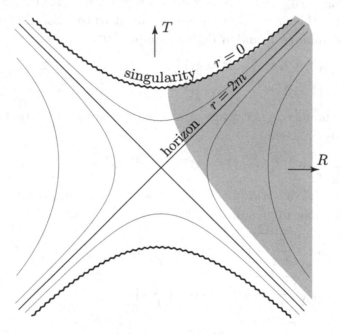

Fig. 8.4 The complete Schwarzschild space-time in Kruskal–Szekeres coordinates (R, T) with θ and ϕ constant. The thin lines represent hypersurfaces on which r is a constant. All (radial) null trajectories in this section propagate at 45° to the vertical. The shaded area indicates that part of the space-time that could represent the external field of a collapsing star forming a black hole, while the remaining part is removed.

The space-time exterior to a star that collapses to form a black hole was previously illustrated in Figure 8.2. For a section of constant ϕ, this region is represented in Kruskal–Szekeres coordinates by the lightly shaded region in Figure 8.4. If a black hole can only be formed by a process that involves the final collapse of a large star, then it is only the shaded region contained in Figure 8.4 that has physical significance. However, the maximally extended space-time may also be investigated.

As illustrated in Figure 8.4, the black hole curvature singularity at $r = 0$ occurs when $T = \sqrt{16m^2 + R^2}$. This is a spacelike hypersurface, which forms a boundary of the physical space-time. In fact, the complete space-time represented by the metric (8.12) has another (naked) curvature singularity when $T = -\sqrt{16m^2 + R^2}$. This corresponds to the white hole singularity that is illustrated in Figure 8.3. In the Kruskal–Szekeres coordinates, the black hole and the white hole singularities are both included, thus manifesting the time symmetry that is inherent in the field equations. Figure 8.4 also illustrates that a white hole is a time-reversed black hole from which light can emerge but which no light can reach.

The region $R > |T|$ in Kruskal–Szekeres coordinates represents the complete static exterior of either a black hole or a white hole outside their horizons. However, there is no need to restrict the range of R to positive values. By considering the full natural range $R \in (-\infty, \infty)$, the metric (8.12) also includes another region with the same properties in which $R < -|T|$. This represents a second complete exterior region. In any spatial section of the complete space-time through $R = 0$ (with $-4m < T < 4m$), *two distinct* exterior static regions are connected via a time-dependent region behind horizons in which a minimum value for r occurs at $R = 0$. Such a spatial section is often described as a "wormhole", or "Einstein–Rosen bridge", in which the two exterior regions are connected by a "throat" which is non-singular for a finite period of time. However, it must be emphasised that no real particles can pass through the throat from one exterior region to the other. These two static regions are causally separated. An observer in either cannot in principle know about the existence of the other.

8.4 Conformal structure

When constructing a compactified conformal diagram of the Schwarzschild space-time, it is convenient to start with the line element in the form (8.10) which involves the double-null coordinates $U, V \in (-\infty, \infty)$. It is then possible to introduce rescaled retarded and advanced null coordinates $\tilde{u}$ and $\tilde{v}$ defined by

$$\tilde{u} = 2 \arctan \left(\frac{U}{4m} \right), \qquad \tilde{v} = 2 \arctan \left(\frac{V}{4m} \right), \qquad (8.14)$$

where

$$-\pi < \tilde{u} < \pi, \qquad -\pi < \tilde{v} < \pi, \qquad -\pi < \tilde{u} + \tilde{v} < \pi.$$

This range of coordinates then covers the complete analytically extended space-time, as illustrated in Figure 8.5. The black hole and the naked white

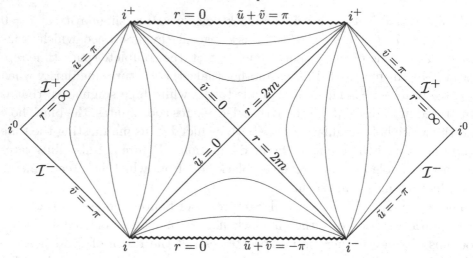

Fig. 8.5 Penrose diagram for the complete Schwarzschild space-time. The θ and ϕ coordinates are suppressed so that each point represents a 2-sphere of radius r. All lines shown are hypersurfaces on which r is a constant.

hole curvature singularities where $r = 0$ are given here by $\tilde{u} + \tilde{v} = \pi$ and $\tilde{u} + \tilde{v} = -\pi$, respectively.

Figure 8.5 clearly illustrates the global structure of the Schwarzschild space-time.[5] It shows all possible regions of the complete analytically extended manifold. In particular, exterior to the horizons there exist two causally separated static regions $r > 2m$ that are asymptotically Minkowski-like. It also demonstrates the spacelike character of the initial and final curvature singularities $r = 0$. The black hole horizon $r = 2m$ is manifested as an event horizon, while the white hole horizon is seen as a Cauchy horizon for initial data on a spacelike surface near the initial big bang-like singularity.

If the Schwarzschild space-time is considered just to represent the exterior

---

[5] Using the compactification given by (8.14), the lines on which r and t are constant approach spacelike and timelike infinities i^0, i^- and i^+ and the "central" point $\tilde{u} = 0 = \tilde{v}$ from a variety of directions as partly shown in Figure 8.5. Such a compactification, however, is not unique and alternatives may be considered. Moreover, this process simply relates the infinite ranges of the coordinates r and t to finite ranges of $\tilde{u}$ and $\tilde{v}$. It does not conformally relate the 3+1-dimensional metric (8.10) to another that has convenient geometrical properties – unlike the conformally flat cases that were explicitly constructed in previous chapters. Nevertheless, the Schwarzschild space-time is asymptotically flat at null and spatial infinity $\mathcal{I}$ and i^0, and it is possible to find a conformal compactification such that the lines on which r and t are constant behave like those of Minkowski space, shown in Figure 3.4, near these locations. In any such construction in which the compactified space-time is better behaved near $\mathcal{I}$ and i^0, however, the singularity at $r = 0$ would not be represented by a straight line (although it would still be spacelike). In Figure 8.5, and in all the conformal diagrams that are given in following chapters, non-null lines on which $r = $ const. are therefore only shown schematically. For a rigorous definition of conformal infinity for non-flat space-times, see Penrose (1963, 1965b), Ashtekar and Hansen (1978), Ashtekar (1980) and Wald (1984).

of a collapsing spherically symmetric object and the resulting black hole, then only the part of the conformal diagram given in Figure 8.6 is relevant.

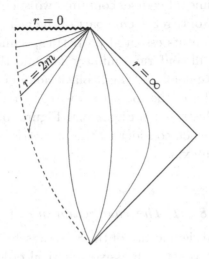

Fig. 8.6 Penrose diagram for the part of the Schwarzschild space-time that represents the exterior of a collapsing spherical object and the resulting black hole, corresponding to the shaded region in Figure 8.4.

For later convenience, it is also helpful to further illustrate the asymptotic properties of the complete exterior region by reintroducing the ϕ coordinate, as given in Figure 8.7. This enables conformal null infinity $\mathcal{I}$ to be represented more clearly as sections of null cones as in Figure 3.4. In this figure, the white hole and black hole horizons are drawn at 45° to the vertical to

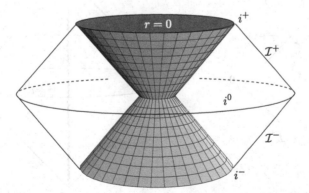

Fig. 8.7 The exterior region of the Schwarzschild space-time with the angular coordinate ϕ reintroduced. The complete space-time contains two such regions that are causally separated from each other. The horizons at $r = 2m$ are drawn at 45° to reflect the fact that they are null, but their area at all times remains constant.

reflect the fact that they are null. However, since they occur at $r = 2m$, their area at any time remains equal to the constant $16\pi m^2$. The complete Schwarzschild space-time, of course, contains two such regions.

Although Figure 8.7 only represents part of the Schwarzschild space-time, it does illustrate some of its essential properties, namely: the character of spacelike infinity i^0, null conformal infinity $\mathcal{I}^\pm$, and the horizons that surround the initial (white hole) and final (black hole) curvature singularities. As $r \to \infty$, the character of the space-time is seen to be the same as that of Minkowski space, as illustrated in Figure 3.4. Figures of this type will in fact be particularly useful when considering the properties of Schwarzschild-like sources in different contexts.

8.4.1 *The case when* $m < 0$

Because of the physical significance of the Schwarzschild solution as describing the space-time exterior to a massive spherical object or black hole, we have concentrated on the case in which $m > 0$. As already emphasised, the parameter m may be interpreted as the mass of the object and, in the real world, this is necessarily positive. However, besides considering the Schwarzschild space-time in regions which do not have this interpretation, this solution with a negative parameter m may also be considered (still assuming that $r \in (0, \infty)$). In this case, there is no horizon and the curvature

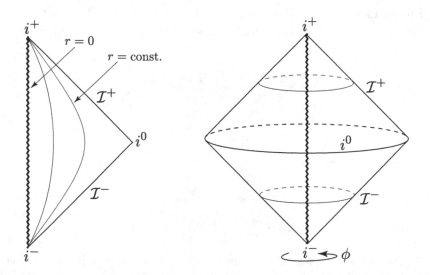

Fig. 8.8 Penrose and conformal diagrams for the Schwarzschild space-time when $m < 0$. In this case, there is a globally naked timelike singularity at $r = 0$.

singularity at $r = 0$ is timelike, globally naked (i.e. the curvature singularity could be "seen" at every point in the space-time) and unstable (see Gleiser and Dotti, 2006).

The Penrose and conformal diagrams for this solution are given schematically in Figure 8.8. It can be seen that the asymptotic structure as $r \to \infty$ is the same as that for Minkowski space as given in Figure 3.4. This case, however, does not correspond to any known physical situation.

8.5 Interior solutions

The vacuum Schwarzschild solution has been interpreted above as representing the exterior field of a spherically symmetric body, such as some kind of star. In fact, it is the unique asymptotically flat vacuum solution around any spherically symmetric object, irrespective of how that body may be composed or how it may evolve in time. It contains no information about its source other than its total mass, and imposes no constraints on its internal structure. However, although this leaves almost total freedom when seeking to model physical bodies that have the required exterior field, a number of physical constraints must be imposed in the construction of realistic stellar models.

In this section, exact solutions will be considered which could represent an idealised star-like material source for an exterior Schwarzschild space-time. For simplicity, attention will be restricted to non-rotating static bodies, and it will be assumed that these are composed of a perfect fluid with density $\rho(r)$ and isotropic pressure $p(r)$. Exact solutions representing such an object must match to the exterior solution on a spherical surface $r = a$, which is necessarily outside the horizon (i.e. $a > 2m$).

Taking the exterior solution in the standard form (8.1), the single parameter m represents the total mass of the source. Inside the body, where $r < a$, it is convenient to generalise this and to introduce a function $\mu(r)$ which represents the total mass inside each spherical surface of constant r whose area is given by $4\pi r^2$. Thus,

$$\mu(r) = \int_0^r 4\pi r^2 \rho(r)\, dr, \tag{8.15}$$

where $\mu(0) = 0$ and $\mu(a) = m$. With this, and continuing to use Schwarzschild coordinates, the line element in the interior region $r < a$ can always be expressed in the static spherically symmetric form

$$ds^2 = -\exp\left(2\Phi(r)\right)dt^2 + \left(1 - \frac{2\mu(r)}{r}\right)^{-1} dr^2 + r^2\left(d\theta^2 + \sin^2\theta\, d\phi^2\right), \tag{8.16}$$

where $\Phi(r)$ is a function that needs to be determined. For this metric, Einstein's field equations imply that

$$\frac{\mathrm{d}\Phi}{\mathrm{d}r} = \frac{\mu + 4\pi r^3 p}{r(r - 2\mu)}. \tag{8.17}$$

In addition, there is the equation for hydrostatic equilibrium. This can be expressed in the form

$$\frac{\mathrm{d}p}{\mathrm{d}r} = -\frac{(\rho + p)(\mu + 4\pi r^3 p)}{r(r - 2\mu)}, \tag{8.18}$$

which is generally known as the Tolman–Oppenheimer–Volkov equation.

The next step in the construction of an interior solution is to specify the equation of state. This may be expressed in the barotropic form $p = p(\rho)$. In this case, there are then two remaining functions, $\rho(r)$ and $\Phi(r)$, which must satisfy (8.17) and (8.18) with (8.15). Starting with some initial value for the central pressure $p(0)$, these equations should be integrated to find the value of r at which the pressure is reduced to zero. This is taken as the value $r = a$ which identifies the surface of the star, i.e. $p(a) = 0$. After integrating (8.17), the arbitrary additive constant in the expression for Φ may then be specified to satisfy the junction condition that $e^{2\Phi(a)} = 1 - 2m/a$. For a more detailed description of this method for constructing stellar models, see Kuchowicz (1971) and Misner, Thorne and Wheeler (1973).

One of the simplest explicit models for the interior of a star, and one of the most useful, is that given by Schwarzschild (1916b). In this case, the density is taken as a constant

$$\rho = \frac{3m}{4\pi a^3}, \qquad \mu(r) = m\frac{r^3}{a^3}, \qquad \text{for} \quad r \leq a.$$

In this case, (8.18) can be integrated and, after fixing the arbitrary constant such that $p = 0$ when $r = a$, this gives

$$p(r) = \frac{3m}{4\pi a^3}\left(\frac{\sqrt{1 - 2mr^2/a^3} - \sqrt{1 - 2m/a}}{3\sqrt{1 - 2m/a} - \sqrt{1 - 2mr^2/a^3}}\right).$$

It can then be seen that this expression diverges, for some $r \in [0, a)$, if $2m/a \geq 8/9$. Thus, for this solution to be physically acceptable, it is necessary that $a > \frac{9}{4}m$. With these expressions, the remaining equation (8.17) can be integrated to give

$$e^{\Phi(r)} = \begin{cases} \frac{3}{2}\sqrt{1 - \frac{2m}{a}} - \frac{1}{2}\sqrt{1 - \frac{2mr^2}{a^3}}, & \text{for } r \leq a, \\ \sqrt{1 - \frac{2m}{r}}, & \text{for } r > a, \end{cases}$$

which is C^1 at $r = a$. For the region in which $r < a$, this is known as the *Schwarzschild interior solution*. The exterior solution is exactly (8.1).

Remarkably, the Schwarzschild interior solution is conformally flat (Buchdahl, 1971). Indeed, it can be shown that any stationary, conformally flat, perfect fluid solution must be the static Schwarzschild interior metric. See also Collinson (1976), Gürses (1977) and Raychaudhuri and Maiti (1979).

Although the pressure and the constant density in the Schwarzschild interior solution are both positive, they do not satisfy a realistic equation of state. Moreover, the solution has the unphysical property that the speed of sound within the interior, defined by $v_s = \sqrt{\mathrm{d}p/\mathrm{d}\rho}$, is greater than the speed of light.

Numerous alternative exact interior solutions have been presented in the extensive literature on this topic. Mostly, but not exclusively, these have been obtained using either Schwarzschild or isotropic coordinates. Many are summarised in the reviews of Delgaty and Lake (1998) and Stephani *et al.* (2003). They have been constructed using various initial ad hoc assumptions, and almost all have at least one unphysical feature. In practice, to obtain realistic stellar models, the equations (8.15), (8.17) and (8.18) have to be integrated numerically.

Time-dependent interior solutions can be similarly constructed. (For a review, again see Stephani *et al.*, 2003.) The simplest of these is that of Oppenheimer and Snyder (1939), in which the interior region is taken to be a spherical section of a FLRW space-time (with zero cosmological constant) as described in Chapter 6. Such a model can represent a collapsing spherical body.

8.6 Exterior solutions

As emphasised above, the Schwarzschild space-time is the unique asymptotically flat vacuum solution representing the exterior of a spherically symmetric body. It is considered to represent the gravitational field of a star. However, the stars that occur in our universe do not exist in an asymptotically flat background. The asymptotic structure is rather that of some cosmological model.

In fact, after a rather complicated change of coordinates, it is possible to match a vacuum Schwarzschild solution to an exterior FLRW solution. Following the early work of Einstein and Straus (1945) and Schücking (1954), this is known as an *Einstein–Straus* model, and is obtained by cutting a sphere out of a FLRW universe (as described in Chapter 6) and replacing it by a Schwarzschild solution for the external field of a spherical body whose

mass is exactly the same as the total mass contained in the part of the
FLRW solution that was removed.[6] (Notice that this is the inverse of the
Oppenheimer–Snyder solution.) Of course, the areal radius of the region
removed has to be expanding in a way that is precisely in accordance with
the general expansion of the corresponding FLRW background.[7]

This model may be used, for example, to argue that planetary orbits
about a star are unaffected by the large-scale expansion of the universe.
However, two features indicate that this conclusion should be treated cau-
tiously. Krasiński (1997) has pointed out that this represents an exceptional situation that is unstable with respect to perturbations of initial data.
In addition, when applied to our solar system, the radius of the vacuum
Schwarzschild region would need to be around 10^3 light-years – a region
that in fact contains many other stars.

Although the conformal structure of each component of this space-time
is known (see Sections 6.4 and 8.4), a construction of a combined Penrose
diagram faces the problem that they are conformally related to different
geometries. Nevertheless, a combined conformal diagram can be obtained.
The details of this have been fully described by Sussman (1985).

It may also be noted that, since the FLRW universe is spherically symmetric about every point, it is possible to remove any number of spherical
regions at random locations within it, and to replace each by spherical regions of the corresponding Schwarzschild space-time. A universe composed
in this way is known as a *Swiss cheese* model. Such a model is particularly useful as it represents a universe which is locally inhomogeneous and
anisotropic, but is still homogeneous and isotropic on a large scale (see Rees
and Sciama, 1968). It describes a universe composed of spherical objects of
constant mass surrounded by spherical vacuum regions that exist in a homogeneous perfect fluid cosmological background. Although such a universe as
a whole has non-zero expansion, the vacuum regions around each body are
static. For further details and applications of this model, see e.g. Kantowski
(1969), Dyer (1976) and Kaiser (1982).

---

[6] It is also possible to include a cosmological constant throughout such a compound space-time.
In this case, the interior solution is part of the Schwarzschild–de Sitter space-time that will
be described in Section 9.4. The combined solution in this case has been given by Balbinot,
Bergamini and Comastri (1988).

[7] The uniqueness of this model within a FLRW background has been discussed by Mars (2001)
and Nolan and Vera (2005).

9

Space-times related to Schwarzschild

In this chapter we will describe a few simple exact solutions that are related to the Schwarzschild space-time, either by changing the value of a discrete parameter that will be introduced below, or by adding a charge parameter or a cosmological constant, or by permitting the parameter m to become a variable. Other one-parameter generalisations of the Schwarzschild metric will be described in Chapters 11, 12 and 14, in which a rotation parameter, a NUT parameter and an acceleration parameter are included, respectively.

9.1 Other A-metrics

In their 1962 review, Ehlers and Kundt identified a family of vacuum space-times which they denoted as A-metrics. These can most generally be represented by the line element

$$ds^2 = -\left(\epsilon - \frac{2m}{r}\right)dt^2 + \left(\epsilon - \frac{2m}{r}\right)^{-1}dr^2 + r^2\frac{2\,d\zeta\,d\bar\zeta}{(1 + \frac{1}{2}\epsilon\zeta\bar\zeta)^2}, \tag{9.1}$$

which includes an arbitrary continuous parameter m and a discrete parameter ϵ, which can either be zero or scaled to the values $+1$ or -1. It may be observed (see Appendix A) that the surfaces on which t and r are constant have positive, zero or negative Gaussian curvature according to whether ϵ is $+1$, 0 or -1, respectively. This parameter may therefore be referred to as the *2-space curvature parameter*. It will in fact occur in many of the solutions described in the following chapters.

Using a suitable null tetrad, which is an obvious generalisation of (8.2), the only non-zero component of the Weyl tensor is given by

$$\Psi_2 = -\frac{m}{r^3}.$$

In all three cases, $\epsilon = +1, 0, -1$, it follows that this is Minkowski space when

$m = 0$. Whenever $m \neq 0$, there exists a curvature singularity at $r = 0$, and it is appropriate to restrict r either to the range $r > 0$, or to $r < 0$.

In the case in which $\epsilon = +1$, it is convenient to use the standard stereographic relation (see Appendix A) $\zeta = \sqrt{2}\tan(\theta/2)e^{i\phi}$, and the line element (9.1) then reduces exactly to that of the Schwarzschild space-time as given in (8.1). This could alternatively be referred to as the *AI* space-time, and has already been described in detail in the previous chapter. It therefore remains to consider the other two cases.

9.1.1 The AII-metrics ($\epsilon = -1$)

For the case in which $\epsilon = -1$, a coordinate singularity occurs when $|\zeta| = \sqrt{2}$. This corresponds to conformal infinity in the 2-space on which t and r are constant, and the coordinate ranges with $|\zeta| < \sqrt{2}$ and $|\zeta| > \sqrt{2}$ are completely equivalent. Taking $|\zeta| < \sqrt{2}$, it is convenient to put (see Appendix A)

$$r = -\sigma \qquad \text{and} \qquad \zeta = \sqrt{2}\,\tanh(\psi/2)\,e^{i\phi},$$

where $\psi \in [0, \infty)$ and $\phi \in [0, 2\pi)$, and the metric (9.1) then becomes

$$\mathrm{d}s^2 = -\left(\frac{2m}{\sigma} - 1\right)\mathrm{d}t^2 + \left(\frac{2m}{\sigma} - 1\right)^{-1}\mathrm{d}\sigma^2 + \sigma^2\left(\mathrm{d}\psi^2 + \sinh^2\psi\,\mathrm{d}\phi^2\right). \quad (9.2)$$

This is the standard form of the *AII*-metric, which contains a single continuous parameter m. Its interpretation, as described below, was given by Gott (1974). Since this space-time has a curvature singularity at $\sigma = 0$, it is convenient to restrict σ to the range $[0, \infty)$, and to permit m to be either positive or negative.

For the case in which $m > 0$, there is a coordinate singularity at $\sigma = 2m$ which is a Killing horizon related to ∂_t, whose properties will be described below. In this case, the metric is static for $0 < \sigma < 2m$,[1] and the curvature singularity at $\sigma = 0$ is timelike. In the region for which $2m < \sigma < \infty$, the metric is time-dependent with σ being the timelike coordinate.

It may immediately be noticed that the properties of this space-time have many things in common with that of Schwarzschild, but a number of features are different. At large distances from the source, the Schwarzschild solution can be considered to represent the gravitational field of a static point mass. In a similar way, at large values of σ, the *AII*-metric can be considered to represent *the gravitational field of a tachyon*: i.e. the source

---

[1] Within the static region, the metric can be cast in Weyl form (see Section 10.1). The Newtonian potential associated with this region of the solution has been determined by Martins (1996) as that of two semi-infinite line sources of specific positive and negative mass per unit length. However, this provides little guidance on the physical source of this space-time.

may be understood as being an object of mass m whose worldline is space-like (see Peres, 1970). Of course, the direction in which such an object may be considered to move across the complete space-time depends on the worldline of the observer, and it may even appear to do so instantaneously. And, whereas in Schwarzschild space-time, surfaces on which $r\,(> 2m)$ and t are constant are 2-spheres centred on the source at any time, in the AII space-time, the surfaces on which $\sigma\,(> 2m)$ and ψ are constant are cylinders centred on this instantaneous spacelike line, some time after the tachyon's occurrence. Moreover, the metric (9.2) only represents those parts of the space-time that correspond to the causal future or past of the tachyon.[2]

These properties can partly be seen in the limit as $m \to 0$, in which the metric (9.2) reduces to Minkowski space in the form

$$\mathrm{d}s^2 = -\mathrm{d}\sigma^2 + \mathrm{d}t^2 + \sigma^2 \left(\mathrm{d}\psi^2 + \sinh^2 \psi \, \mathrm{d}\phi^2\right). \tag{9.3}$$

This is related to the cylindrical form $\mathrm{d}s^2 = -\mathrm{d}T^2 + \mathrm{d}R^2 + R^2 \, \mathrm{d}\phi^2 + \mathrm{d}Z^2$ by the transformation

$$T = \pm\sigma \cosh\psi, \qquad R = \sigma \sinh\psi, \qquad Z = t, \tag{9.4}$$

with ϕ unchanged and $R \in [0, \infty)$. Since $T^2 - R^2 = \sigma^2$, it can be seen that the hypersurface on which $\sigma = 0$ corresponds, not only to the instantaneous spacelike line $T = 0$, $R = 0$ (representing a massless tachyon on the instantaneous Z-axis), but includes the two null hypersurfaces on which $T = \pm R$, which are cylindrical surfaces whose radius contracts and then expands at the speed of light. This null hypersurface corresponds in this limit to the horizon $\sigma = 2m$ in (9.2). It may also be noted that the coordinates of (9.3) only cover the region of Minkowski space with $|T| > R$, which is inside this contracting and expanding cylinder. Within this cylinder, σ is a timelike coordinate, as illustrated in Figure 9.1. (A further analysis of wave surfaces in this limit is given in Section 19.2.1.)

The solution (9.2), however, represents very different behaviour when $m \neq 0$. In this case, the solution includes a curvature singularity and, when $m > 0$, it also includes some kind of horizon. Nevertheless, the weak-field limit described above does reflect the behaviour of the coordinate foliations as $\sigma \to \infty$. In fact, the surfaces on which σ and t are constant are surfaces whose negative Gaussian curvature varies as σ^{-2} (see Appendix A), as in the Minkowski limit. However, this does not imply that the space-time is

---

[2] Gott (1974) has argued that those parts of the space-time for a tachyon that are not causally related to it may be represented by part of the space-time described by the BI-metric which will be considered in Section 16.4.1.

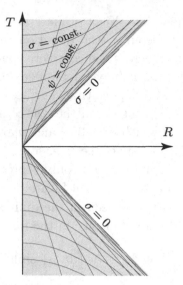

Fig. 9.1 The region of Minkowski space in cylindrical coordinates covered by the weak-field limit (9.3) of the *AII*-metric (9.2) with Z ($= t$) and ϕ constant. The surface $\sigma = 0$ represents a cylinder whose radius decreases to the instant at which the tachyon occurs and then increases at the speed of light.

asymptotically flat, as the source is not spatially bounded and σ is not a measure of distance from the source.

Using similar techniques to those that were employed to analyse the Schwarzschild space-time, the coordinate $\sigma_\star$ defined by

$$\sigma_\star = \int \left(\frac{2m}{\sigma} - 1\right)^{-1} \mathrm{d}\sigma = -\sigma - 2m \log\left|\frac{\sigma}{2m} - 1\right|, \qquad (9.5)$$

may temporarily be introduced. With this, the metric (9.2) becomes

$$\mathrm{d}s^2 = \left(\frac{2m}{\sigma} - 1\right)\left(-\mathrm{d}t^2 + \mathrm{d}\sigma_\star^2\right) + \sigma^2\left(\mathrm{d}\psi^2 + \sinh^2 \psi \, \mathrm{d}\phi^2\right). \qquad (9.6)$$

It is now natural to introduce the null coordinates

$$v = t + \sigma_\star, \qquad u = t - \sigma_\star.$$

These take the metric (9.2) to the alternative Robinson–Trautman (Eddington–Finkelstein-like) forms

$$\begin{aligned}
\mathrm{d}s^2 &= +2\, \mathrm{d}v\, \mathrm{d}\sigma - \left(\frac{2m}{\sigma} - 1\right)\mathrm{d}v^2 + \sigma^2\left(\mathrm{d}\psi^2 + \sinh^2 \psi \, \mathrm{d}\phi^2\right), \\
\mathrm{d}s^2 &= -2\, \mathrm{d}u\, \mathrm{d}\sigma - \left(\frac{2m}{\sigma} - 1\right)\mathrm{d}u^2 + \sigma^2\left(\mathrm{d}\psi^2 + \sinh^2 \psi \, \mathrm{d}\phi^2\right),
\end{aligned} \qquad (9.7)$$

which are clearly nonsingular at $\sigma = 2m$.

In either of the above forms, null geodesics on which ψ and ϕ are constant, are given, respectively, by

$$v = \text{const.}, \quad \text{or} \quad \left(\frac{2m}{\sigma} - 1\right)dv - 2\,d\sigma = 0,$$

and

$$u = \text{const.}, \quad \text{or} \quad \left(\frac{2m}{\sigma} - 1\right)du + 2\,d\sigma = 0.$$

In order to construct Eddington–Finkelstein-like diagrams for the two metric forms given in (9.7), it is convenient to introduce, respectively, the spacelike coordinates

$$\bar{z} = v + \sigma = t - 2m\log\left|\frac{\sigma}{2m} - 1\right|, \qquad \tilde{z} = u - \sigma = t + 2m\log\left|\frac{\sigma}{2m} - 1\right|.$$

Using these, the pairs of null geodesics with ψ, ϕ constant, are given by

$$d\bar{z} = d\sigma, \quad d\bar{z} = -\left(\frac{\sigma + 2m}{\sigma - 2m}\right)d\sigma,$$

and

$$d\tilde{z} = -d\sigma, \quad d\tilde{z} = \left(\frac{\sigma + 2m}{\sigma - 2m}\right)d\sigma.$$

The corresponding Eddington–Finkelstein-like diagrams are those given in Figure 9.2.

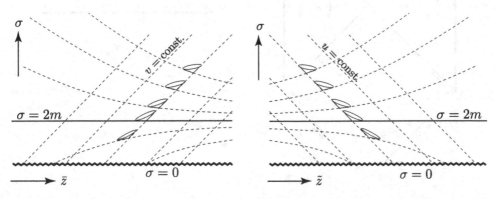

Fig. 9.2 Eddington–Finkelstein-like diagrams for the metrics (9.7) with ψ and ϕ constant. Some null lines are drawn together with some future light cones.

From the diagrams in Figure 9.2, it can be seen that the coordinate singularity at $\sigma = 2m$ does *not* correspond to what is normally considered as an event horizon since events inside the horizon could be seen by observers outside. However, events outside cannot be seen from inside. Of course, it is still a Killing horizon in the sense that the Killing vector ∂_t becomes degenerate on this null surface, and the space-time is static on one side and time-dependent on the other. The character of this horizon will be more transparent using conformal Kruskal–Szekeres-type coordinates.

In fact, it is now straightforward to obtain Kruskal–Szekeres-type coordinates using the exact analogue of the Schwarzschild case in which both the above null coordinates u and v are employed. Hence, the maximally extended AII space-times can be determined and it is possible to construct a conformal diagram for the 2-space in which ψ and ϕ are constants. The result is given in Figure 9.3 (see Gott, 1974, and Louko, 1987) for the two cases with either sign of m. For the case with $m > 0$, there are horizons at $\sigma = 2m$.

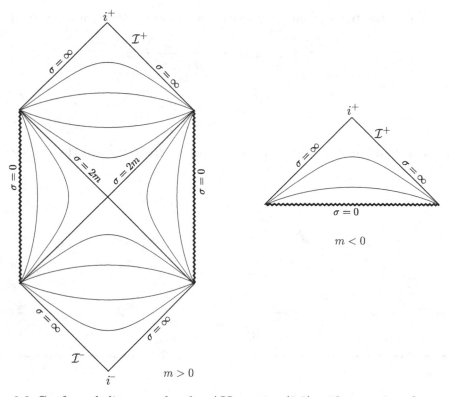

Fig. 9.3 Conformal diagrams for the AII-metrics (9.2) with $m > 0$ and $m < 0$. These represent sections on which ψ and ϕ are constant, so that each point on these diagrams represents a point on an infinite surface of constant negative curvature.

Since the only non-zero component of the curvature tensor is $\Psi_2 = m\sigma^{-3}$, it can be seen that curvature singularities occur at $\sigma = 0$ whenever $m \neq 0$. These are timelike for $m > 0$ and spacelike for $m < 0$. In both cases they are globally naked. It can also be seen that the space-time becomes flat as $m \to 0$, or as $\sigma \to \infty$ with any sign of m. However, σ is not a radial-type coordinate, and the space-time is not globally asymptotically flat.

When $m > 0$, the complete future outside the horizon is covered with $2m < \sigma < \infty$. However, as can be seen from the conformal diagram in Figure 9.3, the analytically extended space-time also includes an equivalent past region. With the inclusion of this identical but time-reversed region, the complete space-time is time symmetric. It can also be seen that it is possible for a test particle to travel on a timelike trajectory from past null infinity to future null infinity, passing through two horizons without encountering the singularity.

The alternative case in which $m < 0$ looks more like a cosmological model in which all timelike and null trajectories start in a (past) spacelike singularity and continue to a Minkowski-like future null infinity. In fact, it can be considered as a vacuum spatially homogeneous but anisotropic cosmological model that is of Bianchi type III, in which σ is a global time coordinate.

9.1.2 The AIII-metrics ($\epsilon = 0$)

Consider, finally, the remaining case of the A-metrics (9.1) in which $\epsilon = 0$. In this case, it is convenient to put $\zeta = \frac{1}{\sqrt{2}}\rho\, e^{i\phi}$, so that the line element becomes

$$ds^2 = \frac{2m}{r}\, dt^2 - \frac{r}{2m}\, dr^2 + r^2\Big(d\rho^2 + \rho^2 d\phi^2\Big). \tag{9.8}$$

Since a scaling freedom has not been applied to ϵ in this case, this freedom can be used here to set the parameter m to any convenient number. In fact, it is conventional to put either $2m = 1$ or $2m = -1$. However, the arbitrary form of the parameter will be retained at this stage so that the weak-field limit as $m \to 0$ can be considered (although it will be absorbed in most of the following transformations).

As for all members of this family of solutions with $m \neq 0$, a curvature singularity occurs at $r = 0$. It is therefore convenient to restrict r to the range $0 < r < \infty$, and to permit m to be either positive or negative. When $m > 0$, r can be seen to be a timelike coordinate. Alternatively, when $m < 0$, r is spacelike and the space-time is static. These two cases describe two distinct space-times which need to be considered separately.

For the time-dependent case in which $m > 0$, t is a spacelike coordinate and r timelike. In this case, the transformation

$$t = \left(\frac{3}{4m}\right)^{1/3} z, \qquad r = \left(\frac{9m}{2}\right)^{1/3} \tau^{2/3}, \qquad \rho\, e^{i\phi} = \left(\frac{2}{9m}\right)^{1/3} (x+iy),$$

puts the line element in the form

$$ds^2 = -d\tau^2 + \tau^{-2/3} dz^2 + \tau^{4/3}(dx^2 + dy^2), \tag{9.9}$$

which is the type D vacuum Kasner solution (22.1) with the exponents $(p_1, p_2, p_3) = (\frac{2}{3}, \frac{2}{3}, -\frac{1}{3})$. This particular solution may be interpreted as a vacuum homogeneous, but anisotropic cosmological model in which τ is a global time coordinate. In this case, the space-time originates in a spacelike big bang-type singularity and expands uniformly in the x and y directions while contracting in the z direction. The conformal diagram for this (and also the static) case is given in Figure 9.4.

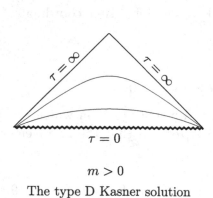

$$m > 0$$

The type D Kasner solution

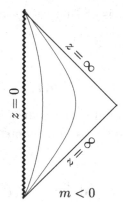

$$m < 0$$

The plane-symmetric static solution

Fig. 9.4 Conformal diagrams for the $AIII$ space-times (9.9) and (9.10) on sections on which x and y are constants. There are curvature singularities at $\tau = 0$ and $z = 0$ in the time-dependent and static cases, respectively.

For the alternative static case in which $m < 0$, one can put

$$t = \left(\frac{3}{4|m|}\right)^{1/3} t', \qquad r = \left(\frac{9|m|}{2}\right)^{1/3} z^{2/3}, \qquad \rho\, e^{i\phi} = \left(\frac{2}{9|m|}\right)^{1/3} (x+iy),$$

giving

$$ds^2 = -z^{-2/3} dt'^2 + dz^2 + z^{4/3}(dx^2 + dy^2), \tag{9.10}$$

which is a vacuum static plane-symmetric solution. Alternatively, putting

$$t = |m|^{-1/3}\, \tilde{t}, \qquad r = 2|m|^{1/3}\, \tilde{z}^{1/2}, \qquad \rho\, e^{i\phi} = \tfrac{1}{2}\, |m|^{-1/3}(\tilde{x} + i\tilde{y}),$$

yields

$$ds^2 = \tilde{z}^{-1/2}(-d\tilde{t}^2 + d\tilde{z}^2) + \tilde{z}(d\tilde{x}^2 + d\tilde{y}^2), \qquad (9.11)$$

which is the more familiar form of this solution as given by Taub in (1951). In fact, it is the only static vacuum solution with plane symmetry.

The physical interpretation of this space-time has been examined by many authors. See particularly Horský and Novotný (1982), Bonnor (1990b), Bedran *et al.* (1997) and references cited in these papers. The main conclusions are the following. The metric (9.11) is static and manifests (local) plane symmetry and a curvature singularity at $\tilde{z} = 0$. It therefore appears to represent the exterior solution to a source covering a static infinite plane at $\tilde{z} = 0$ (a domain wall). However, all timelike geodesics representing test particles appear to be repelled from the singularity. Thus, it seems that the source must have *negative* mass density. And, since an equivalent solution with positive mass density does not appear to exist, this interpretation must therefore be either suspect or unphysical.

Since the curvature tensor is proportional to m, the space-time must reduce to Minkowski space in the limit as $m \to 0$, but this is not obvious from the line element in any of the above forms. However, starting with the line element (9.8) with $m < 0$ and using the transformation

$$r = 2|m|^{-1/3}(1 - m\bar{z})^{1/2}, \qquad t = |m|^{-2/3}\,\bar{t}, \qquad \rho = \tfrac{1}{2}|m|^{1/3}\,\bar{\rho},$$

the metric becomes

$$ds^2 = \frac{1}{\sqrt{1 - m\bar{z}}}\left(-d\bar{t}^2 + d\bar{z}^2\right) + (1 - m\bar{z})\left(d\bar{\rho}^2 + \bar{\rho}^2\,d\phi^2\right), \qquad (9.12)$$

which is valid for $\bar{z} > -|m|^{-1}$. This form clearly reduces to Minkowski space (3.11) when $m = 0$. However, the curvature singularity now occurs when $\bar{z} = -|m|^{-1}$ and thus, in the weak-field limit, the source is located at spatial infinity. This limit therefore only clarifies that the space-time is Minkowski at a large distance from the source, whose character remains unclear. Nevertheless, it does indicate that the source cannot be an infinite plane surface whose constant mass density could become infinitesimally small.

It can also be shown that the metric (9.10) is a special case of the Levi-Civita solution which will be described in Section 10.2. When the coordinate y is periodic, this is normally interpreted as the static field of an infinite line source but, in this case, the gravitational mass per unit length has the specific negative value $\sigma = -\tfrac{1}{2}$ which cannot be varied or made arbitrarily small.

It is also possible to consider the alternative transformation

$$t = \frac{\bar{t}}{|m|^{1/3}}, \qquad r = \frac{|m|^{1/3}}{2}\left(\sqrt{\bar{r}^2 + \bar{z}^2} + \bar{z}\right), \qquad \rho = \frac{\left(\sqrt{\bar{r}^2 + \bar{z}^2} - \bar{z}\right)^{1/2}}{|m|^{1/3}},$$

which takes the metric (9.8) with $m < 0$ to the form

$$ds^2 = -\frac{1}{X}\,d\bar{t}^2 + \frac{2\,X^2}{\sqrt{\bar{r}^2 + \bar{z}^2}}\left(d\bar{r}^2 + d\bar{z}^2\right) + X\,\bar{r}^2\,d\phi^2, \qquad (9.13)$$

where $X = \frac{1}{4}\left(\sqrt{\bar{r}^2 + \bar{z}^2} + \bar{z}\right)$. This has transformed the line element to the static axially symmetric Weyl form that will be considered in Chapter 10. In this context, the source appears to be a semi-infinite rod located on the half-axis $\bar{r} = 0$, $\bar{z} < 0$ (Bonnor, 1990b). However, the mass density of this rod must also have a specific negative value. Both of these interpretations are therefore also suspect.

To the authors' knowledge, a totally satisfactory interpretation of this simple static metric has still not been found.

9.2 The Reissner–Nordström solution

Now consider the case in which the Schwarzschild solution is modified by the addition of an electric charge. An exact solution of the Einstein–Maxwell equations describing such a solution was found independently by Reissner (1916), Weyl (1917) and Nordström (1918), although it is now known only under the names of Reissner and Nordström. This is a static, spherically symmetric, asymptotically flat space-time that can be expressed in the form

$$ds^2 = -\left(1 - \frac{2m}{r} + \frac{e^2}{r^2}\right)dt^2 + \left(1 - \frac{2m}{r} + \frac{e^2}{r^2}\right)^{-1}dr^2 + r^2\left(d\theta^2 + \sin^2\theta\,d\phi^2\right),$$
$$(9.14)$$

where m and e are arbitrary parameters, and it may initially be assumed that $t \in (-\infty, \infty)$, $r \in (0, \infty)$, $\theta \in [0, \pi]$ and $\phi \in [0, 2\pi)$. This is accompanied by an electromagnetic field $F = dA$, with the vector potential given by

$$A = -\frac{e}{r}\,dt,$$

up to a duality rotation. Using Gauss's law, it can be seen that the parameter e may be interpreted as the charge of the source. This solution clearly reduces to that of Schwarzschild when $e = 0$, so that m can be interpreted as the mass of the source, at least in this limit. And, as for the Schwarzschild metric (8.1), r retains its interpretation as an areal coordinate.

In fact, there are charged versions of all the A-metrics which include the discrete 2-space curvature parameter $\epsilon = +1, -1, 0$ (see Section 9.6). However, only the case in which $\epsilon = +1$ will be considered here as this has the greatest physical significance.

It has been seen in the previous chapter that the Schwarzschild solution with $m > 0$ can be interpreted as the gravitational field of a black hole. Similarly, the Reissner–Nordström solution can be interpreted as the field of a *charged black hole*. Moreover, it has been shown to be the unique spherically symmetric, asymptotically flat solution of the Einstein–Maxwell equations (see e.g. Carter, 1987, Ruback, 1988 and Chruściel, 1994). It therefore follows that the exterior of any bounded spherically symmetric distribution of charged mass is described by this metric, which has just the two parameters m and e.

9.2.1 Singularity and horizons

Relative to an appropriately chosen tetrad generalising (8.2), the only non-zero components of the curvature tensor for the Reissner–Nordström metric (9.14) are given by

$$\Psi_2 = -\frac{m}{r^3} + \frac{e^2}{r^4}, \qquad \Phi_{11} = \frac{e^2}{2r^4}. \tag{9.15}$$

From this, it can be seen that, when m and e are non-zero, a curvature singularity occurs at $r = 0$. It is therefore appropriate to continue to restrict the range of r to $(0, \infty)$.

For physical reasons it will normally be assumed that $m > |e|$, as the mass of physical bodies is never negative and charge does not occur without an associated mass. Interestingly, this is also the condition under which additional (coordinate) singularities occur in the metric (9.14) at $r = r_+$ and $r = r_-$ where

$$r_\pm = m \pm \sqrt{m^2 - e^2}. \tag{9.16}$$

These are clearly Killing horizons associated with the Killing vector ∂_t. Their physical significance will be clarified below. In this notation, the relevant metric function $-g_{tt}(r)$ can be expressed as

$$1 - \frac{2m}{r} + \frac{e^2}{r^2} = \frac{(r - r_-)(r - r_+)}{r^2}. \tag{9.17}$$

This differs most significantly from the equivalent expression for the Schwarzschild solution when r is small.

For $r > r_+$, r is a spacelike coordinate, t is timelike, and the solution can be considered as a small perturbation of the Schwarzschild space-time

outside the horizon. The coordinate singularity at $r = r_+$ retains its interpretation as an event horizon. Inside this horizon, for $r_- < r < r_+$, r is
timelike and t spacelike. For the Reissner–Nordström metric, however, the
structure of the space-time nearer the curvature singularity is changed dra-
matically. Not only is there an additional horizon type of singularity at
$r = r_-$, but the innermost region in which $0 < r < r_-$ is again static: the
coordinate r becomes spacelike again and t is timelike. The root $r = r_+$ will
be referred to as the *outer horizon*, while the other root at $r = r_-$ will be
referred to as the *inner horizon*.

In this case, the curvature singularity at $r = 0$ is point-like, just like that
of a point mass in Newtonian theory. Its worldline is timelike. And, at any
time, it can be surrounded by a sphere of arbitrarily small radius. This
structure is in marked contrast with the singularity of the Schwarzschild
black hole, or indeed with that of the (Kerr) rotating black hole that will
be described in Chapter 11.

In order to clarify the character of the two horizons and the global structure of the space-time, it is appropriate to introduce Eddington–Finkelstein-
type coordinates. As for the Schwarzschild solution, it is convenient to temporarily introduce the coordinate

$$r_\star = \int \frac{\mathrm{d}r}{-g_{tt}(r)} = r + \frac{r_+^2}{(r_+ - r_-)} \log\left|\frac{r}{r_+} - 1\right| - \frac{r_-^2}{(r_+ - r_-)} \log\left|\frac{r}{r_-} - 1\right|, \quad (9.18)$$

and the retarded and advanced null coordinates

$$u = t - r_\star, \qquad v = t + r_\star. \qquad (9.19)$$

Using only the advanced null coordinate, the line element (9.14) can be
re-expressed in the form

$$\mathrm{d}s^2 = 2\,\mathrm{d}v\,\mathrm{d}r - \left(1 - \frac{2m}{r} + \frac{e^2}{r^2}\right)\mathrm{d}v^2 + r^2\left(\mathrm{d}\theta^2 + \sin^2\theta\,\mathrm{d}\phi^2\right),$$

which is regular at both horizons $r = r_\pm$. Incoming and outgoing null
geodesics can now be easily identified and an Eddington–Finkelstein-like
diagram can be constructed, as shown in Figure 9.5. This generalises Figure 8.1 and demonstrates the different structure near the singularity. (A
time-reversed version of this, generalising Figure 8.3 and representing a
charged white hole, may also be constructed using the retarded null coordinate u.)

From the light-cone structure that is apparent in Figure 9.5, it may be
seen that any body which passes through the outer (event) horizon r_+ must
also pass through the entire time-dependent region, and then through (or

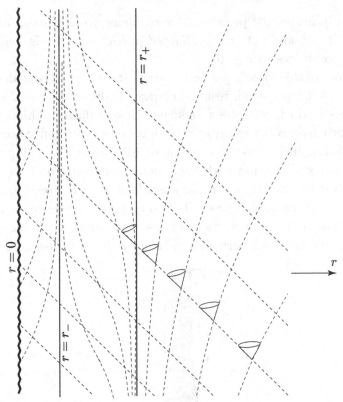

Fig. 9.5 A schematic representation of radial null geodesics in Eddington–Finkelstein-like coordinates with θ and ϕ constant. Ingoing and outgoing null geodesics are represented by the broken lines. Except near the curvature singularity $r = 0$, the vertical axis is taken to be $\bar{t} = t + r_* - r$. However, near the singularity inside the inner horizon, $\bar{t}$ is spacelike and may be replaced by the timelike coordinate t.

possibly just as far as) the inner horizon r_-. Once inside the inner horizon, the space-time is again static and the body need not reach the curvature singularity at $r = 0$.

Now consider the geodesic equations for the motion of an uncharged particle in the space-time described by the metric (9.14). Confining attention to the plane $\theta = \pi/2$, the equations that are equivalent to (8.3) for the Schwarzschild metric are given by

$$\dot{r}^2 + \left(1 - \frac{2m}{r} + \frac{e^2}{r^2}\right)\left(\epsilon_0 + \frac{L^2}{r^2}\right) = E^2, \qquad r^2\dot{\phi} = L,$$

where E and L are constants of the motion, and $\epsilon_0 = 1$ for a timelike geodesic or $\epsilon_0 = 0$ for a null geodesic. This includes the familiar potentials for a Newtonian attraction and the general relativistic correction. However, when

$e \neq 0$, it also indicates the presence of a *repulsive gravitational force* that is proportional to e^2 and acts on *uncharged* particles. This is negligible for large r, but dominates as $r \to 0$.

The presence of this force is particularly significant when considering a collapsing body that has a small residual charge. Unlike the case of a spherical uncharged body which collapses to a spacelike singularity behind a horizon, a charged body may collapse through both its outer and inner horizons, but the unbounded repulsive force as $r \to 0$ generally prevents a collapse to a curvature singularity. The outer surface of such a collapsing body is illustrated in Figure 9.6 using Eddington–Finkelstein-type coordinates (equivalent to Figure 8.2 for the uncharged case). It is important to emphasise, however, that this situation still represents a collapse to a black hole since the body disappears inside an event horizon relative to a distant observer.

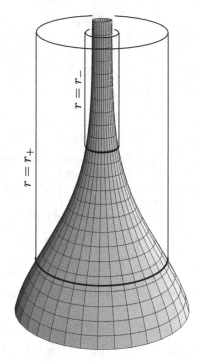

Fig. 9.6 A representation of the outer surface of a collapsing spherical charged body in Eddington–Finkelstein-like coordinates (with $\theta = \pi/2$). If the body collapses through its event horizon $r = r_+$, it will necessarily pass through its inner horizon $r = r_-$ but the outer surface will not reach the curvature singularity $r = 0$.

The occurrence of this short-range repulsive force acting inside the horizons of the Reissner–Nordström space-time has not been widely emphasised in the literature to date. However, the effects of its presence have been noted. A reversal of the collapse of a charged fluid sphere has been noted by Novikov (1966) and Bardeen (1968). However, this reversal occurs inside the event horizon so that it is not observed externally. (For further details of collapsing charged fluid spheres, see e.g. Bekenstein, 1971.) In the same way, de la Cruz and Israel (1967) and Boulware (1973) have noted that a collapsing shell of matter bounces reversibly at non-zero minimal radius. The motion of test particles in this space-time has been further investigated, for example, by Cohen and Gautreau (1979).

In terms of the double-null coordinates (9.19), the metric (9.14) reduces to the form

$$ds^2 = -\frac{(r-r_-)(r-r_+)}{r^2}\, du\, dv + r^2 \left(d\theta^2 + \sin^2\theta\, d\phi^2\right), \tag{9.20}$$

where r is given implicitly in terms of u and v by the above expressions. This metric is badly behaved both at $r = r_+$ and at $r = r_-$. However, Kruskal–Szekeres-like coordinates can be found with which it becomes regular across either of these horizons, although it is not possible to remove both of these coordinate singularities simultaneously.

In order to cover the outer event horizon at $r = r_+$, it is first convenient to introduce new null coordinates U_+ and V_+ defined by

$$U_+ = -\frac{2r_+^2}{(r_+-r_-)} \exp\left(-\frac{(r_+-r_-)}{2r_+^2}\, u\right), \quad V_+ = \frac{2r_+^2}{(r_+-r_-)} \exp\left(\frac{(r_+-r_-)}{2r_+^2}\, v\right),$$

so that

$$U_+ = -\frac{2r_+^2}{(r_+-r_-)} \left|\frac{r-r_+}{r_+}\right|^{1/2} \left|\frac{r-}{r-r_-}\right|^{r_-^2/2r_+^2} \exp\left(-\frac{(r_+-r_-)}{2r_+^2}(t-r)\right),$$

$$V_+ = \frac{2r_+^2}{(r_+-r_-)} \left|\frac{r-r_+}{r_+}\right|^{1/2} \left|\frac{r-}{r-r_-}\right|^{r_-^2/2r_+^2} \exp\left(\frac{(r_+-r_-)}{2r_+^2}(t+r)\right).$$

It is then possible to introduce Kruskal–Szekeres-like coordinates T_+ and R_+ by putting

$$T_+ = \tfrac{1}{2}(V_+ + U_+), \qquad R_+ = \tfrac{1}{2}(V_+ - U_+),$$

in the stationary region, with an equivalent definition in the time-dependent region as in (8.11). With these, the metric (9.20) becomes

$$ds^2 = \frac{r_-r_+}{r^2} \left(\frac{r-r_-}{r_-}\right)^{1+\frac{r_-^2}{r_+^2}} \exp\left(-\frac{(r_+-r_-)}{r_+^2}\, r\right)\left(-dT_+^2 + dR_+^2\right)$$
$$+ r^2 \left(d\theta^2 + \sin^2\theta\, d\phi^2\right),$$

which is continuous across the outer horizon $r = r_+$, although it still has a coordinate singularity at $r = r_-$. In fact, this form of the metric covers the full part of the space-time from the inner horizon at $r = r_-$, through the outer horizon $r = r_+$ and out to $r = \infty$. In particular, it should be noted that the outer horizon corresponds to both null hypersurfaces $U_+ = 0$ (i.e. $T_+ - R_+ = 0$) and $V_+ = 0$ (i.e. $T_+ + R_+ = 0$). These represent the white hole and black hole event horizons, respectively.

For the inner structure, alternative null coordinates can be defined by

$$U_- = \frac{2\,r_-^2}{(r_+-r_-)} \exp\left(\frac{(r_+-r_-)}{2\,r_-^2} u\right), \quad V_- = -\frac{2\,r_-^2}{(r_+-r_-)} \exp\left(-\frac{(r_+-r_-)}{2\,r_-^2} v\right),$$

so that

$$U_- = \frac{2\,r_-^2}{(r_+-r_-)} \left|\frac{r_+}{r-r_+}\right|^{r_+^2/2r_-^2} \left|\frac{r-r_-}{r_-}\right|^{1/2} \exp\left(\frac{(r_+-r_-)}{2\,r_-^2}(t-r)\right),$$

$$V_- = -\frac{2\,r_-^2}{(r_+-r_-)} \left|\frac{r_+}{r-r_+}\right|^{r_+^2/2r_-^2} \left|\frac{r-r_-}{r_-}\right|^{1/2} \exp\left(-\frac{(r_+-r_-)}{2\,r_-^2}(t+r)\right).$$

Then, putting $U_- = T_- + R_-$ and $V_- = T_- - R_-$ in the time-dependent region, the metric (9.20) becomes

$$ds^2 = \frac{r_-r_+}{r^2}\left(\frac{r_+ - r}{r_+}\right)^{1+\frac{r_+^2}{r_-^2}} \exp\left(\frac{(r_+ - r_-)}{r_-^2} r\right)\left(-\,dT_-^2 + dR_-^2\right)$$
$$+ r^2\left(d\theta^2 + \sin^2\theta\,d\phi^2\right),$$

which is continuous across the inner horizon $r = r_-$. This form of the metric covers the region of the space-time from the curvature singularity at $r = 0$, through the inner horizon at $r = r_-$ and as far as the outer event horizon. As in the above case, it should be noted that the inner horizon at $r = r_-$ corresponds to two null hypersurfaces $U_- = 0$ (i.e. $T_- + R_- = 0$) and $V_- = 0$ (i.e. $T_- - R_- = 0$). Moreover, these null hypersurfaces separate four distinct regions. One of these is a time-dependent region that connects with the two distinct exterior static regions at $r = r_+$. It also connects to two distinct static regions inside the inner horizon, which can each be continued as far as timelike singularities at $r = 0$. And these two regions also connect to a second distinct time-dependent region that is between these inner horizons and some new outer horizon.

9.2.2 Conformal structure

Having separately obtained Kruskal–Szekeres-like coordinates for the regions of the Reissner–Nordström space-time in which $r_- < r < \infty$ and $0 < r < r_+$, it is now possible to perform a compactification of these regions in exactly the same way as that performed for the Schwarzschild space-time in (8.14).

This indicates the same asymptotic structure as shown in the conformal diagram in Figure 8.5. However, the structure inside the outer horizon is now significantly different.

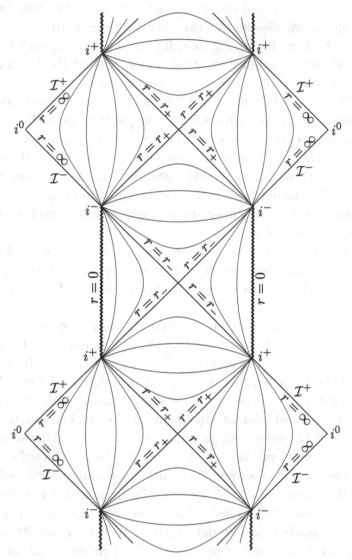

Fig. 9.7 Conformal diagram for the maximally extended Reissner–Nordström space-time is an infinite sequence of components of the type shown. This represents a section on which θ and ϕ are constant. The lines illustrated indicate hypersurfaces on which r is a constant.

After compactifying the coordinates, and by combining different (overlapping) coordinate patches, the initial part of the space-time which is outside the outer horizon can be extended through each horizon to construct a conformal diagram for the complete space-time.[3] This is given in Figure 9.7 following Graves and Brill (1960) (see also Walker, 1970). It shows that a complete maximally extended Reissner–Nordström space-time consists of an infinite sequence of identical structures each containing two exterior regions, inner and outer horizons and timelike curvature singularities.

It can be seen from Figure 9.7 that, for any exterior region $r > r_+$, there simultaneously exists another identical region beyond the horizons. However, a real particle cannot traverse from one to the other on a timelike worldline. This is exactly like the Schwarzschild space-time whose conformal diagram is given in Figure 8.5. A real particle can follow a timelike worldline from an exterior region through the event horizon at $r = r_+$ and then through either branches of the inner horizon at $r = r_-$. Beyond this it would be in a static region containing a naked curvature singularity at $r = 0$. It is possible that the particle's motion will terminate in the singularity but, in view of the repulsive character of the singularity, any uncharged particle following a geodesic will pass through the other $r = r_-$ horizon and emerge through an outer white hole horizon into another asymptotically flat exterior region similar to that from which it started. Various embedding diagrams for hypersurfaces in this space-time have been described by Jacob and Piran (2006).

It can also be seen from Figure 9.7 that the outer horizon at $r = r_+$ has the character of an *event horizon* relative to conformal infinity $\mathcal{I}^+$ or timelike infinity i^+: events inside $r = r_+$ cannot be seen at $\mathcal{I}^+$ or i^+. To consider the character of the inner horizon at $r = r_-$, it should first be noted that the inner region $0 < r < r_-$ is partly bounded by a naked (timelike) singularity at $r = 0$. The causal past of any event in this region can be extended to include both the intermediate region $r_- < r < r_+$ and the singularity itself. Thus, the null hypersurface at $r = r_-$ can be interpreted as a *Cauchy horizon* in the sense that all points outside $r = r_-$ are determined uniquely by Cauchy data given on some prior spacelike section through the complete space-time. However, points inside $r = r_-$ are determined, not only by the same Cauchy data but also by data relating to the singularity itself. The Cauchy horizon is the limit of the unique development of initial data on a complete regular section through the space-time.

Just as part of the Schwarzschild space-time can be taken to represent the exterior field of a collapsing spherical distribution of mass, so part of the Reissner–Nordström space-time can be taken to represent the exterior of a collapsing spherical distribution of charged matter that forms a charged

---

[3] A coordinate system which covers more than one horizon has been given by Israel (1966b) and Klösch and Strobl (1996) (see also Lake, 2006).

black hole. The conformal diagram for this part of the space-time is illustrated in Figure 9.8, at least as far as the Cauchy horizon. This may be the only region of the space-time that has a practical physical interpretation.

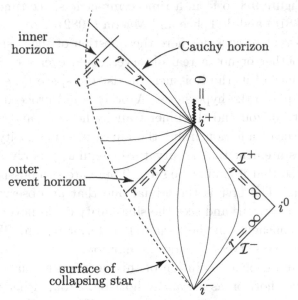

Fig. 9.8 The conformal diagram for that part of the Reissner–Nordström space-time that represents the exterior of a collapsing spherical distribution of charged matter. The external region may be extended through the outer (event) horizon r_+ and, possibly, through the inner horizon r_-, but the surface of the collapsing body does not reach a curvature singularity before meeting a Cauchy horizon. Although the exact solution is extendable through the Cauchy horizon, this horizon is unstable and so may be regarded as part of the boundary of the physical space-time.

Since this topic is clearly important, it is appropriate to include a few additional comments here. Whereas an uncharged spherical body collapses to a singularity in a finite time, this does not seem to happen if some charge is included. In this situation, there is an outer (event) horizon at $r = r_+$. A body whose outer surface collapses through this horizon must continue to collapse through the inner horizon. It cannot stop at some size between these horizons as the r-coordinate there is timelike. However, a distinction needs to be made between the two different types of horizon at which $r = r_-$. One is an inner (Killing) horizon inside which the space-time is locally static. The other is the Cauchy horizon that appears as a result of the conformal compactification. This is the limit of the causal development of the entire infinite external space, and there are good reasons to regard this as the boundary of the physically significant region of space-time (see Poisson and Israel, 1990). Clearly, if a charged body collapses through its event horizon, it must also collapse through the inner Killing horizon. It may then reach a minimum size and start to expand. However, the expansion could be analytically extended beyond the Cauchy horizon, and this could be pictured as extending Figure 9.8 to form a time-symmetric space-time. For details, see Hiscock (1981c) and de Felice and Maeda (1982).

It is also relevant to consider here the question of the cosmic censorship hypothesis: whether or not a real observer could ever see a naked singularity. On the face of it, the Reissner–Nordström space-time could provide a counter-example to the hypothesis. A body could proceed through both the outer event horizon and the inner Cauchy horizon, and would then be able to see (receive radiation from) the curvature singularity. This would contradict the statement that a curvature singularity is always hidden behind a horizon so that it cannot be seen. There are two possible resolutions of this question. The first is the trivial one that an observer who passes through the two horizons and sees the singularity could never communicate his findings to colleagues in the (original) exterior region. The singularity remains hidden to observers outside the horizon.

The second and deeper resolution of this question is related to the character of the inner horizon as a Cauchy horizon. From either of the Figures 9.7 or 9.8, it can be seen that the Cauchy horizon may be considered as a "continuation" of conformal null infinity $\mathcal{I}^+$. Since this encloses a compactification of the entire infinite external region, all information from distant asymptotic regions would finally be seen in a very short time as an observer approached the Cauchy horizon. They would thus appear to be infinitely blue-shifted (see Poisson and Israel, 1990). This would suggest that, in any perturbation of the space-time, the field near the Cauchy horizon would diverge. In fact, it was demonstrated by Simpson and Penrose (1973), Matzner, Zamorano and Sandberg (1979) and Chandrasekhar and Hartle (1982) that the inner horizon is unstable with respect to small perturbations of the space-time. (A simple calculation illustrating this property has been given by Bičák, 2000a.) Thus, in any real space-time, an observer approaching the Cauchy horizon would experience unbounded tidal forces and would therefore not survive to observe the naked singularity. It is for this reason that the physical space-time outside a collapsing charged spherical body has only been illustrated in Figure 9.8 as far as the Cauchy horizon.

9.2.3 (Hyper-)extreme Reissner–Nordström space-times

The above discussion of the Reissner–Nordström space-time has concerned the physically most applicable case in which $m > |e|$. However, the alternative cases in which the arbitrary parameters satisfy different conditions should also be investigated. The curvature tensor components (9.15) imply that the curvature singularity at $r = 0$ remains in all cases with $m, e \neq 0$.

In the *hyperextreme* case, for which $e^2 > m^2$, the metric can be re-expressed in terms of double-null coordinates in the form

$$\mathrm{d}s^2 = - \left(1 - \frac{2m}{r} + \frac{e^2}{r^2} \right) \mathrm{d}u\,\mathrm{d}v + r^2 \left(\mathrm{d}\theta^2 + \sin^2 \theta \,\mathrm{d}\phi^2 \right), \qquad (9.21)$$

where r is given implicitly in terms of $u = t - r_\star$ and $v = t + r_\star$ with $r_\star = \int \frac{1}{-2g_{uv}}\,\mathrm{d}r$. In this case, unlike (9.20), the metric function $g_{uv}(r)$ has no real roots. There are therefore no horizons and the space-time is regular and static for all $r \in (0, \infty)$. For this case, the double-null coordinates can be compactified simply by putting

$$\tilde{u} = 2\arctan\left(\frac{u}{4m} \right), \qquad \tilde{v} = 2\arctan\left(\frac{v}{4m} \right), \qquad (9.22)$$

and the resulting conformal diagram is given in Figure 9.9. It can immediately be seen from this that the (timelike) curvature singularity at $r = 0$ is globally naked.

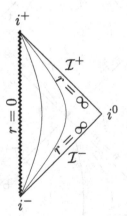

Fig. 9.9 Conformal diagram for the hyperextreme $e^2 > m^2$ Reissner–Nordström space-time with θ and ϕ constant.

In the remaining (intermediate) *extreme* case for which $e^2 = m^2$, the metric function $g_{uv}(r)$ has a single repeated root at $r = m$. This is still a Killing horizon where the norm of ∂_t vanishes. However, the space-time on both sides of this horizon is static. In this case, the extensions across the horizon can easily be obtained using Eddington–Finkelstein-like forms of the metric. Using either a retarded or an advanced coordinate $u = t - r_\star$ or $v = t + r_\star$

where now

$$r_\star = \frac{r(r-2m)}{(r-m)} + 2m \log \left| \frac{r}{m} - 1 \right|, \qquad (9.23)$$

the line element takes either of the forms

$$ds^2 = -2\,du\,dr - \left(1 - \frac{m}{r}\right)^2 du^2 + r^2 \left(d\theta^2 + \sin^2\theta\,d\phi^2\right),$$

or

$$ds^2 = +2\,dv\,dr - \left(1 - \frac{m}{r}\right)^2 dv^2 + r^2 \left(d\theta^2 + \sin^2\theta\,d\phi^2\right),$$

respectively. These are both well behaved across the horizon at $r = m$. As first shown by Carter (1966a), these coordinates can be compactified by again applying the transformation (9.22). The conformal diagram for the extended space-time can then be constructed by applying different coordinate

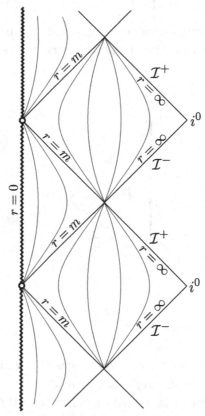

Fig. 9.10 Conformal diagram for the extreme $e^2 = m^2$ Reissner–Nordström space-time with θ and ϕ constant, showing that the curvature singularity at $r = 0$ is globally naked.

patches across the horizons for both the advanced and the retarded coordinates. However, after extending across the horizon to the interior region in each case, further extensions to different exterior regions are also possible. This process can be continued indefinitely, showing that the maximally extended manifold consists of an infinite sequence of domains, as illustrated in Figure 9.10. This also demonstrates that the curvature singularity at $r = 0$ is globally naked.

Clearly the metric for this case can be expressed in terms of double-null coordinates in the form

$$ds^2 = -\left(1 - \frac{m}{r}\right)^2 du\,dv + r^2\left(d\theta^2 + \sin^2\theta\,d\phi^2\right), \qquad (9.24)$$

where $r = r(r_\star(u, v))$. However, as pointed out by Carter (1973), since the expression (9.23) for $r_\star$ contains a pole as well as the familiar logarithmic singularity, it is not possible to express this explicitly in terms of Kruskal–Szekeres-type coordinates, which cover both forms of the horizon. In fact, although it can be shown that this metric form is continuous across the horizons, it is not analytic.

This is the first example of a *degenerate* horizon. It may be considered as a case in which the event horizon and the Cauchy horizon of the case with $m^2 > e^2$ "coincide". Such horizons have vanishing surface gravity. However, in this case, the white hole horizon also corresponds to the Cauchy horizon of a previously existing external region, and the black hole horizon additionally corresponds to the past Cauchy horizon of a different future external region.

Carter (1973) has also demonstrated that the region of the extreme Reissner–Nordström space-time near the degenerate horizon $r = m$ resembles the Bertotti–Robinson space-time (see Section 7.1).

9.3 The Schwarzschild–Melvin solution

A different one-parameter electromagnetic generalisation of the Schwarzschild space-time may now be mentioned – one in which the black hole is immersed in a background electric or magnetic field. This solution of the Einstein–Maxwell equations is a special case of that obtained by Ernst (1976a), see Section 14.2.1, that can be written in the form

$$ds^2 = D^2\left[-\left(1 - \frac{2m}{r}\right)dt^2 + \left(1 - \frac{2m}{r}\right)^{-1}dr^2 + r^2 d\theta^2\right] + D^{-2}r^2\sin^2\theta\,d\phi^2,$$
$$(9.25)$$

where $D = 1 + \frac{1}{4}B^2 r^2 \sin^2\theta$, and the strength of the electric or magnetic field is given by B.

This solution has just two parameters, m and B. When $B = 0$, it is the Schwarzschild solution (8.1) in which m is the mass of the source. Alternatively, when $m = 0$, it is the Melvin solution (7.30) which represents a uniform electric or magnetic field of strength B as described in Section 7.3. There is just a single (event) horizon when $r = 2m$. The space-time is not asymptotically flat when $B \neq 0$. As shown explicitly by Bose and Estaban (1981) and Pravda and Zaslavskii (2005), it is algebraically general (except when m or B vanish, in which case it is of type D).

In view of these properties, the solution (9.25) can be interpreted as a black hole immersed in a uniform magnetic or electric field. It is referred to in the literature as the Schwarzschild–Melvin solution or the magnetised Schwarzschild solution. Wild and Kerns (1980) have shown that the horizon at $r = 2m$ becomes increasingly prolate for an increasing magnetic field,[4] although its area remains constant depending only on m. Moreover, as the horizon is "stretched" in this way, a region around the equator will acquire negative Gaussian curvature if $mB > 1$.

For further details, properties and applications of this space-time, see e.g. Hiscock (1981d), Bičák and Janiš (1985) and Stuchlík and Hledík (1999). Dadhich, Hoenselaers and Vishveshwara (1979) have determined the trajectories of charged particles in this space-time. The chaotic motion of test particles has been reported by Karas and Vokrouhlický (1992).

9.4 The Schwarzschild–de Sitter solution

Now consider the generalisation of the Schwarzschild solution which, in addition to mass parameter m, includes an arbitrary cosmological constant Λ. The metric for this case was discovered by Kottler (1918), Weyl (1919b) and Trefftz (1922), and can be written in the form

$$
ds^2 = -\left(1 - \frac{2m}{r} - \frac{\Lambda}{3}r^2\right)dt^2 + \left(1 - \frac{2m}{r} - \frac{\Lambda}{3}r^2\right)^{-1}dr^2
$$
$$
+ r^2\left(d\theta^2 + \sin^2\theta\, d\phi^2\right),
$$
(9.26)

where, $t \in (-\infty, \infty)$, $r \in (0, \infty)$, $\theta \in [0, \pi]$ and $\phi \in [0, 2\pi)$. This clearly reduces to the Schwarzschild metric (8.1) when $\Lambda = 0$, and to the de Sitter or anti-de Sitter metrics in their spherically symmetric forms (4.9) or (5.6) when $m = 0$. When $m \neq 0$, it has a curvature singularity at $r = 0$.

For small values of r, this solution approximates to the Schwarzschild

---

[4] Similar properties have also been found by Wild, Kerns and Drish (1981) for a rotating generalisation of this space-time (i.e. a Kerr–Melvin solution) first obtained by Ernst and Wild (1976). The magnetic field thus has the opposite effect to that of the rotation.

solution which can be regarded as representing the exterior solution of a spherical body or black hole. For large r, it behaves asymptotically as either the de Sitter solution or the anti-de Sitter solution according to the sign of Λ. The metric (9.26) can therefore be thought of as representing the exterior field of a spherical mass (or black hole) in a de Sitter or anti-de Sitter background. However, as for the Schwarzschild space-time, the complete analytic extension of the space-time may be considered, ignoring its relation to any physical massive source.

To analyse this solution and to perform its global extension, it is convenient initially to concentrate on the metric function

$$g_{tt}(r) = \frac{1}{r}\left(\frac{\Lambda}{3}r^3 - r + 2m\right),\tag{9.27}$$

which is positive for both small and large values of r if $\Lambda > 0$. This indicates that, in these domains, r is a timelike coordinate while t is a spatial coordinate, and the space-time there is time-dependent. Between these domains (and generically for small values of Λm^2), $g_{tt} < 0$ so that, in this "ordinary" region, r is spatial and t temporal and the space-time is static.

It is now necessary to consider separately the various possible cases in which the value of Λm^2 occurs in different ranges.

9.4.1 Black holes in de Sitter space-time

When $0 < 9\Lambda m^2 < 1$, the metric function (9.27) admits two distinct positive real roots at $r = r_b$ and $r = r_c$, ordered such that $r_c > r_b > 0$. The third root is negative since there is no quadratic term in the bracket in (9.27), and thus the sum of the three roots must vanish. The smaller root $r = r_b$ corresponds to a black hole-type event horizon (see Section 8.2), while the larger root $r = r_c$ corresponds to a cosmological de Sitter-type event horizon (see Section 4.3), beyond which the space-time is again time-dependent. The metric is static in the region $r_b < r < r_c$ between the two event horizons.

With this notation, the metric function (9.27) can be re-written as

$$-g_{tt} = \frac{\Lambda}{3r}(r - r_b)(r_c - r)(r + r_b + r_c).$$

This form was used, e.g., by Gibbons and Hawking (1977), Lake and Roeder (1977), and Laue and Weiss (1977) for a qualitative analysis of possible extensions of the metric. As in (8.4), the tortoise coordinate $r_\star = \int(-g_{tt})^{-1}\,dr$ is defined, and the double-null coordinates $u = t - r_\star$, $v = t + r_\star$ are subsequently introduced. The Kruskal–Szekeres-type extension across the two horizons r_b and r_c is then considered and the schematic Penrose diagram which represents the global conformal structure of the Schwarzschild–de Sitter space-time is thus obtained. This is illustrated in Figure 9.11.

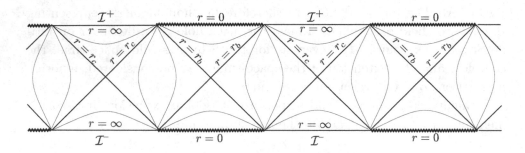

Fig. 9.11 A schematic Penrose diagram for the Schwarzschild–de Sitter space-time when $0 < 9\Lambda m^2 < 1$. This contains an infinite sequence of Schwarzschild-like and de Sitter-like regions. All lines shown have constant values of r.

An explicit analytic extension of the manifold across both horizons has been obtained by Geyer (1980), Bażański and Ferrari (1986), and Bičák and Podolský (1995). This confirms that $r = r_b$ is a Schwarzschild-like black/white hole horizon, and that $r = r_c$ is a de Sitter-like cosmological horizon. The area of the black hole horizon is $4\pi r_b^2$, while that of the cosmological horizon is $4\pi r_c^2$. When all (compactified) Kruskal–Szekeres-like coordinates in different sections of a conformal diagram are combined, the analytic extension of the manifold is obtained. This is bounded by curvature singularities at $r = 0$ and conformal infinities $\mathcal{I}^\pm$ at $r = +\infty$. Moreover, the maximum extended space-time contains an infinite sequence of Schwarzschild-like and de Sitter-like regions, as indicated in Figure 9.11. In the Schwarzschild-like regions, the curvature singularity at $r = 0$ is located behind black and white hole event horizons, and another causally separated cosmological exterior region exists as an analytic extension on any spacelike section that does not reach the singularity. Each exterior cosmological region is de Sitter-like with a spacelike conformal infinity $\mathcal{I}$.

The conformal diagram in Figure 9.11 is clearly spatially periodic. Thus, it is possible to identify corresponding timelike sections. The simplest case is illustrated in Figure 9.12. Alternatively, a space-time could be constructed to include an arbitrary number of Schwarzschild-like black/white holes in an identical number of de Sitter-like backgrounds. However, no observer in such a space-time would ever be able to detect more than a single "period". Moreover, the parameters of each period would be identical.

Geodesics in this space-time have been described by Jaklitsch, Hellaby and

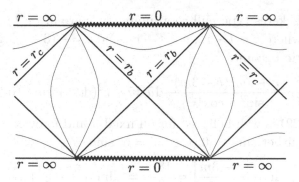

Fig. 9.12 A schematic Penrose diagram for the Schwarzschild–de Sitter space-time in which the spatial edges are identified. This would represent a single Schwarzschild-like black/white hole in a single de Sitter-like universe.

Matravers (1989) and explicitly determined by Hackmann and Lämmerzahl (2008).

An interior solution representing a collapsing sphere of dust in a Schwarzschild–de Sitter background, generalising that of Oppenheimer and Snyder, has been described by Nakao (1992). He also showed that a very large dust sphere of low density does not collapse, as the repulsive effect of the cosmological constant can exceed the mutual gravitational attraction of the dust.

9.4.2 Extreme holes in de Sitter space-time

In the above section, it has been assumed that $0 < 9\Lambda m^2 < 1$. As the value of Λm^2 increases to the upper limit, the position of the black/white hole horizon r_b monotonically increases and the cosmological horizon r_c decreases to the common value $3m$. As pointed out by Hayward, Shirumizo and Nakao (1994), this implies that, for any given value of the cosmological constant, there is a maximum size for the area of a black hole horizon. Here we briefly mention this extreme case of the Schwarzschild–de Sitter space-time (9.26) which is characterised by the condition $9\Lambda m^2 = 1$. In such a situation there exists only one degenerate "double" Killing horizon at the null hypersurface

$$r = r_b = r_c = 3m = 1/\sqrt{\Lambda}.$$

Everywhere else the coordinate r is temporal while t is spatial.

The conformal diagram for this particular case can be constructed explicitly as follows: the tortoise coordinate takes the form

$$r_\star = \frac{9m^2}{r - 3m} + 2m \log \left| \frac{r + 6m}{r - 3m} \right|, \tag{9.28}$$

and Kruskal–Szekeres-type null coordinates $\tilde{u}$, $\tilde{v}$ are given by $u = c\,m \cot \tilde{u}$, $v = c\,m \tan \tilde{v}$, where $u = t - r_\star$, $v = t + r_\star$, and $c = -(3 - 2\log 2) < 0$. With these, the metric becomes

$$ds^2 = -\frac{c^2}{27r}\frac{(r + 6m)(r - 3m)^2}{\sin^2 \tilde{u} \cos^2 \tilde{v}}\,d\tilde{u}\,d\tilde{v} + r^2(d\theta^2 + \sin^2 \theta\,d\phi^2)\,. \tag{9.29}$$

The metric (9.29) is well behaved for all fixed $\tilde{u}$ and $\tilde{v}$, even on the horizons $r = 3m$ where either $\sin \tilde{u} = 0$ or $\cos \tilde{v} = 0$, since

$$\lim_{r \to 3m}\left|\frac{r - 3m}{\sin \tilde{u}}\right| = \frac{18m}{|c|} = \lim_{r \to 3m}\left|\frac{r - 3m}{\cos \tilde{v}}\right|\,. \tag{9.30}$$

The corresponding diagram in the $(\tilde{u}, \tilde{v})$-space is illustrated schematically in Figure 9.13. The part (b) represents a white hole model which evolves from an initial curvature singularity towards a final de Sitter state. Alternatively, by a simple reflection $\tilde{u} \to -\tilde{u}$, $\tilde{v} \to -\tilde{v}$, a black hole space-time (a) with a future singularity may be constructed. The causal structure is evident. Observers "emanate" from the singularity at $r = 0$ (or from the

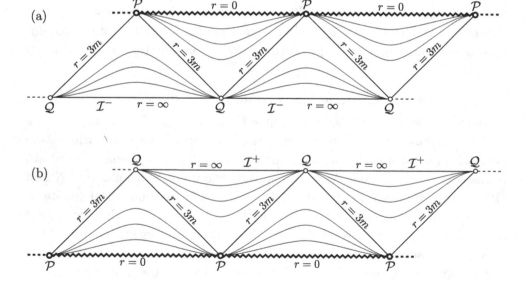

Fig. 9.13 Schematic conformal Penrose diagrams for the extreme Schwarzschild–de Sitter space-time with $9\Lambda m^2 = 1$ in which the horizons are degenerate. The coordinate r is timelike everywhere except at the horizons $r = 3m$ where it is null. (a) The black hole case in which the singularity at $r = 0$ is in the future. The maximal analytic extension of the geometry is obtained by gluing an infinite number of regions shown in the figure, or by joining a finite number of regions via appropriate identifications. (b) The time-reversed diagram ($\tilde{u} \to -\tilde{u}$, $\tilde{v} \to -\tilde{v}$), corresponding to white holes with a past singularity.

"points" $\mathcal{P}$) and, after crossing the horizon, reach the future infinity $\mathcal{I}^+$ ($r = \infty$) or the asymptotic "points" $\mathcal{Q}$. In the black hole case, any timelike geodesic observer falling from the region $r > 3m$ (or the infinity $\mathcal{I}^-$ given by $r = \infty$) will either cross the horizon $r = 3m$ and reach the singularity at $r = 0$, or escape to one of the "asymptotic points" $\mathcal{P}$ given by $u = -\infty$, $v = +\infty$.

Further details concerning this extreme Schwarzschild–de Sitter space-time, such as the character of the asymptotic points $\mathcal{P}$ and $\mathcal{Q}$ investigated by possible geodesic motions, can be found in Podolský (1999) and references therein.

In the remaining (hyperextreme) case for which $9\Lambda m^2 > 1$ there are no horizons. The corresponding Penrose conformal diagram has been given by Geyer (1980) and Bičák and Podolský (1997).

9.4.3 Black holes in anti-de Sitter space-time

For the case when $\Lambda < 0$, it can be seen that the metric function (9.27) must have just a single positive root. This follows from the fact that the sum of the roots must be zero, and that $\frac{\Lambda}{3}r^3 - r + 2m$ is positive when $r = 0$. (We are assuming here that $0 < r < \infty$, and that $m > 0$, so that the space-time has a traditional Schwarzschild limit as $\Lambda \to 0$.) This single root r_b corresponds to a Schwarzschild-like black/white hole horizon.

The schematic conformal diagram is given in Figure 9.14, which indicates that conformal infinity at $r = \infty$ is timelike corresponding to the asymptotic anti-de Sitter structure (see Chapter 5). For an explicit construction of the conformal diagram see, e.g., Lake and Roeder (1977), Geyer (1980), and Bičák and Podolský (1997). Geodesic motions in this space-time have

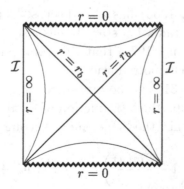

Fig. 9.14 A schematic Penrose diagram for the Schwarzschild–anti-de Sitter space-time. In this case, conformal infinity is timelike.

been described by Cruz, Olivares and Villanueva (2005) and Hackmann and Lämmerzahl (2008).

9.5 Vaidya's radiating Schwarzschild solution

Another important generalisation of the Schwarzschild solution arose from the desire to develop a model of the exterior of a star, which includes its radiation field. Such a "radiating Schwarzschild metric" was found by Vaidya (1943) in the form

$$\mathrm{d}s^2 = -\frac{\dot{m}^2}{f^2}\left(1 - \frac{2m}{r}\right)\mathrm{d}t^2 + \left(1 - \frac{2m}{r}\right)^{-1}\mathrm{d}r^2 + r^2\left(\mathrm{d}\theta^2 + \sin^2\theta\,\mathrm{d}\phi^2\right), \quad (9.31)$$

where $m = m(r, t)$,

$$f(m) = m'\left(1 - \frac{2m}{r}\right),$$

and the dot and prime denote derivatives with respect to t and r, respectively. This space-time has a non-zero energy-momentum tensor of the form (2.25) which corresponds to pure radiation, which is outgoing if $\dot{m} < 0$. The function $m(r, t)$ can be interpreted as the total energy of mass and radiation inside a sphere of areal radius r at time t. It is therefore also appropriate to assume that $m > 0$ and $m' < 0$. The function $f(m)$ may be interpreted as the luminosity of the star in the Newtonian limit.

Vaidya (1951a) has also interpreted this as an exact solution of the Einstein–Maxwell equations, but this would imply the existence of a null and singular radiation of charge. Thus, the source is better interpreted as a central mass which emits a spherical distribution of pure radiation. The interpretation of this solution has been investigated by Raychaudhuri (1953) and Israel (1958). Vaidya (1951b, 1955) has also discussed how to apply boundary conditions relevant to a radiating star. It has also been clarified (Griffiths, 1974; Griffiths and Newing, 1974) that this solution could represent either incoherent photon or neutrino radiation.

In fact, the metric (9.31) can be expressed in a much more useful form (for outgoing radiation) by the introduction of a null coordinate u such that $\mathrm{d}u = -\frac{1}{f(m)}\,\mathrm{d}m$. With this, the line element becomes

$$\mathrm{d}s^2 = -2\,\mathrm{d}u\,\mathrm{d}r - \left(1 - \frac{2m(u)}{r}\right)\mathrm{d}u^2 + r^2\left(\mathrm{d}\theta^2 + \sin^2\theta\,\mathrm{d}\phi^2\right), \quad (9.32)$$

which is clearly a generalisation of the Schwarzschild metric (8.8) in Eddington–Finkelstein-like coordinates for the white hole situation. In this case,

$m(u)$ is now an arbitrary non-increasing function of the retarded null coordinate. This form of the metric was first given by Vaidya (1953). The non-zero components of the curvature tensor are given by

$$\Psi_2 = -\frac{m}{r^3}, \qquad \Phi_{22} = -\frac{m_{,u}}{r^2}.$$

The space-time in this form has been thoroughly analysed by Lindquist, Schwartz and Misner (1965) who have confirmed that $-m_{,u}$ is the luminosity of the star as seen by a distant observer. They have also argued that, since the solution (9.32) corresponds to the white hole situation, the hypersurface $r = 2m(u)$ cannot be realised physically. But this conclusion is in fact misleading and will be clarified below.

For the moment, consider the Vaidya space-time (9.32) without considering any particular physical source (i.e. extend the space-time up to the singularity $r = 0$). In this case, the metric contains a decreasing function $m(u)$, which may be assumed to remain positive. Apart from the presence here of pure radiation, the main difference between this and the Schwarzschild space-time is that the coordinate singularity at $r = 2m(u)$ is not a null hypersurface and therefore cannot be an event horizon. In fact, it is an example of an *apparent horizon*.[5]

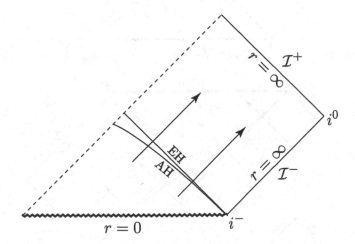

Fig. 9.15 Penrose diagram for the Vaidya space-time representing a white hole in which the radiation propagates outward along null worldlines, indicated by lines with arrows. The apparent horizon (AH) at $r = 2m(u)$ is distinct from the event horizon (EH). An extension to the future, across the broken line, is required.

---

[5] An apparent horizon is technically defined as the hypersurface separating the regions with and without closed trapped surfaces. See Hawking and Ellis (1973) and Wald (1984).

Such an apparent horizon at $r = 2m(u)$ in the Vaidya space-time is space-like whenever m is a decreasing function (it is null when m is constant). Thus, in the presence of radiation, the apparent horizon separates from the past event horizon, which is always null.

The concept of an event horizon is determined by the global asymptotic structure of the space-time and is a null hypersurface separating events that will not be observed at infinity. Similarly, the white hole event horizon is the boundary of the region that can be connected to past timelike infinity i^- by timelike and null lines. The conformal structure for the radiating white hole situation is given in Figure 9.15, following Fayos, Martín-Prats and Senovilla

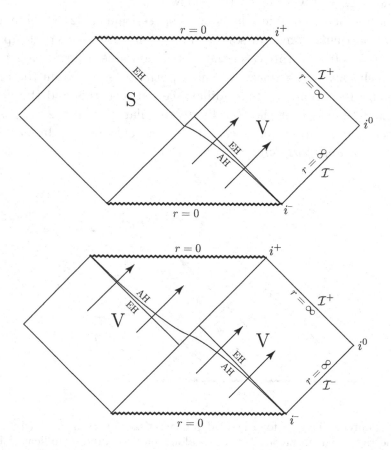

Fig. 9.16 Two possible extensions of the Vaidya space-time. The first is extended with part of the Schwarzschild space-time (S) in which m becomes a constant and a black hole forms. In the second, the Vaidya region (V) is extended with a similar time-reversed region which also forms a black hole, but which contains incoming radiation that appears to originate from the asymptotic region of the alternative exterior.

(1995), in which the apparent horizon (AH) is given by $r = 2m(u)$, but the (past) event horizon (EH) is a null hypersurface from i^-.

It can immediately be seen that the conformal diagram in Figure 9.15 is not complete. It represents the situation in which the parameter m has an initial value on $\mathcal{I}^-$ which decreases to some final positive value. But it is not clear what happens after that. An extension is required. If m subsequently remains constant, an appropriate section of the Schwarzschild space-time can be attached. Israel (1967) was the first to consider such an extension, and Fayos, Martín-Prats and Senovilla (1995) have presented alternative extensions. They have also argued that an analytic extension is not generally to be preferred. Two different extensions are illustrated in Figure 9.16.

For the case in which the function $m(u)$ reduces to zero at some value $u = u_1$, the apparent horizon at $r = 2m(u)$ will eventually reach $r = 0$ at $u = u_1$. However, the extension beyond $u = u_1$ is unclear. In fact, this is more easily discussed in the time-reversed situation for incoming radiation as will be described below in Section 9.5.2.

9.5.1 Radiation from a star

Now consider the exterior field of a radiating star. Using the above metric, this contains a pure radiation field with energy-momentum tensor given by

$$T_{\mu\nu} = -\frac{m_{,u}}{4\pi r^2} k_\mu k_\nu,$$

where $k_\mu = -u_{,\mu}$, so that $\boldsymbol{k} = \partial_r$. The space-time is clearly that of Schwarzschild when m is a constant. The solution can therefore be interpreted as the exterior field of a massive object whose mass is being radiated away in the form of pure radiation. Boundary conditions to an interior solution can be set up on the spherical surface of the body matching it to some appropriate interior solution.

For a body of large mass which emits a small amount of radiation while collapsing to form a black hole, the conformal diagram of the exterior space-time is given in Figure 9.17. It can be seen that this has an identical structure to that of Figure 8.6 for the vacuum case, even though this now contains radiation. The only qualitative difference is that the event horizon (as seen from infinity) is now different from the apparent horizon. Radiation from the star can initially ultimately reach conformal infinity, but in this case the surface of the collapsing body will pass through an event horizon and reach a curvature singularity in a finite time after which the space-time cannot

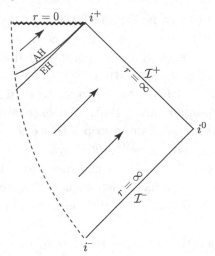

Fig. 9.17 Penrose diagram for the Vaidya space-time for the case when a collapsing radiating object results in a black hole. The space-time contains outgoing null radiation. The apparent horizon (AH) at $r = 2m(u)$ is distinct from the event horizon (EH) which is a null hypersurface from i^+.

be continued. After the surface of the star has passed through its horizon, subsequent radiation from it will only end in the curvature singularity.

On the other hand, one can also consider a situation in which a spherical body will emit so much matter that m will quickly reduce to zero. Since m is interpreted as the mass of the body, and it is assumed that this cannot be negative, radiation will cease when $m = 0$ and the subsequent space-time will be flat Minkowski space. In fact, two possible scenarios arise. If the surface of the body passes through its Schwarzschild horizon, then a black hole is formed and the structure of the space-time will be that illustrated in Figure 9.17. On the other hand, if the outer surface of the body remains outside its Schwarzschild horizon while its volume reduces to zero, then the structure of the space-time will be that illustrated in Figure 9.18 in which a Vaidya region is matched to a subsequent Minkowski region.

Which of these two scenarios occurs in practice will depend on the internal structure of the body and the physical processes which give rise to the radiation. These topics are not considered here.

It should also be noted that the Vaidya solution as described above has been used, with different extensions, to model Hawking radiation from an evaporating black hole in the semi-classical approximation. See Hiscock (1981a,b), Kuroda (1984), Beciu (1984) and Zhao, Yang and Ren (1994).

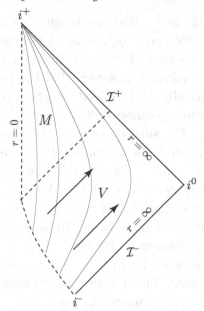

Fig. 9.18 Penrose diagram for the case in which the surface of a collapsing radiating object remains outside its Schwarzschild radius so that no black hole is formed and $m(u)$ reduces to zero. The space-time contains outgoing null radiation in a Vaidya region (V) which is matched to a subsequent Minkowski region (M).

For a recent study which includes a model of the interior of a collapsing body, see Fayos and Torres (2008).

9.5.2 The Vaidya solution for incoming radiation

Now consider the alternative situation in which an initial part of Minkowski space is surrounded by a region of *incoming* null radiation. The Minkowski region will therefore exist prior to a null cone, which forms a boundary to a future region that can be represented by the Vaidya solution with incoming radiation. In other words, the incoming null matter has some initial wavefront across which the Vaidya solution can be matched to a Minkowski interior.

All the incoming null matter will eventually meet at or near $r = 0$. When the particles composing the matter collide, it is possible that they could form a finite-sized spherical body whose radius is greater that its Schwarzschild radius. The size of this body would obviously grow as more matter falls in. The structure of the exterior field would then be similar to the time-inverse of that described in Figure 9.18, although the interior solution would involve different physical processes.

Alternatively, it could occur that the incoming null matter collapses to a point. This would necessarily be inside its Schwarzschild radius and a black hole would thus be formed with a central curvature singularity and an increasing mass. It is found, however, that the precise structure of the space-time depends critically on the initial density of the incoming radiation.

To clarify the situation, assume that the incoming null radiation only occurs for a finite time. The subsequent space-time will be spherically symmetric with an accumulated central mass. This must necessarily be part of a Schwarzschild space-time. The situation thus described can be represented as a compound space-time in which a Vaidya solution is matched to Minkowski space on an initial (inner) null cone, and subsequently to part of a Schwarzschild space-time on a second (outer) null cone. This has been described in detail by Papapetrou (1985).

The line element corresponding to incoming radiation can be obtained from (9.31) by alternatively introducing the advanced null coordinate v defined by $\mathrm{d}v = \frac{1}{f(m)}\,\mathrm{d}m$. This leads to the form

$$\mathrm{d}s^2 = 2\,\mathrm{d}v\,\mathrm{d}r - \left(1 - \frac{2m(v)}{r}\right)\mathrm{d}v^2 + r^2\left(\mathrm{d}\theta^2 + \sin^2\theta\,\mathrm{d}\phi^2\right). \tag{9.33}$$

The compound space-time can then be represented everywhere by this form of the metric, where $m(v)$ is zero in the Minkowski region for which $v < 0$, increases through the Vaidya region where $0 \le v \le v_1$, and is constant in the Schwarzschild region where $v > v_1$. For the Vaidya region, Papapetrou considered the (self-similar) case for which m is a linearly increasing function of v, but other models are also instructive.

Of course, the final singularity in the Schwarzschild region must be hidden behind a horizon. The null hypersurface corresponding to this horizon can then be extended back through the Vaidya region and will reach $r = 0$ after a finite interval. This point cannot be inside the Vaidya region. Since even the first point at which m becomes non-zero will have an effective event horizon, the later Schwarzschild horizon must extend right back at least to this point.

It is also appropriate to consider the future null cone which is attached to the final point $r = 0$ within the Minkowski region. It is possible that this may be contained inside the horizon of the subsequent Schwarzschild region, but it may alternatively extend to future null infinity. Three distinct possible cases arise.

The first case, which is illustrated schematically in Figure 9.19, corresponds to that in which the mass at the centre initially accumulates at a slow rate. In this case, the family of null geodesics which originate at the

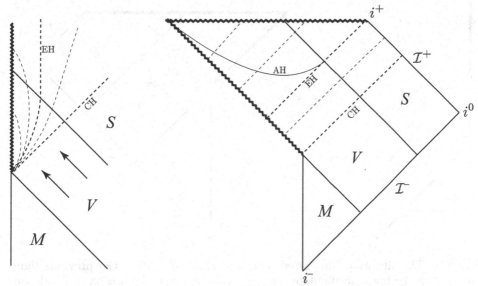

Fig. 9.19 A section V represented by a Vaidya solution with incoming null matter is matched to Minkowski space M on an initial (inner) null cone, and subsequently to part of a Schwarzschild space-time S. (The coordinates θ and ϕ are suppressed.) The figure on the left represents the situation in Eddington–Finkelstein-type coordinates. The figure on the right is the corresponding conformal diagram. The Cauchy horizon (CH), event horizon (EH) and apparent horizon (AH), and some null cones originating at the initial formation of the singularity, are also indicated.

initial point of collapse are not entirely contained inside the event horizon and span the entire causal future. The initial null cone which extends from the singularity to null infinity is a Cauchy horizon. (In this case, the radius of this cone at any time is greater than the Schwarzschild radius associated with the total mass inside it.) Other null cones between this and the extended black hole event horizon extend to future null infinity, while the remainder end back at the curvature singularity either in the Schwarzschild or the Vaidya regions. Since this initial point of collapse could be seen by distant observers outside the subsequent event horizon, it therefore appears as a *globally naked singularity*. This is indicated schematically in the right conformal diagram in Figure 9.19. It may be noted that, since there is a family of outgoing radial null geodesics emanating from the initial point of the singularity, this singularity must be null as indicated in the figure.

In a second case, represented in Figure 9.20, the collapsing null matter has a sufficiently large density that the null cone from the final point of the Minkowski region propagates to a region which is inside the subsequent Schwarzschild horizon. (Or the event horizon of the Schwarzschild region

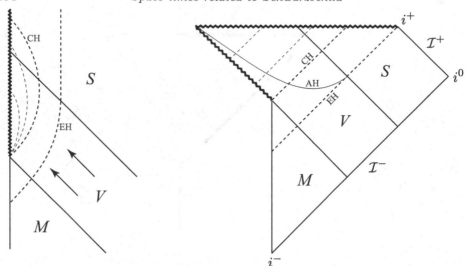

Fig. 9.20 The situation illustrated here is similar to that of the previous figure except that the mass accumulates more quickly at $r = 0$. In this case, the density of the incoming null matter is sufficiently large that the horizon of the subsequent Schwarzschild region extends to $r = 0$ at a point within the Minkowski region.

extends back right through the Vaidya region to the Minkowski region.) Thus the curvature singularity at $r = 0$ in the Vaidya region is entirely hidden behind an event horizon, as would be seen from infinity. Nevertheless, some null geodesics do extend from the initial point of the singularity. (The initial point of the singularity is again null.) Since these geodesics collapse back to the singularity and do not extend to infinity, this situation is referred to as a *locally naked singularity*.

In the third and final case, the collapsing null matter has such a large density that null rays cannot escape at all. Even null geodesics from the final point of the Minkowski region are immediately captured by the singularity. Thus the curvature singularity at $r = 0$ in the Vaidya region is entirely censored. Such a situation is illustrated schematically in Figure 9.21.

The appearance of the naked singularity which occurs in this context in the first case (as illustrated in Figure 9.19) was first pointed out by Steinmüller, King and Lasota (1975). And since its occurrence contradicts the cosmic censorship hypothesis, it has subsequently received significant detailed attention. Most of this has been based on the Papapetrou (1985) self-similar model (see e.g. Nolan, 2001). Dwivedi and Joshi (1989, 1991) have examined the growth of curvature along causal geodesics, and Nolan and Waters (2005) and Nolan (2007) have shown that the Cauchy horizon for this model is stable with respect to non-spherically symmetric perturbations

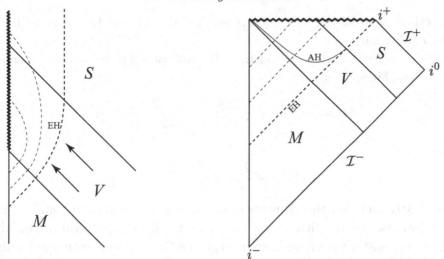

Fig. 9.21 In the situation illustrated here, the mass accumulates so quickly at $r = 0$ that no null rays can emerge from the singularity. The apparent horizon in the Vaidya region V necessarily meets the event horizon at the junction with the Schwarzschild region S and the origin of the Minkowski region M.

of the space-time, while the event horizon is unstable. Further work to clarify this point is in progress.

The easy answer to the apparent contradiction with the cosmic censorship hypothesis, however, is simply that, when null matter collapses, it may realistically be expected to initially form a solid spherical lump of finite size (greater than its Schwarzschild horizon). This lump would grow through the Vaidya region, and a curvature singularity just does not arise. However, it may also be pointed out that, even if the null radiation collapses to a point, the naked singularity is just the single event at which the singularity initially forms. The null cones from this event span the entire causal future. Thus, even if such a situation were to exist, any (test) radiation from this event would have negligible amplitude when seen by a distant observer.

It should be mentioned finally that the problem of expressing the Vaidya metric in double-null coordinates has been considered in detail by Waugh and Lake (1986) and Girotto and Saa (2004).

9.6 Further generalisations

The above generalisations of the Schwarzschild space-time have been obtained either by introducing new parameters (ϵ, e, B or Λ) or by permitting m to vary. These have been considered one at a time in the previous subsections, but all the cases described can also be considered together in any combination.

For the case in which m is constant, the additional parameters ϵ, e and Λ may be combined to give

$$ds^2 = -Q\,dt^2 + Q^{-1}dr^2 + r^2\frac{2\,d\zeta\,d\bar{\zeta}}{(1 + \frac{1}{2}\epsilon\zeta\bar{\zeta})^2}, \tag{9.34}$$

where

$$Q = \epsilon - \frac{2m}{r} + \frac{e^2}{r^2} - \frac{\Lambda}{3}r^2.$$

This clearly includes the A-metrics, the Reissner–Nordström solution and the Schwarzschild–de Sitter solution. This family of space-times contains surfaces of constant positive, zero or negative Gaussian curvature according to whether $\epsilon = +1, 0, -1$, respectively. As $r \to \infty$ (or when m and e vanish), they approach the de Sitter or anti-de Sitter spaces using the different foliations that are described in detail below in Section 19.2.2.

When $\epsilon = +1$, these space-times are spherically symmetric, and it is convenient to put $\zeta = \sqrt{2}\tan(\theta/2)e^{i\phi}$. The metric (9.34) may then be described as the Reissner–Nordström–de Sitter solution. It represents a black hole-type solution, which generally possess three horizons according to the non-zero values of the parameters m, e and Λ. These correspond to outer (event) and inner[6] (Cauchy) black hole horizons and a cosmological horizon. It may therefore be interpreted as the space-time of a charged black hole in a de Sitter or anti-de Sitter universe. Its extensions and conformal structures have been analysed by Laue and Weiss (1977) and Brill and Hayward (1994).

In particular, it may be noted that these three horizons still occur in the special case in which $e = m$ if $\Lambda > 0$. In this case, the black hole horizons do not degenerate as they do in the extreme Reissner–Nordström solution with $\Lambda = 0$, as described in Subsection 9.2.3. Such space-times are said to describe *lukewarm black holes* because the surface gravity on the event horizon is the same as that on the cosmological horizon. Solutions in which the inner and outer black hole horizons coincide are called *cold*. Those with a triply degenerate horizon are called *ultracold*. For further details of these properties see Mellor and Moss (1989), Romans (1992), Kastor and Traschen (1993) and Brill *et al.* (1994).

Apart from the first section on the A-metrics, this chapter has concentrated on solutions for which $\epsilon = +1$. Such space-times are spherically symmetric and admit surfaces of constant positive curvature. However, solu-

---

[6] For a discussion of the stability of the inner Cauchy horizon in Reissner–Nordström–de Sitter space-times, see Mellor and Moss (1990, 1992), Brady and Poisson (1992) and Chambers (1997).

tions with alternative values of ϵ have also been investigated and a few brief comments on these are now included. Such solutions are still of algebraic type D and possess a curvature singularity at $r = 0$ if m and/or e are non-zero. If $\Lambda \neq 0$, they may possess up to three horizons which correspond to generalised forms of inner and outer black hole horizons and a cosmological horizon according to the values (and signs) of m, e and Λ.

Solutions of this type for which $\epsilon = 0$ have surfaces of zero curvature. Their metric can be expressed in the form

$$ds^2 = -Q\,dt^2 + Q^{-1}dr^2 + r^2(dx^2 + dy^2),$$

where $Q = -2mr^{-1} + e^2r^{-2} - \frac{1}{3}\Lambda r^2$. If it is assumed that $x, y \in (-\infty, \infty)$, then these space-times have plane symmetry. However, the coordinates x and y could also be considered to have finite range and the hypersurfaces with their limiting values could be identified. In this way space-times with the different topologies of planes ($R^1 \times R^1$), cylinders ($R^1 \times S^1$) or tori ($S^1 \times S^1$) could be constructed. Such space-times with a negative cosmological constant are often referred to as *black membrane solutions, black string solutions* or *toroidal black holes*, respectively. Originally given by Kar (1926), their properties have been investigated by Lemos (1995a,b), Huang and Liang (1995), Cai and Zhang (1996), Bañados (1998) and others. Moreover, Lemos (1995b) and Lemos and Zanchin (1996) have constructed a rotating, charged cylindrical black hole, or black string solution, by performing a coordinate transformation before making the topological identification, and have investigated some of its properties.

The case in which $\epsilon = -1$ that has surfaces of constant negative curvature may also be considered. Such space-times are often referred to as being quasi-spherical. Their metric can be expressed in the form

$$ds^2 = -Q\,dt^2 + Q^{-1}dr^2 + r^2(d\psi^2 + \sinh^2\psi\,d\phi^2),$$

where $Q = -1 - 2mr^{-1} + e^2r^{-2} - \frac{1}{3}\Lambda r^2$. These solutions are sometimes known as *topological black holes*. However, in view of the interpretation given to the *AII*-metric in Section 9.1.1 (and supported by the description of the foliation of the background given in Section 19.2.2), these could more appropriately be interpreted as representing the field of a tachyon in a de Sitter or anti-de Sitter background. Particular attention has been focused on the case in which $\Lambda < 0$, for which the region with large r is static. For conformal diagrams of appropriate sections and other properties of these space-times, see Brill, Louko and Peldán (1997), Holst and Peldán (1997), Mann (1997), Smith and Mann (1997) and Vanzo (1997).

To include the Vaidya solution in this general family of space-times, it is appropriate to introduce a null coordinate u and to consider a generalised form of the metric (9.32) which can be expressed as

$$ds^2 = -2\,du\,dr - Q\,du^2 + r^2 \frac{2\,d\zeta\,d\bar{\zeta}}{(1 + \frac{1}{2}\epsilon\zeta\bar{\zeta})^2}, \tag{9.35}$$

where

$$Q = \epsilon - \frac{2m(u)}{r} + \frac{e^2(u)}{r^2} - \frac{\Lambda}{3}r^2. \tag{9.36}$$

These solutions are all contained in the family of Robinson–Trautman space-times of type D that will be described in Chapter 19, where the foliations of the background and the particular case of the Vaidya–(anti-)de Sitter solution (19.51) will be explicitly mentioned. Space-times with the metric (9.35) with (9.36) generally contain null charged radiation (see Frolov, 1974a), as well as a cosmological constant.

Finally, it must be mentioned that some other one-parameter generalisations of the Schwarzschild space-time also exist. The introduction of a rotation parameter leads to the Kerr space-time that will be discussed in Chapter 11. A NUT parameter can also be introduced as described in Chapter 12. It is also possible to introduce an acceleration parameter, leading to a space-time that will be considered in detail in Chapter 14.

In fact, it is possible to combine all these parameters in a single form of the metric. This is given by the general family of Plebański–Demiański solutions. These are all of algebraic type D and include seven physical parameters. They will be described in detail, including certain important subcases, in Chapter 16.

10

Static axially symmetric space-times

Very shortly after Einstein published his field equations for general relativity, Weyl (1917) and Levi-Civita (1918) considered static and axially symmetric solutions. These will be briefly reviewed in this chapter with emphasis on a few important special cases, particularly those that are asymptotically flat at large distances either from the axis or from a finite number of sources.

It will be seen that, in spite of their obvious and simple symmetry properties, the physical interpretation of these solutions is generally far from trivial. For example, problems sometimes arise from the identification of points associated with the periodic coordinate. Without such identifications, these space-times with both spatial and temporal symmetries may be alternatively interpreted as being plane symmetric. Further problems arise because, in the natural coordinates for the assumed symmetry, the behaviour of any curvature singularities generally appears to depend on the direction from which they are approached. This occurs for singularities on the axis and elsewhere.

10.1 The Weyl metric

A general line element possessing both the timelike and axial symmetries, generated by the commuting Killing vector fields ∂_t and ∂_ϕ, can be expressed in the canonical form

$$ds^2 = -e^{2U}dt^2 + e^{-2U}\left[e^{2\gamma}(d\eta^2 + d\xi^2) + \rho^2 d\phi^2\right], \qquad (10.1)$$

where U, γ and ρ depend only on the coordinates η and ξ. The vacuum field equations with $\Lambda = 0$ imply that $\rho(\eta, \xi)$ is an arbitrary harmonic function satisfying the two-dimensional Laplace equation $\rho_{,\eta\eta} + \rho_{,\xi\xi} = 0$. It is then convenient to adopt *Weyl's canonical coordinates* ρ and z, where $z(\eta, \xi)$ is the harmonic conjugate to ρ.[1] Although these are sometimes referred to as "cylindrical coordinates", they do not necessarily have such an interpretation in the full four-dimensional space-time. In terms of these coordinates, the metric is expressed in the form

$$ds^2 = -e^{2U} dt^2 + e^{-2U} \left[e^{2\gamma}(d\rho^2 + dz^2) + \rho^2 d\phi^2 \right], \qquad (10.2)$$

where U and γ are now functions of ρ and z, and it is usually assumed that $t \in (-\infty, \infty)$, $\rho \in [0, \infty)$, $z \in (-\infty, \infty)$ and $\phi \in [0, 2\pi)$.

With $\Lambda = 0$, the vacuum field equations for the metric (10.2) imply that

$$U_{,\rho\rho} + \frac{1}{\rho} U_{,\rho} + U_{,zz} = 0. \qquad (10.3)$$

This may be recognised as Laplace's equation $\nabla^2 U = 0$ for an axially symmetric function in an unphysical Euclidean 3-space in cylindrical polar coordinates, though the coordinates ρ, z, ϕ here have a different meaning.

The basic equation (10.3) is a *linear* equation whose general solution can be expressed in a variety of forms. For any particular solution for U, the remaining metric function γ can be obtained, at least in principle, by evaluating the quadratures

$$\gamma_{,\rho} = \rho \left(U_{,\rho}^{\;2} + U_{,z}^{\;2} \right), \qquad \gamma_{,z} = 2\rho U_{,\rho} U_{,z}. \qquad (10.4)$$

These equations are automatically integrable in view of (10.3). In this sense, all the vacuum solutions of this class are formally known.

In the linear approximation, the metric (10.2) implies that $g_{tt} \sim -1 - 2U$, and U satisfies Laplace's equation (10.3). It is therefore natural to regard U as the analogue of a Newtonian potential. This prompts the following solution-generating procedure:

- Start with the exact Newtonian potential for some classical axially symmetric system in a flat space expressed in terms of standard cylindrical ρ, z coordinates.
- Take this as the appropriate expression for U in the metric (10.2) and calculate the remaining metric function γ using (10.4).
- Interpret the solution as the gravitational field for the analogous Newtonian source.

Unfortunately, as will be seen, the last step does not straightforwardly lead to the correct physical interpretation of the space-time derived. In fact, the

---

[1] Weyl's canonical coordinates ρ and z are thus both harmonic functions satisfying Laplace's equation such that $\rho + iz$ is analytic.

sources of the Weyl solutions generally possess a number of subtleties such that most are still not fully understood.

It can be seen from (10.2) that the "axis" on which $\rho = 0$ is regular if, and only if, $\gamma \to 0$ as $\rho \to 0$.[2] If this condition is not satisfied for some value or range of z, then some kind of singularity occurs at these points.

Another important feature that should be noticed immediately about the metric (10.2) is that the corresponding space-time is necessarily static for arbitrary values of the functions U and γ. In these coordinates, the metric cannot therefore be extended through any horizons that may occur. It can only describe static regions.

Normally, solutions of Weyl's class are algebraically general. However, static regions of all solutions of type D can also be expressed in terms of the line element (10.2). Some of these will be given explicitly below, either in the following sections of this chapter or in appropriate sections of later chapters.

10.1.1 Flat solutions within a Weyl metric

Minkowski space in cylindrical polar coordinates (3.11) is obviously included in the Weyl metric (10.2) when

$$U = 0, \qquad \gamma = 0.$$

However, the solution

$$U = \log \rho, \qquad \gamma = \log \rho,$$

leads to the metric $\mathrm{d}s^2 = -\rho^2\,\mathrm{d}t^2 + \mathrm{d}\rho^2 + \mathrm{d}z^2 + \mathrm{d}\phi^2$, which also describes (part of) Minkowski space-time, in the form (3.14) which is expressed in terms of coordinates that are adapted to uniform acceleration, as described in Section 3.5. In this case, ρ, z and ϕ are all Cartesian-like coordinates, and trajectories with ρ, z and ϕ constant have uniform acceleration in the ρ-direction. As will be shown in the following section, this is also a special case of the Levi-Civita metric (10.9), with (10.10), whose Newtonian interpretation of the potential function U corresponds to an infinite line source along the axis, for which the mass per unit length is given by $\sigma = \frac{1}{2}$.

In addition, Gautreau and Hoffman (1969) have shown that there is just one further solution, namely

$$U = \tfrac{1}{2}\log\left(\sqrt{\rho^2 + z^2} + z\right), \qquad \gamma = \tfrac{1}{2}\log\left(\frac{\sqrt{\rho^2 + z^2} + z}{2\sqrt{\rho^2 + z^2}}\right), \qquad (10.5)$$

---

[2] For a discussion of conditions for the existence of an axis of symmetry, see Van den Bergh and Wils (1985).

which also represents a flat space-time. In this case, the Newtonian interpretation of the potential U is that of a semi-infinite line source located on the negative z-axis, and whose mass per unit length is again $\sigma = \frac{1}{2}$ (for details, see Bonnor and Martins, 1991). In this case, the transformation $\rho = \tilde{z}\tilde{\rho}$, $z = \frac{1}{2}(\tilde{z}^2 - \tilde{\rho}^2)$ leads to $ds^2 = -\tilde{z}^2 dt^2 + d\tilde{z}^2 + d\tilde{\rho}^2 + \tilde{\rho}^2 d\phi^2$, which is again the uniformly accelerated metric (3.14) but with acceleration now in the $\tilde{z}$-direction. It may also be seen that the singularity corresponding to the "source" of this space-time, and which is located on the negative z-axis, now occurs on the "plane" $\tilde{z} = 0$, which corresponds to the acceleration horizon.

It has thus been demonstrated that three different Newtonian potentials give rise to (parts of) Minkowski space in different coordinate systems. Two of these correspond to the fictitious forces that are associated with accelerated coordinates, rather than genuine gravitational fields. Moreover, as emphasised by Gautreau and Hoffman (1969) and many others, the sum of a number of Newtonian potentials does not usually correspond to the potential of the superposition of the associated physical systems. And the interpretation of the Newtonian potential in the unphysical flat space is not necessarily a reliable guide to the physical interpretation of the four-dimensional space-time.

10.1.2 The Weyl solutions

An important family of exact, asymptotically flat, axially symmetric vacuum solutions can be obtained by first transforming to spherical coordinates. Putting $\rho = r \sin\theta$ and $z = r\cos\theta$, the metric (10.2) takes the form

$$ds^2 = -e^{2U} dt^2 + e^{-2U} \left[e^{2\gamma}(dr^2 + r^2 d\theta^2) + r^2 \sin^2\theta\, d\phi^2 \right],$$

and the field equation (10.3) becomes

$$r^2 U_{,rr} + 2r U_{,r} + U_{,\theta\theta} + \cot\theta\, U_{,\theta} = 0.$$

The asymptotically flat solutions of this equation can be expressed as

$$U = -\sum_{n=0}^{\infty} a_n\, r^{-(n+1)}\, P_n(\cos\theta), \tag{10.6}$$

where $P_n(\cos\theta)$ are Legendre polynomials. The corresponding expression for γ is given explicitly by

$$\gamma = -\sum_{l=0}^{\infty}\sum_{m=0}^{\infty} a_l a_m \frac{(l+1)(m+1)}{(l+m+2)} \frac{(P_l P_m - P_{l+1} P_{m+1})}{r^{l+m+2}}.$$

These are known as the *Weyl solutions*.

The simplest member of this family is obtained by putting $a_0 = m$ and $a_i = 0$ for $i \geq 1$, giving

$$U = -\frac{m}{r}, \qquad \gamma = -\frac{m^2 \sin^2 \theta}{2r^2}. \tag{10.7}$$

This solution, which was first used by Curzon (1924) and Chazy (1924), has a spherically symmetric Newtonian potential. However, although this potential is that of a point particle located at $r = 0$, the space-time itself is not spherically symmetric. Moreover, the singularity at $r = 0$ has a very interesting but complicated structure. The interpretation of this space-time will be described in Section 10.5.

In classical Newtonian theory, the coefficients a_n in the expansion (10.6) are the sequence of multipole moments, a_0 being the monopole moment, a_1 the dipole moment, and so on. In the Weyl coordinates of (10.2), they can be expanded in the form

$$U = -\sum_{n=0}^{\infty} a_n \frac{P_n\left(z/\sqrt{\rho^2 + z^2}\right)}{(\rho^2 + z^2)^{(n+1)/2}}.$$

However, it must be emphasised that this multipole expansion in the unphysical flat space does not directly correspond to the actual multipole expansion of a physical source in space-time. For a review of the definitions of the multipole moments of a source in an asymptotically flat space-time, see Quevedo (1990).

At this point, it may be noted only that the spherically symmetric Schwarzschild solution arises from a different separation of equation (10.3) which involves Legendre functions of both the first and second kinds. This has been described by Erez and Rosen (1959), who have also considered the perturbation of the Schwarzschild space-time which arises from the next term in this expansion.

10.2 Static cylindrically symmetric space-times

To consider the special case of the vacuum Weyl space-times in which there exists an additional spacelike Killing vector ∂_z, it is interesting to start with a different form of the metric,[3] namely

$$ds^2 = -x_1^{2p_0}\, dx_0^2 + dx_1^2 + x_1^{2p_2}\, dx_2^2 + x_1^{2p_3}\, dx_3^2, \tag{10.8}$$

---

[3] This is the general solution of the vacuum field equations for a diagonal metric with $g_{11} = 1$, and the remaining coefficients being functions of x_1 only. It can be formally derived from the Kasner solution that is given here in (22.1) by applying a complex transformation which effectively interchanges the roles of x_0 and x_1.

where the constants p_0, p_2 and p_3 satisfy the constraints $p_0 + p_2 + p_3 = 1$ and $p_0^2 + p_2^2 + p_3^2 = 1$. It is convenient here to express these constants in terms of a single parameter σ in the form

$$p_0 = 2\sigma\Sigma^{-1}, \qquad p_2 = 2\sigma(2\sigma - 1)\Sigma^{-1}, \qquad p_3 = (1 - 2\sigma)\Sigma^{-1},$$

where $\Sigma = 1 - 2\sigma + 4\sigma^2$. The coordinate transformation

$$x_0 = k^{-p_0}\Sigma^{p_0}t, \qquad x_1 = k\Sigma^{-1}\rho^{\Sigma}, \qquad x_2 = k^{1-p_2}\Sigma^{p_2}z, \qquad x_3 = k^{-p_3}\Sigma^{p_3}\phi,$$

then puts the metric in the form

$$\mathrm{d}s^2 = -\rho^{4\sigma}\mathrm{d}t^2 + k^2\rho^{4\sigma(2\sigma-1)}\left(\mathrm{d}\rho^2 + \mathrm{d}z^2\right) + \rho^{2(1-2\sigma)}\mathrm{d}\phi^2, \tag{10.9}$$

in which an additional scaling parameter k has been introduced, and it is now generally assumed that $\phi \in [0, 2\pi)$ with $\phi = 2\pi$ identified with $\phi = 0$.

It is obvious that the metric (10.9) is a particular case of (10.2) with (10.3), in which the metric functions

$$U = 2\sigma \log \rho, \qquad \gamma = 4\sigma^2 \log \rho + \log k, \tag{10.10}$$

are independent of z. In this form, the corresponding space-time appears to be both static and cylindrically symmetric. However, except for some special values of the parameters that will be identified below, it does not possess a regular axis. In fact, this vacuum solution was first found by Levi-Civita (1919) and (10.9) is known as the *Levi-Civita metric*. However, it was a very long time before its physical interpretation was clarified by Marder (1958), Gautreau and Hoffman (1969), Bonnor (1979a) and Herrera *et al.* (2001) among others.

It may be noted that the metric function $U = 2\sigma \log \rho$ is formally the Newtonian potential for an infinite uniform line source, for which σ is the mass per unit length. In this case, the line source is represented by the singularity on the axis. By examining this singularity in detail, Israel (1977) has confirmed that σ may be interpreted as the effective gravitational mass per unit proper length. When $\sigma > 0$, the source has positive pressures and a negative energy density. In addition, the behaviour of g_{tt} shows that it is infinitely red-shifted. However, as will be clarified below, this interpretation appears to be appropriate only when σ is small.

It is often convenient to introduce an alternative pair of scaling parameters B and C such that the metric (10.9) takes the form

$$\mathrm{d}s^2 = -\rho^{4\sigma}\mathrm{d}t^2 + \rho^{4\sigma(2\sigma-1)}\left(\mathrm{d}\rho^2 + B^2\mathrm{d}z^2\right) + C^2\rho^{2(1-2\sigma)}\mathrm{d}\phi^2. \tag{10.11}$$

If z or ϕ are non-periodic coordinates, they may always be rescaled to remove the constants B or C, respectively. However, if either of these is regarded as an angular coordinate with a given period, then the corresponding additional parameter remains and will determine any angular deficit.

Using the null tetrad

$$k = \tfrac{1}{\sqrt{2}} \left(\rho^{-2\sigma} \partial_t + \rho^{2\sigma(1-2\sigma)} \partial_\rho \right),$$

$$l = \tfrac{1}{\sqrt{2}} \left(\rho^{-2\sigma} \partial_t - \rho^{2\sigma(1-2\sigma)} \partial_\rho \right),$$

$$m = \tfrac{1}{\sqrt{2}} \left(\tfrac{1}{B} \rho^{2\sigma(1-2\sigma)} \partial_z + \tfrac{i}{C} \rho^{2\sigma-1} \partial_\phi \right),$$

the components of the Weyl tensor are

$$\Psi_0 = \Psi_4 = -\sigma(1 - 4\sigma^2)\rho^{-2(1-2\sigma+4\sigma^2)},$$

$$\Psi_1 = \Psi_3 = 0,$$

$$\Psi_2 = -\sigma(1 - 2\sigma)^2 \rho^{-2(1-2\sigma+4\sigma^2)},$$

which shows that these space-times are normally algebraically general. To determine the physical significance of these components, it is convenient to re-introduce the coordinate $x_1 = \Sigma^{-1}\rho^\Sigma$. In fact, this is the *proper radial distance* from the axis $\rho = 0$, defined such that $\rho^{2\sigma(2\sigma-1)}\mathrm{d}\rho = \mathrm{d}x_1$. With this, the scalar curvature invariant (2.13), for example, becomes

$$I = \frac{4\sigma^2(1 - 2\sigma)^2(1 - 2\sigma + 4\sigma^2)}{\rho^{4(1-2\sigma+4\sigma^2)}} = \frac{4\sigma^2(1 - 2\sigma)^2}{(1 - 2\sigma + 4\sigma^2)^3 \, x_1^4}. \qquad (10.12)$$

It can then be seen that the space-time is flat when $\sigma = 0$, $\tfrac{1}{2}$ or ∞. For all other values of this parameter, there is a curvature singularity at $\rho = 0$. The space-time is of type D when $\sigma = \tfrac{1}{4}$, 1 or $-\tfrac{1}{2}$.[4]

For the case in which $\sigma = 0$, the metric (10.11) simply represents part of Minkowski space in which ρ, z and ϕ are cylindrical polar coordinates and the parameter B can be removed. With this, the metric takes the form (3.12) which describes a space-time with a cosmic string along the axis for which, with $\phi \in [0, 2\pi)$, the deficit angle $2\pi\delta$ is given by $\delta = 1 - C$.

When $\sigma = \tfrac{1}{2}$, however, (10.11) again represents part of Minkowski space, but with both z and ϕ now being non-periodic coordinates so that the parameters B and C can both be removed. In this case, the metric takes the form (3.14), which is the uniformly accelerated metric in which trajectories with ρ, z and ϕ constant have acceleration ρ^{-1} in the (Cartesian) ρ-direction, as described in Section 3.5 (and also Section 10.1.1 for the case with $U = \log \rho$).

It can also be seen from (10.12) that the space-time becomes (locally) flat

---

[4] This follows since $\Psi_1 = 0 = \Psi_3$ and, for these values, $\Psi_2 \neq 0$ and either $\Psi_0 = 0 = \Psi_4$ or $\Psi_0\Psi_4 = 9\Psi_2{}^2$ (see Section 2.1.2).

as $\rho \to \infty$. However, it is not asymptotic to Minkowski space in cylindrical coordinates unless $\sigma = 0$. Indeed, for $\sigma > 0$,

$$\lim_{\rho \to \infty} \left[\sqrt{g_{\phi\phi}(\rho + \delta\rho)} - \sqrt{g_{\phi\phi}(\rho)} \right] = \lim_{\rho \to \infty} \left[(1 - 2\sigma)C\rho^{-2\sigma}\delta\rho \right] = 0,$$

for any finite change $\delta\rho$ in the coordinate ρ. Thus, an increase in the coordinate ρ does not lead to an increase in the circumference of an asymptotically large circle on which t, z and ρ are constant even though the proper radius becomes unbounded.

Such non-standard behaviour in the asymptotics of this space-time are not unexpected as even the classical Newtonian potential $U = 2\sigma \log \rho$ diverges at infinity. Indeed, for the metric (10.11), the geodesic equations are

$$\dot{t} = E\rho^{-4\sigma}, \qquad \dot{z} = P\rho^{4\sigma(1-2\sigma)}, \qquad \dot{\phi} = L\rho^{2(2\sigma-1)},$$

$$\rho^{8\sigma^2}\dot{\rho}^2 + \epsilon_0 \rho^{4\sigma} + C^2 L^2 \rho^{-2(1-4\sigma)} + B^2 P^2 \rho^{8\sigma(1-\sigma)} = E^2,$$

where E, P and L are constants and $\epsilon_0 = 1, 0$ for timelike and null geodesics, respectively. This indicates that, as in the classical case, unbounded orbits are not permitted for timelike orbits with $\sigma > 0$. The gravitational field of this infinite line source is so strong, even for small σ, that test particles cannot escape to infinity.

It is also of interest to note that the metric (10.11) is invariant under the transformation

$$\rho = (a\rho')^{4\sigma'^2}, \qquad t = a^{-2\sigma'}t', \qquad z = \phi', \qquad \phi = z', \qquad (10.13)$$

with

$$2\sigma = 1/2\sigma', \qquad B = a^{2\sigma'-1}C', \qquad C = a^{2\sigma'(1-2\sigma')}B',$$

where $a = (2\sigma')^{-2/(1-2\sigma'+4\sigma'^2)}$. This indicates that, for the metric (10.11), the parameter range $\sigma \in (0, \frac{1}{2}]$ is exactly equivalent to the range $\sigma' \in [\frac{1}{2}, \infty)$ and vice versa, but with the roles of z and ϕ interchanged. Thus, (10.11) may be considered to represent a cylindrically symmetric space-time with periodic coordinate ϕ when σ is small. As σ increases, however, the cylindrical surfaces on which t and ρ are constant appear to open out and become infinite planes when $\sigma = \frac{1}{2}$. As σ increases beyond that, these surfaces again become cylinders but with z now being the periodic coordinate.

For this space-time to be interpreted as representing the gravitational field around an infinite line mass, there is clearly a bound to the permitted range of the parameter σ. In fact, Gautreau and Hoffman (1969) have shown that circular timelike geodesics only exist when $0 < \sigma < \frac{1}{4}$, and they then occur for all constant values of ρ. When $\sigma = \frac{1}{4}$, the circular geodesics (for all values of ρ) become null, supporting the conclusion that this is some kind of limit. For larger values of σ, it had been suggested that a different, probably non-axially symmetric, interpretation is required.

Considering the metric (10.9) to represent the exterior of an infinite cylinder of finite radius, Marder (1958), Evans (1977), Bonnor (1979a), da Silva *et al.* (1995c) and Bičák *et al.* (2004) have constructed corresponding interior perfect fluid solutions. Such an interpretation is only appropriate when σ is small, and they appear to confirm the upper limit of $\sigma = \frac{1}{4}$. Lathrop and Orsene (1980) and Bičák and Žofka (2002) have obtained the same upper limit for the exterior of a cylinder of counter-rotating dust and a thin cylindrical shell, respectively.

In fact, for the limiting case in which $\sigma = \frac{1}{4}$, the metric (10.11) becomes

$$\mathrm{d}s^2 = \rho \left(-\mathrm{d}t^2 + C^2 \,\mathrm{d}\phi^2\right) + \rho^{-1/2}(\mathrm{d}\rho^2 + \mathrm{d}z^2). \qquad (10.14)$$

This is a non-standard form of the $BIII$-metric (i.e. the so-called B-metric with $\epsilon = 0$ that will be identified in Section 16.4.1), which is of type D but has no known interpretation. (The equivalent case with $\sigma = 1$, which is the asymptotic form of the Melvin universe (7.21) for large ρ, can also be transformed to this form using (10.13), and with the free parameter B replacing C.)

However, it may be noted that Bonnor and Davidson (1992), Philbin (1996) and Haggag and Desokey (1996) have found interior perfect fluid solutions that are valid for the extended range $0 < \sigma < \frac{1}{2}$. These have been further interpreted by Bonnor (1999) who has shown that, as σ increases towards $\frac{1}{2}$, the proper radius of the cylindrical source tends to infinity. Thus, the boundary of the cylindrical source tends locally to that of a plane. This strongly suggests that the exterior with $\sigma = \frac{1}{2}$ has a plane source – a result that is consistent with the fact that this is Minkowski space adapted to uniform acceleration in the Cartesian-like ρ-direction.

For the case in which $\sigma > \frac{1}{2}$, Philbin (1996) has suggested wall sources. However, as pointed out above, the range $\sigma \in [\frac{1}{2}, \infty)$ is equivalent to the range $(0, \frac{1}{2}]$ when the roles of z and ϕ are interchanged. Negative values of σ may also be considered.[5]

A generalisation of the Levi-Civita metric (10.9) to include a non-zero cosmological constant was obtained by Linet (1986) and Tian (1986). In fact, this results in a dramatic modification of the geometry of the space-time as

---

[5] For the particular case in which $\sigma = -\frac{1}{2}$, the metric (10.9) is that of the plane symmetric Taub (1951) solution that can be transformed to either of the forms (9.10) or (9.11). This is the static case of the type D $AIII$-metric described in Section 9.1.2. It has been interpreted as part of a cylindrically symmetric space-time by Jensen and Kučera (1994).

analysed by da Silva *et al.* (2000) and Žofka and Bičák (2008). The limit of this solution as $\sigma \to 0$ has been investigated by Bonnor (2008) and shown to be a particular non-expanding type D Plebański–Demiański solution which is included in the class given by (16.27) with $\alpha = 0 = \gamma$. The Linet–Tian solution with a positive cosmological constant has been investigated in more detail by Griffiths and Podolský (2010) who have shown that it can be matched to a toroidal region of the Einstein static universe.

10.3 The Schwarzschild solution

The Schwarzschild metric described in Chapter 8 is the unique vacuum spherically symmetric space-time, which is necessarily asymptotically flat. It is given in the standard form by (8.1), in which the parameter m is usually understood as being the mass of the source. On applying the transformation

$$r = m(x + 1), \qquad \cos\theta = y,$$

to prolate spheroidal coordinates x, y, the standard line element becomes

$$ds^2 = -\left(\frac{x-1}{x+1}\right) dt^2 + m^2 \left(\frac{x+1}{x-1}\right) dx^2 + m^2 \frac{(x+1)^2}{(1-y^2)} dy^2$$
$$+ m^2(x+1)^2(1-y^2)\, d\phi^2,$$

where, to retain Lorentzian signature, it is necessary that $|y| < 1$. The static region $r > 2m$ now occurs when $x > 1$. Within this region, it is possible to make the further transformation (see Figure 13.1)

$$\rho = m\sqrt{(x^2-1)(1-y^2)}, \qquad z = m\, x\, y,$$

which takes the line element to the Weyl form (10.2) with

$$e^{2U} = \frac{R_+ + R_- - 2m}{R_+ + R_- + 2m}, \qquad e^{2\gamma} = \frac{(R_+ + R_-)^2 - 4m^2}{4R_+R_-}, \tag{10.15}$$

where $R_\pm^2 = \rho^2 + (z \pm m)^2$. In this case, the function U can alternatively be expressed in the form

$$U = \tfrac{1}{2}\log\left(\frac{R_- + z - m}{R_+ + z + m}\right), \tag{10.16}$$

which is formally the Newtonian potential for a finite rod, located along the part of the axis $\rho = 0$ for which $|z| < m$, whose mass per unit length is $\sigma = \tfrac{1}{2}$. Thus the "rod" has length $2m$ and its total mass is m.

Thus the static region of a spherically symmetric and asymptotically flat gravitational field can be expressed in terms of Weyl coordinates, with the corresponding Newtonian potential being that of a finite rod whose mass per unit length is exactly $\frac{1}{2}$. Moreover, the horizon of the Schwarzschild space-time at $r = 2m$ is located on that part of the axis $\rho = 0$ that corresponds to the rod. Relative to Schwarzschild coordinates, the different parts of the axis of the Weyl coordinates have the following meanings:

$$
\begin{array}{llllll}
\text{half-axis} & \theta = \pi, & r \geq 2m & \leftrightarrow & \rho = 0, & z \leq -m, \\
\text{horizon} & 0 < \theta < \pi, & r = 2m & \leftrightarrow & \rho = 0, & -m < z < m, \\
\text{half-axis} & \theta = 0, & r \geq 2m & \leftrightarrow & \rho = 0, & m \leq z.
\end{array}
$$

This illustrates very nicely that the classical interpretation of the associated Newtonian potential is not a good guide to the geometry of the associated space-time. Moreover, since it is limited to static regions, it gives no clue to possible extensions through horizons.

It is also interesting to observe that the potential (10.16) may be formally regarded as the superposition of those for *two* semi-infinite rods. The first of these would have mass per unit length $\sigma = \frac{1}{2}$ and be located on that part of the axis for which $z < m$. The second semi-infinite rod component would have mass per unit length $\sigma = -\frac{1}{2}$ and be located on the part of the axis for which $z < -m$, thus cancelling out the source on this part. It may be noted that the first of these on its own corresponds to the potential that is given in (10.5) and that leads to Minkowski space in accelerated coordinates. However, the second component with negative mass per unit length $\sigma = -\frac{1}{2}$ can be seen to correspond to that which leads to the plane symmetric Taub (1951) solution in the form (9.13). This is the type D $AIII$-metric that has been described in Section 9.1.2. It is counter-intuitive that this particular combination of potentials would give rise to the simple spherically symmetric Schwarzschild space-time.

10.4 The Zipoy–Voorhees solution

As shown above, the Schwarzschild solution in Weyl form corresponds to the Newtonian potential (10.16) for a rod whose mass per unit length is $\sigma = \frac{1}{2}$. It is clearly possible to generalise this to solutions with arbitrary values of σ. Denoting the length of the rod by 2ℓ, located at $z \in (-\ell, \ell)$, the corresponding potential is

$$
U = \sigma \log \left(\frac{R_- + z - \ell}{R_+ + z + \ell} \right),
$$

where $R_\pm = \sqrt{\rho^2 + (z \pm \ell)^2}$. Of course, the mass of the rod is now given by $m = 2\sigma\ell$ and, in terms of the two parameters m and ℓ, the vacuum solution for the metric (10.2) can be expressed in the form

$$\mathrm{e}^{2U} = \left(\frac{R_+ + R_- - 2\ell}{R_+ + R_- + 2\ell}\right)^{m/\ell}, \qquad \mathrm{e}^{2\gamma} = \left(\frac{(R_+ + R_-)^2 - 4\ell^2}{4\,R_+\,R_-}\right)^{m^2/\ell^2}.$$

This family of solutions was initially discovered by Bach and Weyl (1922) and Darmois (1927). However, it is now generally referred to either as the Zipoy–Voorhees solution or as the γ-metric. It is usually described in terms of two parameters, namely m, the mass of the corresponding rod, and δ, which is known as the *deformation parameter* and is defined such that

$$\delta = 2\sigma = m/\ell.$$

Properties of these solutions have been investigated further by Zipoy (1966), Gautreau and Anderson (1967), Bonnor and Sackfield (1968), Voorhees (1970), Esposito and Witten (1975) and many others.

Clearly, this solution reduces to the Schwarzschild solution when $\sigma = \frac{1}{2}$ or $\delta = m/\ell = 1$. It also reduces to the Curzon–Chazy solution in the limit when $\ell \to 0$. Herrera, Paiva and Santos (1999) have shown that it also reduces to the Levi-Civita metric (10.9) in the limit as $\ell \to \infty$. Moreover, this value of σ, where the Zipoy–Voorhees metric reduces to the Schwarzschild metric, is exactly the same as when the Levi-Civita metric becomes flat with a possible topological defect.

Except for the flat case in which σ, m and thus δ vanish, the finite line segment $\rho = 0$, $-\ell < z < \ell$ in Weyl coordinates is singular. This corresponds to a curvature singularity in the space-time except when $\sigma = \frac{1}{2}$ or $\delta = m/\ell = 1$, when it is the horizon of the Schwarzschild space-time.

It can be seen that the proper circumference of a small circle around the axis for any constant $z \in (-\ell, \ell)$, with t constant, is given approximately by $2^{1+\delta}\pi(\ell^2 - z^2)^{\delta/2}\rho^{1-\delta}$ for $\rho \ll 1$. As $\rho \to 0$, this clearly vanishes for $\delta < 1$, is equal to $4\pi\sqrt{\ell^2 - z^2}$ for the Schwarzschild case $\delta = 1$ in which $\ell = m$, and diverges for $\delta > 1$. Moreover, within this range, the circumference of such circles increases as ρ decreases when $\delta > 1$.

Kodama and Hikida (2003) have shown that this line segment is geometrically pointlike for $\delta < 0$ (although this has a negative Komar mass), rod-like for $0 < \delta < 1$, and ring-like for $\delta > 1$. These singularities are always naked. In addition, the end points of this segment, namely $\rho = 0$, $z = \pm\ell$ are generally directional singularities (at least when $\delta \geq 2$) in that the curvature tensor vanishes when these points are approached from the direction of the regular part of the axis. Some properties of this directional singularity in the Zipoy–Voorhees solution have been described by Hoenselaers (1978) and Papadopoulos, Stewart and Witten (1981). An interior for it has been constructed by Stewart *et al.* (1982).

In terms of prolate spheroidal coordinates x and y such that

$$\rho = \ell \sqrt{(x^2 - 1)(1 - y^2)}, \qquad z = \ell\, x\, y,$$

the metric functions of the Zipoy–Voorhees space-times simply become

$$e^{2U} = \left(\frac{x-1}{x+1}\right)^{\delta}, \qquad e^{2\gamma} = \left(\frac{x^2 - 1}{x^2 - y^2}\right)^{\delta^2},$$

and the metric takes the form

$$ds^2 = -e^{2U} dt^2 + \Sigma^2 \left(\frac{dx^2}{x^2 - 1} + \frac{dy^2}{1 - y^2}\right) + R^2 d\phi^2,$$

where

$$\Sigma^2 = \ell^2 (x+1)^{\delta(\delta+1)}(x-1)^{\delta(\delta-1)}(x^2 - y^2)^{1-\delta^2},$$

$$R^2 = \ell^2 (x+1)^{1+\delta}(x-1)^{1-\delta}(1 - y^2).$$

These solutions are the static limit of the stationary axially symmetric solutions of Tomimatsu and Sato (1972, 1973) (see Section 13.2.2) which are only known explicitly for integer values of the deformation parameter δ, and will not be considered in this book. Alternatively, the Tomimatsu–Sato solutions may be regarded as the rotating generalisations of the Zipoy–Voorhees solutions.

10.5 The Curzon–Chazy solution

As already mentioned in Section 10.1.2, the simplest of the Weyl solutions is that given by (10.7). The sources of such space-times are usually referred to as Curzon–Chazy particles since they were used (for two such sources) by Curzon (1924) and Chazy (1924). Re-expressed in Weyl coordinates, the metric functions for the line element (10.2) are specifically

$$U = -\frac{m}{\sqrt{\rho^2 + z^2}}, \qquad \gamma = -\frac{m^2 \rho^2}{2(\rho^2 + z^2)^2}. \tag{10.17}$$

This formally arises by taking U to be the Newtonian potential of a point mass located at $\rho = 0$, $z = 0$. However, it can be seen that the resulting space-time is not spherically symmetric. Moreover, there is a curvature singularity at $\rho = 0$, $z = 0$ that is not surrounded by a horizon and is therefore "naked". Nevertheless, any "light" emitted from it becomes infinitely red-shifted, so that it is effectively invisible.

This curvature singularity has a complicated directionally dependent structure (Stachel, 1968; Gautreau and Anderson, 1967). In fact, Cooperstock and Junevicus (1974) have shown that it is not encountered by certain families of curves which approach the origin along the axis of symmetry. This suggests that the singularity may not be "point-like", and that such curves may be extendable through it. Such a possibility was first investigated in detail by Szekeres and Morgan (1973) who have shown that almost all null geodesics approach the z-axis very rapidly in Weyl coordinates as $\rho^2 + z^2 \to 0$. Their conclusion is that such geodesics can be extended through the singularity in a way that is C^∞ but not analytic.

It is also of interest to note that the circumference of the circular lines on which t, ρ and z are constant is given by $2\pi\rho\, e^{m/\sqrt{\rho^2+z^2}}$. Although this behaves as expected like $2\pi\rho$ for $\rho \gg 1$, and generally vanishes on the axis $\rho = 0$, it actually diverges in the plane $z = 0$.

These results suggest that there is some kind of "hole" through the singularity that is aligned with the z-axis. It was subsequently confirmed by Scott and Szekeres (1986a,b) that the curvature singularity has the structure of a ring. They have shown that geodesics that approach the origin along the z-axis can be extended through it. However, having passed through the origin, such geodesics do not necessarily re-emerge from it, but may generally be presumed to enter new regions of space-time.

In fact, the work of Scott and Szekeres (1986a,b) shows that the Weyl coordinates are not well suited to describing the structure of the singularity in this space-time. The apparent point $\rho = 0$, $z = 0$ in Weyl coordinates actually covers regular points and a spacelike infinity as well as a curvature singularity. This structure can be clarified by making a more appropriate choice of coordinates. Scott and Szekeres have analysed the global structure of this space-time using approximate asymptotic coordinates. Taylor (2005) has preferred to use the Kretschmann scalar (denoted as α) and its harmonic conjugate (denoted as β). However, it is not easy to re-express the metric in terms of either of these coordinate systems. Nevertheless, the ring structure of the curvature singularity is clearly shown in these new coordinates, as illustrated schematically in Figure 10.1 for a region in which $z \geq 0$ and $\sqrt{\rho^2 + z^2}$ less than some constant R_s. This figure clearly demonstrates that the axis at $\rho = 0$ does not meet the singularity. Moreover, there are trajectories that can go between the singularity and the axis and reach a new spatial infinity. This extended region is shown as partly bounded by an "edge" that is also at infinite proper distance, but not reached by any geodesics.

Scott and Szekeres (1986b) have shown that, within the Curzon–Chazy

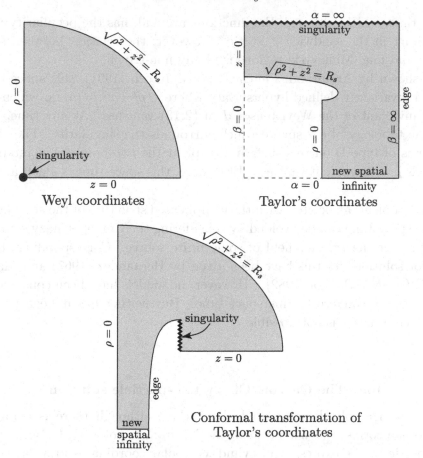

Fig. 10.1 Equivalent regions of a spatial section of the Curzon–Chazy space-time on which t and ϕ are constant, shown in Weyl coordinates and those of Taylor in which α is the Kretschmann scalar. The third figure is a conformal transformation of that in Taylor's coordinates and is equivalent to that of Scott and Szekeres. This demonstrates the character of the ring singularity and shows that trajectories can pass through it to reach a new spatial infinity.

space-time but in the extended region through the ring, there exists a null hypersurface across which the space-time can be smoothly extended to part of a Minkowski space. This may be considered to form an event horizon since events behind it could never be seen by observers near the Curzon–Chazy infinity. In fact, it is possible for an observer to start in a Minkowski space, cross this null hypersurface into a static region described by the Curzon–Chazy solution, possibly to pass through the ring singularity, but anyway to possibly cross another null hypersurface into a further Minkowski region. Such a possibility may be considered to represent a kind of "sandwich" space-time, which, although the junctions are null, has the peculiarity that the filling in the sandwich is static. However, the relation between such initial and final Minkowski regions is highly non-trivial.

As shown by Arianrhod, Fletcher and McIntosh (1991), this space-time has an invariantly defined hypersurface where $\sqrt{\rho^2 + z^2} = m$, on which the cubic invariant of the Weyl tensor J in (2.13) vanishes. At any time, this has the topology of a 2-sphere which surrounds the singularity. The Weyl tensor is of type D on this surface, except at the two points that intersect the axis $\rho = 0$, where it vanishes. Elsewhere, this space-time is algebraically general.

The problems associated with the complicated structure of the singularity in this space-time can be avoided by considering the Curzon–Chazy solution to represent the *exterior* field of some finite source. Corresponding static interior solutions for this have been given by Hernandez (1967) and Marek (1967) (see also Bonnor, 1982).[6] However, no models have been constructed that actually collapse to this space-time. But neither has it been proved that such a model is not possible.

10.6 The Curzon–Chazy two-particle solution

It has been amply demonstrated above that, although there is a formal one-to-one correspondence between the Newtonian potential for an axially symmetric source expressed in cylindrical polar coordinates in a flat space and the gravitational field of the corresponding four-dimensional space-time, there is no equivalent correspondence in the physical interpretations of these fields.

In principle, static axially symmetric solutions can be generated by superposing any number of Newtonian potentials for such cases in the expression for U. However, although the equation (10.3) for U is linear, a nonlinearity enters through the equations (10.4) for γ. The physical interpretations of such superpositions need to be treated with considerable care. Nevertheless, some interesting and important results can be obtained in this way.

According to such a procedure, it is clearly possible to construct solutions corresponding to a superposition of two Curzon–Chazy particles. In fact, this was the solution given initially by Curzon (1924) and Chazy (1924) in

---

[6] The Hernandez (1967) solution approaches the Schwarzschild interior solution near the centre, but the Lichnerowicz junction conditions are not satisfied on the boundary. These junction conditions are satisfied by the Marek (1967) solution, but the normal energy conditions are only manifestly satisfied in the weak-field limit.

which

$$U = -\frac{m_1}{\sqrt{\rho^2 + (z-a)^2}} - \frac{m_2}{\sqrt{\rho^2 + (z+a)^2}}.$$

This appears to represent two Newtonian point sources of "mass" m_1 and m_2 located at the points $z = a$ and $z = -a$, respectively. It would thus seem to describe the unphysical situation of two separate gravitating bodies in a static configuration (Silberstein, 1936). However, on evaluating the metric function γ, Einstein and Rosen (1936) found that there is a conical singularity along the axis between these two points. This singularity may be interpreted as a kind of structure known as a *Weyl strut* which holds the two "sources" apart in a static configuration.

For this case, it is found (Katz, 1968) that the stress in the Weyl strut, at the lowest order, is approximately equal to $m_1 m_2 / 4a^2$, which is exactly as expected from Newtonian theory. Moreover, Israel (1977) has shown that, although the stress is that given above, the strut also has a negative energy density of equal magnitude. Thus, the Weyl strut has no effective

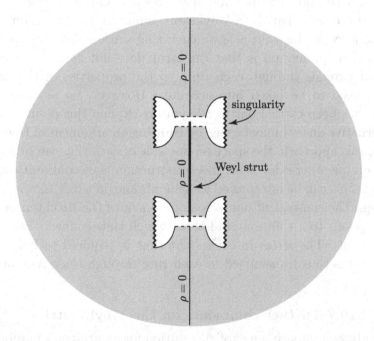

Fig. 10.2 (Taylor, 2005) A spatial section through the double Curzon space-time on which t and ϕ are constant (with $\phi + \pi$ included), showing schematically the character of the ring singularities through which the space-time can be extended in either direction.

gravitational mass – a result which is consistent with the fact that it does not contribute to the gravitational potential U.

In fact, it can be seen from (10.4) that the function γ necessarily takes constant values on the axis $\rho = 0$. However, it may take different values on different sections of the axis that are separated by particular sources, and this allows for the existence of Weyl struts. This also implies that the stress along any particular strut must be constant along its length.

The structure of the space-time near the singularities in the above double Curzon–Chazy particle solution has been nicely described by Taylor (2005). The ring-like structure of each source is illustrated in Figure 10.2. It is interesting in this case that the Weyl strut appears on the axis between the two sources. However, geodesics that approach the ring singularities can pass through them into different regions of space-time according to the direction in which they approach the ring, and to which ring they approach. It follows that the Weyl struts do not generally terminate near the rings but pass through both of them reaching infinity in some new covering space. Each regular part of the axis may similarly be extended through the corresponding ring to new regions of space-time.

If the above solution for two Curzon–Chazy particles is interpreted as two sources that are held apart by a Weyl strut, an interesting question arises as to how the force in the strut acts on each ring source. The answer implied by the previous paragraph is that the strut does not actually reach these sources but extends through each ring, so the properties of the extended space also need to be taken into account. However, no analysis of such extensions has been carried out to adequately explain this point.

An alternative answer may be suggested using an argument of Hoenselaers (1995). In this approach, the space on one side of each ring can be identified with that on the other side. With this construction, a discontinuity occurs in the metric. This can be interpreted as a membrane on a disc that is bounded by the ring. The consequent non-zero component of the Ricci tensor on this disc corresponds to a thin material sheet which thus connects each ring to the Weyl strut. The stress in each strut that is required to hold the two sources apart is thus transmitted to each ring through these membranes.

10.7 Further comments on the Weyl metrics

Using the above approach, an explicit solution for an arbitrary number of co-linear Schwarzschild-like bodies was constructed by Israel and Khan (1964). Such sources are represented as a set of non-overlapping rods along the axis of the Weyl coordinates. In general, such solutions contain conical singularities on the axis. However, a static solution with a regular axis outside the sources can be obtained if the number of sources is odd and if alternate ones have negative mass. Israel and Khan described the case of three such sources in detail. In this case, the central body has negative mass and its rod-like source in Weyl coordinates corresponds to the point-like naked singularity of the negative mass Schwarzschild solution. They also described an interesting limit of this that will be further referred to in the context of accelerating bodies in chapter 15.

In addition, Szekeres (1968) has demonstrated that static, two-body solutions can exist without the need for an intervening strut. He has given examples of such situations in which at least one of the sources has a multipole structure. If this is chosen in certain ways, it allows equilibrium solutions to be achieved for which the axis is regular everywhere except at two sources. However, it must be recalled that a dipole component consists of a positive and a negative mass in very close proximity. Such a combined object would spontaneously accelerate under the action of its internal gravitational fields. In fact, an exact solution describing such a situation is given in Section 15.3.2. In Szekeres' examples, the force that would tend to produce a spontaneous acceleration is exactly balanced by an attraction towards a second body. The interesting point, noticed by Szekeres, is that equilibrium can be achieved between (multipole) particles even when both bodies have positive monopole components.

Weyl solutions may also have singularities at points *off the axis*. Because of the axial symmetry, these will form more obvious kinds of ring singularities. The first such solution was given by Bach and Weyl (1922), but they generally have properties that are difficult to visualise. It is known in some cases that, for a small circle around the axis $\rho = 0$, these have infinite circumference for a finite radius. In fact, this would appear to be the generic case (see Scott, 1989). Moreover, it is generally found that the value of the Kretschmann scalar at this ring will depend upon the direction from which it is approached: i.e. it is another directional singularity. In fact, only special geodesics reach the ring, and these do so only from certain preferred directions (Hoenselaers, 1990).

A further ring solution was constructed by Thorne (1975). This appears to be unique among toroidal solutions in that the singularity is locally cylindrically symmetric, being equivalent to a section of the Levi-Civita solution for an infinite line mass.

An interesting situation arises if the solution for a ring singularity is combined with one for a particle located at a different point on the axis. A static solution of this type will only be possible if a Weyl strut is present to hold them apart. And the question again arises as to how the ring is connected to the strut – a question that was specifically answered by Hoenselaers (1995). As in the above explanation for the double Curzon solution, the point is that the space on one side of the ring contains a conical singularity (representing the Weyl strut), while that on the other side does not. If these two spaces are identified, a discontinuity in the metric occurs on a membrane bounded by the ring. Again the consequent non-zero component of the Ricci tensor on this membrane corresponds to a material sheet connecting the ring to the Weyl strut. Further examples of such structures have been described by Letelier and Oliviera (1998).

Another significant family of static, axially symmetric solutions can be constructed by taking any solution that is regular in the half space $z \geq 0$, but which contains arbitrary sources and singularities in $z < 0$, and matching the half space $z \geq 0$ to its image about $z = 0$. Such a space-time would be continuous across the junction, but there would be discontinuities in the derivatives of the metric functions. These would give rise to distributional components in the Ricci tensor, which could be interpreted as being caused by a thin axially symmetric sheet of matter. Although such a disc source would generally be infinite in extent, its total mass could well be finite, and it could provide an approximate model for the field of some astrophysical objects such as some thin galaxies. However, the source must be considered to be composed of equal streams of particles propagating in circular orbits in opposite directions but without collisions. Generally, differential pressures would also occur.

This method for constructing the gravitational field of a *static disc of counter-rotating particles* was introduced in general relativity by Morgan and Morgan (1969). It was further extended and applied by Lynden-Bell and Pineault (1978), Lemos (1988), Bičák, Lynden-Bell and Pichon (1993) and Bičák, Lynden-Bell and Katz (1993). This approach has been used to describe thin discs around black holes and other astrophysically significant structures. See for example Lemos and Letelier (1994), Semerák, Zellerin and Žáček (1999), and Semerák (2002) for a thorough review.

10.8 Axially symmetric electrovacuum space-times

For static, axially symmetric Einstein–Maxwell fields, the introduction of the electromagnetic field introduces a nonlinearity into the field equation which generalises (10.3).[7] Specific space-times are therefore more difficult

---

[7] It is assumed that the electromagnetic field, like the gravitational field, is both static and axially symmetric.

to obtain. Nevertheless, many interesting solutions of this type are known and will be briefly reviewed here.

It was pointed out by Bonnor (1954) that, to every electrostatic solution, there is a corresponding magnetostatic solution. Thus, different physical situations may be represented by the same metric, and the different electromagnetic fields are related by a duality transformation of the Maxwell field. An example of this is the axially symmetric Melvin solution, for which $U = \log(1 + \frac{1}{4}B^2\rho^2)$, that has already been described in Section 7.3, and which may be interpreted in terms of either an electric or a magnetic field.

For solutions of this type, Weyl (1917) identified an electrovacuum class in which the equipotential surfaces of the gravitational and electromagnetic fields are the same. In this case, the main field equations become linear and the superposition of solutions is possible exactly as in the vacuum case. Specifically, the metric function (gravitational potential) e^{2U} and the electromagnetic (Ernst) potential Φ are functionally related by the condition $e^{2U} = 1 - 2c^{-1}\Phi + \Phi^2$, where c is an arbitrary complex constant.[8]

It is significant that the Reissner–Nordström solution is included in this class. Using the transformation to Weyl coordinates

$$r = m + \tfrac{1}{2}(R_+ + R_-), \qquad \cos\theta = \frac{1}{2d}(R_+ - R_-),$$

where $R_\pm^2 = \rho^2 + (z \pm d)^2$, and $d^2 = m^2 - e^2$, the metric (9.14) can be re-expressed in the form

$$ds^2 = -\frac{(R_+ + R_-)^2 - 4d^2}{(R_+ + R_- + 2m)^2}\,dt^2 + \frac{(R_+ + R_- + 2m)^2}{4R_+R_-}(d\rho^2 + dz^2)$$
$$+ \frac{(R_+ + R_- + 2m)^2}{(R_+ + R_-)^2 - 4d^2}\,\rho^2\,d\phi^2, \tag{10.18}$$

with

$$\Phi = \frac{2e}{R_+ + R_- + 2m},$$

(see Gautreau, Hoffman and Armenti, 1972). It is also possible, due to the ability to superpose solutions, to obtain that for two Reissner–Nordström sources at different points on the axis. Such a possibility will be discussed below. However, it may be noted that an equilibrium configuration with a regular axis is obtained when each source is an extreme object whose charge has the same magnitude as its mass (so that the mutual gravitational

---

[8] The particular form of this condition arises from the assumptions that the space-time is static and asymptotically flat, that $e^{2U} = -g_{tt}$ and Φ are functionally related, and that the electromagnetic field vanishes at infinity. It will be derived for the more general stationary case in (13.28).

attraction between the sources is exactly balanced by their electrostatic repulsion).

Because of its occurrence in such static configurations, it is appropriate to consider this case first in greater detail. When $e = \pm m$ in (10.18), $d = 0$ and $R_+ = R_-$, the metric for the extreme Reissner–Nordström solution becomes

$$ds^2 = -\left(1 + \frac{m}{R}\right)^{-2} dt^2 + \left(1 + \frac{m}{R}\right)^2 \left(d\rho^2 + dz^2 + \rho^2 d\phi^2\right), \qquad (10.19)$$

where $R^2 = \rho^2 + z^2$ and $\Phi = \pm m(R + m)^{-1}$. This is clearly regular everywhere except at $R = 0$. For such a case, Das (1962) has found an interior solution, which is a sphere of charged dust particles whose mass density has the same magnitude as its charge density, and which can be matched to (10.19) across a surface of constant R. However, as pointed out by Hartle and Hawking (1972), a different analytic extension through the horizon is also possible.

In fact, the "point" $R = 0$, which is the origin $\rho = 0$, $z = 0$ of the Weyl coordinates in this case, is not a point at all, but a surface having area $4\pi m^2$. And, since $r - m = R = \sqrt{\rho^2 + z^2}$, it can be seen that this actually corresponds to the degenerate horizon $r = m$ of the extreme Reissner–Nordström solution (see Subsection 9.2.3). As shown by Hartle and Hawking (1972), any trajectory which approaches this "point" can be extended through it into the static interior region of the extreme Reissner–Nordström space-time. In fact, the analytically extended region is given by the metric

$$ds^2 = -\left(1 - \frac{m}{R}\right)^{-2} dt^2 + \left(1 - \frac{m}{R}\right)^2 \left(d\rho^2 + dz^2 + \rho^2 d\phi^2\right). \qquad (10.20)$$

This is effectively the extension of (10.19) to negative values of R. Alternatively, this "interior" metric may be regarded as that with a negative value for m. In any approach, however, it is important to note that this "interior" region has a curvature singularity at $R = m$. The trajectory of a test particle which falls into this singularity is illustrated in Figure 10.3.

Of course, it is also possible that, having passed through the horizon into the region represented by (10.20), the test particle could return again to the origin corresponding to the horizon rather than continuing to the singularity. In such a case, the particle would subsequently emerge into a different exterior region which is also illustrated in the conformal diagram of Figure 9.10.

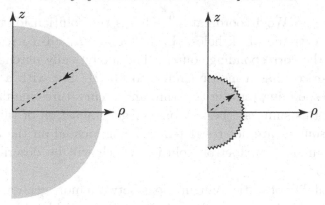

Fig. 10.3 The trajectory of a particle which falls radially into an extreme Reissner–Nordström black hole is represented by the broken line. It starts in the exterior region, given by the metric (10.19) and shown in the left figure, and moves toward the origin of Weyl coordinates which represents the horizon. It then continues in the right-hand figure, representing the interior region given by the metric (10.20), and proceeds from the horizon at the origin to the curvature singularity at $\rho^2 + z^2 = m^2$.

10.8.1 Equilibrium configurations with distinct sources

Within the above class of solutions in which the metric function e^{2U} and the electromagnetic potential Φ are functionally related, it is possible to superpose potentials to construct static space-times representing configurations with distinct sources.

For example, it is possible to obtain the solution for two charged Curzon–Chazy particles of masses m_1 and m_2 and charges e_1 and e_2, respectively. Such a solution was obtained explicitly by Cooperstock and de la Cruz (1979), who showed that there is generally a strut between them. However, the axis between the sources is regular if $m_1{}^2 = e_1{}^2$ and $m_2{}^2 = e_2{}^2$. This corresponds to the particular classical case where the gravitational attraction between the sources is exactly balanced by an electrostatic repulsion for which each source itself is critically charged.

More importantly, a superposition of an arbitrary number (n say) of extreme Reissner–Nordström solution sources located along a common axis of symmetry has been described by Azuma and Koikawa (1994). In this case, the exterior metric is given by

$$ds^2 = -\Big(1 + \sum_i \frac{m_i}{R_i}\Big)^{-2} dt^2 + \Big(1 + \sum_i \frac{m_i}{R_i}\Big)^2 \big(d\rho^2 + dz^2 + \rho^2 d\phi^2\big), \quad (10.21)$$

with $\Phi = (1 + \sum \frac{m_i}{R_i})^{-1}$, where $i = 1, \ldots n$, $R_i{}^2 = \rho^2 + (z - z_i)^2$, and each source has mass m_i and charge $e_i = m_i$ (or $e_i = -m_i$), and is located at

$\rho = 0$, $z = z_i$ in Weyl coordinates.[9] This is the solution for the region exterior to n extreme black holes. Each "point" $R_i = 0$ corresponds to the horizon of the corresponding source. The analytically extended region inside this horizon has a metric similar to (10.21) but with a negative sign in the corresponding term in the sum, and a curvature singularity then occurs where this sum vanishes. A further generalisation of this situation in which the sources are not restricted to being located on the axis is known as the Majumdar–Papapetrou solution which will be described briefly in Section 22.5.

The possibility of static configurations between more general (non-extreme) Reissner–Nordström sources has also received considerable attention. In spite of many initial contrary indications, the plausibility of such solutions has been demonstrated by Bonnor (1981, 1993) (see also Chandrasekhar and Xanthopoulos, 1989) and further examined by Perry and Cooperstock (1997), Bretón, Manko and Aguilar Sánchez (1998) and Emparan and Teo (2001).

An explicit solution describing such a situation was eventually given by Alekseev and Belinski (2007), who found an equilibrium configuration, in the absence of struts, which also depends on the distance between the sources.[10] Specifically, the condition for equilibrium is given by

$$m_1 m_2 = (e_1 - \nu)(e_2 + \nu), \qquad \nu = \frac{m_2 e_1 - m_1 e_2}{\ell + m_1 + m_2},$$

where ℓ is the "distance" between the centres of the sources in terms of Weyl coordinates. It can be shown that this condition cannot be satisfied for two separate non-extreme black holes. (This is consistent with a general result of Chruściel and Tod, 2007.) Neither can it be satisfied for two naked singularities (hyperextreme sources), or if just one of the sources is extreme. However, an equilibrium configuration is possible, with a specific separating distance ℓ, either if both sources are extreme, as described above, or if one of the sources is a black hole and the other is a naked singularity. Moreover, in this case, the charge of the black hole can even vanish, so a neutral (Schwarzschild-like) black hole could exist in equilibrium with a naked singularity. This is possible due to the repulsive nature of gravity near

---

[9] Notice that, for a single particle, this expression for the electromagnetic potential differs from that following (10.19) only by the addition of a constant.

[10] Manko (2007) has subsequently shown that the Alekseev–Belinski solution is included in the family of solutions given by Bretón, Manko and Aguilar Sánchez (1998). Actually, it is also "contained" in a number of other general families of solutions. However, these were not expressed in terms of appropriate "physical" parameters, the condition for a regular axis was not given in a compact form, and their physical properties were not interpreted in detail: i.e. this particularly interesting case was not explicitly identified.

a naked Reissner–Nordström singularity. Further details of this solution, with an investigation of its gravitational and electric fields, have been given by Paolino and Pizzi (2008).

A further asymptotically flat, static, axially symmetric solution that describes a magnetic (or electric) dipole was given by Bonnor (1966a). This was further analysed by Emparan (2000), who argued that it represents two extreme Reissner–Nordström sources with equal mass and opposite (magnetic) charge, which he calls a *black dihole*. The two sources are held apart with a fixed distance between their horizons either by two semi-infinite cosmic strings or by a strut, represented by conical singularities. It was subsequently observed (see the note in Emparan and Teo, 2001) that this solution had also been obtained in a different form by Chandrasekhar and Xanthopoulos (1989). Emparan (2000) has also shown that the conical singularities could be eliminated by an appropriately chosen additional magnetic field.

Finally, it may be noted that all the solutions that have been described in this chapter are static. They represent the external fields of non-rotating axially symmetric systems whose generalisation to include rotation will be described in Chapter 13. However, the space-times represented by the Weyl solutions are generally not restricted to static situations. For example, the Schwarzschild solution, which is included in this class, is time-dependent inside its horizon. The point to be emphasised is that the metric (10.2) can only describe static regions of any such space-time. Whenever horizons exist in these solutions, the space-time can and should be extended through them to time-dependent regions. Some further solutions of this class which represent accelerating sources, and so also have non-static regions, will be described in Chapters 14 and 15.

11

Rotating black holes

The vacuum exterior of a spherically symmetric body is represented by the Schwarzschild space-time. If such a body collapses, it could form a black hole as described in Section 8.2. However, most astrophysically significant bodies are rotating. And, if a rotating body collapses, the rate of rotation will speed up, maintaining constant angular momentum. This is a most important process to model, the details of which are extremely complicated. Nevertheless, the end result could be expected to be a stationary rotating black hole. The field representing such a situation will be described in this chapter.

The Schwarzschild solution is the unique vacuum spherically symmetric space-time. It is also asymptotically flat, static, non-radiating (of algebraic type D), and includes an event horizon. The simplest rotating generalisation of this would be expected to be a stationary, axially symmetric and asymptotically flat space-time. To describe a rotating black hole, it should also include an event horizon. In fact, there is only one solution (Carter, 1971b) that satisfies all these properties, and that is the one obtained by Kerr (1963). The main geometrical properties of this very important type D solution will be described here. Other stationary, axially symmetric space-times that do not have horizons but could well represent the exteriors of some rotating sources will be described in Chapter 13.

11.1 The Kerr solution

Although it was not originally discovered in this form,[1] it is convenient to present this solution here in terms of the spheroidal-like coordinates of

---

[1] The metric of this important class of space-times was initially discovered in the Kerr–Schild form $g_{\mu\nu} = \eta_{\mu\nu} - 2S\,k_\mu k_\nu$, where $\eta_{\mu\nu}$ is the Minkowski metric (in some coordinate system), S is a scalar function, and k is a null vector with respect to both $g_{\mu\nu}$ and $\eta_{\mu\nu}$. The history of its discovery has been described by Kerr (2008).

Boyer and Lindquist (1967). In these coordinates, the Kerr metric can be expressed as

$$ds^2 = -\frac{\Delta_r}{\varrho^2}\left(dt - a\sin^2\theta\,d\phi\right)^2 + \frac{\varrho^2}{\Delta_r}\,dr^2 + \varrho^2\,d\theta^2 + \frac{\sin^2\theta}{\varrho^2}\left(a\,dt - (r^2 + a^2)d\phi\right)^2,$$

(11.1)

where

$$\varrho^2 = r^2 + a^2\cos^2\theta, \qquad \Delta_r = r^2 - 2mr + a^2,$$

and m and a are arbitrary constants. For this metric, relative to a null tetrad that is based on the repeated principal null directions, the only non-zero component of the curvature tensor is

$$\Psi_2 = -\frac{m}{(r + i\,a\cos\theta)^3},$$

(11.2)

confirming that, when $m \neq 0$, the space-time is of type D.

It may immediately be observed that the metric (11.1) reduces to that of Schwarzschild in the standard spherical form (8.1) when $a = 0$. In this limit, m is the *mass* of the source, and it may be assumed that this parameter will still describe some kind of mass even when $a \neq 0$. It will usually be assumed to be positive. The parameter a is interpreted as a *rotation parameter* and is found to measure the twist of the repeated principal null congruence. These are exactly the parameters that are required to represent the field of a source that has the independent physical quantities of mass and angular momentum. Since the Schwarzschild space-time has already been described in Chapter 8, it is assumed in the remainder of this chapter that $a \neq 0$.

The metric (11.1) is invariant under the transformation $a \to -a$ with either $\phi \to -\phi$ or $t \to -t$. Any of these replacements on their own would lead to a space-time for which the rotation is in the opposite direction. The space-time is also symmetric about the equatorial hyperplane $\theta = \pi/2$.

It can also be seen that the metric (11.1) is singular when $\Delta_r = 0$ or $\rho = 0$. However, in view of (11.2), a curvature singularity only occurs when $r = 0$, $\theta = \pi/2$, and $m \neq 0$. And, since $r = 0$ is not generally a singularity, the space-time may normally be extended to negative values of r. To understand this space-time, these singularities need to be investigated in detail.

These and some other properties will be described briefly below. Further properties, such as explicit Killing vectors, geodesics, physical processes, the metric in other useful coordinate systems, and perturbations, have been thoroughly described in the books specifically devoted to this topic by Chandrasekhar (1983), Frolov and Novikov (1998), O'Neill (1995) and Wiltshire, Visser and Scott (2008).

11.1.1 Horizons

When considering the character of the singularity at $\Delta_r = 0$, it may initially be observed from (11.1) that the coordinate r is spacelike when $\Delta_r > 0$ and timelike when $\Delta_r < 0$. The boundaries $\Delta_r = 0$ between these regions are thus Killing horizons associated with ∂_t. They are not curvature singularities, as can immediately be seen from (11.2).

In fact, provided $m^2 > a^2$, Δ_r has two roots at $r = r_+$ and $r = r_-$, where

$$r_{\pm} = m \pm \sqrt{m^2 - a^2}. \tag{11.3}$$

In the Schwarzschild limit in which $a \to 0$, the root $r = r_+$ corresponds to the event horizon at $r = 2m$, while the other root $r = r_-$ coincides with the curvature singularity at $r = 0$.

It may also be calculated that the determinant of the metric on the 3-surface on which r is constant is given by

$$\det(g_{\mu\nu})\Big|_{r=\text{const.}} = -\varrho^2 \Delta_r \sin^2 \theta.$$

This confirms that these surfaces are timelike when $\Delta_r > 0$, spacelike when $\Delta_r < 0$, and null when $\Delta_r = 0$. Indeed, the expressions (11.3) are remarkably similar to the expressions (9.16) for the locations of the inner and outer horizons of the Reissner–Nordström space-time, with the rotation parameter a here replacing the charge parameter e. It may therefore be concluded that the two null surfaces on which $\Delta_r = 0$ must be some kind of horizons, between which the space-time is time-dependent.

It is now convenient to introduce the constant quantities

$$\Omega_{\pm} = \frac{a}{2mr_{\pm}} = \frac{a}{r_{\pm}^2 + a^2}.$$

On the horizons, where $r = r_{\pm}$, the particular combination of Killing vectors given by $L_{\pm} = \partial_t + \Omega_{\pm} \partial_\phi$ satisfies the condition $g_{\mu\nu} L_{\pm}^{\mu} L_{\pm}^{\nu} = 0$. It follows that the family of curves $(t, r_{\pm}, \theta_0, \phi_0 + \Omega_{\pm} t)$, where θ_0 and ϕ_0 are constants, are a set of null lines on these horizons which rotate about the axis of symmetry with angular velocity $\Omega_{\pm}$. In contrast with the Schwarzschild and Reissner–Nordström cases, both horizons of the Kerr space-time at $r = r_{\pm}$ thus appear to rotate about the axis, though with different angular velocities.

Just as for the Schwarzschild space-time in standard coordinates (8.1), the Boyer–Lindquist coordinates of (11.1) are badly behaved near the horizons. In particular, for any infalling photon or particle, $t \to \infty$ as $r \to r_+$, as in the Schwarzschild case. However, in this case, it is also seen that $\phi \to \infty$, which corresponds to an infinite twisting of worldlines around the horizon.

Thus, to analyse the character of the Kerr space-time near the horizon, it is necessary to transform both the t and ϕ coordinates.

Following the method described by Misner, Thorne and Wheeler (1973), it is possible to construct Eddington–Finkelstein-type coordinates which cover these coordinate singularities by introducing an advanced[2] null coordinate v and an "untwisted" angular coordinate $\tilde{\phi}$ such that

$$\mathrm{d}v = \mathrm{d}t + \frac{r^2 + a^2}{\Delta_r}\,\mathrm{d}r, \qquad \mathrm{d}\tilde{\phi} = \mathrm{d}\phi + \frac{a}{\Delta_r}\,\mathrm{d}r.$$

With these, the metric (11.1) may be rewritten in the form

$$\mathrm{d}s^2 = -\left(1 - \frac{2mr}{\varrho^2}\right)\left(\mathrm{d}v - a\sin^2\theta\,\mathrm{d}\tilde{\phi}\right)^2 \tag{11.4}$$

$$+ 2\left(\mathrm{d}v - a\sin^2\theta\,\mathrm{d}\tilde{\phi}\right)\left(\mathrm{d}r - a\sin^2\theta\,\mathrm{d}\tilde{\phi}\right) + \varrho^2\left(\mathrm{d}\theta^2 + \sin^2\theta\,\mathrm{d}\tilde{\phi}^2\right),$$

which was originally obtained by Kerr (1963). In this form, the metric is clearly non-singular when $\Delta_r = 0$. In particular, it may be seen from (11.4) that incoming and outgoing trajectories on which θ and $\tilde{\phi}$ are constant, can be pictured in the same way as that described in Sections 8.2 and 9.2.1. It follows that the general character of the Kerr space-time with $m > a$ may be considered to be qualitatively similar to aspects of the Reissner–Nordström space-time as described in Section 9.2, and illustrated particularly in Figures 9.5 and 9.6.

The outer horizon at $r = r_+$ retains its character as an event horizon, exactly as in the non-rotating Schwarzschild and Reissner–Nordström cases: particles can pass through it in the inward direction, but not emerge from it into the exterior region. And a particle falling in through this horizon reaches the second (inner) horizon at $r = r_-$ in a finite time before possibly encountering the curvature singularity.

Between the horizons, the space-time is time-dependent but, inside the inner horizon $r < r_-$, it is again stationary. The conformal structure of this space-time will be considered below. But, from its similarity with the Reissner–Nordström space-time, it may be provisionally assumed that the inner horizon must behave as a Cauchy horizon. The character of the inner region, however, will need to be considered in more detail.

Although the general structure of their horizons may be similar, the Reissner–Nordström space-time is spherically symmetric, while the Kerr

---

[2] The label "advanced" may be misleading because, with $a \neq 0$, the hypersurfaces of constant v are actually timelike. A foliation of the space-time by null hypersurfaces that are asymptotic to Minkowski light cones at infinity has been given by Pretorius and Israel (1998), see (11.7).

space-time is not. Nevertheless, the surfaces on which t and r are constant (including the horizons) remain topologically 2-spheres.

The geometry of the surfaces on which $r = r_\pm$ and t is constant have been investigated in detail by Smarr (1973) (see also Visser, 2008). The area of each horizon is given by

$$\mathcal{A}_\pm = \int_0^{2\pi} \int_0^\pi \sqrt{g_{\theta\theta}\, g_{\phi\phi}} \Bigg|_{\substack{r=r_\pm \\ t=\text{const.}}} \mathrm{d}\theta \, \mathrm{d}\phi = 8\pi\, m\, r_\pm \, .$$

For the event horizon, this approximates to $\mathcal{A}_+ \approx 16\pi m^2 - 4\pi a^2$ to first order in a^2/m^2.

In particular, Smarr has shown that, as the rate of rotation increases, these surfaces become more oblate: i.e. the equatorial circumference of the horizon increases, while the polar circumference decreases. This is consistent with intuitive expectations based on the behaviour of material rotating bodies. And, for larger values of the rotation before the extreme limit is reached (specifically when $3a^2 > r_+^2$, i.e. for $\frac{3}{4}m^2 < a^2 < m^2$), the Gaussian curvature at the two polar regions of the event horizon becomes negative. In this case, this surface cannot be globally embedded in a Euclidean 3-space. The Gaussian curvature of the inner horizon is always negative near the two polar regions.

To conclude this section, it must of course be noted that, if $m^2 < a^2$, no horizons occur and the curvature singularity is naked. This case is referred to as a *hyperextreme* Kerr space-time. The limiting case in which $m^2 = a^2$, and which has a degenerate horizon, is referred to as an *extreme* Kerr space-time.

11.1.2 Curvature singularity and central disc

It follows from (11.2) that, provided $m \neq 0$, a curvature singularity occurs when $r = 0$, $\theta = \pi/2$. This occurs in the stationary region that, for the case when $m^2 > a^2$, is inside the inner horizon $r = r_-$.

The space-time is clearly flat when $m = 0$. In this limit, the metric (11.1) reduces to

$$\mathrm{d}s^2 = -\mathrm{d}t^2 + (r^2 + a^2\cos^2\theta)\left(\frac{\mathrm{d}r^2}{r^2 + a^2} + \mathrm{d}\theta^2\right) + (r^2 + a^2)\sin^2\theta\, \mathrm{d}\phi^2, \quad (11.5)$$

which can be obtained from the standard Cartesian form of the Minkowski metric (3.1) using the transformation

$$x = \sqrt{r^2 + a^2}\,\sin\theta\cos\phi,$$
$$y = \sqrt{r^2 + a^2}\,\sin\theta\sin\phi,$$
$$z = r\cos\theta.$$

Within this flat space, the coordinate lines on which r or θ are constant in the plane $\phi = 0, \pi$ are illustrated in Figure 11.1. Considering the full range of $\phi \in [0, 2\pi)$, it can be seen that, in this weak-field limit, the surface on which $r = 0$ is a *disc* of radius a in the plane $z = 0$. Moreover, the singularity at $r = 0$, $\theta = \pi/2$ is located on the circumference of this disc. However, this limit corresponds to the hyperextreme case. For this, the source of the space-time is a naked *ring singularity*.

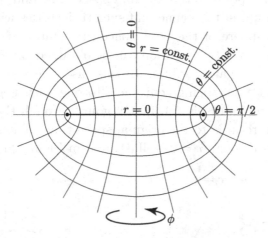

Fig. 11.1 Coordinate lines on which r or θ are constant in the plane $\phi = 0, \pi$ with $t = $ const. in the space (11.5) which represents the weak-field limit of the hyperextreme Kerr space-time. Lines with constant r are confocal ellipses, while the locally orthogonal lines of constant θ are hyperbolae.

For a rotating black hole solution which possesses outer and inner horizons, however, the structure of the singularity is more subtle. In this case, it is necessary that $a^2 < m^2$. The weak-field limit can therefore give no information as, when this constraint remains satisfied, the outer horizon shrinks to a point and all details of the interior structure are lost. Nevertheless, it is clear that $r = 0$ is not point-like. Only curves that approach it in the equatorial plane $\theta = \pi/2$ will encounter a curvature singularity. All other curves that approach $r = 0$ only reach a regular point.

For any constant t, the surface on which $r = 0$ in the space-time (11.1) has the metric

$$\mathrm{d}s_2^2 = a^2 \cos^2 \theta \, \mathrm{d}\theta^2 + a^2 \sin^2 \theta \, \mathrm{d}\phi^2.$$

This may be considered to represent a disc of radius a in the flat 2-space $\mathrm{d}s_2^2 = \mathrm{d}x^2 + \mathrm{d}y^2$, where $x = a \sin \theta \cos \phi$, $y = a \sin \theta \sin \phi$. Of this disc, it is only the rim on which $\theta = \pi/2$ that corresponds to the curvature singularity. All other points are regular. Thus, even when it is hidden inside a horizon, the curvature singularity of the Kerr space-time has the structure of a ring, and the behaviour of the coordinates near it may be assumed to be similar to that illustrated in Figure 11.1. Further aspects of the geometry of the hypersurface at $r = 0$, and the properties of the space-time near the ring singularity, have been described in detail by Punsly (1985, 1990).

All trajectories that reach the disc at $r = 0$ for values other than $\theta = \pi/2$, only reach a regular point of this vacuum space-time and may be extended through it. It is simplest to assume that such trajectories will re-emerge into the opposite hemisphere of the same space-time, but such an extension is not C^1. In fact, an analytic extension can be obtained simply by continuing the coordinate r to negative values. In this case, the extended space-time differs from the original, being equivalent to one with negative mass (for details, see Hawking and Ellis, 1973). On going through the disc again, the analytic extension returns to the original space-time. The structure of this analytic extension through $r = 0$ is illustrated in Figure 11.2.

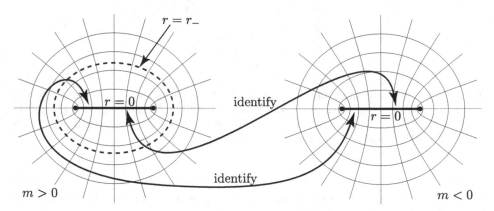

Fig. 11.2 The maximum analytic extension of the Kerr space-time through the disc at $r = 0$ is obtained by taking two space-times with positive and negative values for m, and identifying the upper surface of one disc with the lower surface of the other and vice versa. The space-time with $m > 0$ contains outer and inner horizons at $r = r_\pm$; that with $m < 0$ does not.

In the extended space-time in which mr is negative, curves near the ring singularity on which t, r and θ are constant are closed timelike curves. Carter (1968a) has pointed out that these can in fact be deformed to pass through any point of the analytically extended space-time so that causality is completely violated in this region. He has also shown that, although the extended space-time is geodesically incomplete at the ring singularity, the only timelike and null geodesics that reach it occur in the equatorial plane in the side with positive r.

11.1.3 Ergoregions

The coefficient g_{tt} in the metric (11.1) is $-(r^2 - 2mr + a^2 \cos^2\theta)/\varrho^2$, and this changes signs on the spacelike surfaces on which $r = r_{e\pm}$, where

$$r_{e\pm} = m \pm \sqrt{m^2 - a^2 \cos^2\theta}. \tag{11.6}$$

Thus, curves with constant values of r, θ and ϕ (i.e. curves aligned with the Killing vector ∂_t) are only timelike outside $r = r_{e+}$ or inside $r = r_{e-}$. Between these limits, such curves cannot be worldlines of real test particles, and all timelike trajectories with fixed r and θ must orbit about the centre in the same direction in which the source rotates. The outer limit $r = r_{e+}$ is thus known as the *stationary limit*. It is also the surface of infinite redshift since $g_{tt} = 0$ at $r = r_{e+}$.

By comparing (11.6) with (11.3), it can be seen that the surface r_{e+} is out-

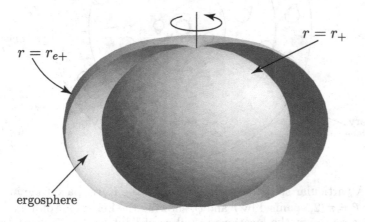

Fig. 11.3 The ergosphere is a region $r_+ < r < r_{e+}$ of the Kerr space-time with $m^2 > a^2$ that is just outside the oblate outer (event) horizon, inside which nothing can be seen. It is thickest at the equator and vanishes at the poles. The outer boundary of the ergosphere has a deficit angle at the poles. This illustration is schematic, not an embedding.

side the outer horizon r_+, which it touches at the poles $\theta = 0, \pi$. It is furthest from this horizon in the equatorial plane $\theta = \pi/2$, where $r_{e+} = 2m$. In the region for which $r_+ < r < r_{e+}$, the coordinates r and t are both spacelike. This region, which is known as the *ergosphere*, is illustrated schematically in Figure 11.3. Within it, it is not possible to "stand still", but it is possible to escape to infinity.

The significance of the ergosphere can be seen in the illustration given in Figure 11.4. This represents the equatorial plane $\theta = \pi/2$ of a particular spacelike section through the Kerr space-time. (This section will be clarified in Figure 11.7.) The position of light emitted from the points indicated by heavy dots is represented a short time later by circles. These circles can be seen to rotate about the centre as well as being attracted towards it. Inside the stationary limit surface, the points are outside their corresponding circles, indicating that curves with constant values of r, θ and ϕ (and hence $\tilde{\phi}$) cannot be timelike in these regions. It is also evident from Figure 11.4 that, within the ergosphere, non-rotating stationary observers are forced to co-rotate with the black hole.

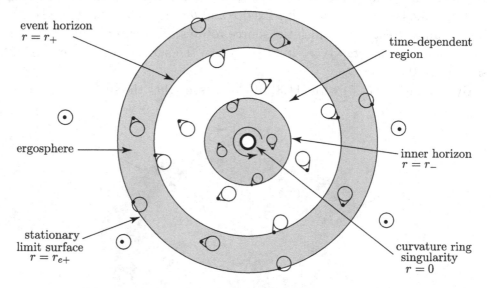

Fig. 11.4 A particular spacelike section through the Kerr space-time in the equatorial plane $\theta = \pi/2$, spanned by r and $\tilde{\phi}$, showing the behaviour of light cones. The small circles represent the locations of pulses of light emitted a short time before at the points that are represented by heavy dots. The ergosphere and the inner ergoregion are shaded. In this equatorial plane, stationary points are not possible anywhere inside the stationary limit surface.

Properties of the stationary limit surface $r = r_{e+}$ have been evaluated in detail by Pelavas, Neary, and Lake (2001). In particular they have shown that this surface remains topologically spherical and has a surface area given approximately by $16\pi m^2 + 4\pi a^2$ to first order in a^2/m^2. They have also shown that this surface has a deficit angle on the axis.

There is also an equivalent ergoregion inside the inner horizon which, on the equatorial plane $\theta = \pi/2$, actually reaches as far in as the curvature singularity where $r = 0$. This is illustrated schematically in Figure 11.5.

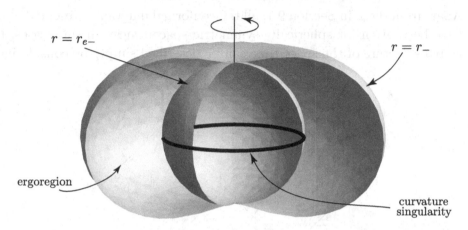

Fig. 11.5 The stationary region inside the inner horizon $r = r_-$ of the Kerr space-time is illustrated schematically at any time t. The inner horizon has negative curvature near the poles (though the axis remains regular). The curvature singularity at $r = 0$, $\theta = \pi/2$ is a ring of radius a. The surface on which $r = 0$ is the disc bounded by this ring. Those trajectories that approach the disc $r = 0$ with $\theta \neq \pi/2$ encounter regular points. The complementary ergoregion $0 \leq r_{e-} < r < r_-$ reaches from the inner horizon to the singularity.

For hyperextreme Kerr space-times that do not contain horizons, the er-

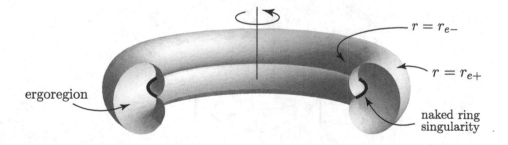

Fig. 11.6 The stationary limits of the ergoregion for a hyperextreme Kerr space-time can be represented explicitly in the weak-field limit given by (11.5). This region extends to the naked curvature singularity which has the structure of a ring.

gosphere has the topology of a toroidal region about the axis whose inner circle is the naked ring singularity. Using the metric of the Minkowski weak-field limit (11.5), the inner and outer boundaries of the ergoregion (11.6) can be represented explicitly. These are illustrated in Figure 11.6.

11.1.4 Conformal diagrams

Apart from those in Section 9.1, all the conformal diagrams presented above have been given for spherically symmetric space-times. In such cases, the global structure of the space-time can be described simply by considering a

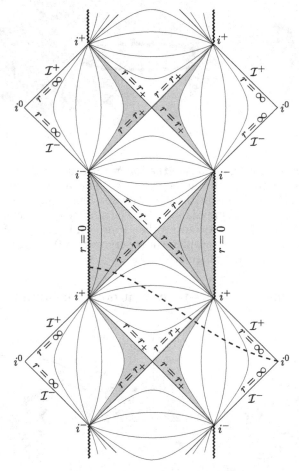

Fig. 11.7 The conformal diagram for the Kerr space-time in the equatorial plane $\theta = \pi/2$, with $\tilde{\phi}$ constant. The dashed line indicates the type of spacelike section that was illustrated in Figure 11.4. All other lines are those on which r is constant. The shaded areas indicate the outer and inner ergoregions.

worldsheet on which θ and ϕ are constants, and every point on the resulting diagram may be considered to represent a 2-sphere. However, the Kerr space-time is not spherically symmetric: its global structure is θ-dependent.

In view of the existence of Eddington–Finkelstein-like coordinates as previously used in (11.4), it is clear that there is no difficulty in extending the space-time across the horizons. This was first achieved explicitly, together with the appropriate compactification, by Carter (1966b, 1968a).

In fact, in the equatorial plane in which $\theta = \pi/2$, the conformal diagram for the maximally extended Kerr space-time is formally identical to that for

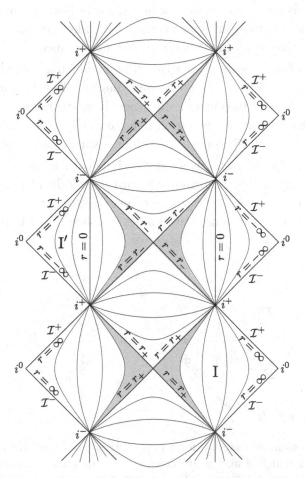

Fig. 11.8 The conformal diagram for the maximally extended Kerr space-time for a worldsheet on which θ and $\tilde{\phi}$ are constant, with $\theta \neq \pi/2$. In this general case, the space-time can be analytically extended through $r = 0$ to a new region I′ with r negative. This extended region includes no horizons and may be extended to conformal infinity $\mathcal{I}$. The shaded areas indicate the outer and inner ergoregions.

the Reissner–Nordström space-time given in Figure 9.7. This is repeated in Figure 11.7, which also indicates the type of spacelike section that was used in Figure 11.4. However, if $\theta \neq \pi/2$, then $r = 0$ is not a curvature singularity and the space-time can be analytically extended through it to negative values of r, and to a new conformal infinity at $r = -\infty$, as shown in Figure 11.8.

It can now be seen that, relative to an exterior region labelled I in Figure 11.8, the horizons at $r = r_+$ have the character of a black or white hole event horizon: events inside the black hole horizon cannot be seen from i^+. Also $r = r_-$ has the character, either of an inner Killing horizon inside which the space-time is again stationary, or of a Cauchy horizon just as described for the Reissner–Nordström space-time. Extended regions, such as I′ in which r is negative, contain no horizons but are acausal.

For the case of an *extreme* Kerr space-time in which $a^2 = m^2$, the horizons (11.3) degenerate. Again, the conformal diagram for the maximal extension resembles that for the extreme Reissner–Nordström space-time given in Figure 9.10 except that, for $\theta \neq \pi/2$, it is generally possible to extend the space-time through $r = 0$ to negative values, exactly as above.

Finally, for a *hyperextreme* Kerr space-time with $a^2 > m^2$, there are no horizons. And again the conformal diagram is exactly like that for the hyperextreme Reissner–Nordström space-time of Figure 9.9, but with a general extension through $r = 0$ to negative values, as shown in Figure 11.9.

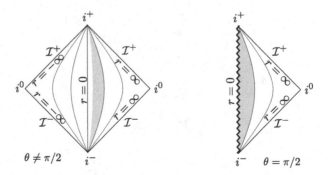

Fig. 11.9 The conformal diagram for a hyperextreme $a^2 > m^2$ Kerr space-time for a worldsheet on which θ and $\tilde{\phi}$ are constant. Generally, the space-time can be analytically extended through $r = 0$ to a new region with r negative. However, when $\theta = \pi/2$, a curvature singularity occurs at $r = 0$. The shaded area represents the ergoregion which, when $\theta = \pi/2$, extends to $r = 0$.

11.1.5 Further comments on the Kerr solution

The gravitational field of any black hole may be investigated by considering the orbits of test particles (timelike geodesics) around it. In fact, as for all vacuum type D space-times, the geodesic equations for the Kerr metric are separable and can be integrated explicitly. This was first achieved by Carter (1968a,b) who expressed them in terms of quadratures.

At large distances, the field is effectively Newtonian and approximately Keplerian elliptic orbits could occur. However, corrections are needed for orbits that pass closer to the outer horizon. Wilkins (1972) has investigated bound geodesics, and Boyer and Lindquist (1967) and Bardeen (1973) have analysed geodesics in the equatorial plane. For descriptions of geodesics using Boyer–Lindquist coordinates, see Chandrasekhar (1983) and Plebański and Krasiński (2006). (See also the references in Sharp, 1979.) A modern study that includes a possible cosmological constant has been undertaken by Kraniotis (2004, 2005, 2007) – see also references therein.

By a careful examination of geodesics in the equatorial plane in the weak-field limit of a Kerr space-time, Boyer and Price (1965) have shown that a test particle experiences the equivalent of a Coriolis-type force due to a rotating body with angular momentum ma. This interpretation has been confirmed by Cohen (1968) using integral methods.

Interestingly, the geodesic equations also reveal that the central disc on which $r = 0$ is repulsive (Carter, 1968a). Moreover, the repulsive effects appear to be most effective on the axis of symmetry and vanish in the equatorial plane. These properties have been further investigated by de Felice and Bradley (1988), who identified the region around this disc in which the gravitational field has a net repulsive effect. Specifically, the boundary between the net attractive and net repulsive regions is the hypersurface $r = a \cos \theta$ which, at any time, has the topology of a 2-sphere with equator corresponding to the curvature singularity. The existence of such a repulsive component in the gravitational field which appears to be aligned with the axis is also manifested in the spin-spin interaction between aligned rotating bodies that will be described in Section 13.2.2.

An axially symmetric null foliation of this space-time that is useful in a number of contexts has been constructed by Pretorius and Israel (1998). By introducing a tortoise-like coordinate $r_\star$ and a latitudinal coordinate λ, they have re-expressed the Kerr metric (11.1) in the form

$$\mathrm{d}s^2 = \frac{\Delta_r}{R^2}(-\mathrm{d}t^2 + \mathrm{d}{r_\star}^2) + \frac{L^2}{R^2}\,\mathrm{d}\lambda^2 + R^2 \sin^2 \theta (\mathrm{d}\phi - \omega_B\,\mathrm{d}t)^2. \qquad (11.7)$$

Unfortunately, this has the drawback that the metric functions R, L and

ω_B are not expressible in an explicit elementary form because the new coordinates $r_\star$ and λ are elliptic functions of the Boyer–Lindquist coordinates. Using this, they have constructed a foliation of the Kerr space-time by null hypersurfaces $t \pm r_\star = $ const. that are asymptotic to standard light cones of Minkowski space at $\mathcal{I}^+$. These have the important property that they are free of caustics for all positive values of the coordinate r.

Because of its importance as space-time with appropriate properties for the exterior of a rotating source, many people have tried unsuccessfully to find a possible perfect fluid interior to the Kerr space-time. It is now known that a rigidly rotating interior fluid solution cannot be matched to an exterior Kerr metric. The search for a perfect fluid interior is continuing.

The possible collapse of a non-rotating spherically symmetric body to form a black hole has already been outlined in Sections 8.2 and 9.2, and illustrated in Figures 8.2 and 9.6. It is now appropriate to include some general comments on whether an initially rotating body could collapse to form a black hole whose exterior is described by the Kerr metric.

Consider a body that has some small initial rotation, and whose exterior field is at least approximately given by the Kerr metric. If it begins to collapse, then, as its size reduces, the speed of its rotation would increase in order to conserve angular momentum. As this happens, centrifugal forces arise that could possibly halt the collapse or change its structure. Interestingly, many large astrophysical objects have an angular momentum parameter a that is currently significantly greater than their mass m. However, the parameters a and m for compact objects such as neutron stars are of the same order. It may thus be concluded that, as a large body collapses, some mechanism must exist by which angular momentum is reduced. Such processes include the fragmentation of the body, the radiation of matter and gravitational waves and the effects of magnetism. Only after such processes could a body collapse to form a rotating black hole rather than a naked singularity. (For details of this and other aspects of the physics of black holes, see Frolov and Novikov, 1998.) On the other hand, if a body progresses through a sequence of quasistationary rotating states, Meinel (2004) has shown that the only possible limit is the extreme Kerr space-time.

The source of a Kerr space-time could alternatively be a thin disc. A solution for such a case could be constructed by making two symmetrical spacelike cuts of the Kerr space-time either side of the singularity, discarding the middle region, and pasting together the remaining surfaces. Although the metric can be kept continuous across the junction, discontinuities in the derivatives of the metric on this surface will introduce distributional components in the Ricci tensor, which could be interpreted in terms of a thin disc of counter-rotating particles. However, unlike the static Weyl solutions of this type, the two sets of counter-rotating particles are not symmetric. Such a situation has been analysed in detail by Bičák and Ledvinka (1993), who have shown that as the "cuts" approach the event horizon, the speed of the particles can approach that of light, and the space-time can include a toroidal ergoregion. Although such a source extends to infinity, it has finite mass and exhibits a number of interesting properties. (For similar discs with non-zero pressure, see Pichon and Lynden-Bell, 1996.)

11.2 The Kerr–Newman solution

The charged version of the Kerr space-time, that is still of type D and for which the electromagnetic field is aligned, is given by the metric

$$ds^2 = -\frac{\Delta_r}{\varrho^2}\left(dt - a\sin^2\theta\, d\phi\right)^2 + \frac{\varrho^2}{\Delta_r}\, dr^2 + \varrho^2\, d\theta^2 + \frac{\sin^2\theta}{\varrho^2}\left(a\, dt - (r^2+a^2)d\phi\right)^2,$$

(11.8)

with

$$\varrho^2 = r^2 + a^2\cos^2\theta, \qquad \Delta_r = r^2 - 2mr + a^2 + e^2,$$

where m and a have the same meaning as above and the additional parameter e denotes the charge of the source. This is accompanied by a non-zero electromagnetic field $F = dA$, where the vector potential is

$$A = -\frac{e\, r}{r^2 + a^2\cos^2\theta}\left(dt - a\sin^2\theta\, d\phi\right).$$

(11.9)

The metric (11.8) is known as the Kerr–Newman solution. It was first presented explicitly by Newman *et al.* (1965) and clearly includes both the Kerr solution (11.1) when $e = 0$ and the Reissner–Nordström solution (9.14) when a vanishes.

Apart from the inclusion of an electromagnetic field, the gravitational properties of this space-time are qualitatively the same as those of the Kerr space-time described above (see Carter, 1968a). The locations of the horizons in this case are determined by the roots of the quadratic function $\Delta_r = r^2 - 2mr + a^2 + e^2$ and, provided $m^2 > a^2 + e^2$, these are qualitatively the same as those for the special cases in which either a or e are zero. Thus, the conformal diagrams for the Kerr–Newman family of space-times are analogous to those for the uncharged Kerr space-times or, in the equatorial plane, to those of the non-rotating Reissner–Nordström solutions. This clarifies why the conformal diagrams for these two cases are so similar.

It is a most remarkable observation that, if a material body of arbitrary structure were to collapse inside an event horizon, then details of its structure must be radiated away and the final exterior field must almost certainly reduce to a Kerr–Newman space-time that is characterised by just three parameters – its total mass m, its total charge e and total angular momentum am. All other features of its original structure are lost. This is known as the *no-hair conjecture*. See Carter (1987) and references therein for a classical discussion of this theorem, and Chruściel (1994, 1996) and Heusler (1998) for more critical reviews.

11.3 The Kerr–Newman–(anti-)de Sitter solution

The above family of solutions can also be further generalised to include a non-zero cosmological constant Λ. Such a case is usually referred to as the Kerr–Newman–(anti-)de Sitter space-time. Its metric can be expressed in standard Boyer–Lindquist-type coordinates (Gibbons and Hawking, 1977) in the form

$$
\begin{aligned}
ds^2 = -\frac{\Delta_r}{\Xi^2 \varrho^2}\Big(dt - a\sin^2\theta\,d\phi\Big)^2 + \frac{\varrho^2}{\Delta_r}\,dr^2 + \frac{\varrho^2}{\Delta_\theta}\,d\theta^2 \\
+ \frac{\Delta_\theta \sin^2\theta}{\Xi^2 \varrho^2}\Big(a\,dt - (r^2 + a^2)\,d\phi\Big)^2,
\end{aligned}
\tag{11.10}
$$

where

$$
\begin{aligned}
\varrho^2 &= r^2 + a^2\cos^2\theta, \\
\Delta_r &= (r^2 + a^2)(1 - \tfrac{1}{3}\Lambda r^2) - 2mr + e^2, \\
\Delta_\theta &= 1 + \tfrac{1}{3}\Lambda a^2\cos^2\theta, \\
\Xi &= 1 + \tfrac{1}{3}\Lambda a^2,
\end{aligned}
$$

and the potential for the electromagnetic field is the same as in (11.9). Formally, it is possible to remove the constant Ξ by applying a rescaling to the coordinates t and ϕ. However, this constant is included here so that the metric has a well-behaved axis both at $\theta = 0$ and $\theta = \pi$ with $\phi \in [0, 2\pi)$. It should also be noted, however, that the metric (11.10) only retains Lorentzian signature for all $\theta \in [0, \pi]$ provided $\tfrac{1}{3}\Lambda a^2 > -1$.

The main qualitative differences between this solution and (11.8) without the cosmological constant concern its behaviour at large distances. It can be seen that the function $\Delta_r(r)$ is now a quartic, and this may possess four (or two, or no) real roots. When roots exist, two will typically represent the outer and inner black hole horizons around the singularity. These correspond to the event and Cauchy horizons of the Kerr black hole. And, if $\Lambda > 0$, there must exist two roots which represent cosmological horizons. One of these will occur in the region with positive r, while the other appears in the analytically extended region in which r is negative. In addition, the character of conformal infinity $\mathcal{I}$ will be spacelike or timelike according to whether Λ is positive or negative (see Chapters 4 and 5, respectively).

Fig. 11.10 The schematic conformal structure of the analytically extended Kerr–Newman–de Sitter space-time with $\theta \neq \pi/2$. Regions for which $r < 0$ are indicated in a darker shading. (If $\theta = \pi/2$, these areas are omitted as $r = 0$ is a curvature singularity.) In the outwardly extended regions, de Sitter-like conformal infinities $\mathcal{I}^+$ and $\mathcal{I}^-$ each occur for both $r = \infty$ and $r = -\infty$. The outer event horizons at $r = r_+$ are indicated by $\mathcal{H}_{\mathrm{o}}$, while the inner Cauchy horizons at $r = r_-$ are labelled $\mathcal{H}_{\mathrm{i}}$. Cosmological horizons in regions with both positive and negative r are denoted by $\mathcal{H}_{\mathrm{c}}$. Extensions through different asymptotic regions in which r has different signs are multiply connected. They do not necessarily reach the same regions, and appropriate cuts are indicated by the broken lines.

The structure of the analytically extended Kerr–Newman–de Sitter space-time, for which $\Lambda > 0$, is represented in the conformal diagram given in Figure 11.10 (see Gibbons and Hawking, 1977). Cosmological horizons exist in this case. Thus, as well as an infinite number of possible extensions in the timelike direction through successive outer and inner horizons, an indefinite number of spacelike extensions are also possible through each cosmological horizon. However, although extensions taken in a different order through successive horizons could lead to the same region, this is not necessary, and multiply connected extensions may be assumed to occur.

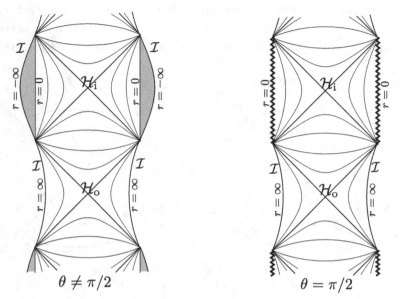

Fig. 11.11 The conformal structure of the analytically extended Kerr–Newman–anti-de Sitter space-time for which $\mathcal{I}$ is timelike. If $\theta = \pi/2$, there is a curvature singularity at $r = 0$. However, when $\theta \neq \pi/2$, the space-time can be extended to the shaded regions in which $r < 0$. The outer event horizons at $r = r_+$ are indicated by $\mathcal{H}_{\mathrm{o}}$, and the inner Cauchy horizons at $r = r_-$ are labelled $\mathcal{H}_{\mathrm{i}}$.

In the equatorial plane, a curvature singularity occurs at $r = 0$. However, when $\theta \neq \pi/2$, the space-time can be extended to negative values of r and then through an alternative cosmological horizon, and subsequently to conformal infinity or to another region with $r > 0$. This leads to additional opportunities for multiple connectivity.

The global structure of the analytically extended Kerr–Newman–anti-de Sitter space-time, for which $\Lambda < 0$, is much simpler as conformal infinity $\mathcal{I}$ is then timelike and cosmological horizons do not occur. It is represented in the conformal diagrams given in Figure 11.11. A curvature singularity only occurs for the section on which $\theta = \pi/2$. For all other values of θ, the space-time can be extended through $r = 0$ to regions in which r is negative and out to a new $\mathcal{I}$.

Further generalisations of the Kerr space-time, which are still of algebraic type D, will be described in Chapter 16. As well as charge and the cosmological constant, these include a number of additional parameters.

12

Taub–NUT space-time

In this chapter, we describe what is widely known as the Taub–NUT solution. This was first discovered by Taub (1951), but expressed in a coordinate system which only covers the time-dependent part of what is now considered as the complete space-time. It was initially constructed on the assumption of the existence of a four-dimensional group of isometries so that it could be interpreted as a possible vacuum homogeneous cosmological model.

This solution was subsequently rediscovered by Newman, Tamburino and Unti (1963) as a simple generalisation of the Schwarzschild space-time. And, although they presented it with an emphasis on the exterior stationary region, they expressed it in terms of coordinates which cover both stationary and time-dependent regions. In addition to a Schwarzschild-like parameter m which is interpreted as the mass of the source, it contained two additional parameters – a continuous parameter l which is now known as the NUT parameter, and the discrete 2-space curvature parameter which is denoted here by ϵ. It is only the case in which $\epsilon = +1$, which includes the Schwarzschild solution, that was obtained by Taub. The cases with other values of ϵ are generalisations of the other A-metrics.

We will follow the usual convention of referring to the case in which $\epsilon = +1$ as the Taub–NUT solution. However, there are two very different interpretations of this particular case. Both of these have unsatisfactory aspects in terms of their global physical properties. In one interpretation, the space-time contains a semi-infinite line singularity, part of which is surrounded by a region that contains closed timelike curves. The other interpretation, which is due to Misner (1963), contains no singularities. These are removed only at the expense of introducing a periodic time coordinate throughout the stationary region. However, this also has other undesirable features, such that Misner was led to conclude that the complete space-time has no reasonable physical interpretation.

Nevertheless, the Taub–NUT solution is a simple non-radiating exact solution of Einstein's field equations, and it is important to investigate its properties, if only to determine the kind of structures and pathologies that could appear within this theory. Moreover, this solution is nowadays also being reconsidered in the context of higher-dimensional theories of semi-classical quantum gravity. Its properties therefore need to be fully understood at the basic classical level.

In recent years, it has become common to refer to the NUT parameter l as the *magnetic mass* or the *gravitomagnetic monopole moment*. This interpretation is based on an analogue of one aspect of a property of one of the possible interpretations, but is not relevant in the context of the alternative global interpretation of the space-time.

It may finally be remarked here that, in the second (Misner's) interpretation, the Taub–NUT solution has a number of properties that are usually considered to be undesirable in any reasonable representation of a space-time. This is so much the case that Misner (1967) has presented it as "a counter-example to almost anything". These properties will be described below. They need to be understood because they clarify the need for some of the precise requirements that should be imposed on a space-time and particularly, for example, in conditions for singularity theorems.

12.1 The NUT solution

The NUT solution is often quoted in the form

$$\mathrm{d}s^2 = -f(r)\left(\mathrm{d}\bar{t} - 2l\cos\theta\,\mathrm{d}\phi\right)^2 + \frac{\mathrm{d}r^2}{f(r)} + \left(r^2 + l^2\right)\left(\mathrm{d}\theta^2 + \sin^2\theta\,\mathrm{d}\phi^2\right), \quad (12.1)$$

where

$$f(r) = \frac{r^2 - 2mr - l^2}{r^2 + l^2}, \quad (12.2)$$

and m and l are constants, of which l is known as the NUT parameter.[1] It is always possible to take $m \geq 0$ since, if m were negative, its sign could be reversed by the coordinate transformation $r \to -r$.

When $l = 0$ and $m \neq 0$, the metric reduces to the Schwarzschild solution (8.1) in which m is the familiar parameter representing the mass of the source. Newman, Tamburino and Unti (1963) have shown that, when l is small, the inclusion of this parameter induces a small additional advance in

---

[1] It may be noted that the NUT solution as originally obtained by Newman, Tamburino and Unti (1963) also includes the additional 2-space curvature parameter ϵ. This is given in Section 12.3. The convention that only the form in which $\epsilon = +1$ is known as the NUT solution follows that of Misner (1963).

the perihelion of approximately elliptic orbits in the stationary region of the space-time. However, when $l \neq 0$, the space-time has very different global properties to that of Schwarzschild, as will be shown below.

Because of the $l \cos \theta$ term in the metric (12.1), the space-time does not have a well-behaved axis at both $\theta = 0$ and $\theta = \pi$. However, it is possible to introduce a regular axis at $\theta = 0$ by applying the transformation $\bar{t} = t + 2l\phi$. This takes the line element (12.1) to

$$\mathrm{d}s^2 = -f(r)\left(\mathrm{d}t + 4l\sin^2 \tfrac{1}{2}\theta\,\mathrm{d}\phi\right)^2 + \frac{\mathrm{d}r^2}{f(r)} + (r^2 + l^2)\left(\mathrm{d}\theta^2 + \sin^2\theta\,\mathrm{d}\phi^2\right),$$

$$(12.3)$$

which is the form that will generally be adopted below.

In terms of a natural tetrad with $\boldsymbol{k} = \frac{1}{f(r)}\partial_t + \partial_r$ and $\boldsymbol{l} = \frac{1}{2}\partial_t - \frac{f(r)}{2}\partial_r$, some of the spin coefficients are given by $\rho = -\frac{1}{r+\mathrm{i}l}$, $\kappa = \sigma = \lambda = \nu = 0$ and $\mu = -\frac{f(r)}{2(r+\mathrm{i}l)}$, and the only non-zero component of the Weyl tensor is

$$\Psi_2 = -\frac{m + \mathrm{i}l}{(r + \mathrm{i}l)^3}.$$

From this it is clear that the space-time is of Petrov type D, and that both repeated principal null congruences are geodesic and shear-free but have non-zero expansion and twist. Moreover, the twist of each congruence is proportional to the NUT parameter l. This parameter may therefore initially be regarded as a *twist parameter*, although the possible sources of this twist need to be investigated.

It may also be observed that this space-time is asymptotically flat in the sense that the Riemann tensor decays as r^{-3} as $r \to \pm\infty$. However, the metric (12.3) clearly contains some kind of singularity along the half axis $\theta = \pi$. Thus the solution cannot be *globally* asymptotically flat.

Also, in contrast to the Schwarzschild solution, all components of the Riemann tensor are analytic functions of r over any range when $l \neq 0$: *no scalar polynomial curvature singularity exists* for any value of r. It is therefore natural for r to cover the full range $r \in (-\infty, \infty)$. However, a singularity of some kind clearly occurs whenever $f(r) = 0$. In fact, $f(r)$ given by (12.2) has two distinct roots with positive and negative values of r given by

$$r_\pm = m \pm \sqrt{m^2 + l^2}.$$

$$(12.4)$$

These clearly correspond to two distinct Killing horizons associated with the symmetry generated by ∂_t. Their properties will be considered in detail below. For $r < r_-$ and $r > r_+$, $f(r) > 0$, so that the coordinate r is spacelike and the metric is stationary. These will be referred to as the NUT regions, namely NUT$_-$ and NUT$_+$, respectively. These are separated by a time-dependent (Taub) region $r_- < r < r_+$ in which r is timelike and t spacelike.

It is, of course, possible to regard this solution as the exterior field of some object with radius greater than r_+. Indeed, interior solutions of this type have been obtained by Bradley *et al.* (1999). These contain a rotating perfect fluid which does not have a unique axis of rotation, but appears to rotate about all radial directions. With such an interior, there is no need to consider the time-dependent and the other stationary regions, but the singular half-axis at $\theta = \pi$ still needs to be explained in this context. In the following subsections, however, the properties of this space-time will continue to be explored over the complete natural range of the coordinates.

12.1.1 *The NUT regions in Weyl coordinates*

Within the stationary NUT regions in which $f(r) > 0$, the solution can be expressed in terms of Weyl's coordinates with the metric in the Lewis–Papapetrou form for stationary axisymmetric systems (see Chapter 13): i.e.

$$ds^2 = -e^{2U}(d\bar{t} + A\,d\phi)^2 + e^{-2U}\left[e^{2\gamma}(d\rho^2 + dz^2) + \rho^2 d\phi^2\right]. \tag{12.5}$$

In particular, the NUT solution is given by

$$e^{2U} = \frac{(R_+ + R_-)^2 - 4(m^2 + l^2)}{(R_+ + R_- + 2m)^2 + 4l^2}, \qquad A = -\frac{l}{\sqrt{m^2 + l^2}}\,(R_+ - R_-),$$

$$e^{2\gamma} = \frac{(R_+ + R_-)^2 - 4(m^2 + l^2)}{4R_+ R_-}, \qquad R_\pm^2 = \rho^2 + \left(z \pm \sqrt{m^2 + l^2}\right)^2,$$

$$\tag{12.6}$$

(see Gautreau and Hoffman, 1972; and Stephani *et al.*, 2003). When $l = 0$, these expressions reduce to the corresponding ones (10.15) for the Schwarzschild solution. However, when $l \neq 0$, the function A is non-zero. As pointed out in Stephani *et al.* (2003), this indicates that the NUT solution can be considered as an exterior field of a *rotating* source in which the NUT parameter l is a measure of the angular momentum.

With the expressions (12.6), the metric (12.5) can be seen to reduce exactly to the form (12.1) by putting

$$\rho = \sqrt{r^2 - 2mr - l^2}\,\sin\theta, \qquad z = (r - m)\cos\theta,$$

so that $R_\pm = r - m \pm \sqrt{m^2 + l^2}\cos\theta$. It can immediately be seen from this that the axis $\rho = 0$ of the Weyl coordinates corresponds to the horizon $r = r_\pm$ as well as the axis $\theta = 0, \pi$. For the NUT$_+$ region, the different parts of the axis have meanings that are equivalent to those of the Schwarzschild solution, namely:

$$\text{half axis} \quad \theta = \pi \quad \leftrightarrow \quad \rho = 0, \qquad z \leq -\sqrt{m^2 + l^2},$$

$$\text{horizon} \quad r = r_+ \quad \leftrightarrow \quad \rho = 0, \quad -\sqrt{m^2 + l^2} < z < \sqrt{m^2 + l^2},$$

$$\text{half axis} \quad \theta = 0 \quad \leftrightarrow \quad \rho = 0, \qquad \sqrt{m^2 + l^2} \leq z.$$

For the NUT$_-$ region, the order is reversed. With the expressions (12.6), both parts of the axis are singular. But either of these could be made regular by applying the transformation $\bar{t} = t \pm 2l\phi$ or by adding $\pm 2l$ to A. It can also be seen that the usual potential function $U(\rho, z)$ can be expressed as

$$e^{2U} = \frac{R_+ + R_- - 2m - \frac{4l^2}{R_+ + R_- + 2m}}{R_+ + R_- + 2m + \frac{4l^2}{R_+ + R_- + 2m}},$$

which indicates that the effect of NUT parameter l vanishes asymptotically for large values of $R_\pm$, i.e. for $\rho \gg \sqrt{m^2 + l^2}$.

Circular orbits of test particles in this space-time have been investigated by Kagramanova and Ahmedov (2006) using these coordinates. Their results are consistent with those described in Section 12.1.3 below.

12.1.2 The Taub region

In the region $r_- < r < r_+$, in which $f(r) < 0$, the coordinate r in the metric (12.3) is timelike and t is spacelike. In fact, it was this region, which will be referred to here as the Taub region, that was first presented in a different form by Taub (1951) as a vacuum homogeneous cosmological model.[2]

In this region, it is sometimes convenient to rewrite the metric in an alternative form. First put $r = \tau$, $f(r) = -V(\tau)$, and $t = 2l(\psi' - \phi)$, so that (12.3) becomes

$$ds^2 = -\frac{d\tau^2}{V(\tau)} + (2l)^2 V(\tau)(d\psi' - \cos\theta \, d\phi)^2 + (\tau^2 + l^2)(d\theta^2 + \sin^2\theta \, d\phi^2), \quad (12.7)$$

where

$$V(\tau) = \frac{l^2 + 2m\tau - \tau^2}{l^2 + \tau^2}, \qquad (12.8)$$

and $r_- < \tau < r_+$. Then, with the further Eddington–Finkelstein-like trans-

---

[2] Part of this region may also be interpreted as the interaction region of some specific colliding plane wave space-times. This will be described in Section 21.5.3.

formation $\psi' = \psi + \frac{1}{2l} \int V^{-1}(\tau) d\tau$, this line element takes the form

$$ds^2 = 4l(d\psi - \cos\theta\, d\phi)d\tau + V(\tau)(2l)^2(d\psi - \cos\theta\, d\phi)^2$$
$$+(\tau^2 + l^2)(d\theta^2 + \sin^2\theta\, d\phi^2). \tag{12.9}$$

This form is often used, but it has the obvious disadvantage that it does not possess a nice limit as $l \to 0$.

Within the Taub region, the timelike coordinate r varies between the two limits r_- and r_+. This would indicate that trajectories remain within this region for a finite proper time given by

$$\int_{r_-}^{r+} \frac{d\tau}{\sqrt{V(\tau)}} = \int_{r_-}^{r+} \frac{\sqrt{r^2 + l^2}}{\sqrt{r - r_-}\sqrt{r_+ - r}}\, dr.$$

It is interesting that this expression is independent of all other coordinates. Thus all trajectories which enter the Taub region, remain within it for the same amount of proper time. It is therefore possible to regard the Taub region as a vacuum homogeneous cosmological model (it is of Bianchi type IX) which exists for a finite time but for which the initial and final states are not curvature singularities.[3]

12.1.3 The singularity on the axis and a gravitomagnetic interpretation

It may be seen that, when $l \neq 0$ in the metric (12.3), the half of the symmetry axis on which $\theta = 0$ is regular. It is therefore appropriate to consider ϕ as a periodic coordinate in which $\phi = 2\pi$ is identified with $\phi = 0$. However, with this identification, the other half of the axis on which $\theta = \pi$ is singular.[4] (Of course, this could be reversed, or the entire axis made singular by a transformation of the coordinate t.) And, since the curvature tensor components remain bounded everywhere, this is a quasi-regular singularity rather than a scalar polynomial curvature singularity.

One approach to this singular half-axis has been advocated by Bonnor (1969a) (see also Sackfield, 1971; and Dowker, 1974). This treats the singularity at $\theta = \pi$ as the natural idealisation of a semi-infinite massless source

---

[3] Wheeler (1980) has constructed a specific case of the Taub universe which will exist for a period as long as that of the Friedman closed dust model.

[4] It is possible to take a spacelike section through the non-singular part of the axis (outside the horizon) and glue the resulting nonsingular part to its mirror image. This produces a stationary solution with a regular axis, but introduces non-zero components of the Ricci tensor on the junction between the two parts. This can be interpreted physically as a thin disc of counter-rotating particles. Such a model has been constructed by González and Letelier (2000), in which they have shown that such a model for the NUT solution includes a non-zero heat flow.

of angular momentum: i.e. the source of the NUT metric is considered as a thin semi-infinite spinning rod. This is justified as, in the weak-field limit in which l is small and $m = 0$, the metric perturbation corresponds to that of a semi-infinite spike of angular momentum.

An interesting electromagnetic analogue of this solution has been described by Lynden-Bell and Nouri-Zonoz (1998). (For a brief summary, see Bičák, 2000a.) This represents it as a general relativistic analogue of the Newtonian field in which the equation of motion of a particle is

$$m_1 \ddot{\mathbf{r}} = -V'(r)\, \hat{\mathbf{r}} + e_1\, \dot{\mathbf{r}} \times \mathbf{B},$$

where $\mathbf{B} = -Q\, r^{-2}\, \hat{\mathbf{r}}$. This describes the motion of a charged particle under the action of a central electric force and a magnetic force which is perpendicular to the instantaneous plane of motion. The classical orbits lie on cones. The only force acting in the instantaneous plane of the orbit is the central force, and so the position vector sweeps out equal areas in equal times. Moreover, if the central force is an inverse square law force with potential $V = -mm_1/r$ and the cone is slit along some radial line and flattened, then the orbit is exactly a conic section. This could correspond to an elliptic orbit in a plane with a wedge missing, and the removal of the wedge would result in an advance of the perihelion in successive orbits.

As shown by Lynden-Bell and Nouri-Zonoz (1998) and Bičák (2000a), the stationary (NUT) regions of the Taub–NUT solution possess many similar properties. It is also consistent with the conclusion of Newman, Tamburino and Unti (1963) that, when l is small, its presence induces a small additional advance in the perihelion of approximately elliptic orbits in this region. Although this is a purely gravitational solution, trajectories in this space-time appear to describe the motion of a test particle subject to the action of both a central force, and a force which behaves like a magnetic force acting on a charged particle. However, the magnetic field $\mathbf{B}$ in the Newtonian analogue is purely radial. Thus the origin in this representation behaves as though it were something like a charged magnetic monopole which interacts with a test mass as though it were a charged particle. Making the analogy between a gravitational and an electrostatic attraction, the parameter m can be regarded as the "mass" of the source, consistent with its Schwarzschild limit. The other magnetic-like force is proportional to the NUT parameter l, which may therefore be referred to as the *magnetic mass* or the *gravitomagnetic monopole moment* of the source.

Lynden-Bell and Nouri-Zonoz (1998) also proved that, as in the Newtonian case, all geodesics of NUT space lie on spatial cones. This has interesting consequences for the gravitational lensing properties of the space-time. Light rays are not merely bent as they pass the origin, but are also twisted. The differential twisting produces a characteristic spiral shear that is peculiar to such gravitomagnetic monopoles.

Demiański and Newman (1966) and also Dowker (1974) had previously drawn the analogy between the NUT solution and Dirac's theory of magnetic monopoles. They regarded the source of the NUT metric at $r = 0$ as an ordinary mass together with a gravitomagnetic monopole (sometimes called a magnetic mass or dyon). Of course, magnetic monopoles do not occur in isolation in the real world. In the theory of such objects two monopoles with opposite polarity are considered to be connected by what is referred to as a Dirac string. In this case, the semi-infinite line singularity at $\theta = \pi$ is regarded as the gravitational analogue of a Dirac string. However, as pointed out by Bonnor (1992), whereas in electromagnetic theory the string has no effect, the singularity along the half-axis in the NUT solution does affect space-time. (See also Zimmerman and Shahir, 1989.)

As already observed, the action of a central electromagnetic force on a charged particle may be considered to be analogous to the action of a gravitational force on a test mass in the NUT space-time, and the parameter m can be interpreted as the gravitational mass of the source, at least in the limit as $l \to 0$. However, there is no classical gravitational analogue of the effect of a magnetic monopole on a charged particle. Thus the interpretation of this solution as containing a gravitomagnetic monopole is not fully satisfactory. Also, since there is no central curvature singularity when $l \neq 0$, it is not clear where the source of the field is located.

It may alternatively be pointed out that the Newtonian equation expressed above can also be written in the form

$$m_1 \ddot{\mathbf{r}} = -V'(r)\,\hat{\mathbf{r}} - 2m_1\,\boldsymbol{\Omega} \times \dot{\mathbf{r}},$$

in which the term $2m_1\,\boldsymbol{\Omega} \times \dot{\mathbf{r}}$ is the Coriolis force associated with the local rotation of the frame of reference. Thus, the NUT solution may alternatively be interpreted in terms of a central gravitational force, together with a Coriolis force in which the angular velocity $\boldsymbol{\Omega}$ of the local coordinates is radial and varies as r^{-2}. And, since a rotating field cannot be distributed over the surface of a sphere without the occurrence of at least one singularity, the existence of the singularity on the axis is not unexpected.

Such an approach is more consistent with Bonnor's interpretation, described above, in which the source of the NUT metric is considered as a thin semi-infinite spinning rod injecting angular momentum into the space-time. This seems to be a more natural interpretation of this solution than that of gravitomagnetism (unless this is what is meant by that term). But it may be recalled that the effects of a Coriolis force are usually interpreted in general relativity in terms of "frame dragging".

12.1.4 The occurrence of closed timelike curves

Following the above approach of using the line element (12.3) over the full natural range of the coordinates, the half-axis $\theta = 0$ is regular and the coordinate ϕ is periodic. However, there is a semi-infinite (quasiregular) singularity along the half-axis $\theta = \pi$. Also, curves on which t, r and θ are constants have

$$\mathrm{d}s^2 = -\Big(4l^2 f(r)(1 - \cos\theta)^2 - (r^2 + l^2)\sin^2\theta\Big)\mathrm{d}\phi^2.$$

Such intervals are always spacelike if $f(r) \leq 0$. However, it can be seen that they become timelike in those parts of the NUT regions, where $f(r) > 0$, for which

$$\cos\theta < -\frac{r^2 + l^2 - 4l^2 f}{r^2 + l^2 + 4l^2 f}.$$

Moreover, since ϕ is a periodic coordinate, they clearly form *closed timelike curves* (although they are not geodesics). It can be seen that such curves occur in finite regions surrounding the singular half-axis within each of the NUT regions. Here, the hypersurfaces $t = $ const. cease to be spacelike. Moreover, the boundaries of these regions do not correspond to a coordinate singularity. They cannot therefore be interpreted as an axis or as any kind of physical barrier. Thus, there appears to be nothing that would prevent the more acceptable regions of the space-time from being extended to these regions which include closed timelike curves.

Bonnor (2001) has suggested that, in such cases, the singular half-axis acts like a *torsion singularity* which injects angular momentum into the space-time.[5] Further, it is suggested that the existence of closed timelike curves is associated with aspects of *classical spin*. Such an interpretation has not yet achieved wide recognition. Most relativists still regard the existence of regions containing closed timelike curves with suspicion. Yet it would appear that they occur naturally and regularly within many families of exact solutions of Einstein's equations.

---

[5] Actually, this is equivalent to what is often referred to as a spinning cosmic string, as described for a flat space-time in Section 3.4.1.

12.1.5 Conformal structure

For the NUT solution in the form (12.3), purely radial lines with $\theta = $ const. and $\phi = $ const. have

$$ds^2 = -f(r)\,dt^2 + \frac{dr^2}{f(r)}, \qquad \text{where} \qquad f(r) = \frac{r^2 - 2mr - l^2}{r^2 + l^2}. \qquad (12.10)$$

From this it is obvious that the hypersurfaces $r = $ const. are timelike when $f > 0$ (i.e. in the NUT regions), and spacelike when $f < 0$ (i.e. in the Taub region in which $r_- < r < r_+$ "between" two NUT regions). The horizons connecting these regions are given by $r = r_\pm = m \pm \sqrt{m^2 + l^2}$. In order to describe the global structure of this space-time, it is necessary to investigate the continuation of the space-time across these horizons.

To achieve this, it is first appropriate to temporarily introduce the tortoise coordinate

$$r_\star = \int \frac{dr}{f(r)} = r + r_- \log\left|\frac{r}{r_-} - 1\right| + r_+ \log\left|\frac{r}{r_+} - 1\right|.$$

The null coordinates $u = t - r_\star$ and $v = t + r_\star$ then reduce (12.10) to the double-null form

$$ds^2 = -\frac{(r - r_-)(r - r_+)}{r^2 + l^2}\,du\,dv.$$

In the outer NUT region

$$e^{-u} = \left(\frac{r-r_+}{r_+}\right)^{r_+} \left(\frac{r-r_-}{r_-}\right)^{r_-} e^{r-t}, \qquad e^{v} = \left(\frac{r-r_+}{r_+}\right)^{r_+} \left(\frac{r-r_-}{r_-}\right)^{r_-} e^{r+t}.$$

In order to cover the horizon $r = r_+$, it is convenient to introduce new null coordinates U_+ and V_+ defined by

$$U_+ = -e^{-u/2r_+} \quad = -\left(\frac{r-r_+}{r_+}\right)^{1/2} \left(\frac{r-r_-}{r_-}\right)^{r_-/2r_+} e^{(r-t)/2r_+},$$

$$V_+ = e^{v/2r_+} \quad = \left(\frac{r-r_+}{r_+}\right)^{1/2} \left(\frac{r-r_-}{r_-}\right)^{r_-/2r_+} e^{(r+t)/2r_+}.$$

Then, it is possible to introduce Kruskal–Szekeres-like coordinates T_+ and R_+ by putting

$$T_+ = \tfrac{1}{2}(V_+ + U_+) = \left(\frac{r-r_+}{r_+}\right)^{1/2} \left(\frac{r-r_-}{r_-}\right)^{r_-/2r_+} e^{r/2r_+} \sinh(t/2r_+),$$

$$R_+ = \tfrac{1}{2}(V_+ - U_+) = \left(\frac{r-r_+}{r_+}\right)^{1/2} \left(\frac{r-r_-}{r_-}\right)^{r_-/2r_+} e^{r/2r_+} \cosh(t/2r_+).$$

$$(12.11)$$

With these, the two-dimensional line element (12.10) becomes

$$ds^2 = \frac{4\,r_+^3\,r_-}{r^2 + l^2} \left(\frac{r-r_-}{r_-}\right)^{1-r_-/r_+} e^{-r/r_+} \left(-dT_+^2 + dR_+^2\right),$$

which is well behaved across the horizon $r = r_+$. This construction explicitly demonstrates that the singularity in the metric (12.3) at $r = r_+$ is a removable coordinate singularity which corresponds to a horizon in the space-time.

Inversely, the transformation (12.11) can be expressed as

$$\left(\frac{r-r_+}{r_+}\right)\left(\frac{r-r_-}{r_-}\right)^{r_-/r_+} e^{r/r_+} = R_+{}^2 - T_+{}^2,$$

$$t = 2r_+ \operatorname{arctanh}(T_+/R_+). \tag{12.12}$$

In the T_+, R_+-coordinates, the horizon $r = r_+$ corresponds to the pair of null lines $T_+ = \pm R_+$, and the hypersurfaces given by $r = $ const. are hyperbolae. The hypersurfaces $t = $ const. in the T_+, R_+-plane are simply given by the straight lines $T_+ \propto R_+$. It may be noted, however, that these coordinates are singular at $r = r_-$. They only cover the range $r_- < r < \infty$. On the other hand, it may be observed that the T_+ and R_+ coordinates can each be either positive or negative. Thus, these coordinates cover two separate Taub and two NUT$_+$ spaces, as illustrated in Figure 12.1.

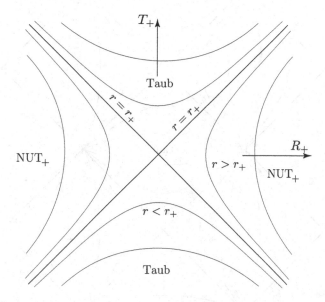

Fig. 12.1 With $\theta = $ const. and $\phi = $ const., the T_+, R_+-space covers two separate Taub and two NUT$_+$ regions and is regular over the horizon $r = r_+$. In this space, the hypersurfaces $r = $ const. are a family of hyperbolae as shown. They are timelike in the NUT$_+$ regions $r_+ < r < \infty$, and spacelike in the Taub regions $r_- < r < r_+$.

It is similarly possible to introduce the alternative Kruskal–Szekeres-like coordinates

$$T_- = \left(\frac{r_+ - r}{r_+}\right)^{r_+/2r_-} \left(\frac{r_- - r}{r_-}\right)^{1/2} e^{r/2r_-} \sinh(t/2r_-),$$

$$R_- = \left(\frac{r_+ - r}{r_+}\right)^{r_+/2r_-} \left(\frac{r_- - r}{r_-}\right)^{1/2} e^{r/2r_-} \cosh(t/2r_-),$$

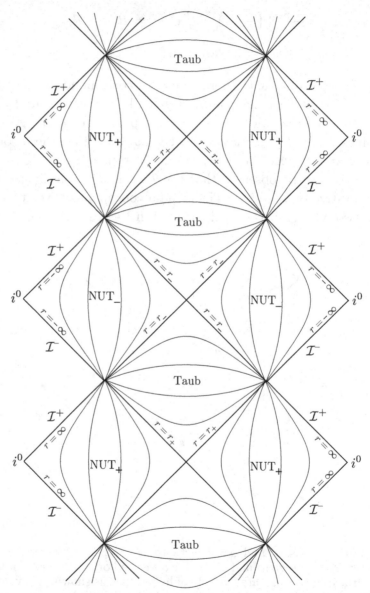

Fig. 12.2 The conformal diagram for a complete analytic extension of the r, t-surface ($\theta = $ const., $\phi = $ const.) for the Taub–NUT space-time when r and t can take any values $r, t \in (-\infty, \infty)$. The lines drawn have constant r.

with which the two-dimensional line element becomes

$$ds^2 = \frac{4\,r_-^3\,r_+}{r^2 + l^2} \left(\frac{r_+ - r}{r_+}\right)^{1-r_+/r_-} e^{-r/r_-} \left(-\,dT_-^2 + dR_-^2\right),$$

which is continuous across the other horizon $r = r_-$. These coordinates cover the range $-\infty < r < r_+$, which duplicates the range within the Taub region which is also covered by the T_+, R_+-coordinates. Since these coordinates overlap, it is possible to patch together the various regions to obtain a complete analytic extension of the Taub region.

With an additional conformal compactification, these pictures can be patched together, as indicated schematically in Figure 12.2, to form a complete analytic extension of the (totally geodesic) r, t-section of the space-time on which θ and ϕ are constant. This has been described by Miller, Kruskal and Godfrey (1971). The complete space-time contains an infinite sequence of Taub regions separated alternately by NUT$_+$ and NUT$_-$ regions.

With a combination of the above transformations, it may be noted that the metric (12.3) can be expressed as

$$ds^2 = F\left[-\,dT_\pm^2 + dR_\pm^2 - \tfrac{4l}{r_\pm}(R_\pm dT_\pm - T_\pm dR_\pm)\sin^2 \tfrac{1}{2}\theta\,d\phi\right.$$
$$\left. + \tfrac{4l^2}{r_\pm^2}(T_\pm^2 - R_\pm^2)\sin^4 \tfrac{1}{2}\theta\,d\phi^2\right] + (r^2 + l^2)(d\theta^2 + \sin^2\theta\,d\phi^2),$$

where

$$F = -\frac{4r_\pm^2 f}{U_\pm V_\pm} = \frac{4\,r_\pm^3\,r_\mp}{(r^2 + l^2)}\left|\frac{r - r_\mp}{r_\mp}\right|^{1-r_\mp/r_\pm} e^{-r/r_\pm}.$$

These two metrics (with alternate signs) can be used to represent all regions of the analytically extended space-time, as illustrated in Figure 12.2. That with upper signs represents domains in which $r_- < r < \infty$, while that with lower signs represents domains with $-\infty < r < r_+$. However, there is still a singularity at $\theta = \pi$ which is surrounded in the NUT regions by a smaller region which contains closed timelike curves.

12.2 Misner's construction of a singularity-free solution

An alternative approach to the interpretation of the Taub–NUT solution has been suggested by Misner (1963). This involves a different structure in which the axis is forced to be *completely regular*. It can be obtained by introducing two coordinate patches, one in each hemisphere. The metric is taken in the form (12.3) for $0 < \theta < \pi/2$ only. The transformation $t \rightarrow \tilde{t} = t - 4l\phi$ is then applied to the same metric for the other hemisphere $\pi/2 < \theta < \pi$.

The two parts are then joined at $\theta = \pi/2$. However, since ϕ is an axial coordinate in which $\phi = 0$ and $\phi = 2\pi$ have to be identified, this transformation can only be achieved consistently by making the t coordinate periodic with period $8\pi l$ – a procedure which also ensures that the two coordinate patches join smoothly.[6] Thus, although this procedure completely removes the singularity from the axis (indeed the solution is everywhere singularity-free), it does so at the expense of requiring that all particles moving along coordinate time lines within the NUT regions do so along *closed timelike curves*.

Within the time-dependent Taub region, this procedure is not so problematic. In this case, it is the coordinate $r \in (r_-, r_+)$ that is timelike. The coordinate t is spacelike and also periodic, indicating that the spatial part of the space-time here is closed: i.e. the Taub region is compact. In fact, the metric in this region is most conveniently expressed in the form (12.9) in which $r \equiv \tau$, namely

$$ds^2 = 4l(d\psi - \cos\theta\, d\phi)d\tau + V(\tau)(2l)^2(d\psi - \cos\theta\, d\phi)^2$$
$$+ (\tau^2 + l^2)(d\theta^2 + \sin^2\theta\, d\phi^2),$$

where $\theta \in [0, \pi]$, $\phi \in [0, 2\pi)$, $\psi \in [0, 4\pi)$. By comparing this form of the metric with that of the 3-sphere given in the Appendix in (B.13), it can immediately be seen that the sections on which $\tau (\equiv r) = $ const. have the same topology as the 3-sphere. (See also Misner, 1963; and Miller, Kruskal and Godfrey, 1971.) It follows that the space-time in this region is spatially homogeneous, with a topology that is homeomorphic to $R^1 \times S^3$.

In fact, the coordinates of (12.9) cover the complete manifold for which $\tau \in (-\infty, \infty)$, including both the Taub region and the NUT regions in which $V(\tau) < 0$. The topology $R^1 \times S^3$, therefore, applies to the complete space-time. However, in the NUT regions, R^1 is spacelike and the 3-sphere Lorentzian with one timelike dimension, consistent with the existence there of closed timelike curves. Thus, although the Taub region is globally hyperbolic, the complete space-time is not.

Within the Taub region, sections of constant τ are topological 3-spheres with proper volume proportional to $V(\tau)(\tau^2 + l^2) = (\tau_+ - \tau)(\tau - \tau_-)$. This clearly vanishes at the horizons where it connects to the NUT regions. Thus, as $\tau (\equiv r)$ increases from τ_-, the proper volume of the 3-sphere increases to a maximal value and then decreases again to vanish at $\tau = \tau_+$. The null hypersurfaces across which these regions are joined are Cauchy horizons and have interesting topological properties. These are referred to as *Misner bridges* and have been described by Misner and Taub (1968) and Ryan and

---

[6] Dowker (1974) has argued that the introduction of a periodic time is the analogue of the Dirac quantisation.

Shepley (1975). They are smooth, compact null hypersurfaces diffeomorphic to S^3, whose generators are closed null geodesics.

It has already been noted that the Taub solution in the form (12.9) does not have a non-degenerate space-time limit as $l \to 0$. The Misner construction of imposing the topology $R^1 \times S^3$ similarly has no meaning in this limit, as the period of the closed timelike curves in the NUT regions is proportional to l, and this is meaningless in the limit as $l \to 0$.

Finally, it should be emphasised that the construction described in this section represents just one possible interpretation of the solution (12.3). By introducing the topology $R^1 \times S^3$, the time-dependent region acquires a reasonable interpretation as the closed Taub universe which exists for a finite time. However, the associated stationary NUT regions, which have closed timelike curves, are more difficult to interpret. Moreover, the junctions between the two regions are problematic. In fact, Misner has remarked that such a space-time has no reasonable interpretation.

Usually, it is the space-time with this interpretation which is referred to as the Taub–NUT solution. The alternative interpretation, as described in Section 12.1, contains a semi-infinite topological singularity, part of which is surrounded by a region which contains closed timelike curves. With different ranges and identifications of the coordinates, these are two alternative interpretations of the same line element.

It should also be emphasised that, in this (Misner's) interpretation, there is no justification for regarding the NUT parameter l as a "magnetic mass" or "gravitomagnetic monopole moment". In this context, l appears as a measure of the periodic time of the closed timelike curves in the NUT regions, or of the proper volume of the spatial sections of the Taub universe.

Interestingly, the interior solutions of Bradley *et al.* (1999) can still be applied in this context, but the time coordinate is still periodic.

12.2.1 Conformal structure in Misner's interpretation

Now consider the conformal diagram for the nonsingular Taub–NUT space-time described above. In this approach, the t-coordinate is periodic and the NUT regions contain closed timelike curves everywhere.

It must immediately be emphasised that the construction of a space-time with topology $R^1 \times S^3$ is incompatible with the conformal structure illustrated in Figure 12.2. In this case, various points on each hypersurface $r = $ const. at different values of t would have to be identified, and these identifications cannot be achieved consistently throughout the space-time. In particular, the central points at which the distinct NUT regions are pictured as touching should be removed from the space-time. This has been analysed in detail by Miller, Kruskal and Godfrey (1971).

The difficulties involved arise because, in the Misner construction, t is a periodic coordinate. Thus, points on any line $t = $ const. have to be identified with points on other lines with $t = $ const. $+ 8n\pi l$ for any integer n. And it can be seen from the Kruskal–Szekeres-like transformation (12.11) and its inverse (12.12) that these lines have a degeneracy at the points $T_\pm = 0$, $R_\pm = 0$. Although the Kruskal–Szekeres-like coordinates given by (12.11) seem to cover these points perfectly adequately, as is illustrated in Figures 12.1 and 12.2, the periodic identifications imply that these points (where $r = r_\pm$) have to be removed from the space-time. In fact, these behave as quasiregular singular points (see Konkowski, Helliwell and Shepley, 1985). They can also be considered to arise because the two possible extensions from the Taub region given by $t = 2l\psi \pm \int V^{-1}(\tau)\mathrm{d}\tau$ cannot be made simultaneously. With these points missing, the resulting manifold is non-Hausdorff (see Hájíček, 1970).

Within each of the three distinct regions, it is possible to concentrate on a single period of the t-coordinate given by $t_0 \leq t \leq t_0 + 8\pi l$. This is illustrated in Figure 12.3. Within the Taub region, the periodic identification of the spacelike t-coordinate simply reflects the fact that the space is bounded (equivalent to S^3). However, in the two NUT regions, it reflects the exis-

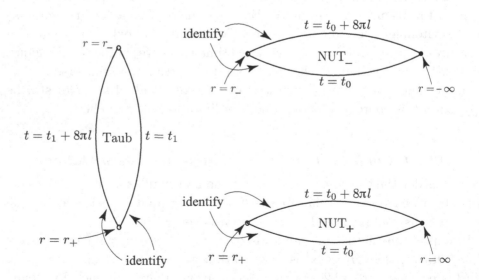

Fig. 12.3 A schematic conformal diagram for the r, t-surface (with θ and ϕ constants) for the Taub and NUT space-times using the Misner periodic identifications. The coordinate t is spacelike and r is timelike in the Taub space-time.

tence of closed timelike curves. Specifically, a particle on a timelike line in the NUT region progresses to $t_0 + 8\pi l$ and then returns to t_0. It does not necessarily progress to the singular point (where $r = r_\pm$) and thence into the Taub region. In fact, in any period, it may move along this region either toward the singular point or toward $r = \infty$. Null geodesics with θ and ϕ constant will also either move to the point at infinity or spiral to the singular point, reaching it after a finite affine distance.

With the above periodic identifications, the whole space-time is non-singular in the sense that the curvature invariants are everywhere well behaved, but is singular in the sense used in the "singularity theorems", namely that some timelike or null geodesic cannot be continued to infinite values of the affine parameter. (A null geodesic with decreasing r will spiral to the quasiregular singular point, which it reaches after a finite affine distance and can be extended no further.) This pathological behaviour of "incomplete geodesics imprisoned in a compact neighbourhood of the horizon" was inspirational in the rigorous definition of singularities. This is an example in which geodesic incompleteness is not connected with a strong gravitational field. There is no physical singularity. The geodesics come to a point at the edge of a coordinate patch at which the basis vectors are no longer linearly independent. It can be shown, however, that this situation is unstable (Misner and Taub, 1968). After the addition of even the smallest amount of matter, this pathological behaviour will not occur and a curvature singularity will arise.

12.3 The wider family of NUT solutions

A more general form of the NUT solution may now be considered which contains the additional parameter that was included in the original paper of Newman, Tamburino and Unti (1963). This can most generally be represented by the line element

$$ds^2 = -f(r)\left[dt + l\frac{\mathrm{i}(\zeta \mathrm{d}\bar\zeta - \bar\zeta \mathrm{d}\zeta)}{1 + \frac{1}{2}\epsilon\zeta\bar\zeta} \right]^2 + \frac{\mathrm{d}r^2}{f(r)} + (r^2 + l^2)\frac{2\,\mathrm{d}\zeta\,\mathrm{d}\bar\zeta}{(1 + \frac{1}{2}\epsilon\zeta\bar\zeta)^2}, \quad (12.13)$$

where,

$$f(r) = \frac{\epsilon(r^2 - l^2) - 2mr}{r^2 + l^2}. \quad (12.14)$$

In addition to the continuous parameters m and l, this form also includes the discrete 2-space curvature parameter ϵ, which can be scaled to any of the values $+1$, -1 or 0. It can be shown that, using a suitable null tetrad, the only non-zero component of the curvature tensor is

$$\Psi_2 = -\frac{m + i\,\epsilon l}{(r + i\,l)^3}.$$ (12.15)

As above, it may be assumed that $m \geq 0$.

When $l = 0$ and $m \neq 0$, these solutions reduce to the family of A-metrics, as discussed in Section 9.1, which all contain curvature singularities at $r = 0$. However, with $l \neq 0$, these space-times contain no curvature singularities and, generally, $r \in (-\infty, \infty)$. Interestingly, the alternative limit in which $m = 0$ and $l \neq 0$ does not seen to differ greatly from the general case.

In the case in which $\epsilon = +1$, it is convenient to put $\zeta = \sqrt{2}\tan(\theta/2)e^{i\phi}$, and the line element (12.13) then reduces exactly to (12.3) in which (12.14) becomes (12.2). This case clearly corresponds to the form of the Taub–NUT solution that has been described in detail in the previous parts of this chapter. The remaining two cases will now be described.

12.3.1 The case $\epsilon = -1$

For the case in which $\epsilon = -1$, the coordinate singularity which occurs when $|\zeta| = \sqrt{2}$ corresponds to conformal infinity in the 2-space on which t and r are constant (see Appendix A). Taking $|\zeta| \leq \sqrt{2}$, it is convenient to put $\zeta = \sqrt{2}\tanh(\psi/2)e^{i\phi}$, so that the line element (12.13) reduces to the form

$$\mathrm{d}s^2 = -f(r)\big(\mathrm{d}t + 4l\sinh^2\tfrac{1}{2}\psi\,\mathrm{d}\phi\big)^2 + \frac{\mathrm{d}r^2}{f(r)} + (r^2 + l^2)\big(\mathrm{d}\psi^2 + \sinh^2\psi\,\mathrm{d}\phi^2\big),$$ (12.16)

where

$$f(r) = \frac{l^2 - 2mr - r^2}{l^2 + r^2}.$$

The coordinate singularity which occurs at $\psi = 0$ clearly has the character of an axis which is everywhere regular, so one may take $\psi \in [0, \infty)$ and $\phi \in [0, 2\pi)$.

The solution (12.16) is stationary when $f > 0$. And, since the numerator of f is proportional to a quadratic which has two roots $r_\pm = -m \pm \sqrt{m^2 + l^2}$, there is a single stationary region where $r_- < r < r_+$, and two time-dependent regions $r < r_-$ and $r > r_+$. In either of the time-dependent regions, the metric may be considered as a vacuum cosmological solution of Bianchi type VIII.

Since the only non-zero component of the Weyl tensor is that given by (12.15), it can be seen that the space-time must be asymptotically flat as

$r \to \pm\infty$, which occurs in the time-dependent regions where r is a time-like coordinate. Moreover, the metric (12.16) can be considered as a one-parameter generalisation of the AII-metric which can be interpreted at a large distance as representing the gravitational field of a tachyon (see Section 9.1.1). In the asymptotic time-dependent region where $r \gg r_+$, this solution can therefore be similarly interpreted as the distant field of a tachyon with some kind of additional structure.

The existence of closed timelike curves

Consider the family of lines with constant values of r, ψ and t. A distance along these lines can be determined by the line element

$$ds^2 = \left[(r^2 + l^2)\sinh^2 \psi - 4l^2 f(r)(\cosh \psi - 1)^2\right]d\phi^2$$
$$= \left[(r^2 + l^2 - 4l^2 f)\cosh \psi + (r^2 + l^2 + 4l^2 f)\right](\cosh \psi - 1)d\phi^2.$$

The total length of these circular lines for $\phi \in [0, 2\pi)$ is zero when $\psi = 0$, which clearly corresponds to an axis. However, when $4l^2 f > r^2 + l^2$ (which occurs in part of each stationary region), it is also zero when

$$\cosh \psi = \frac{4l^2 f + r^2 + l^2}{4l^2 f - r^2 - l^2}. \tag{12.17}$$

This cannot represent an axis as it does not correspond to a coordinate singularity of the metric. In fact, it must represent a *closed null line* around the axis. Moreover, when

$$\frac{4l^2 f(r)}{r^2 + l^2} > \frac{\cosh \psi + 1}{\cosh \psi - 1} = \left(\coth \tfrac{\psi}{2}\right)^2,$$

these curves become timelike. And, since ϕ remains periodic, these are *closed timelike curves*. Domains which contain such curves occur within each stationary region for small values of r, large values of ψ and any non-zero value of l. There is no physical reason to prevent the space-time from being extended to such domains.

Regions containing closed timelike curves occur whenever $4l^2 f > r^2 + l^2$: i.e. when

$$(r^4 + 6l^2 r^2 - 3l^4) + 8ml^2 r < 0.$$

It can immediately be seen that, for any values of the arbitrary parameters, there must always occur at least one range of r in which this inequality is satisfied. This will identify a distinct region within the stationary region(s) of the space-time away from the horizons where $f = 0$. Moreover, if more than one range of r exists in which the above inequality is satisfied, these would identify distinct regions containing closed timelike curves. The boundaries of these regions are closed null hypersurfaces identified as the roots of (12.17).

The conformal structure

Siklos (1976) has given the conformal diagram for maximally extended spaces of this type. These relate to the 2-space in which ψ and ϕ are constants, which has the metric

$$ds^2 = -f(r)dt^2 + \frac{dr^2}{f(r)}, \qquad \text{where} \qquad f(r) = -\frac{r^2 + 2mr - l^2}{r^2 + l^2},$$

and $r \in (-\infty, \infty)$. Horizons occur at $r = r_\pm = -m \pm \sqrt{m^2 + l^2}$. The hypersurfaces $r = $ const. are timelike when $f > 0$, which occurs when $r_- < r < r_+$. They are spacelike when $f < 0$, which occurs either when $r < r_-$ or when $r > r_+$. Thus, there is a stationary region between distinct time-dependent regions. The conformal diagram of the maximally extended space-time on these sections is as illustrated in Figure 12.4.

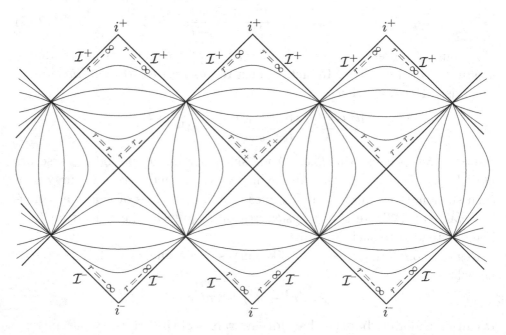

Fig. 12.4 A conformal diagram for a complete analytic extension of the r, t-surface ($\psi = $ const., $\phi = $ const.) for the metric (12.16).

As for the previous case, it is again possible to introduce null coordinates $v = t + \int \frac{dr}{f(r)}$ and $u = t - \int \frac{dr}{f(r)}$ so that, for the region in which $r > r_+$,

$$e^v = \left(\frac{r-r_+}{r_+}\right)^{r_+} \left(\frac{r-r_-}{r_-}\right)^{r_-} e^{r+t}, \qquad e^{-u} = \left(\frac{r-r_+}{r_+}\right)^{r_+} \left(\frac{r-r_-}{r_-}\right)^{r_-} e^{r-t},$$

and the 2-metric becomes

$$ds^2 = \frac{(r-r_-)(r-r_+)}{r^2 + l^2} \, du\, dv.$$

In order to cover the horizon $r = r_+$, it is first convenient to introduce new null coordinates $V_+ = e^{v/2r_+}$ and $U_+ = -e^{-u/2r_+}$. Then, introducing Kruskal–Szekeres-like coordinates $T_+ = \frac{1}{2}(U_+ + V_+)$ and $R_+ = \frac{1}{2}(V_+ - U_+)$, the two-dimensional line element becomes

$$ds^2 = \frac{4\, r_+^3\, r_-}{r^2 + l^2} \left(\frac{r-r_-}{r_-}\right)^{1-r_-/r_+} e^{-r/r_+} \left(-dT_+{}^2 + dR_+{}^2\right),$$

which can be extended across the horizon $r = r_+$. In addition, it can be seen that the horizon $r = r_+$ corresponds to the null lines $T_+ = \pm R_+$, and the lines $r = $ const. are hyperbolae. A similar construction can be obtained to extend the metric across the other horizon $r = r_-$. This approach leads to the conformal diagram of the maximally extended space-time that is illustrated in Figure 12.4.

12.3.2 The case $\epsilon = 0$

For the case in which $\epsilon = 0$ in the line element (12.13), it is convenient to put $\zeta = \frac{1}{\sqrt{2}} \rho e^{i\phi}$, so that the metric reduces to the form

$$ds^2 = -f(r)\left(dt + l\rho^2 d\phi\right)^2 + \frac{dr^2}{f(r)} + \left(r^2 + l^2\right)\left(d\rho^2 + \rho^2 d\phi^2\right), \qquad (12.18)$$

where

$$f(r) = -\frac{2mr}{r^2 + l^2}.$$

This space-time is stationary and axially symmetric whenever $f(r) > 0$ (i.e. for $r < 0$ since it is assumed that $m \geq 0$), and time-dependent when $r > 0$. When $l = 0$, it reduces to the $AIII$-metric (9.8), which is the Kasner type D solution when $r > 0$, or Taub's plane symmetric solution when $r < 0$.

Irrespective of the sign of r, the coordinate singularity at $\rho = 0$ behaves exactly like an axis for which ϕ is the associated periodic coordinate, and it is possible to generally take $\rho \in [0, \infty)$ and $\phi \in [0, 2\pi)$. A horizon occurs when $r = 0$. Also, the space-time is asymptotically flat when $r \to \pm\infty$ in the sense that the curvature tensor vanishes in this limit.

On lines for which r, t and ρ are constant, it can be seen that

$$ds^2 = \left((r^2 + l^2) + \frac{2ml^2\rho^2 r}{(r^2 + l^2)} \right) \rho^2 d\phi^2,$$

and this will be timelike if

$$2ml^2\rho^2 r < -(r^2 + l^2)^2.$$

Thus, since ϕ is a periodic coordinate, there will exist closed timelike curves for large negative values of $ml^2\rho^2 r$. Provided both m and l are non-zero, these occur away from the axis $\rho = 0$ in the stationary region of the space-time for which r is negative. Thus, although this solution is well behaved near the axis $\rho = 0$, the space-time is not globally causal.

Conformal diagrams for sections of the maximally extended spaces with the metric (12.18) have also been given by Siklos (1976). These correspond to the 2-space in which ρ and ϕ have constant values: i.e. to

$$ds^2 = \frac{2mr}{r^2 + l^2} dt^2 - \frac{r^2 + l^2}{2mr} dr^2,$$

where the surfaces $r = $ const. are timelike when $r < 0$ and spacelike when $r > 0$. The conformal diagram for these sections is given in Figure 12.5.

To investigate the properties of the space-time near the Killing horizon at $r = 0$, it is convenient to introduce the null coordinates

$$v = t + \frac{1}{2m} \left(\frac{1}{2}r^2 + l^2 \log\left|\frac{r}{l}\right| \right), \qquad u = t - \frac{1}{2m} \left(\frac{1}{2}r^2 + l^2 \log\left|\frac{r}{l}\right| \right),$$

so that the line element can be written as

$$ds^2 = \frac{2mr}{r^2 + l^2} du\, dv.$$

When $r > 0$, new null coordinates U and V can now be defined such that

$$V = e^{mv/l^2} = \sqrt{\frac{r}{l}}\, e^{(r^2 + 4mt)/4l^2}, \qquad U = -e^{-mu/l^2} = -\sqrt{\frac{r}{l}}\, e^{(r^2 - 4mt)/4l^2}.$$

Kruskal–Szekeres-type coordinates T, Z can then be introduced by putting

$$T = \tfrac{l^2}{2m}(V + U) = \tfrac{l}{m}\sqrt{lr}\, e^{r^2/4l^2} \cosh(\tfrac{m}{l^2}t),$$

$$Z = \tfrac{l^2}{2m}(V - U) = \tfrac{l}{m}\sqrt{lr}\, e^{r^2/4l^2} \sinh(\tfrac{m}{l^2}t),$$

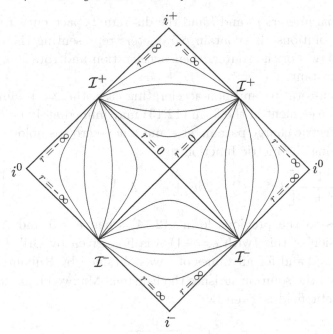

Fig. 12.5 The conformal structure of sections of the NUT space with $\epsilon = 0$ and ρ and ϕ constant. Lines given by $r = $ const. are timelike in the regions in which $r < 0$, and are spacelike when $r > 0$. The null hypersurfaces at $r = 0$ are Killing horizons corresponding to ∂_t, but not event horizons or Cauchy horizons.

for which the inverse transformation is given by

$$\frac{l^3}{m^2} r \, \mathrm{e}^{r^2/2l^2} = T^2 - Z^2, \qquad t = -\frac{l^2}{m} \operatorname{arctanh}\left(\frac{Z}{T}\right).$$

With these, the two-dimensional line element becomes

$$\mathrm{d}s^2 = \frac{2ml}{r^2 + l^2}\, \mathrm{e}^{-r^2/2l^2}(\mathrm{d}T^2 - \mathrm{d}Z^2),$$

which is continuous across $r = 0$. The horizon $r = 0$ corresponds to the pair of null lines $T = \pm Z$ through which the space-time can be continued. However, although these are Killing horizons, they are not event horizons or Cauchy horizons.

12.4 Further generalisations of the Taub–NUT solutions

The Taub–NUT solution is contained in the large Plebański–Demiański family of solutions that will be discussed in Chapter 16. In addition to the continuous parameters m and l and the discrete 2-space curvature parameter ϵ, these solutions also contain parameters representing the electric and magnetic charges of the source, their acceleration and rotation, and a cosmological constant.

For the non-rotating and non-accelerating case, the NUT family of solutions with line element in the form (12.13) may immediately be extended to include an electric charge parameter e and a non-zero cosmological constant Λ by modifying the metric function to

$$f(r) = \frac{1}{r^2 + l^2}\left[\epsilon(r^2 - l^2) - 2mr + e^2 - \Lambda\left(\tfrac{1}{3}r^4 + 2l^2r^2 - l^4\right)\right]. \quad (12.19)$$

This reduces to the previous form (12.14) when $e = 0$ and $\Lambda = 0$. The charged version of this (with $\epsilon = +1$) was first given by Brill (1964). The case with $\Lambda \neq 0$ and for all values of ϵ was obtained by Ruban (1972).

In this case, the solution satisfies the Einstein–Maxwell equations and the electromagnetic field is given by

$$\Phi_1 = \frac{e}{2(r^2 + l^2)}.$$

According to the value of the discrete parameter, one can make the following coordinate transformations (see Appendix A):

$$\epsilon = +1: \qquad \text{put} \qquad \zeta = \sqrt{2}\tan\tfrac{1}{2}\theta\, e^{i\phi},$$

$$\epsilon = 0: \qquad \text{put} \qquad \zeta = \tfrac{1}{\sqrt{2}}\rho\, e^{i\phi},$$

$$\epsilon = -1: \qquad \text{put} \qquad \zeta = \sqrt{2}\tanh\tfrac{1}{2}\psi\, e^{i\phi}.$$

With these choices, the line element (12.13) can be written in the form

$$ds^2 = -f(r)\left[dt + l\left\{\begin{matrix}\dfrac{4\sin^2\frac{1}{2}\theta}{\rho^2}\\[4pt]4\sinh^2\frac{1}{2}\psi\end{matrix}\right\}d\phi\right]^2 + \frac{dr^2}{f(r)}$$

$$+(r^2 + l^2)\left[\left\{\begin{matrix}d\theta^2\\d\rho^2\\d\psi^2\end{matrix}\right\} + \left\{\begin{matrix}\sin^2\theta\\\rho^2\\\sinh^2\psi\end{matrix}\right\}d\phi^2\right], \qquad (12.20)$$

where $f(r)$ is given by (12.19). These are clearly generalisations of the NUT metric in the three cases (12.3), (12.18) and (12.16), respectively.

When $l = 0$, this solution reduces to the generalised family of Reissner–Nordström–de Sitter space-times as described in Section 9.6. Thus, when $\epsilon = +1$, the constant parameters m and e can be interpreted as denoting the mass and the charge of the source, at least in this limit. These solutions clearly become asymptotically de Sitter or anti-de Sitter, according to the sign of Λ.

The greatest difference in including these parameters, however, is due to the fact that, with a non-zero cosmological constant, the numerator of the function $f(r)$ becomes a quartic rather than a quadratic. Thus there will generally exist up to five distinct regions in which the sign of f alternates, and four roots of $f(r) = 0$, which correspond to two horizons of the type considered above and an additional pair of cosmological-type horizons.

13

Stationary, axially symmetric space-times

The static region of an axially symmetric space-time, representing the field of non-rotating sources outside any event horizon, was described in Chapter 10 using the Weyl metric (10.2). For the more general case in which the sources are rotating but their fields remain stationary, the line element may be extended to what is known as the Weyl–Lewis–Papapetrou form (see Lewis, 1932; Levy and Robinson, 1963; Papapetrou, 1966)

$$ds^2 = -f\,(dt + A\,d\phi)^2 + f^{-1}\left[e^{2\gamma}(d\rho^2 + dz^2) + \rho^2 d\phi^2\right], \qquad (13.1)$$

where f, γ and A are functions of the spatial coordinates $\rho \in [0, \infty)$ and $z \in (-\infty, \infty)$ only. The metric (13.1) represents stationary, axially symmetric space-times since there exist timelike and spacelike Killing vectors ∂_t and ∂_ϕ, and ϕ is taken to be periodic with $\phi \in [0, 2\pi)$. It is convenient here to replace the expression e^{2U} in (10.2) by f, although this is still generally assumed to be positive. The additional function A, which appears here, represents some kind of rotation about the axis at $\rho = 0$.

For the metric (13.1), the field equations for vacuum and electrovacuum solutions are nonlinear but are known to be integrable. They can be expressed in terms of the Ernst equations (Ernst, 1968a,b) which will be introduced in Sections 13.2 and 13.3. Associated with this are various solution-generating techniques that can be employed to obtain particular families of solutions from any suitable "seed" solution, though these methods will not be described here. The purpose of this chapter is simply to describe the physical properties of some basic space-times of this type.

13.1 Cylindrically symmetric solutions

For simplicity, it is natural to initially consider space-times of this type for which there exists an additional Killing vector ∂_z, so that the metric functions depend on ρ only. Assuming the above periodicity in ϕ, these are accordingly interpreted as stationary solutions with cylindrical symmetry. For such space-times, it is convenient to make the substitution $\gamma = \mu + \frac{1}{2}\log f$ so that the metric (13.1) takes the form

$$\mathrm{d}s^2 = -f(\mathrm{d}t + A\,\mathrm{d}\phi)^2 + \mathrm{e}^{2\mu}(\mathrm{d}\rho^2 + \mathrm{d}z^2) + \rho^2 f^{-1}\mathrm{d}\phi^2. \tag{13.2}$$

The reason for this is that it facilitates extensions to regions in which f becomes negative, although such extensions do not share the properties of being stationary and axially symmetric.

13.1.1 The Lewis family of vacuum solutions

The general family of exact stationary cylindrically symmetric vacuum solutions was obtained independently by Lanczos (1924) and Lewis (1932). It can be given by the metric (13.2) with

$$f = \rho\left(a\rho^{-n} - \frac{c^2}{n^2 a}\rho^n\right),$$

$$\mathrm{e}^{2\mu} = k^2\rho^{(n^2-1)/2}, \tag{13.3}$$

$$A = \frac{c}{na}\rho^{n+1}f^{-1} + b,$$

which contain five constant (not independent) parameters n, a, b, c and k.[1] This particular line element is known as the *Lewis metric*. It contains *two distinct classes* of space-times according to whether n is real or imaginary. When n is real, the remaining parameters also have to be real. However, when n is imaginary, they have to be chosen so that the metric is real.

If it is required that the metric be regular on an axis at $\rho = 0$, then, as shown by Davies and Caplan (1971), only the flat case with $n = 1$ is permitted. It will therefore be assumed that some kind of source surrounds this infinite axis. The symmetries of these solutions also support their possible interpretation as exteriors for rotating cylindrical sources. However, identifying the physical meaning of the parameters for these space-times has proved to be significantly more difficult than for the static case.

In fact, MacCallum and Santos (1998) have shown that the Lewis metric has just *three independent parameters*. One of these, namely n, characterises the local gravitational field and appears in the curvature scalars. The remaining parameters (a, b, c and k, which are not independent) give information about the topological identification made to produce cylindrical

---

[1] It is usual practice to use a coordinate rescaling to put $k = 1$. However, this is not applied here to facilitate the interpretation of the van Stockum solutions described below.

symmetry. In particular, when $k = 1$, b and c might be related to the vorticity of the source and to the topological frame dragging. However, the general interpretation of these parameters still remains unclear, and the constraint relating them is unknown.

The Weyl class of Lewis metrics

When the parameter n is *real*, solutions of the above Lewis metric are referred to as belonging to the Weyl class. In this case, a, b, c and k must also be real. In addition, and for this case only, the metric can be diagonalised (see Frehland, 1971; Tipler, 1974; Som, Teixeira and Wolk, 1976; and Bonnor, 1980).

Specifically, for the case in which $a > 0$, the transformation

$$t = \frac{1}{\sqrt{a}}\left(1 - \frac{bc}{n}\right)\tau - b\sqrt{a}\,\tilde{\phi}, \qquad \phi = \frac{c}{n\sqrt{a}}\tau + \sqrt{a}\,\tilde{\phi},$$

takes the metric (13.2) with (13.3) to the form

$$\mathrm{d}s^2 = -\rho^{1-n}\,\mathrm{d}\tau^2 + k^2\rho^{(n^2-1)/2}(\mathrm{d}\rho^2 + \mathrm{d}z^2) + \rho^{n+1}\,\mathrm{d}\tilde{\phi}^2. \tag{13.4}$$

This may be recognised as the static Levi-Civita solution (10.9) in which $n = 1 - 4\sigma$, to which it also reduces when $a = 1$, $b = 0$ and $c = 0$. And it may be recalled that σ may be interpreted, in the Newtonian limit, as the mass per unit length of an infinite line source located on the axis.

For the alternative case in which $a < 0$, the transformation

$$t = b\sqrt{-a}\,\tau + \frac{1}{\sqrt{-a}}\left(1 - \frac{bc}{n}\right)\tilde{\phi}, \qquad \phi = -\sqrt{-a}\,\tau + \frac{c}{n\sqrt{-a}}\,\tilde{\phi},$$

takes the metric (13.2) with (13.3) to the form

$$\mathrm{d}s^2 = -\rho^{1+n}\,\mathrm{d}\tau^2 + k^2\rho^{(n^2-1)/2}(\mathrm{d}\rho^2 + \mathrm{d}z^2) + \rho^{1-n}\,\mathrm{d}\tilde{\phi}^2, \tag{13.5}$$

which is also the static Levi-Civita solution (10.9), but now with $n = 4\sigma - 1$ (i.e. with n replaced by $-n$).

It must therefore be concluded that this family of space-times is *locally* isomorphic to the static Levi-Civita solutions (10.9). They are consequently referred to as belonging to the locally static Weyl class. However, since $\phi = 0$ is identified with $\phi = 2\pi$, the time coordinate τ is periodic unless $b = 0$. Space-times of this class with $b \neq 0$ therefore contain closed timelike curves. As emphasised by da Silva *et al.* (1995b), this also implies that these space-times are topologically different from the Levi-Civita solution.

Moreover, after transforming to the Levi-Civita form (13.4) or (13.5), the roles of the coordinates ϕ and z can be interchanged, for a corresponding change in the range of the parameter n, by further applying the transformation (10.13).

It may be noticed that, if $na > 0$, f is positive in the range $0 < \rho < \rho_1$, where $\rho_1 = |\frac{na}{c}|^{1/n}$, and negative for $\rho_1 < \rho < \infty$. Alternatively, if $na < 0$, the opposite occurs. When $f < 0$, the t coordinate is spacelike. However, since the metric can always be transformed to Levi-Civita form for all values of ρ, the Killing horizon at $\rho = \rho_1$ has no *local* significance. It arises due to the rotation of the coordinates and has properties that resemble those of the boundary of the ergosphere in the Kerr space-time.

In addition, the coefficient of $d\phi^2$ in the metric (13.2) with (13.3) is given by $\frac{1}{a}(1 - \frac{bc}{a})^2 \rho^{n+1} - b^2 a \rho^{1-n}$. Thus, ϕ is a spacelike coordinate only if $an > 0$ and $\rho_2 < \rho < \infty$ where $\rho_2 = |ab/(1 - \frac{bc}{n})|^{1/n}$, or $an < 0$ and $0 < \rho < \rho_2$. Otherwise it is timelike. Thus, closed timelike curves occur asymptotically when $na < 0$, or near the axis when $na > 0$.

General properties of these space-times that are globally stationary but locally static have been described by Stachel (1982). In particular, he has argued that they may be considered as manifesting the gravitational analogue of a kind of classical electromagnetic Aharonov–Bohm effect (by which, for example, a locally electrostatic field can be globally magnetostatic).

Since space-times of this class are locally isomorphic to the Levi-Civita solutions, they must be flat when $n = \pm 1, \pm\infty$, and of type D when $n = 0, \pm 3$. However, it is not obvious that these space-times possess cylindrical symmetry. A curvature singularity occurs at $\rho = 0$ except in the flat cases, and with $n > 0$, f is positive when $\rho < \rho_1$ and negative when $\rho > \rho_1$. (When $a > 0$ and $n < 0$, or $a < 0$ and $n > 0$, it is the other way round.) Moreover, the assumed cylindrical symmetry is achieved only by formally identifying $\phi = 0$ with $\phi = 2\pi$, but it is this identification that introduces the existence of closed timelike curves. These space-times may be appropriate as matched exteriors to infinite cylindrical sources. But although they become locally flat as $\rho \to \infty$, like the Levi-Civita solution, they are not asymptotic to a flat cylindrically symmetric space-time if $n \neq 1$ with $a > 0$.

The Lewis class of Lewis metrics

The functions (13.3) represent a different class of space-times when the parameter n is *purely imaginary*. In this case, a and b have to be complex in a way that ensures that the metric is real. Such solutions are referred to as belonging to the Lewis class. They can be expressed in terms of the real parameters $\tilde{n}$, a_1, b_1, a_2 and b_2 where $a_1 b_2 - a_2 b_1 = 1$, and

$$n = i\tilde{n}, \qquad\qquad a = \tfrac{1}{2}(a_1^2 - b_1^2) + i a_1 b_1,$$

$$b = \frac{a_1 a_2 + b_1 b_2 + i}{a_1^2 + b_1^2}, \qquad\qquad c = \tfrac{1}{2}\tilde{n}(a_1^2 + b_1^2).$$

With these, the metric functions (13.3) take the forms

$$f = \rho\left[(a_1^2 - b_1^2)\cos(\tilde{n}\log\rho) + 2a_1 b_1\sin(\tilde{n}\log\rho)\right],$$

$$e^{2\mu} = k^2 \rho^{-(\tilde{n}^2+1)/2}, \tag{13.6}$$

$$A = \rho f^{-1}\left[(a_1 a_2 - b_1 b_2)\cos(\tilde{n}\log\rho) + (a_1 b_2 + a_2 b_1)\sin(\tilde{n}\log\rho)\right].$$

da Silva *et al.* (1995a) have shown that, in spite of their formal similarities with the Weyl class, space-times in this family are *not* locally isomorphic to the Levi-Civita solutions. And neither do they include any locally flat space-times. They are asymptotically flat as $\rho \to \infty$ only when $\tilde{n}^2 < 3$. On the other hand, if $\tilde{n}^2 > 3$, there is a curvature singularity at $\rho = \infty$, which occurs at a finite proper radial distance from the axis.

13.1.2 The van Stockum solutions

The first exact solutions in which rigidly rotating, infinitely long cylinders of fluid were matched to the above Lewis family of vacuum exteriors were those obtained by van Stockum (1937). Properties of these solutions have been analysed by Tipler (1974) and more extensively by Bonnor (1980, 1992).

In this case, the *interior* solution is described by just a single parameter ω. In terms of the metric (13.2), it is given by

$$f = 1, \qquad \mu = -\tfrac{1}{2}\omega^2\rho^2, \qquad A = -\omega\rho^2, \qquad \varrho = \tfrac{1}{2\pi}\omega^2 e^{\omega^2\rho^2},$$

where $\varrho(\rho)$ here denotes the matter density. It consists of rigidly rotating dust in which the net gravitational attraction towards the axis is exactly balanced by the centrifugal force due to the rotation.[2] The coordinates are corotating with the fluid, so that $u^\mu = \partial_t$.

The speed of each particle of the fluid increases with distance from the axis. In the Newtonian analogue of such a situation, in any rigid rotation there will exist a certain finite radius beyond which the particles move at speeds greater than that of light. Of course, such a situation is not possible

---

[2] In fact, van Stockum (1937) identified a complete family of rigidly rotating axially symmetric dust solutions. Confusingly, these are known as the van Stockum class of solutions. Properties of these space-times have been analysed by Bonnor (1977, 2005) and Bratek, Jałocha and Kutschera (2007). This section only considers the cylindrically symmetric member of this class which van Stockum matched to vacuum exteriors.

in a relativistic theory. Instead, it may be noticed that the coefficient of $d\phi^2$ in the metric is $\rho^2(1 - \omega^2\rho^2)$. Thus, when $|\omega\rho| > 1$, ϕ is a timelike coordinate, and circles around the axis at constant t, ρ and z are *closed timelike curves*.

Bonnor (1980) has shown that the angular velocity of the fluid with respect to a locally non-rotating frame is given by $\omega(1 - \omega^2\rho^2)^{-1}$, so that ω may be interpreted as the angular velocity of the fluid on the axis. Geodesics within the fluid have been considered by Opher, Santos and Wang (1996). In particular, they have shown that all timelike and null geodesics are bounded within the region $\omega^2\rho^2 \leq 1$ if the fluid extends beyond this into the region that contains closed timelike curves. These properties of a rigidly rotating fluid are shared with those of the Gödel universe, which will be described in Section 22.6.

In view of the above properties, physically acceptable solutions (without closed timelike curves in the interior) may be constructed by limiting the fluid region to a cylinder of finite width $\rho \leq R$, such that $\omega^2 R^2 < 1$, and by taking the exterior to be vacuum. Such a family of solutions is then characterised by the two parameters ω and R.

In fact, this family of solutions contains three physical parameters, namely the circumference of the cylindrical source, its angular velocity, and its mass per unit length. However, since the fluid in the source is rotating in such a way that the gravitational attraction towards the axis is exactly balanced by the centrifugal force due to the rotation, the second and third of these physical parameters are related and can be represented by the single parameter ω.

For $0 < \omega R < 1$, the mass m and angular momentum L of the source per unit length were given by Bonnor (1980) in the form

$$m = \int u^\mu T_\mu{}^\nu n_\nu \, d^3x = 1 - \sqrt{1 - \omega^2 R^2},$$

$$L = \int \xi^\mu T_\mu{}^\nu n_\nu \, d^3x = \tfrac{2}{3}\omega^{-1}\left(1 - \sqrt{1 - \omega^2 R^2}\right) - \tfrac{1}{3}\omega R^2\sqrt{1 - \omega^2 R^2},$$

where $u^\mu = \partial_t$ and $\xi^\mu = \partial_\phi$ are the Killing vectors, n_μ is the unit normal to the spacelike surface of integration, d^3x is the three-dimensional volume element, and the integration is over the interior of the cylinder.

For this interior, the matched *exterior* solution (13.2) has three distinct forms which depend on the specific value of ωR.

Case I: $0 < \omega R < \frac{1}{2}$

$$f = \frac{\rho}{R} \frac{\sinh(\epsilon - \theta)}{\sinh \epsilon},$$

$$e^{2\mu} = e^{-\omega^2 R^2} \left(\frac{R}{\rho}\right)^{2\omega^2 R^2},$$

$$A = \omega R^2 \frac{\sinh(\epsilon + \theta)}{\sinh(\epsilon - \theta)},$$

where $\theta = \sqrt{1 - 4\omega^2 R^2} \log(\rho/R)$ and $\operatorname{sech} \epsilon = 2\omega R$. This is a member of the Weyl class of Lewis solutions (13.3) for which $n = \sqrt{1 - 4\omega^2 R^2}$, $a = \frac{1+n}{2n} R^{-1+n}$, $b = -(\frac{1-n}{1+n})\omega R^2$, $c = \omega$, and $k^2 = R^{2\omega^2 R^2} e^{-\omega^2 R^2}$. In this vacuum region, ϕ is spacelike everywhere, so there are no closed timelike curves around the axis. However, f becomes negative for large values of ρ, so that t there becomes spacelike. This occurs because the coordinates are rotating, having been initially defined in the interior region to be corotating with the fluid.

In this external region, this solution is locally isomorphic to the Levi-Civita solution (10.9) with $\sigma = \frac{1}{4}(1 - \sqrt{1 - 4\omega^2 R^2})$, so that σ is within the usual range $\sigma \in (0, \frac{1}{4})$. Thus, as clarified by Bonnor (1980), these solutions are locally, but not globally, static. It is of interest to note that σ is here equal to m (the mass per unit length) only to first order in ωR.

Case II: $\omega R = \frac{1}{2}$

$$f = \frac{\rho}{R} \left[1 - \log(\rho/R)\right],$$

$$e^{2\mu} = e^{-1/4}(R/\rho)^{1/2},$$

$$A = \frac{R}{2} \left(\frac{1 + \log(\rho/R)}{1 - \log(\rho/R)}\right),$$

This is the special type D case of the Weyl class of Lewis solutions (13.3) in which $n = 0$ that is locally isomorphic to the critical Levi-Civita solution with $\sigma = \frac{1}{4}$, namely (10.14), which is a non-standard form of the *BIII*-metric.

Case III: $\omega R > \frac{1}{2}$

$$f = \frac{\rho}{R} \frac{\sin(\epsilon - \theta)}{\sin \epsilon},$$

$$e^{2\mu} = e^{-\omega^2 R^2} \left(\frac{R}{\rho}\right)^{2\omega^2 R^2},$$

$$A = \omega R^2 \frac{\sin(\epsilon + \theta)}{\sin(\epsilon - \theta)},$$

where $\theta = \sqrt{4\omega^2 R^2 - 1} \log(\rho/R)$ and $\sec \epsilon = 2\omega R$. This belongs to the Lewis class of Lewis solutions (13.6) for which $\tilde{n} = \sqrt{4\omega^2 R^2 - 1}$, but the identification of the parameters a_1, b_1, a_2 and b_2 is rather complicated.

In this case, the external solution contains closed timelike curves around the cylinder. In fact, as pointed out by Tipler (1974), any two events can be connected by a closed timelike curve. Causality is therefore violated globally. In fact, Steadman (2003) has shown that closed timelike geodesics exist in this case. Steadman (1998) has also demonstrated that null geodesics are radially bounded.

Bonnor (1980) showed that this space-time is asymptotically flat as $\rho \to \infty$ provided $\omega^2 R^2 < 1$. However, if $\omega^2 R^2 > 1$, there is a curvature singularity at $\rho = \infty$ which is at a finite proper radial distance from $\rho = R$. For the particular case in which $\omega R = 1$, Bonnor (1979b) has also shown that the exterior region is locally isometric to Petrov's homogeneous vacuum space-time.[3]

13.1.3 Other cylindrically symmetric solutions

da Silva *et al.* (2002) have helped to clarify the physical meaning of the parameters in the Weyl class of Lewis metrics by considering the source to be a rotating cylindrical shell, with the interior taken to be flat. They have found that the exterior, which in this case is locally isomorphic to the Levi-Civita space-time, does not include closed timelike curves either if $a > 0$ and $\sigma \in [0, \frac{1}{4}]$, or if $a < 0$ and $\sigma \in [\frac{1}{4}, \frac{1}{2}]$.

Other interiors may also be considered. A cylindrical source of differentially rotating dust has been constructed by Vishveshwara and Winicour (1977), and perfect fluid sources with non-zero pressure have been given by da Silva *et al.* (1995b), Davidson (2000, 2001) and Ivanov (2002). In fact, there is a vast literature on rotating cylindrically symmetric perfect fluid solutions which could act as possible interior solutions for the Lewis metric. However, of the family given by Krasiński (1975), very few have so far been matched appropriately.

The extension of the Lewis metric to include a cosmological constant was obtained by Santos (1993) and discussed further by MacCallum and Santos (1998). In fact, this introduces a dramatic modification of the geometry of

---

[3] Petrov (1962) presented a vacuum solution of Einstein's equations with the metric

$$\mathrm{d}s^2 = e^{kx} \cos(\sqrt{3}kx)(-\mathrm{d}t^2 + \mathrm{d}z^2) - 2e^{kx} \sin(\sqrt{3}kx)\,\mathrm{d}t\,\mathrm{d}z + \mathrm{d}x^2 + e^{-2kx}\mathrm{d}y^2,$$

where k is an arbitrary constant. This is the only vacuum solution that is homogeneous in the sense that it admits a simply transitive four-parameter group of motions.

this space-time. Bonnor, Santos and MacCallum (1998) matched this space-time as an exterior to a cylinder whose interior is the Gödel (1949) solution, which represents a homogeneous universe filled with rigidly rotating dust (see Section 22.6). This matching has been given more explicitly by Griffiths and Santos (2010).

13.2 Vacuum, axially symmetric solutions

From this point on, the metric functions are considered to depend on both ρ and z, and it is most convenient to take the line element in the form (13.1) in which it is assumed that $f > 0$. It is then possible with this metric to represent the complete family of stationary axially symmetric space-times. These include the static regions of the space-times described in Chapters 8 and 10, as well as the stationary regions of the rotating black hole space-time and the NUT solution as described in Chapters 11 and 12.

13.2.1 The Ernst equation

The vacuum field equations for the metric (13.1) can be expressed in terms of two sets of nonlinear equations. First, there are the main equations that involve only the functions f and A, and which can be written in the form

$$
f\left(f_{,\rho\rho} + f_{,zz} + \rho^{-1}f_{,\rho}\right) - f_{,\rho}^{2} - f_{,z}^{2} + \rho^{-2}f^{4}(A_{,\rho}^{2} + A_{,z}^{2}) = 0,
$$
$$
\left(\rho^{-1}f^{2}\,A_{,\rho}\right)_{,\rho} + \left(\rho^{-1}f^{2}\,A_{,z}\right)_{,z} = 0. \tag{13.7}
$$

And then there are the secondary equations

$$
\gamma_{,\rho} = \tfrac{1}{4}\rho f^{-2}(f_{,\rho}^{2} - f_{,z}^{2}) - \tfrac{1}{4}\rho^{-1}f^{2}(A_{,\rho}^{2} - A_{,z}^{2}),
$$
$$
\gamma_{,z} = \tfrac{1}{2}\rho f^{-2}f_{,\rho}f_{,z} - \tfrac{1}{2}\rho^{-1}f^{2}\,A_{,\rho}A_{,z}, \tag{13.8}
$$

which are automatically integrable in view of (13.7). Thus, once the main equations (13.7) are solved for f and A, the remaining metric function γ can always be found in principle by integrating the secondary equations (13.8) as simple quadratures. (For more details see Islam, 1985.)

It may be noticed that the second equation in (13.7) implies the existence of a "potential" function $\varphi(\rho, z)$ such that

$$
\varphi_{,\rho} = -\rho^{-1}f^{2}\,A_{,z}, \qquad \varphi_{,z} = \rho^{-1}f^{2}\,A_{,\rho}. \tag{13.9}
$$

Eliminating A between these equations, and substituting for A in (13.7), leads to two equations that can be expressed in the form

$$f \, \nabla^2 f - f_{,\rho}^{\,2} - f_{,z}^{\,2} + \varphi_{,\rho}^{\,2} + \varphi_{,z}^{\,2} = 0,$$

$$f \, \nabla^2 \varphi - 2 f_{,\rho} \, \varphi_{,\rho} - 2 f_{,z} \, \varphi_{,z} = 0,$$

where $\nabla^2 = \rho^{-1} \partial_\rho (\rho \, \partial_\rho) + \partial_z^{\,2}$ is the two-dimensional Laplacian operator. It is then convenient to introduce a complex function $\mathcal{E}(\rho, z)$ defined as

$$\mathcal{E} \equiv f + \mathrm{i}\,\varphi, \tag{13.10}$$

in terms of which the above pair of equations can be expressed in the form

$$(\mathcal{E} + \bar{\mathcal{E}}) \, \nabla^2 \mathcal{E} = 2(\nabla \mathcal{E})^2. \tag{13.11}$$

In fact, this form of the equations is coordinate invariant, where the square of the gradient is $(\nabla \mathcal{E})^2 = g^{\mu\nu} \mathcal{E}_{,\mu} \mathcal{E}_{,\nu}$, and the Laplacian (or generalised d'Alembertian) is given by

$$\nabla^2 \mathcal{E} = (g^{\mu\nu} \mathcal{E}_{,\nu})_{;\mu} = \frac{1}{\sqrt{-g}} \left(\sqrt{-g} \, g^{\mu\nu} \mathcal{E}_{,\nu} \right)_{,\mu}.$$

In view of the symmetries of these space-times, this expression only involves derivatives with respect to two remaining independent coordinates.

The equation (13.11) was first given by Ernst (1968a). The function $\mathcal{E}$ is known as the *Ernst potential*, and (13.11) is known as the *Ernst equation*. In fact, this equation applies in any space-time that has two commuting Killing vectors. More significantly, it happens to be integrable, and can therefore be solved by a variety of solution-generating techniques.

For any solution of the equation (13.11), the metric function f in (13.1) is the real part of $\mathcal{E}$, and A and then γ can be obtained from (13.9) and (13.8), or their equivalent forms in different coordinate systems. In terms of the Ernst function, it may be noted that the secondary quadratures (13.8) can be written in the form

$$\gamma_{,\rho} = \tfrac{1}{4}\rho f^{-2} \left(\mathcal{E}_{,\rho} \bar{\mathcal{E}}_{,\rho} - \mathcal{E}_{,z} \bar{\mathcal{E}}_{,z} \right),$$
$$\gamma_{,z} = \tfrac{1}{4}\rho f^{-2} \left(\mathcal{E}_{,\rho} \bar{\mathcal{E}}_{,z} + \mathcal{E}_{,z} \bar{\mathcal{E}}_{,\rho} \right). \tag{13.12}$$

An alternative form of the Ernst equation, which was also given by Ernst (1968a), can be obtained by putting

$$\mathcal{E} = \frac{\xi - 1}{\xi + 1}. \tag{13.13}$$

With this, (13.11) becomes

$$(\xi \bar{\xi} - 1) \, \nabla^2 \xi = 2\bar{\xi} \, (\nabla \xi)^2, \tag{13.14}$$

which is more suitable in some contexts.

For many interesting stationary axially symmetric space-times, it is appropriate to introduce prolate spheroidal coordinates (x, y) as in Sections 10.3 and 10.4. As illustrated in Figure 13.1, these are related to Weyl's canonical coordinates as

$$\left.\begin{array}{l} \rho = \sqrt{x^2 - 1}\,\sqrt{1 - y^2}, \\[2mm] z = x\,y, \end{array}\right\} \qquad \left\{\begin{array}{l} x = \tfrac{1}{2}(R_+ + R_-), \\[2mm] y = \tfrac{1}{2}(R_+ - R_-), \end{array}\right. \qquad (13.15)$$

where $R_\pm^2 = \rho^2 + (z \pm 1)^2$. It may be noted in passing that, although there is only a correspondence between these coordinate systems for $x \in (1, \infty)$ and $y \in (-1, 1)$, as the Weyl coordinates only cover the stationary region of any space-time, it is sometimes of importance to consider an extended region in which $x \le 1$. In terms of these coordinates, the metric (13.1) becomes

$$ds^2 = -f(dt + A\,d\phi)^2 + f^{-1}e^{2\gamma}(x^2 - y^2)\left(\frac{dx^2}{x^2 - 1} + \frac{dy^2}{1 - y^2}\right)$$
$$+ f^{-1}(x^2 - 1)(1 - y^2)\,d\phi^2. \qquad (13.16)$$

For this, the alternative form of the Ernst equation (13.14) is explicitly

$$(\xi\bar{\xi} - 1)\left[(x^2 - 1)\xi_{,xx} + 2x\,\xi_{,x} + (1 - y^2)\xi_{,yy} - 2y\,\xi_{,y}\right]$$
$$= 2\bar{\xi}\left[(x^2 - 1)\xi_{,x}^2 + (1 - y^2)\xi_{,y}^2\right], \qquad (13.17)$$

and the equations (13.9) for A are

$$A_{,x} = (1 - y^2)f^{-2}\varphi_{,y}, \qquad A_{,y} = -(x^2 - 1)f^{-2}\varphi_{,x}, \qquad (13.18)$$

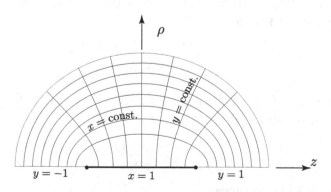

Fig. 13.1 Prolate spheroidal coordinates illustrated in terms of Weyl's canonical coordinates using (13.15).

while those for γ, namely (13.8) or (13.12), can be expressed as

$$
\begin{aligned}
\gamma_{,x} = \frac{(1-y^2)}{2(x^2-y^2)f^2}\bigg[& x(x^2-1)\left(f_{,x}^2+\varphi_{,x}^2\right) \\
& -x(1-y^2)\left(f_{,y}^2+\varphi_{,y}^2\right)-2y(x^2-1)\left(f_{,x}f_{,y}+\varphi_{,x}\varphi_{,y}\right)\bigg], \\
\gamma_{,y} = \frac{(x^2-1)}{2(x^2-y^2)f^2}\bigg[& y(x^2-1)\left(f_{,x}^2+\varphi_{,x}^2\right) \\
& -y(1-y^2)\left(f_{,y}^2+\varphi_{,y}^2\right)+2x(1-y^2)\left(f_{,x}f_{,y}+\varphi_{,x}\varphi_{,y}\right)\bigg].
\end{aligned}
\tag{13.19}
$$

13.2.2 Some exact solutions

Since the Ernst equation is integrable, a huge number of solutions are now known and others can straightforwardly be constructed (see for example Belinski and Verdaguer, 2001, and Stephani *et al.*, 2003). However, the purpose of this section is not to review all known solutions, but simply to comment on the physical interpretation of a few particularly important cases.

Among the simplest and most important of these solutions is that which corresponds to the *Kerr solution* for a rotating black hole which was described in detail in Chapter 11. Using the alternative form of the Ernst equation (13.17), the solution which gives rise to this space-time is

$$
\xi = p\,x + i\,q\,y,
\tag{13.20}
$$

where p and q are real constants which satisfy the constraint $p^2+q^2=1$.[4] It therefore follows from (13.10) and (13.13) that

$$
f = \frac{p^2x^2+q^2y^2-1}{(px+1)^2+q^2y^2}, \qquad \varphi = \frac{2\,q\,y}{(px+1)^2+q^2y^2},
$$

and from (13.18) that

$$
A = -\frac{2\,q\,(1-y^2)(px+1)}{p\,(p^2x^2+q^2y^2-1)},
$$

where the arbitrary constant of integration has been set to zero. The remaining equations (13.19) can then be integrated to obtain

$$
e^{2\gamma} = m^2\,\frac{p^2x^2+q^2y^2-1}{x^2-y^2},
$$

in which the constant of integration has been written as $\log m$. To express

---

[4] It may be noted that the Kerr–NUT solution is obtained by taking $\xi = e^{i\beta}(p\,x+i\,q\,y)$ where p, q and β are real constants satisfying $p^2+q^2=1$.

this in the standard Boyer–Lindquist form (11.1) of the Kerr space-time, it is first necessary to put $q = a/m$ (so that $p = \sqrt{m^2 - a^2}/m$) and then to make the coordinate transformation

$$x = \frac{r - m}{\sqrt{m^2 - a^2}}, \qquad y = \cos\theta.$$

For details see Chandrasekhar (1983) or Islam (1985).

It may be noted that, in these prolate spheroidal coordinates, the outer (event) horizon of the Kerr space-time occurs at $x = 1$, while $y = \pm 1$ correspond to poles on the axis. These coordinates also cover the time-dependent region inside the event horizon, for which $x \leq 1$, that is not covered by Weyl's canonical coordinates. In spite of this, however, the metric form (13.16) is not well adapted to the study of horizons, or indeed for many other aspects that are required for the physical interpretation of these space-times.

A different solution of the Ernst equation (13.14) was obtained by *Tomimatsu and Sato* (1972) which can be written in the form

$$\xi = \frac{p^2 x^4 + q^2 y^4 - 1 - 2\mathrm{i}\,pqxy(x^2 - y^2)}{2px(x^2 - 1) - 2\mathrm{i}\,qy(1 - y^2)}, \tag{13.21}$$

where p and q are constants satisfying $p^2 + q^2 = 1$. Tomimatsu and Sato (1973) have also obtained further sequences of solutions for which the expressions for ξ can be expressed as a rational fraction of polynomials. These correspond to different integer values of the deformation parameter δ, for which the Kerr solution (13.20) has $\delta = 1$ and (13.21) has $\delta = 2$. These are asymptotically flat space-times representing the fields of some kinds of rotating sources which generalise the Zipoy–Voorhees solution that was described in Section 10.4. However, they generally contain naked ring singularities. And, as for the Kerr solution, no appropriate internal solutions covering the singularity are known. For details of these and further properties, see Gibbons and Russell-Clark (1973), Kinnersley and Kelley (1974), Ernst (1976b), Economou (1976), Yamazaki (1977) and Hoenselaers and Ernst (1983).

Using various solution-generating techniques, a *double Kerr solution* can be constructed. This was first obtained by Kramer and Neugebauer (1980) and expressed in a more concise form by Yamazaki (1983). It has two distinct sources on the axis in Weyl coordinates, and reduces to the Kerr space-time if either of these sources vanishes. Normally, an asymptotically flat solution containing two separate sources would also posses a conical singularity on the axis between them, corresponding to a strut holding them apart. However, as pointed out by Bonnor and Steadman (2004), in the general family of solutions obtained by Kramer and Neugebauer, each source generally also possesses a non-zero NUT parameter (see also Kramer, 1986) which leads to an associated (torsion) singularity involving closed timelike curves near the axis, in addition to a conical singularity. In particular, Bonnor and Steadman have identified two cases that are asymptotically flat and for which a conical singularity only occurs between the sources. One of these does not include closed timelike curves.

In fact, it is possible to find a special choice of parameters in this particular family of solutions which can ensure that the axis is regular everywhere except at the sources themselves. The existence of such a stationary configuration can only be understood if some kind of interactive force arises due to the angular momentum of each source, and which exactly balances the gravitational attraction between them (for a discussion of this, see Bonnor, 2002). Dietz and Hoenselaers (1985), however, have shown that such a balance can only occur between hyperextreme sources. They have thus established that a stationary asymptotically flat space-time containing two rotating black holes (i.e. sources with horizons) in equilibrium cannot occur. (See also Manko and Ruiz, 2001.)

The repulsive character of a spin-spin interaction is also illustrated in a solution which represents two rotating Curzon–Chazy particles. In this case it is again possible to determine parameters such that the space-time is asymptotically flat and the axis between the particles is regular. For details, see Dietz and Hoenselaers (1982).

As mentioned at the end of Section 10.7, for static axially symmetric space-times, it is possible to construct solutions with infinite discs of counter-rotating particles by cutting an axially symmetric solution on a transverse plane and joining it to its mirror image. Such a procedure is also possible in the stationary case. Using Weyl's canonical coordinates and the metric (13.1), for any exact solution which has plane-reflection symmetry relative to $z = 0$, the hypersurface $z = b$ can be identified with $z = -b$. With such an identification, the region between these hypersurfaces is discarded. This produces non-zero components of the Ricci tensor on the junction that can be related to a surface energy-momentum tensor using the methods developed by Israel (1966a) and Barrabès and Israel (1991). The special choice $b = $ const., independent of the radial coordinate, leads to a corresponding thin disc of matter that is free of radial pressure. If the original metric represents an asymptotically flat axially symmetric space-time of certain positive total mass and angular momentum, then, for sufficiently large b, it is possible to interpret the disc source as being composed of counter-rotating streams of particles which move along their respective circular geodesics. The difference between the densities of both streams is the source of the angular momentum. If the identification is performed in a stronger field, this interpretation may not be possible, e.g. because timelike geodesics no longer exist in the plane of the disk, or because a negative proper mass density has to be assigned to one of the streams. Such an interpretation for the Kerr solution has been given by Bičák and Ledvinka (1993) (see Section 11.1.5), and for more general space-times by Ledvinka (1998). However, no solutions have yet been found by this method for which the source has purely differentially rotating matter with no counter-rotating material.

Most solution-generating techniques associated with the Ernst equation (13.11) produce new solutions from some initial "seed" without any reference to their physical interpretation (although many do preserve the condition of asymptotic flatness). However, solutions with particular properties can be generated by the matrix Riemann–Hilbert problem method developed initially by Hauser and Ernst (1980). This requires initial data to be specified on the boundary of the space-time, namely the axis, the surface of any source and at infinity. Normally, the integrals involved in solving this problem are impossibly complicated but, remarkably, Neugebauer and Meinel (1993, 1994, 1995) have used this approach to derive the exact solution for a finite rigidly rotating disc of dust. See also Neugebauer, Kleinwächter and Meinel (1996).

This analytic approach has been extended to the construction of exact solutions representing counter-rotating discs of dust by Klein (2001) and Frauendiener and Klein (2001).

13.3 Axially symmetric electrovacuum space-times

Many static axially symmetric solutions of the Einstein–Maxwell equations are known. These generalise the Weyl solutions and were described in Section 10.8. However, the introduction of an electromagnetic field already introduces an essential nonlinearity into the field equations, and a further generalisation to stationary solutions is straightforward. In fact, the main field equations are given by the electrovacuum Ernst equations that will be described below. After presenting these equations, some important exact solutions will briefly be described, particularly those that describe the fields of rotating charged sources.

13.3.1 The Ernst electrovacuum equations

The integrable Ernst equation (13.11) was generalised by Ernst (1968b) to include a non-zero electrovacuum field. To achieve this, it may first be recalled that, for a stationary axially symmetric solution with the metric (13.1), the electromagnetic four-potential A_μ only has non-zero components A_t and A_ϕ that are themselves functions of ρ and z only. The component A_t may be regarded as the electric potential, and it can be shown that a magnetic potential A'_ϕ exists such that

$$A'_{\phi,\rho} = \rho^{-1}f(A_{\phi,z} - A A_{t,z}), \qquad A'_{\phi,z} = -\rho^{-1}f(A_{\phi,\rho} - A A_{t,\rho}), \quad (13.22)$$

where $fA = -g_{t\phi}$ as in (13.1). A complex potential Φ can then be defined such that its real and imaginary parts are the electric and magnetic potentials, respectively: i.e.

$$\Phi \equiv A_t + i\, A'_\phi. \tag{13.23}$$

Thus a duality rotation is simply given by the replacement $\Phi \to e^{i\alpha}\Phi$.

It is then convenient to introduce a new complex function $\mathcal{E}$, generalising (13.10), such that

$$\mathcal{E} \equiv f - \Phi\bar{\Phi} + i\,\varphi, \tag{13.24}$$

in which $f = -g_{tt}$, and φ is a modified potential for the metric function A that is related to the angular momentum of the source. It is given by

$$\begin{aligned}
\varphi_{,\rho} &= -\rho^{-1}f^2\, A_{,z} + i\left(\bar{\Phi}\Phi_{,\rho} - \Phi\bar{\Phi}_{,\rho}\right), \\
\varphi_{,z} &= \rho^{-1}f^2\, A_{,\rho} + i\left(\bar{\Phi}\Phi_{,z} - \Phi\bar{\Phi}_{,z}\right),
\end{aligned} \tag{13.25}$$

which generalises (13.9).

With these potential functions defined as above, Ernst (1968b) has shown that the main Einstein–Maxwell field equations reduce to the pair of complex equations

$$\begin{aligned}
\left(\mathcal{E} + \bar{\mathcal{E}} + 2\Phi\bar{\Phi}\right)\nabla^2\mathcal{E} &= 2\left(\nabla\mathcal{E} + 2\bar{\Phi}\nabla\Phi\right)\cdot\nabla\mathcal{E}, \\
\left(\mathcal{E} + \bar{\mathcal{E}} + 2\Phi\bar{\Phi}\right)\nabla^2\Phi &= 2\left(\nabla\mathcal{E} + 2\bar{\Phi}\nabla\Phi\right)\cdot\nabla\Phi.
\end{aligned} \tag{13.26}$$

These coordinate-invariant forms clearly reduce to (13.11) when the electromagnetic field vanishes.

The equations given above are equivalent to the main Einstein–Maxwell field equations. However, to obtain a complete solution, it is still necessary to solve the quadratures for the remaining metric function γ. Using Weyl's canonical coordinates, these may be written in the form

$$
\gamma_{,\rho} = \tfrac{1}{4}\rho f^{-2}\Big[(\mathcal{E}_{,\rho} + 2\bar{\Phi}\Phi_{,\rho})(\bar{\mathcal{E}}_{,\rho} + 2\Phi\bar{\Phi}_{,\rho}) - (\mathcal{E}_{,z} + 2\bar{\Phi}\Phi_{,z})(\bar{\mathcal{E}}_{,z} + 2\Phi\bar{\Phi}_{,z})\Big]
$$
$$
-\rho f^{-1}(\Phi_{,\rho}\bar{\Phi}_{,\rho} - \Phi_{,z}\bar{\Phi}_{,z}),
$$

$$
\gamma_{,z} = \tfrac{1}{4}\rho f^{-2}\Big[(\mathcal{E}_{,\rho} + 2\bar{\Phi}\Phi_{,\rho})(\bar{\mathcal{E}}_{,z} + 2\Phi\bar{\Phi}_{,z}) + (\mathcal{E}_{,z} + 2\bar{\Phi}\Phi_{,z})(\bar{\mathcal{E}}_{,\rho} + 2\Phi\bar{\Phi}_{,\rho})\Big]
$$
$$
-\rho f^{-1}(\Phi_{,\rho}\bar{\Phi}_{,z} + \Phi_{,z}\bar{\Phi}_{,\rho}),
$$

$$\tag{13.27}$$

which generalises (13.12).

13.3.2 Some exact solutions

For static solutions, it is possible to take $A = 0$ and $\arg\Phi = \text{const.}$ so that φ also vanishes and hence $\mathcal{E}$ is real. For example, the Ernst potentials for the Reissner–Nordström solution can be expressed in prolate spheroidal coordinates as

$$
\mathcal{E} = \frac{x - m}{x + m}, \qquad \Phi = \frac{e}{x + m},
$$

and hence $f = (x^2 - m^2 + e^2)/(x + m)^2$, where m and e are the usual parameters. Consistent with (10.18) using Weyl's canonical coordinates, these potentials are given by

$$
\mathcal{E} = \frac{R_+ + R_- - 2m}{R_+ + R_- + 2m}, \qquad \Phi = \frac{2e}{R_+ + R_- + 2m},
$$

where $R_\pm^2 = \rho^2 + (z \pm d)^2$, and $d^2 = m^2 - e^2$.

For stationary solutions, it is first appropriate to consider the case in which $\mathcal{E}$ is an analytic function of Φ. In this case, the Ernst equations (13.26) imply that $\mathcal{E}$ must be a linear function of Φ. If, in addition, it is assumed that the space-time is asymptotically flat and the electromagnetic field vanishes at infinity (i.e. $\mathcal{E} \to 1$ and $\Phi \to 0$ at infinity), then

$$
\mathcal{E} = 1 - 2c^{-1}\Phi, \tag{13.28}
$$

where c is a complex constant. In this case, Ernst (1968b) has shown that, if ξ_0 is a solution of the alternative form (13.14) of the vacuum Ernst equation, then

$$
\mathcal{E} = \frac{\sqrt{1 - c\bar{c}}\,\xi_0 - 1}{\sqrt{1 - c\bar{c}}\,\xi_0 + 1}, \qquad \Phi = \frac{c}{\sqrt{1 - c\bar{c}}\,\xi_0 + 1}, \tag{13.29}
$$

is a solution of the electrovacuum equations (13.26). Thus, the Ernst potential for any vacuum equation can be used to generate a corresponding electrovacuum solution.

Using (13.29), Ernst (1968b) has then shown that the Kerr–Newman solution for a rotating charged black hole may be obtained from the seed potential (13.20) which leads to the vacuum Kerr solution, namely $\xi_0 = p\,x + \mathrm{i}\,q\,y$ in prolate spheroidal coordinates, where p and q are real constants which satisfy the constraint $p^2 + q^2 = 1$. The *Kerr–Newman* solution is then obtained by putting

$$q = \frac{a}{\sqrt{m^2 - e^2}}, \qquad p = \frac{\sqrt{m^2 - a^2 - e^2}}{\sqrt{m^2 - e^2}}, \qquad |c| = \frac{e}{m}.$$

The metric (11.8), which is discussed in Section 11.2 in terms of Boyer–Lindquist coordinates, is finally obtained by putting

$$x = \frac{r - m}{\sqrt{m^2 - a^2 - e^2}}, \qquad y = \cos\theta.$$

Magnetised generalisations of the Kerr solution have been given by Kramer (1984), Manko and Sibgatullin (1992a, 1992b) and Manko (1993). The latter solution is a special case of that for a rotating, charged, magnetised mass of Manko and Sibgatullin (1993). It represents a spinning black hole with a magnetic dipole field. It is asymptotically flat, but is significantly more complicated than the charged (Kerr–Newman) case as it is not symmetric about the equatorial plane and is not of algebraic type D. However, the global structure and physical properties of this space-time do not yet appear to have been analysed in detail.

Static electrovacuum space-times with axial symmetry have already been considered in Section 10.8. It was shown there that any number of charged Curzon–Chazy particles or Reissner–Nordström-type sources could be superposed along the axis. These generally require Weyl struts to hold the sources apart. However, the electrostatic repulsion can balance the gravitational attraction either for extremely charged objects, or for a hyperextreme source and a black hole at a particular distance. Since it has been shown in the previous section that a spin-spin interaction can also be repulsive, it is appropriate to consider if this could contribute to the existence of an equilibrium configuration between two charged and rotating black holes: i.e. one in which both sources have horizons.

A stationary generalisation of such an equilibrium configuration for *two extreme sources* has long been known. In fact, it is possible to consider a random distribution of any number of such sources, as their mutual electromagnetic repulsions exactly balance their gravitational attractions. These solutions, known as the *PIW metrics*, were given independently by Perjés (1971) and Israel and Wilson (1972). They are a generalisation of the static

Majumdar–Papapetrou solutions that will be described in Section 22.5. Bonnor and Ward (1973) have shown that the axisymmetric solution containing two such sources includes a topological singularity on the axis between them (or connecting each to infinity). This comprises both a deficit angle and a "torsion" singularity, possibly corresponding to a rotating cosmic string. The stationarity of these solutions arises from the introduction of both a rotation and a NUT parameter associated with each source. Further properties of these solutions have been described by Bonnor and Steadman (2005), who have considered each source to be charged, rotating, magnetic objects in equilibrium and have pointed out that, as for NUT sources, regions between the sources which contain closed timelike curves necessarily exist. In fact, the charge of each source is equal to its mass, and its magnetic moment is equal to its angular momentum. Explicit solutions for the case in which the angular momentum of each source is not aligned have been given by Ward (1973).

These PIW space-times have also been analysed by Hartle and Hawking (1972). In particular they have shown that, for a single source, the singular semi-infinite axis can be removed by employing Misner's construction (see Section 12.2) which introduces a periodic time coordinate. They also conjectured that the only solutions of this family that are asymptotically flat and non-singular outside horizons are the static Majumdar–Papapetrou solutions. Under certain reasonable conditions, this conjecture has been established by Chruściel, Reall and Tod (2006).

Because of the essential nonlinearity of the field equations, it is far from obvious how to obtain an exact solution that represents a superposition of two non-extreme rotating sources. Nevertheless, Bičák and Hoenselaers (1985) have used the transformation (13.29), which generates the Kerr–Newman solution from the Kerr solution, and applied this to the vacuum solution representing two equal Kerr black holes spinning around the same axis. In this case, it gives rise to a solution that represents two equal Kerr–Newman black holes. They have found a condition by which the conical singularities along the axis are removed, while simultaneously keeping the total mass of the system positive. Unfortunately, however, although the axis can be made regular, this solution also contains at least one naked ring singularity in the plane of symmetry $z = 0$ away from the axis.

Alternative solutions representing a pair of charged and magnetised rotating sources with an aligned axis can be generated by various inverse-scattering techniques, although these may also introduce unwanted features. Bretón, Manko and Aguilar Sánchez (1999) have reviewed such solutions and have found that conditions for an asymptotically flat equilibrium configuration can be obtained explicitly when at least one of the sources is hyperextreme.

Further solutions representing a pair of charged rotating black holes have been given by Manko, Martín and Ruiz (1994). Also, Sod-Hoffs and Rodchenko (2007) have analysed the solutions obtained by Ernst, Manko and Ruiz (2006) and have shown that two counter-rotating equal Kerr–Newman–NUT objects can be in equilibrium when a certain condition is satisfied. However, two counter-rotating equal masses endowed with arbitrary magnetic and electric dipole moments cannot exist in equilibrium under any choice of the parameters.

14

Accelerating black holes

This chapter considers the vacuum solution that was referred to as the C-metric in the classic review of Ehlers and Kundt (1962) – a label that has generally been used ever since. In fact, the static form of this solution was originally found by Levi-Civita (1918) and Weyl (1919a), and has subsequently been rediscovered many times. Its basic properties were first interpreted by Kinnersley and Walker (1970) and Bonnor (1983). Specifically, it was shown that, with its analytic extension, this solution describes a pair of causally separated black holes which accelerate away from each other due to the presence of strings or struts that are represented by conical singularities.

The C-metric is a generalisation of the Schwarzschild solution which includes an additional parameter that is related to the acceleration of the black holes. In fact, generalisations to "accelerating" versions of all three A-metrics have been described by Ishikawa and Miyashita (1983). These include what may be called the CI, CII and $CIII$-metrics. However, it is only the CI-metric, which describes accelerating black holes, that will be considered in the present chapter.

General properties of space-times such as this, which admit boost and rotation symmetries, were described by Bičák (1968). Asymptotic and other properties of the C-metric were further investigated by Farhoosh and Zimmerman (1980a), Ashtekar and Dray (1981), Dray (1982), Bičák (1985), Cornish and Uttley (1995a) and Sládek and Finley (2010). See also Pravda and Pravdová (2000) and Griffiths, Krtouš and Podolský (2006), on which the present chapter is based and from where the figures are taken.

This chapter will conclude with a consideration of various further generalisations of the C-metric which include parameters representing, respectively, an electric charge, a rotation and a cosmological constant.

14.1 The C-metric

It is appropriate to start with the line element for the C-metric in the well-known form

$$ds^2 = \frac{1}{A^2(x+y)^2}\left(-F\,dt^2 + \frac{dy^2}{F} + \frac{dx^2}{G} + G\,d\phi^2\right), \qquad (14.1)$$

where G and F are cubic functions of x and y, respectively:

$$G = 1 - x^2 - 2MAx^3, \qquad F = -1 + y^2 - 2MAy^3. \qquad (14.2)$$

This solution contains two constant parameters M and A. It reduces to a specific form of Minkowski space when $M = 0$, but has no obvious limit as $A \to 0$.

It may be observed that the functions G(x) and F(y) are related by the condition that $F(w) = -G(-w)$. It follows that these functions must have the same structural properties such as the same number of (related) roots. In particular, G and F will each possess three distinct real roots if $27M^2A^2 < 1$. Since this corresponds to the physically most interesting situation, it will generally be assumed that this condition is satisfied.

It may also be noted that, in the line element (14.1) with (14.2), a coordinate freedom has been used to remove the linear terms in G and F. Although this form of the metric has been extensively used in the literature, a significant simplification was recently introduced by Hong and Teo (2003). Their innovation was to use the freedom in applying a linear transformation to the coordinates x and y, not to remove the linear components in the cubic functions as in (14.2) but, rather, to use the transformation in such a way that the root structure of the cubics is expressed in a very simple form. (A rescaling of the other coordinates t and ϕ, and the parameters M and A, is also required, and the new coordinates and parameters are denoted as $\{x, y, \tau, \varphi\}$, m and α, respectively.) Specifically, assuming the existence of three roots with the coefficient of the cubic term being negative, transformations are applied to fix the roots of the metric function in x at $+1$, -1 and $-1/(2\alpha m)$. In order to preserve the order of the roots, it is necessary to assume that $0 < 2\alpha m < 1$. In this way, Hong and Teo have re-expressed the line element (14.1) in the more convenient form

$$ds^2 = \frac{1}{\alpha^2(x+y)^2}\left(-F\,d\tau^2 + \frac{dy^2}{F} + \frac{dx^2}{G} + G\,d\varphi^2\right), \qquad (14.3)$$

where

$$G = (1 - x^2)(1 + 2\alpha mx), \qquad F = -(1 - y^2)(1 - 2\alpha my). \qquad (14.4)$$

To maintain a Lorentzian signature of the metric (14.3), it is necessary that $G > 0$, which implies that the coordinate x must be constrained to lie between appropriate roots of the cubic function G. On the other hand, there is no constraint on the sign of F (static regions occur when $F > 0$). It can also be seen from (14.3) that $x + y = 0$ corresponds to conformal infinity. Thus a physical space-time must satisfy either $x + y > 0$ or $x + y < 0$. With these constraints on the coordinates x, y, the metric (14.3) can still describe four qualitatively different space-times depending on particular choices of the coordinate ranges. We will here restrict attention to the physically most important case $x \in [-1, 1]$ and $x + y > 0$, in which the metric describes a space-time that contains accelerating black holes.

In the form of the metric (14.3), the roots of G and F have very simple explicit expressions, while still satisfying the condition that $F(w) = -G(-w)$. In fact, their simplified root structure is of considerable assistance both when performing calculations, and in the physical interpretation of this solution.

14.1.1 The C-metric in spherical-type coordinates

Although the x, y coordinates of (14.3) and (14.4) are useful when performing calculations and in analysing the global structure of the space-time, their physical interpretation can be clarified by introducing the closely related coordinates r, θ, which play the role of "spherical" coordinates around the black holes. Since x is restricted to the range $-1 \le x \le 1$, it is natural to introduce the coordinate transformation

$$x = -\cos\theta, \qquad y = \frac{1}{\alpha r}, \qquad \tau = \alpha t, \qquad (14.5)$$

where $\theta \in [0, \pi]$. With (14.5), the metric becomes

$$ds^2 = \frac{1}{(1 - \alpha r \cos\theta)^2} \left(-Q \, dt^2 + \frac{dr^2}{Q} + \frac{r^2 \, d\theta^2}{P} + P \, r^2 \sin^2\theta \, d\varphi^2 \right), \quad (14.6)$$

where

$$P = 1 - 2\alpha m \cos\theta, \qquad Q = \left(1 - \frac{2m}{r}\right)(1 - \alpha^2 r^2). \qquad (14.7)$$

This is a two-parameter family of solutions which, in contrast to (14.3), has the important property that it reduces precisely to the familiar form of the spherically symmetric Schwarzschild solution (8.1) when $\alpha = 0$.

Relative to a null tetrad which is naturally adapted to (14.6), the only non-zero component of the curvature tensor is that given by

$$\Psi_2 = -m \left(\frac{1}{r} - \alpha \cos\theta \right)^3 = -m\alpha^3 (x + y)^3. \qquad (14.8)$$

When $m \neq 0$, the space-time is therefore of algebraic type D and a curvature singularity occurs at $r = 0$, so that one may take $r > 0$. This expression also shows that the roots of Q at $r = 2m$ and $r = 1/\alpha$ must be coordinate singularities which are clearly Killing horizons associated with the Killing vector ∂_t. Notice also that the space-time becomes flat for $x + y \to 0$, i.e. near conformal infinity.

Since this solution is a one-parameter generalisation of the Schwarzschild space-time, it is natural to continue to interpret the constant m (which is assumed to be positive) as the parameter characterising the mass of the source, and r as a Schwarzschild-like radial coordinate with a *black hole horizon* $\mathcal{H}_\mathrm{o}$ at $r = 2m$. The remaining question therefore concerns the character of the new parameter α (which must also be positive) and the horizon $\mathcal{H}_\mathrm{a}$ at $r = 1/\alpha$. It will be argued that this solution describes an accelerating black hole in which α can be interpreted as the acceleration parameter. Several reasons for this interpretation will be given. The first is that the horizon

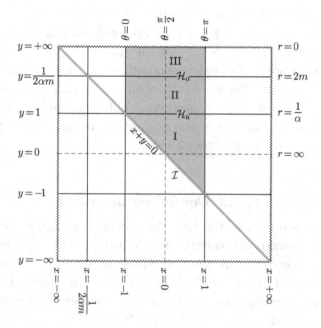

Fig. 14.1 The full ranges of the x, y-coordinates are indicated together with the roots of the cubic functions $G(x)$ and $F(y)$ given by (14.4). The diagonal $x + y = 0$ corresponds to conformal infinity $\mathcal{I}$. The shaded region represents the part of the space-time that has most physical significance and is discussed here. In the r, θ-coordinates, $\theta = 0$ and $\theta = \pi$ correspond to coordinate poles. For $\theta = $ const., r increases from zero at the singularity through the black hole horizon $\mathcal{H}_\mathrm{o}$ and the acceleration horizon $\mathcal{H}_\mathrm{a}$, and then either to conformal infinity or to the coordinate limit $r = \infty$, through which the space-time can be extended.

at $r = 1/\alpha$ has the character of an *acceleration horizon* associated with a boost symmetry. It will also be shown that α is exactly the acceleration of the corresponding source in the weak-field limit.

The relation between the r, θ and x, y coordinates, the roots of Q, and the constraint representing conformal infinity are illustrated in Figure 14.1. As indicated, the horizons divide the space-time into three distinct regions labelled I, II, and III. In region II, the Killing vector ∂_t is timelike and the space-time static. The coordinate t is temporal here, while r is spatial. In regions I and III, the character of the coordinates t and r is interchanged.

The C-metric space-time also admits a second Killing vector ∂_φ which is always spatial and corresponds to a rotational symmetry. It can be seen from the metric (14.6) that the roots of the metric coefficient $g_{\varphi\varphi}$, where the norm of the Killing vector ∂_φ vanishes, correspond to poles $\theta = 0$ and $\theta = \pi$ of the coordinate θ. These identify the *axis of symmetry* of the space-time. The coordinate φ is thus taken to be periodic with the range $\varphi \in [0, 2\pi C)$, where $\varphi = 2\pi C$ is identified with $\varphi = 0$, and the significance of the positive parameter C will be discussed in Section 14.1.3.

As already mentioned, $x + y = 0$ corresponds to conformal infinity and it has been assumed that $y > -x$ (see the shaded region illustrated in Figure 14.1). In the above spherical-type coordinates, the region near conformal infinity is problematic. For $\theta \in [0, \frac{\pi}{2}]$, conformal infinity is simply given by $r = (\alpha \cos \theta)^{-1}$. For $\theta \in (\frac{\pi}{2}, \pi]$, however, conformal infinity is not reached even for $r \to \infty$. To fully cover this asymptotic region of the space-time, it is better to revert to the x, y-coordinates.

14.1.2 Conformal diagrams

It has already been pointed out that the coordinate singularities at $r = 1/\alpha$ and at $r = 2m$ correspond to horizons. Coordinates that are regular across either of these horizons, but not both simultaneously, have been explicitly given by Griffiths, Krtouš and Podolský (2006). The method for obtaining such coordinates and then compactifying them to obtain conformal diagrams will now be outlined.

It is first convenient to introduce the double-null coordinates

$$u = t - r_\star, \qquad v = t + r_\star, \tag{14.9}$$

where $r_\star = \int Q^{-1} \mathrm{d}r$ is a tortoise coordinate given explicitly as

$$\alpha r_\star = k_{\mathrm{c}} \log |1 + \alpha r| + k_{\mathrm{a}} \log |1 - \alpha r| + k_{\mathrm{o}} \log \left| 1 - \frac{r}{2m} \right|, \tag{14.10}$$

where

$$k_{\rm c} = \frac{1}{2(1+2\alpha m)}, \quad k_{\rm a} = -\frac{1}{2(1-2\alpha m)}, \quad k_{\rm o} = \frac{2\alpha m}{1-4\alpha^2 m^2}. \quad (14.11)$$

The null coordinates u, v diverge near both the acceleration and black hole horizons. To analyse the behaviour near these, it is therefore necessary to make the rescaling

$$U = (-1)^i(-\operatorname{sign} k)\,\exp\left(-\frac{\alpha u}{2k}\right), \quad V = (-1)^j(-\operatorname{sign} k)\,\exp\left(\frac{\alpha v}{2k}\right),$$
$$(14.12)$$

where $k = k_{\rm a}$ or $k = k_{\rm o}$, respectively, and where each distinct domain is identified by two appropriately chosen integers i and j. These domains may be of type I, II or III as identified in Figure 14.1. Kruskal–Szekeres-type coordinates

$$T = \tfrac{1}{2}(V+U), \quad R = \tfrac{1}{2}(V-U), \quad (14.13)$$

then cover the corresponding horizon, which is given by $T = \pm R$. Exactly as for the horizons described in previous chapters, these coordinates, for any i, j, cover four distinct domains of two different types.

It is natural to start the procedure at an asymptotic time-dependent domain of type I which is labelled by the integers $(i, j) = (0, 0)$ (see Figure 14.2). This can be extended across the acceleration horizon $\mathcal{H}_{\rm a}$ at $r = 1/\alpha$, using $k = k_{\rm a}$, to two static regions of type II, with integer labels $(0, -1)$ and $(-1, 0)$, which are causally separated by $\mathcal{H}_{\rm a}$. These domains can also be extended to a further (prior) non-static domain of type I with $(i, j) = (-1, -1)$. In terms of these coordinates, the metric is regular across this horizon. These four domains are indicated by the darker shaded regions in Figure 14.2.

Each of the static regions $(0, -1)$ or $(-1, 0)$ is already defined up to their corresponding black hole horizons $\mathcal{H}_{\rm o}$ at $r = 2m$. And they can now be further extended across these horizons using (14.12) with $k = k_{\rm o}$. Just as for other black hole event horizons, these regions are each attached to two other time-dependent domains of type III, which are interpreted as the interiors of the corresponding black and white holes and are given integer labels $(1, -1)$ and $(0, -2)$, or $(-1, 1)$ and $(-2, 0)$. These are also connected to additional static domains of type II labelled $(1, -2)$ or $(-2, 1)$. This process can be repeated across any subsequent horizon. The space-time can thus be continued into a maximally extended manifold which consists of (infinitely) many distinct domains of types I, II or III, which are connected at the corresponding horizons.

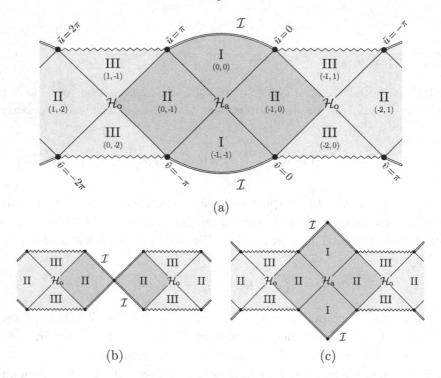

(a)

(b)   (c)

Fig. 14.2 Conformal diagrams for the complete C-metric space-time on sections on which $\theta, \varphi = $ const., for different values of θ. The entire space-time is composed of distinct domains of types I, II and III, labeled here by the integers (i, j). It is bounded by curvature singularities at $r = 0$ and by conformal infinity $\mathcal{I}$, as indicated in (a). Diagrams (b) and (c) indicate the limiting cases on the axes $\theta = 0$ and $\theta = \pi$, respectively.

Notice, however, that the range of the coordinates U and V in the domains of type III is restricted by the presence of the curvature singularity at $r = 0$. In these domains, this is given by the condition $UV = 1$ using $k = k_\text{o}$. Similarly, the domains of type I are restricted by the presence of conformal infinity $\mathcal{I}$ which, for $\theta \in [0, \frac{\pi}{2}]$, is given by $r = |\alpha \cos \theta|^{-1}$. To cover these domains fully, it is better to revert to the coordinates x and y using (14.5) in which conformal infinity is given by $y = -x$, as this also permits y to be negative. Using either coordinates, the important point to note is that the location of conformal infinity $\mathcal{I}$ depends on the value of θ (or x). According to the definitions (14.9)–(14.12), this gives an upper limit (dependent on θ) on the product UV, using $k = k_\text{a}$.

In order to explicitly construct conformal diagrams, corresponding to sections in which θ and φ are constant, it is necessary to compactify the U, V

coordinates. In fact, this can be achieved globally by putting

$$\tan\frac{\tilde{u}}{2} = (-\operatorname{sign} k)\ U^{-\operatorname{sign} k} = (-1)^i\ \exp\!\Big(\frac{\alpha u}{2|k|}\Big), \qquad \tilde{u} \in \Big(i\pi, (i+1)\pi\Big),$$

$$\tan\frac{\tilde{v}}{2} = (-\operatorname{sign} k)\ V^{-\operatorname{sign} k} = (-1)^j\ \exp\!\Big(-\frac{\alpha v}{2|k|}\Big), \qquad \tilde{v} \in \Big(j\pi, (j+1)\pi\Big).$$

$$(14.14)$$

With the choices $k = k_{\mathrm{a}}$ or $k = k_{\mathrm{o}}$, the metric is smooth across either the acceleration or the black hole horizons, respectively. On the horizons, either of the coordinates $\tilde{u}$ or $\tilde{v}$ are integer multiples of π. Different domains of the coordinates u, v, which are separated by horizons $\mathcal{H}_{\mathrm{a}}$ and $\mathcal{H}_{\mathrm{o}}$, thus correspond to different blocks of the coordinates $\tilde{u}$, $\tilde{v}$ and are naturally labelled by the integers (i, j). The whole space-time is covered by $\tilde{u}, \tilde{v} \in (-\infty, \infty)$, constrained by conditions at conformal infinity and the singularity. These conditions imply that the sum of the integers $i + j$ takes one of the values 0, -1 or -2. The resulting two-dimensional conformal diagrams are given in Figure 14.2.

The above extensions across the horizons, and the compactification into two-dimensional conformal diagrams, were obtained by transforming the coordinates t and r only; the angular coordinates θ and φ remained unchanged. However, the location of conformal infinity in the two-dimensional conformal diagrams has been shown to depend on the value of θ. For general θ, the location of $\mathcal{I}$ in these diagrams is given by a spacelike line as typically shown in Figure 14.2(a), and it is not clear a priori why this is not manifestly null as one would expect for a vacuum, asymptotically flat, solution of the Einstein equations without a cosmological constant. This will be clarified in Sections 14.1.4 and 14.1.5. It may also be recalled that the coordinates r, θ do not reach as far as $\mathcal{I}$ for $\theta \in (\frac{\pi}{2}, \pi]$. For this range, it is preferable to revert to the x, y-coordinates (14.5). With this, conformal infinity on the inner axis $x = 1$ ($\theta = \pi$) is seen to be null, as shown in Figure 14.2(c). However, this axis is generally singular as will be discussed below. In addition, on the outer axis $x = -1$ ($\theta = 0$), conformal infinity "coincides" with the horizon at $r = 1/\alpha$, as shown in Figure 14.2(b), but it will be shown that this axis also generally corresponds to a topological singularity.

14.1.3 Some geometrical and physical properties

The C-metric in either of the forms (14.3) or (14.6) is expressed in terms of *three* positive real parameters m and α (satisfying $2\alpha m < 1$) and C, which is hidden in the range of the coordinate $\varphi \in [0, 2\pi C)$. The purpose of this section is to relate these parameters to certain invariant geometrical and physical quantities.

First, consider the regularity of the axis of symmetry. For a small circle around the half-axis $\theta = 0$, with t and r constant and with the above range of φ,

$$\frac{\text{circumference}}{\text{radius}} = \lim_{\theta \to 0} \frac{2\pi CP \sin \theta}{\theta} = 2\pi C(1 - 2\alpha m). \qquad (14.15)$$

In general, this is not exactly 2π, thus implying the existence of a conical singularity. Similarly, around the other half axis $\theta = \pi$,

$$\frac{\text{circumference}}{\text{radius}} = \lim_{\theta \to \pi} \frac{2\pi CP \sin \theta}{\pi - \theta} = 2\pi C(1 + 2\alpha m), \qquad (14.16)$$

which implies the existence of a conical singularity with a *different* conicity. The deficit or excess angles of either of these two conical singularities can be removed by an appropriate choice of C, but not both simultaneously. In general, the constant C can thus be seen to determine the balance between the deficit/excess angles on the two parts of the axis. In particular, one natural choice is to remove the conical singularity at $\theta = \pi$ by setting $C = (1 + 2\alpha m)^{-1}$. In this case, the deficit angles at the poles $\theta = 0, \pi$ are, respectively,

$$\delta_\pi = 0, \qquad \delta_0 = \frac{8\pi \alpha m}{1 + 2\alpha m} . \qquad (14.17)$$

The surfaces on which t and r are constant, showing the conical singularity at $\theta = 0$, are illustrated schematically in Figure 14.3.

The conical singularity with constant deficit angle along the half-axis $\theta = 0$ can be interpreted as representing a *semi-infinite cosmic string* under tension (see Section 3.4). This extends from the source at $r = 0$ right out to conformal infinity $\mathcal{I}$. The metric (14.6) can therefore be understood as representing a Schwarzschild-like black hole that is being accelerated along the axis $\theta = 0$ by the action of a force which corresponds to the tension in a cosmic string.

It is often assumed that the range of the rotational coordinate is 2π. This can be achieved by the simple rescaling

$$\varphi = C \phi,$$

where $\phi \in [0, 2\pi)$. For the above natural choice, where a cosmic string occurs

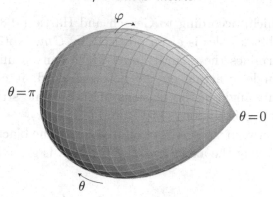

Fig. 14.3 The surface of constant t and r, illustrated as an embedding in an artificial Euclidean space in such a way that it has the correct inner geometry induced by the metric. This is regular at $\theta = \pi$, but there is a conical singularity at $\theta = 0$ corresponding to the deficit angle $\delta_0 = \frac{8\pi\alpha m}{1+2\alpha m}$. However, this is not the correct embedding in the real curved space-time as the coordinates t, r, θ, φ of the C-metric (14.6) are orthogonal, so, in the real space-time, there is no sharp vertex.

only along the half-axis $\theta = 0$, the line element takes the form

$$
\mathrm{d}s^2 = \frac{1}{(1 - \alpha r \cos\theta)^2} \left(-Q\,\mathrm{d}t^2 + \frac{\mathrm{d}r^2}{Q} + \frac{r^2\,\mathrm{d}\theta^2}{P} + \frac{P\,r^2 \sin^2\theta}{(1 + 2\alpha m)^2}\,\mathrm{d}\phi^2 \right),
$$
(14.18)

where P and Q are still given by (14.7).

The parameters m and α naturally occur in the specific forms of the metric (14.3) and (14.6). It is therefore useful to relate these to certain invariant geometrical quantities. Some of the physically most interesting ones are now given.

The area of the black hole horizon $\mathcal{H}_\mathrm{o}$ at $r = 2m$ is given by

$$
\mathcal{A} = \int_0^{2\pi C} \int_0^\pi \sqrt{g_{\theta\theta}\, g_{\varphi\varphi}} \, \bigg|_{\substack{r=2m \\ t=\mathrm{const.}}} \mathrm{d}\theta\,\mathrm{d}\varphi = \frac{16\pi C m^2}{1 - 4\alpha^2 m^2}.
$$
(14.19)

The surface gravity of the horizons, defined by $\kappa^2 = -\frac{1}{2}\chi_{\mu;\nu}\chi^{\mu;\nu}$, is unique up to a normalisation of the Killing vector $\chi^\mu = \partial_t$. An invariant quantity is thus given by the ratio of the surface gravity on the black hole and acceleration horizons $\mathcal{H}_\mathrm{o}$ and $\mathcal{H}_\mathrm{a}$. These are given by

$$
\kappa_\mathrm{o} = \frac{1 - 4\alpha^2 m^2}{4m}, \qquad \kappa_\mathrm{a} = \alpha(1 - 2\alpha m),
$$
(14.20)

which are, respectively, related to the Hawking and Unruh temperatures of the black hole and accelerating frame. It may also be observed that

$\kappa_{\rm o}\mathcal{A} = 4\pi Cm$ which, according to Geroch and Hartle (1982), implies that the mass of the black holes is given precisely by Cm. Notice also that the ratio $\kappa_{\rm o}/\kappa_{\rm a}$ determines the quantity $m\alpha$, which can be understood as the force per unit angle φ acting on the holes. Indeed, it corresponds to a difference per unit angle in the tensions of the cosmic strings on the two parts of the axis (see Equations (14.15) and (14.16)).

The distance between the points of bifurcation of the black hole and acceleration horizons along the axis $\theta = \pi$ at $t = $ const. is given by the complete elliptical integral

$$\mathcal{L} = \int_{2m}^{1/\alpha} \sqrt{g_{rr}}\,\Big|_{\substack{\theta=\pi \\ t=\text{const.}}} \mathrm{d}r = \frac{1}{\alpha\sqrt{1+2\alpha m}}\,\mathbf{E}\left(\frac{1-2\alpha m}{1+2\alpha m}\right). \qquad (14.21)$$

(Notice that this does not depend on the choice of t.) Similarly, the proper time from the point of bifurcation of the black hole horizon to the singularity along the axis $\theta = \pi$ and $t = $ const. is given by

$$\mathcal{T} = \int_0^{2m} \sqrt{-g_{rr}}\,\Big|_{\substack{\theta=\pi \\ t=\text{const.}}} \mathrm{d}r = \frac{1}{\alpha\sqrt{1+2\alpha m}}\left[\mathbf{K}\left(\frac{4\alpha m}{1+2\alpha m}\right) - \mathbf{E}\left(\frac{4\alpha m}{1+2\alpha m}\right)\right],$$
$$(14.22)$$

which is finite and reduces to πm when $\alpha = 0$. It may be observed that the quantities $\mathcal{L}/m$ and $\mathcal{T}/m$ depend only on the product αm.

14.1.4 The Minkowski limit

In order to understand the meaning of the parameter α, it is natural to consider the weak-field limit $m \to 0$ in which the curvature singularity disappears and the black hole source is replaced by a test particle located at $r = 0$. In this limit, the metric (14.6) becomes

$$\mathrm{d}s^2 = \frac{1}{(1 - \alpha r \cos\theta)^2}\left[-(1 - \alpha^2 r^2)\mathrm{d}t^2 + \frac{\mathrm{d}r^2}{(1 - \alpha^2 r^2)} + r^2(\mathrm{d}\theta^2 + \sin^2\theta\,\mathrm{d}\varphi^2)\right].$$
$$(14.23)$$

Since the curvature tensor (14.8) vanishes in this limit, the metric (14.23) must represent (at least a part of) Minkowski space. In fact, one can apply the transformation

$$\hat{\zeta} = \frac{\sqrt{|1 - \alpha^2 r^2|}}{\alpha(1 - \alpha r \cos\theta)}, \qquad \hat{\rho} = \frac{r \sin\theta}{1 - \alpha r \cos\theta}, \qquad (14.24)$$

with $\tau = \alpha t$, leaving φ unchanged. With this, the metric (14.23), in the static region II in which $r < 1/\alpha$, takes the form (3.14) which is adapted to uniform acceleration in cylindrical coordinates

$$\mathrm{d}s^2 = -\hat{\zeta}^2\,\mathrm{d}\tau^2 + \mathrm{d}\hat{\zeta}^2 + \mathrm{d}\hat{\rho}^2 + \hat{\rho}^2\mathrm{d}\varphi^2\,. \tag{14.25}$$

This can be put into the standard form $\mathrm{d}s^2 = -\mathrm{d}\hat{T}^2 + \mathrm{d}\hat{Z}^2 + \mathrm{d}\hat{\rho}^2 + \hat{\rho}^2\mathrm{d}\varphi^2$ of the Minkowski metric in cylindrical coordinates by the transformation $\hat{T} = \pm\hat{\zeta}\sinh\tau$, $\hat{Z} = \pm\hat{\zeta}\cosh\tau$, where $\hat{\zeta} \in [0,\infty)$ and $\tau \in (-\infty,\infty)$ and the choice of signs (giving two copies of the region II) in both expressions is the same. In the region I in which $r > 1/\alpha$, the metric (14.23) can be similarly transformed using (14.24) followed by $\hat{T} = \pm\hat{\zeta}\cosh\tau$, $\hat{Z} = \pm\hat{\zeta}\sinh\tau$, where again the choice of signs is the same for the two copies of this region.

The complete Minkowski space is thus covered by four regions, of either type I or II, that are separated by the null hypersurfaces $\hat{T} = \pm\hat{Z}$ on which $r = 1/\alpha$. These are acceleration horizons $\mathcal{H}_\mathrm{a}$ which form the boundaries of the regions described by the uniformly accelerated metric (14.25). In either of the regions II, points with constant values of r, θ and φ have uniform acceleration: i.e. they follow the worldline $\hat{Z}^2 - \hat{T}^2 = \frac{1-\alpha^2 r^2}{\alpha^2(1-\alpha r\cos\theta)^2}$. In particular, *the acceleration of a test particle located at the origin $r = 0$ is exactly α* in the positive or negative $\hat{Z}$ direction, according to the sign of $\hat{Z}$, as illustrated in Figure 14.4(a).

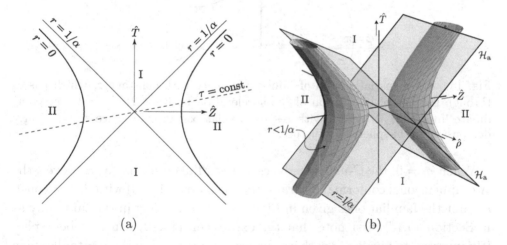

(a)   (b)

Fig. 14.4 In the weak-field limit of the C-metric, a causally separated pair of test particles at $r = 0$ accelerate away from each other in opposite spatial directions, as illustrated in (a). The acceleration horizon at $r = 1/\alpha$ is given by $\hat{Z} = \pm\hat{T}$. For $r < 1/\alpha$, surfaces $r = $ const. wrap around the accelerated particles at $r = 0$, as depicted in (b) with φ suppressed.

Now consider a smooth spatial section $\tau = $ const. through the regions II of the above Minkowski space. This passes through $\hat{T} = 0 = \hat{Z}$, which is where the acceleration horizon $\mathcal{H}_\mathrm{a}$ bifurcates. Each half-section, $\hat{Z} > 0$ or $\hat{Z} < 0$, is covered by the coordinates r, θ, φ with $r \in [0, 1/\alpha)$. With φ ignored, the coordinate lines for $\{r, \theta\}$ (or alternatively $\{y, x\}$) are drawn in Figure 14.5 using axes which correspond to the uniformly accelerated coordinates $\hat{\zeta}$ and $\hat{\rho}$ of (14.24). It can be seen that these lines have *bi-polar structure* (or bi-spherical structure if the angular coordinate φ around the axis $\hat{\rho} = 0$ is included) with origins at $r = 0$. The coordinates r, θ, φ thus play the role of spherical coordinates near the origins, i.e. around the two accelerated test particles, as shown in Figure 14.4(b). However, they deform as $r \to 1/\alpha$.

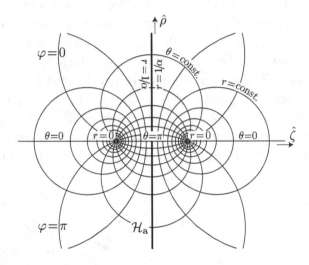

Fig. 14.5 On a spatial section of Minkowski space with constant τ, which passes through the line of bifurcation of the acceleration horizons $\hat{T} = 0 = \hat{Z}$, the coordinate lines of $\{r, \theta\}$ drawn with respect to flat background space $\{\hat{\zeta}, \hat{\rho}\}$ clearly demonstrate a bi-spherical structure.

When $m = 0$, the C-metric reduces to flat Minkowski space. However, the two-dimensional conformal diagrams for the metric (14.23) with $\theta, \varphi = \mathrm{const.}$ are not the familiar ones given in Figure 3.4. Proceeding in the same way as in Section 14.1.2, first note that the expression (14.10) can now be explicitly inverted to give $\alpha r = \coth \alpha r_\star$ in regions of type I and $\alpha r = \tanh \alpha r_\star$ in regions of type II. Double-null coordinates u, v are defined again by (14.9) and their compactified versions $\tilde{u}, \tilde{v}$ are given by (14.14) with $2|k| = 1$. The whole space-time is covered by two domains of type I and two domains of type II, labelled by $(i, j) = (0, 0), (-1, -1)$ and $(i, j) = (-1, 0), (0, -1)$, respectively. These are separated by the horizon $\mathcal{H}_\mathrm{a}$ at $\tilde{u} = 0$ or $\tilde{v} = 0$ corresponding to $r = 1/\alpha$. In the domains II, the coordinate range is restricted by the condition that $r > 0$, and $r = 0$ represents the trajectory of the test particles corresponding to the limit of the black holes.

Introducing the coordinates $\tilde{u} = \eta + \tilde{\chi}$, $\tilde{v} = \eta - \tilde{\chi}$, the above transformations combine in both regions I and II to give

$$\alpha t = \frac{1}{2} \log \left| \frac{\sin \eta - \sin \tilde{\chi}}{\sin \eta + \sin \tilde{\chi}} \right|, \qquad \alpha r = \frac{\cos \tilde{\chi}}{\cos \eta}, \qquad (14.26)$$

which convert the metric (14.23) to the form

$$ds^2 = \frac{1}{\alpha^2 (\cos \eta - \cos \tilde{\chi} \cos \theta)^2} \left(-d\eta^2 + d\tilde{\chi}^2 + \cos^2 \tilde{\chi} \left(d\theta^2 + \sin^2 \theta \, d\varphi^2 \right) \right). \qquad (14.27)$$

This is explicitly conformally related to the metric of the Einstein static universe (3.8) with conformal factor $\Omega = \cos \eta - \cos \tilde{\chi} \cos \theta$, $R_0 = \alpha^{-1}$ and $\tilde{\chi} = \frac{\pi}{2} - \chi$. Conformal infinity $\mathcal{I}$ occurs where $\Omega = 0$, and this obviously depends on the choice of θ. Two-dimensional conformal diagrams on which θ and φ are constant are illustrated in Figure 14.6. These are unfamiliar figures which, however, exactly correspond to particular sections of the familiar three-dimensional (double-cone) representation of conformal Minkowski space, as given in Figure 3.4.

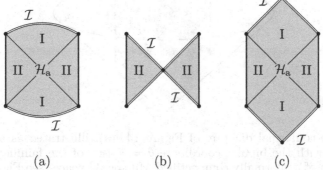

(a)   (b)   (c)

Fig. 14.6 Conformal diagrams of typical sections $\theta, \varphi = \text{const.}$ in the weak-field limit of the C-metric, with axes given by coordinates η and $\tilde{\chi}$ for (a) general θ, (b) $\theta = 0$ and (c) $\theta = \pi$. The accelerating test particles at $r = 0$ are located on the vertical boundaries of region II, and conformal infinity $\mathcal{I}$ is illustrated by the double-line boundaries of regions I.

Indeed, the above non-standard form of the compactified Minkowski space (14.27) is related to the familiar one given in (3.10), in which the conformal factor is $\cos \eta + \cos \chi$, via the transformation

$$\left. \begin{array}{l} \sin \tilde{\chi} = \sin \chi \cos \vartheta, \\[2mm] \tan \theta = \tan \chi \sin \vartheta, \end{array} \right\} \Leftrightarrow \left\{ \begin{array}{l} \cos \chi = -\cos \tilde{\chi} \cos \theta, \\[2mm] \tan \vartheta = -\cot \tilde{\chi} \sin \theta, \end{array} \right. \qquad (14.28)$$

with $R_0{}^2 = 1/\alpha^2$ and ϑ and φ replacing θ and ϕ. The coordinates η, χ are those used for the construction of the usual Minkowski conformal diagrams, in which the conformal boundary $\mathcal{I}$ is manifestly null. By contrast, using the coordinates $\eta, \tilde{\chi}$ of (14.27), $\mathcal{I}$ is manifestly θ-dependent.

The non-standard diagrams in Figure 14.6 arise because the metric (14.23) involves *bipolar* coordinates r, θ that are adapted to the *pair* of uniformly accelerating test particles. Suppressing φ, the spatial section $\eta = 0$ ($\tau = 0$) through the two regions of type II is now a "conformally compactified" version of the surface with bipolar coordinates, illustrated in Figure 14.5. The conformal diagrams of Figure 14.6 correspond to vertical (cylindrical) sections of the familiar three-dimensional "double-cone" diagram of the Minkowski space for the corresponding constant value of θ, as illustrated in Figure 14.7.

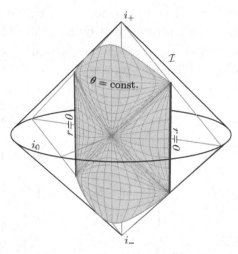

Fig. 14.7 The conformal diagram of Figure 14.6(a), illustrated as a specific vertical section, with the bipolar coordinate $\theta = $ const., of the familiar double-cone representation of conformally compactified Minkowski space (see Figure 3.4).

The two uniformly accelerating test particles at $r = 0$ are represented by two vertical lines in the conformal diagrams and also in Figure 14.7. The surfaces $\theta, \varphi = $ const., corresponding to the family of conformal diagrams in Figure 14.6, are vertical cylindrical surfaces spanned between the worldlines of the origins $r = 0$ and which intersect conformal infinity at an *apparently* spacelike line. This clarifies the fact that conformal infinity $\mathcal{I}$ in the two-dimensional conformal diagram in Figure 14.6(a) generally has a "spatial" character. In addition, the fact that the acceleration horizon is clearly indicated in each of these conformal diagrams enables the location of this horizon to be identified throughout the three-dimensional picture. This is illustrated in Figure 14.8.

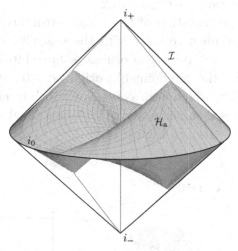

Fig. 14.8 The location of the acceleration horizon $\mathcal{H}_\mathrm{a}$ in the conformally compactified Minkowski space.

14.1.5 Global structure of the C-metric

The full C-metric in the form (14.6) with $m \neq 0$ has already been extended across the horizons in Section 14.1.2, where conformal diagrams were constructed. The purpose here is to combine these two-dimensional sections to construct a three-dimensional representation of the complete space-time (with the trivial φ direction suppressed) outside the black hole horizons, in a similar way to that for the weak-field limit obtained in the previous section.

By comparing the conformal diagrams in Figure 14.2 with those of the corresponding weak-field limit in Figure 14.6, it may immediately be observed that the asymptotic structure of region I is similar in both these cases. Thus, the asymptotic structure of the C-metric space-time close to $\mathcal{I}$ must be basically the same as that illustrated in Figure 14.7. Differences will occur, however, nearer to the sources at $r = 0$.

From the diagrams in Figure 14.2, it can be seen that each asymptotic domain of type I extends through different acceleration horizons to two black holes that are causally separated by the acceleration horizon $\mathcal{H}_\mathrm{a}$. Each of these has an internal structure which is qualitatively similar to that of the familiar Schwarzschild black hole. The inner structure of a single black hole is well understood, and only the exterior of the black holes outside their horizons $\mathcal{H}_\mathrm{o}$ need be considered. Accordingly, it is convenient to concentrate here on four adjacent domains, namely two asymptotic domains of type I and two static domains of type II, as indicated by the darker shaded regions in the conformal diagrams of Figure 14.2.

To construct a representation of the causal structure of this part of the space-time, it is convenient to start with the spacelike surface $t = 0$ in the static regions II on which the r, θ coordinates have bipolar structure, as in Figure 14.5. However, the two point-like origins $r = 0$ must now be replaced by two circles $r = 2m$, which each correspond to bifurcations of the black hole horizons $\mathcal{H}_o$. It is then necessary to include a timelike dimension in the vertical direction, as shown in Figure 14.9(a).

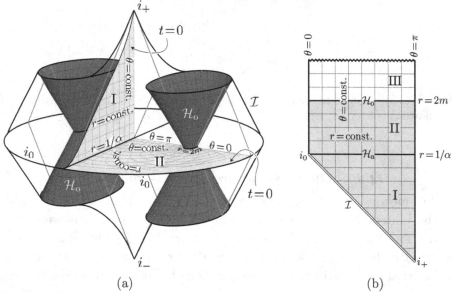

(a)   (b)

Fig. 14.9 The framework for constructing a representation of the global structure of the C-metric space-time outside the black hole horizons $\mathcal{H}_o$ (i.e. regions of types I and II) is given in (a) with the φ direction suppressed. The corresponding part of Figure 14.1 is shown as the shaded region in (b). The coordinate lines $r = $ const., $\theta = $ const. on the surface $t = 0$, which correspond to those shown in Figure 14.5, are shown in both (a) and (b).

The global structure of the Minkowski limit of the C-metric was illustrated in Figure 14.7, where each vertical cylindrical surface $\theta = $ const. corresponds to a two-dimensional conformal diagram of the family pictured in Figure 14.6. In a similar way, an analogous figure describing the global structure of the full C-metric outside the black hole horizons can be constructed. This is achieved by raising cylindrical surfaces vertically above the lines $\theta = $ const. in the plane $t = 0$. These surfaces correspond exactly to the conformal diagrams of Figure 14.2 for the same constant θ. Typical examples are shown in Figure 14.10. The black hole horizons $\mathcal{H}_o$ are indicated as two null conical surfaces.

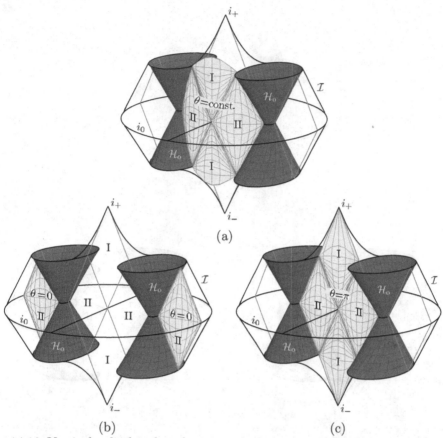

Fig. 14.10 Vertical cylindrical surfaces on which $\theta = $ const. are here the same sections as the conformal diagrams in Figure 14.2. These surfaces can be drawn for $\theta \in [0, \pi]$ and two antipodal values of φ (e.g., $\varphi = 0, \pi$). Together, they lead to a three-dimensional representation of the conformally compactified C-metric spacetime, which shows its global structure outside the black hole horizons $\mathcal{H}_o$. The special cases with $\theta = 0, \pi$, which generally include conical singularities, are shown in (b) and (c).

By taking the complete family of such surfaces for all $\theta \in [0, \pi]$, an illustrative three-dimensional representation of the conformally compactified C-metric space-time may be constructed, which captures its global properties and causal structure. This construction explicitly identifies coordinate labels at all points in a three-dimensional space. This not only allows various coordinate surfaces to be drawn, but also identifies those that have particular physical significance, as already anticipated in Figure 14.9(a). Specifically, surfaces representing conformal infinity $\mathcal{I}$, the acceleration horizon $\mathcal{H}_a$ and the black hole horizons $\mathcal{H}_o$ are pictured in Figure 14.11.

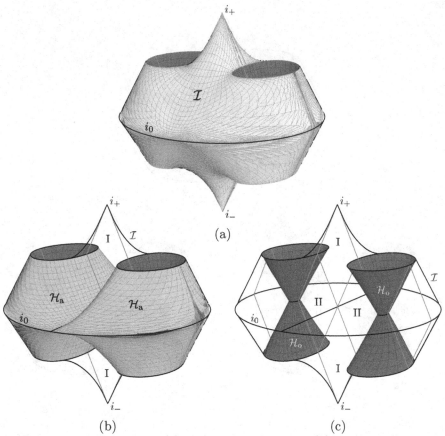

Fig. 14.11 Representations of (a) conformal infinity $\mathcal{I}$, (b) the acceleration horizon $\mathcal{H}_a$ and (c) the two black hole horizons $\mathcal{H}_o$.

Figure 14.11(a) shows the conformal infinity $\mathcal{I}$ of the space-time. It is a null surface which resembles Minkowski-like $\mathcal{I}$, but with two pairs of "holes" which represent the locations where the black hole horizons $\mathcal{H}_o$ intersect conformal infinity. Figure 14.11(b) illustrates the structure of the acceleration horizon $\mathcal{H}_a$. We see that this horizon "touches" infinity only in two pairs of null generators on the axis $\theta = 0$, as in Figure 14.2(b). It also completely encloses the static regions of type II around the two black holes. The black hole horizons $\mathcal{H}_o$, shown in Figure 14.11(c), are two null cones. They are generalisations of the vertical origins $r = 0$ in Figure 14.7 which describes the weak-field limit. (These and other pictures are given by Griffiths, Krtouš and Podolský (2006), and some instructive animations of this space-time and its generalisations are available on the website http://utf.mff.cuni.cz/~krtous/physics/index.html.)

14.1.6 Boost-rotation symmetric form

The C-metric can also be expressed in a form in which the boost and rotation symmetries are explicitly manifested. This is yet another argument indicating that the space-time represents a pair of accelerated sources. It is achieved by first applying the transformation (which generalises (14.24)):

$$
\zeta = \frac{\sqrt{|1-\alpha^2 r^2|}}{\alpha(1-\alpha r\cos\theta)}\sqrt{1-2\alpha m\cos\theta}, \qquad
\rho = \frac{r\sin\theta}{1-\alpha r\cos\theta}\sqrt{1-\frac{2m}{r}},
$$
(14.29)

with $\tau = \alpha t$, and φ unchanged. From this, the following correspondences can immediately be observed:

acceleration horizon $\mathcal{H}_\mathrm{a}$: $\quad r = \frac{1}{\alpha} \quad \leftrightarrow \quad \zeta = 0, \ 0 < \rho < \infty,$

inner axis: $\qquad\qquad\quad \theta = \pi \quad \leftrightarrow \quad \rho = 0, \ 0 \leq \zeta \leq \frac{1}{\alpha}\sqrt{1-2\alpha m},$

black hole horizon $\mathcal{H}_\mathrm{o}$: $\quad r = 2m \quad \leftrightarrow \quad \rho = 0, \ \begin{aligned}\tfrac{1}{\alpha}\sqrt{1-2\alpha m} < \zeta \\ < \tfrac{1}{\alpha}\sqrt{1+2\alpha m},\end{aligned}$

outer axis: $\qquad\qquad\quad \theta = 0 \quad \leftrightarrow \quad \rho = 0, \ \frac{1}{\alpha}\sqrt{1+2\alpha m} \leq \zeta.$

Notice, however, that the coordinates $\{r,\theta\}$ and $\{\zeta,\rho\}$ each cover only half of the section $\tau = $ const. corresponding to a single domain of type II. To cover the regions around both black holes, a second copy of these coordinates is required.

Within region II, an application of the transformation (14.29) takes the metric (14.6) to the form

$$
\mathrm{d}s^2 = -e^{\mu}\zeta^2\mathrm{d}\tau^2 + e^{\lambda}(\mathrm{d}\zeta^2 + \mathrm{d}\rho^2) + e^{-\mu}\rho^2\mathrm{d}\varphi^2 \,,
$$
(14.30)

where the functions μ and λ are given by

$$
e^{\mu} = \frac{1-2m/r}{1-2\alpha m\cos\theta}\,,
$$

$$
e^{-\lambda} = \left(1-\frac{2m}{r}\right)(1-2\alpha m\cos\theta) + \frac{m^2}{r^2}(1-\alpha^2 r^2)\sin^2\theta \,.
$$
(14.31)

Notice that the metric (14.30) reduces to the uniformly accelerated metric (14.25) when $\mu, \lambda \to 0$. Moreover, since μ and λ are independent of τ and φ, it can be put explicitly into the boost-rotation symmetric form (see Bičák and Schmidt, 1989b)

$$ds^2 = \frac{1}{Z^2 - T^2} \left[-e^\mu (Z dT - T dZ)^2 + e^\lambda (Z dZ - T dT)^2 \right] + e^\lambda d\rho^2 + e^{-\mu} \rho^2 d\varphi^2,$$
(14.32)

by performing the transformation $T = \pm \zeta \sinh \tau$, $Z = \pm \zeta \cosh \tau$ in the static regions II (and $T = \pm \zeta \cosh \tau$, $Z = \pm \zeta \sinh \tau$ in the domains of type I, with the same signs chosen in each case for each domain). This explicitly covers all four domains of types I and II that are separated by the acceleration horizons $T = \pm Z$.

Of course the metric functions (14.31) need to be re-written in terms of the variables ζ and ρ. For this purpose, it is convenient to introduce three auxiliary functions $R_i > 0$, $i = 1, 2, 3$, which will be given explicitly in (14.38). In terms of these, the metric functions can be expressed in the forms

$$e^\mu = \frac{R_1 + R_2 - 2m}{R_1 + R_2 + 2m}, \qquad e^\lambda = \frac{\left((1 - 2\alpha m) R_1 + (1 + 2\alpha m) R_2 + 4\alpha m R_3 \right)^2}{4 (1 - 4\alpha^2 m^2)^2 \, R_1 R_2}.$$
(14.33)

14.1.7 Weyl and other coordinates in the static region

The C-metric has already been described here using several sets of coordinates, specifically $\{x, y\}$ and $\{r, \theta\}$, which are simply related by (14.5), and $\{\zeta, \rho\}$ given by (14.29). Each of these parametrise the surfaces $\tau, \varphi = $ const. in the static domains of type II. However, within these domains, the C-metric is also a member of Weyl's class of solutions that were introduced in Chapter 10. It is therefore appropriate to also express this solution in terms of the Weyl coordinates $\{\bar{z}, \bar{\rho}\}$.

The relation between Weyl coordinates and the boost-rotation symmetric coordinates ζ, ρ is given by hyperbolic and parabolic orthogonal transformations, which are well known from the flat two-dimensional space (see Bičák and Schmidt, 1989b), namely

$$\left. \begin{array}{l} \bar{z} = \frac{\alpha}{2}(\zeta^2 - \rho^2) - \frac{1}{2\alpha}, \\[2mm] \bar{\rho} = \alpha \zeta \rho, \end{array} \right\} \Leftrightarrow \left\{ \begin{array}{l} \sqrt{\alpha}\, \zeta = \sqrt{\sqrt{\bar{\rho}^2 + (\bar{z} + \frac{1}{2\alpha})^2} + (\bar{z} + \frac{1}{2\alpha})}, \\[3mm] \sqrt{\alpha}\, \rho = \sqrt{\sqrt{\bar{\rho}^2 + (\bar{z} + \frac{1}{2\alpha})^2} - (\bar{z} + \frac{1}{2\alpha})}. \end{array} \right.$$
(14.34)

In this case, for constant t, the axis $\bar{\rho} = 0$ of the Weyl coordinates corresponds to the entire boundary of a static domain of type II (see Figure 14.1).

It covers both the axis of symmetry and the black hole and acceleration horizons. The different parts of the axis have the following meanings:

$$
\begin{array}{llll}
\text{acceleration horizon } \mathcal{H}_\text{a} & \leftrightarrow & \bar{\rho} = 0, & \bar{z} < -\frac{1}{2\alpha}, \\
\text{inner axis} & \leftrightarrow & \bar{\rho} = 0, & -\frac{1}{2\alpha} \le \bar{z} \le -m, \\
\text{black hole horizon } \mathcal{H}_\text{o} & \leftrightarrow & \bar{\rho} = 0, & -m < \bar{z} < m, \\
\text{outer axis} & \leftrightarrow & \bar{\rho} = 0, & m \le \bar{z}.
\end{array}
$$

With (14.34) and $\tau = \alpha t$, the metric (14.30) takes the Weyl form (10.2), namely

$$
ds^2 = -e^{2U} dt^2 + e^{-2U} \left[e^{2\gamma} (d\bar{\rho}^2 + d\bar{z}^2) + \bar{\rho}^2 d\varphi^2 \right], \tag{14.35}
$$

where U and γ are functions of $\bar{\rho}$ and $\bar{z}$, which are given in terms of the auxiliary functions R_i by

$$
e^{2U} = \alpha \, \frac{R_1 + R_2 - 2m}{R_1 + R_2 + 2m} \left(R_3 + \bar{z} + \tfrac{1}{2\alpha} \right),
$$

$$
e^{-2U+2\gamma} = \frac{\left((1 - 2\alpha m)R_1 + (1 + 2\alpha m)R_2 + 4\alpha m R_3 \right)^2}{8\alpha \, (1 - 4\alpha^2 m^2)^2 \, R_1 R_2 R_3}. \tag{14.36}
$$

In this case, by introducing the points $\bar{z}_1 = m$, $\bar{z}_2 = -m$, $\bar{z}_3 = -1/2\alpha$ on the Weyl axis $\bar{\rho} = 0$, the auxiliary functions $R_i > 0$, $i = 1, 2, 3$ can simply be expressed, respectively, as

$$
R_i = \sqrt{\bar{\rho}^2 + (\bar{z} - \bar{z}_i)^2} \,. \tag{14.37}
$$

As already mentioned in Chapter 10, the coordinates $\{\bar{\rho}, \bar{z}\}$ can be formally interpreted as cylindrical polar coordinates in an unphysical flat 3-space. In this case, following Bonnor (1983), Hong and Teo (2003) and others, the functions R_i can be understood as distances from the points $\bar{z} = \bar{z}_i$, $\bar{\rho} = 0$, as illustrated in Figure 14.12. Moreover, the function U can be interpreted as a superposition of the potential for a rod (producing a spherical black hole) as in (10.15), and that for a uniform acceleration as in (10.5), with a shift in the point of application.

In fact, the functions R_i can be expressed explicitly using (14.34) and

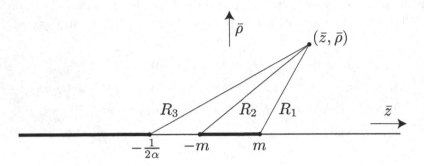

Fig. 14.12 In Weyl coordinates, the axis $\bar{\rho} = 0$ combines the acceleration horizon for $\bar{z} < -1/2\alpha$, the black hole horizon for $\bar{z} \in (-m, m)$, and the inner and outer parts of the axis of symmetry $\theta = 0, \pi$. The auxiliary functions R_i can be understood as distances from the edges of these horizons, measured in the unphysical Euclidean space in which Weyl coordinates $\bar{z}, \bar{\rho}$ would be standard cylindrical coordinates.

(14.29) in each of the following equivalent coordinate forms as

$$R_{1,2} = \sqrt{\bar{\rho}^2 + (\bar{z} \mp m)^2}$$

$$= \sqrt{(1 \pm 2\alpha m)\rho^2 + \frac{1}{4\alpha^2}\left(\alpha^2(\zeta^2 + \rho^2) - (1 \pm 2\alpha m)\right)^2}$$

$$= \frac{r - m(1 \mp \alpha r \pm \cos\theta + \alpha r \cos\theta)}{1 - \alpha r \cos\theta},$$

$$R_3 = \sqrt{\bar{\rho}^2 + \left(\bar{z} + \frac{1}{2\alpha}\right)^2}$$

$$= \tfrac{1}{2}\alpha\,(\rho^2 + \zeta^2)$$

$$= \frac{(1 + \alpha r \cos\theta) - 2\alpha m(\cos\theta + \alpha r)}{2\alpha(1 - \alpha r \cos\theta)}.$$

$$(14.38)$$

By substituting (14.29) into (14.34), it is also found that the direct transformation between the original metric (14.6) and the Weyl form (14.35) is given by

$$\bar{z} = \frac{(\cos\theta - \alpha r)\left(r - m(1 + \alpha r \cos\theta)\right)}{(1 - \alpha r \cos\theta)^2}, \qquad \bar{\rho} = \frac{r \sin\theta \sqrt{PQ}}{(1 - \alpha r \cos\theta)^2}. \qquad (14.39)$$

In terms of the auxiliary functions (14.38), the more complicated inverse transformation to (14.29) or (14.39) can be written explicitly as

$$r = \frac{(1 - 2\alpha m)R_1 + (1 + 2\alpha m)R_2 + 4\alpha m R_3}{(1 - 4\alpha^2 m^2) - \alpha(1 - 2\alpha m)R_1 + \alpha(1 + 2\alpha m)R_2 + 2\alpha R_3},$$

$$\cos\theta = \frac{-(1 - 2\alpha m)(1 + \alpha m)R_1 + (1 + 2\alpha m)(1 - \alpha m)R_2 + 4\alpha^2 m^2 R_3}{m[(1 - 4\alpha^2 m^2) - \alpha(1 - 2\alpha m)R_1 + \alpha(1 + 2\alpha m)R_2 + 2\alpha R_3]},$$

$$(14.40)$$

or equivalently

$$\cos\theta = \frac{-(1 - 4\alpha^2 m^2) - \alpha(1 - 2\alpha m)R_1 + \alpha(1 + 2\alpha m)R_2 + 2\alpha R_3}{\alpha[(1 - 2\alpha m)R_1 + (1 + 2\alpha m)R_2 + 4\alpha m R_3]}.$$

It may finally be noted that the C-metric can also be expressed in Bondi coordinates (see Bonnor, 1990a).

14.2 Accelerating charged black holes

A charged version of the C-metric occurs as a very simple extension of the solution discussed in the previous section. In this case, the appropriate parameters are the mass m, charge e, acceleration α and the hidden parameter C. The metric has the same form as in (14.6) where $\varphi \in [0, 2\pi C)$, but now

$$P = 1 - 2\alpha m \cos\theta + \alpha^2 e^2 \cos^2\theta, \qquad Q = \left(1 - \frac{2m}{r} + \frac{e^2}{r^2}\right)(1 - \alpha^2 r^2), \quad (14.41)$$

and the vector potential for the electromagnetic field is given by

$$A = -\frac{e}{r}\,dt.$$

This clearly reduces to the Reissner–Nordström solution (9.14) in the limit in which $\alpha \to 0$, and to the C-metric when $e = 0$. As before, there is a curvature singularity at $r = 0$, and the roots of Q correspond to horizons. In this case, there exists both an acceleration horizon at $r = \alpha^{-1}$ and inner and outer black hole horizons at $r = r_\pm = m \pm \sqrt{m^2 - e^2}$. The main difference between this case and that discussed in the previous section is in the appearance of the inner (Cauchy) horizon and the consequent change in the character of the curvature singularity. In this case, the singularity will be timelike and further extensions through the inner horizon will occur, so that the internal structure of the black holes will be the same as that of the Reissner–Nordström solution discussed in Section 9.2. This has been described in detail by Cornish and Uttley (1995b) and Hawking and Ross (1997).

Again, deficit or excess angles occur on both parts of the symmetry axis, one of which may be removed by an appropriate choice of the parameter C. For example, the half-axis $\theta = \pi$ is regular if $C = (1 + 2\alpha m + \alpha^2 e^2)^{-1}$. With this choice, the deficit angle along the other half-axis $\theta = 0$ is given by

$$\delta_0 = \frac{8\pi\alpha m}{1 + 2\alpha m + \alpha^2 e^2},$$

generalising (14.17). With this, the space-time given by (14.6) with (14.41) can be interpreted as that of a charged black hole which is being accelerated along the axis of symmetry by the action of a conical singularity representing a cosmic string. Details of this space-time can be calculated exactly as in the previous section.

14.2.1 The Ernst solution

The above comments relate to a charged version of the C-metric, which has virtually the same interpretation as its uncharged limit: the sources are accelerated by forces that are interpreted as tensions in cosmic strings that are represented by conical singularities along parts of the axis of symmetry. Once a charge has been introduced, however, another possibility arises: the forces causing the acceleration could arise from an external electric or magnetic field. A solution representing exactly such a case was found by Ernst (1976a). This can be written in the form

$$ds^2 = \frac{1}{(1 - \alpha r \cos\theta)^2} \left[D^2 \left(-Q\,dt^2 + \frac{dr^2}{Q} + \frac{r^2\,d\theta^2}{P} \right) + \frac{P}{D^2} r^2 \sin^2\theta\,d\varphi^2 \right],$$

(14.42)

where P and Q are given by (14.41) and

$$D = \left(1 - \tfrac{1}{2}eB\cos\theta\right)^2 + \frac{B^2}{4} \frac{r^2 \sin^2\theta\, P}{(1 - \alpha r \cos\theta)^2},$$

in which B is an additional parameter representing the strength of the electromagnetic field.

It can clearly be seen that this solution reduces to the Schwarzschild–Melvin solution (9.25) when $\alpha = 0$ and $e = 0$. When m also vanishes, this is just the Melvin universe (7.30). Moreover, it is the charged C-metric described above when $B = 0$. It is therefore reasonable to interpret this solution as representing a pair of accelerating charged black holes in the presence of an external electric field of strength B.

For this solution, the deficit angles on the axis of symmetry $\theta = 0$ and

$\theta = \pi$ are given by

$$\delta_0 = 2\pi \left(1 - C \frac{1 - 2\alpha m + \alpha^2 e^2}{(1 - \frac{1}{2}eB)^4}\right), \qquad \delta_\pi = 2\pi \left(1 - C \frac{1 + 2\alpha m + \alpha^2 e^2}{(1 + \frac{1}{2}eB)^4}\right),$$

respectively. It follows from this that the conical singularities on either half of the axis can be made to vanish by making an appropriate choice of the constant C. However, unlike the previous cases, it is now possible to remove the deficit angles on both parts of the axis *simultaneously*. This is achieved by choosing the value of the electromagnetic field B such that

$$eB = 2\frac{K-1}{K+1}, \qquad \text{where} \qquad K^4 = \frac{1 + 2\alpha m + \alpha^2 e^2}{1 - 2\alpha m + \alpha^2 e^2}.$$

With this choice, the deficit angles are the same (i.e. $\delta_0 = \delta_\pi$) and, for a particular choice of C, they can both be made zero.

By analogy with the above case in which an external electromagnetic field has been introduced, it has also been suggested that an external gravitational field could similarly be inserted. Exact solutions describing such situations have been explicitly constructed by Ernst (1978) and Bonnor (1988), the first of which was further investigated by Dray and Walker (1980). Obviously, these solutions are not asymptotically flat.

14.3 Accelerating and rotating black holes

Within the family of type D electrovacuum solutions presented by Plebański and Demiański (1976), there exist particular cases which represent black hole-like sources that both accelerate and rotate. This complete family of solutions is described in detail in Chapter 16. However, it is appropriate here to explicitly mention the case that directly generalises the charged C-metric to include a non-zero rotation. (For a more detailed discussion, see Griffiths and Podolský, 2005 and 2006a).

In fact, when some of the parameters of the Plebański–Demiański solutions are set to zero, a solution is obtained that has come to be known as the "spinning C-metric". This has been studied by many authors (e.g. Farhoosh and Zimmerman, 1980b,c; Bičák and Pravda, 1999; Letelier and Oliveira, 2001; Pravda and Pravdová, 2002). However, it was subsequently shown by Hong and Teo (2005) that a different choice of parameters, which removes the specific properties associated with a non-zero NUT parameter, is more appropriate to represent an accelerating and rotating pair of black holes. This is clarified in Chapter 16, where it will be explained why the Plebański–Demiański parameter n should not generally be identified with the NUT parameter l.

Specifically, the special case of (16.24) which describes a pair of accelerating and rotating charged black holes without any NUT-like properties is given by the metric

$$ds^2 = \frac{1}{\Omega^2} \left\{ -\frac{Q}{\varrho^2} \left(dt - a\sin^2\theta\,d\varphi\right)^2 + \frac{\varrho^2}{Q}\,dr^2 + \frac{\varrho^2}{P}\,d\theta^2 \right.$$
$$\left. + \frac{P}{\varrho^2}\sin^2\theta \left[adt - (r^2 + a^2)d\varphi\right]^2 \right\}, \tag{14.43}$$

where

$$\Omega = 1 - \alpha r\cos\theta, \qquad P = 1 - 2\alpha m\cos\theta + \alpha^2(a^2 + e^2)\cos^2\theta,$$
$$\varrho^2 = r^2 + a^2\cos^2\theta, \qquad Q = (a^2 + e^2 - 2mr + r^2)(1 - \alpha^2 r^2).$$

The vector potential for the electromagnetic field is given by

$$A = -\frac{e\,r}{r^2 + a^2\cos^2\theta}\left(dt - a\sin^2\theta\,d\varphi\right),$$

and the only non-zero components of the curvature tensor relative to a natural null tetrad are

$$\Psi_2 = \left[-m(1 - i\,\alpha a) + e^2\left(\frac{1 + \alpha r\cos\theta}{r - i\,a\cos\theta}\right)\right]\left(\frac{1 - \alpha r\cos\theta}{r + i\,a\cos\theta}\right)^3,$$

$$\Phi_{11} = \frac{e^2}{2}\frac{(1 - \alpha r\cos\theta)^4}{(r^2 + a^2\cos^2\theta)^2}.$$

These indicate the presence of a Kerr-like ring singularity at $r = 0$, $\theta = \frac{\pi}{2}$.

It can easily be seen that this solution explicitly contains the C-metric (14.6) with (14.7) when both a and e vanish, the accelerating Reissner–Nordström solution (14.6) with (14.41) when $a = 0$ and the Kerr–Newman metric (11.8) when $\alpha = 0$. In view of these limits, and provided $m^2 > a^2 + e^2$, the metric (14.43) may be interpreted as representing accelerating and rotating charged black holes. However, it is still necessary to check that unwelcome features are not also included.

The metric (14.43) corresponds precisely to that of Hong and Teo (2005). It contains four arbitrary real parameters – the mass of the black hole m, its charge e, its acceleration α and its rotation parameter a. Part of conformal infinity occurs when $r = 1/|\alpha\cos\theta|$ for $\theta \in [0, \frac{\pi}{2}]$, exactly as for the C-metric to which it approximates asymptotically. If $m^2 \geq a^2 + e^2$, the expression for Q factorises as

$$Q = (r_- - r)(r_+ - r)(1 - \alpha^2 r^2),$$

where

$$r_\pm = m \pm \sqrt{m^2 - a^2 - e^2}.$$

The expressions for $r_\pm$ are identical to those for the locations of the outer (event) and inner (Cauchy) horizons of a non-accelerating Kerr–Newman black hole. However, in this case, there is another horizon at $r = 1/\alpha$ which corresponds to an acceleration horizon. Generally, there are conical singularities on the axis of symmetry which depend on the parameter C where $\varphi \in [0, 2\pi C)$. If C is chosen to make either the axis $\theta = 0$ or $\theta = \pi$ regular, the conical singularity remaining on the other part of the axis can be interpreted as a semi-infinite cosmic string or a strut between the sources.[1] These may be considered to cause the black holes to accelerate, and the corresponding deficit or excess angles are given, respectively, by (16.21) or (16.22) with $a_3 = 2\alpha m$ and $a_4 = -\alpha^2(a^2 + e^2)$. Most importantly, it can be seen that there are no closed timelike curves near the axis. This confirms that the solution (14.43) does not possess any NUT-like properties. As argued by Hong and Teo (2005), it is therefore the most appropriate metric for describing a pair of accelerating and rotating black holes.

The metric (14.43) can be analytically extended through either the black hole horizons or the acceleration horizon by generalising the Kruskal–Szekeres-like transformations given above. After a further compactification, the resulting conformal diagram for sections with θ and φ constant is a natural combination of that for the C-metric in asymptotic regions, with that for the Kerr or Reissner–Nordström solution in the interior black hole regions.

Alternatively, the metric can be analytically extended through the acceleration horizon by transforming it to the boost-rotation symmetric form which generalises (14.32) to include rotating sources (see Bičák and Pravda, 1999)

$$
ds^2 = -\frac{e^\mu}{Z^2 - T^2} \left[(Z dT - T dZ) + A(Z^2 - T^2) d\varphi \right]^2
$$
$$
+ e^\lambda \left[\frac{(Z dZ - T dT)^2}{Z^2 - T^2} + d\rho^2 \right] + e^{-\mu} \rho^2 \frac{\gamma^2}{\beta^2} d\varphi^2.
\tag{14.44}
$$

Here μ, λ and A are specific functions of ρ and $Z^2 - T^2$, and β and γ are constants that have to be carefully chosen to identify the cosmic strings causing the acceleration. This has been given explicitly (for the more general case which also includes a NUT parameter) by Griffiths and Podolský (2006a).

---

[1] As in the Ernst solution that was described in the previous subsection, an electromagnetic field can be introduced to remove the conical singularities. This has been given explicitly by Bičák and Kofroň (2010).

14.4 The inclusion of a cosmological constant

The general family of electrovacuum solutions of algebraic type D presented by Plebański and Demiański (1976) also includes a cosmological constant. They must therefore contain particular cases which describe accelerating and rotating charged black holes in asymptotically de Sitter or anti-de Sitter universes. In their original forms, these special cases were not clearly identified, nor were they well suited for physical interpretation. It is much more convenient to consider them in the general form described in detail in Chapter 16, and for this specific case by Podolský and Griffiths (2006).

For this general family of accelerating and rotating black holes with a non-zero cosmological constant Λ but with no NUT-like properties, the metric can be written exactly as in (14.43) with Ω and ϱ^2 unchanged but with the expressions for P and Q replaced by

$$
\begin{aligned}
P &= 1 - 2\alpha m \cos\theta + \left[\alpha^2(a^2 + e^2) + \tfrac{1}{3}\Lambda a^2\right]\cos^2\theta, \\
Q &= \left(a^2 + e^2 - 2mr + r^2\right)\left(1 - \alpha^2 r^2\right) - \tfrac{1}{3}\Lambda(a^2 + r^2)r^2.
\end{aligned}
\tag{14.45}
$$

This solution contains five arbitrary parameters: the mass of the black hole m, its charge e, its rotation a, its acceleration α and the cosmological constant Λ.

One of the most significant differences between this case and that described above, in which $\Lambda = 0$, arises from the fact that the expression for Q no longer has roots that are simply related to the above distinct parameters. Nevertheless, this expression is still a quartic and, for the physically significant solutions, it must generally (for $\alpha^2 + \tfrac{1}{3}\Lambda > 0$) have up to three positive real roots. Such roots would correspond to inner (Cauchy) and outer black hole (event) horizons and to an acceleration horizon. However, the explicit expressions for these roots will now be much more involved. As above, analytic extensions through the acceleration horizon reveal that the complete space-time represents a pair of causally separated black holes which accelerate away from each other in opposite spatial directions.

The other significant difference that arises in this case is that, asymptotically, the space-time has the character of de Sitter or anti-de Sitter space. There remains a curvature (Kerr-like ring) singularity at $r = 0$, but conformal infinity $\mathcal{I}$ is now spacelike or timelike, respectively.

The case in which $\alpha^2 + \tfrac{1}{3}\Lambda \leq 0$, however, has a very different character. This condition implies that there is no acceleration horizon, so that the space-time represents at most a single black hole with inner (Cauchy) and outer black hole (event) horizons. Moreover, such a source is accelerating in an asymptotically anti-de Sitter background. In this case, however, it is necessary that $1 + \alpha^2(a^2 + e^2) + \frac{1}{3}\Lambda a^2 > |2\alpha m|$ to retain the full range of the azimuthal coordinate $\theta \in [0, \pi]$. (For details of these different cases, see e.g. Krtouš, 2005.)

As will be described for the more general case in Section 16.3, conical singularities generally occur on the axis of symmetry. Parts of this axis may be made regular by specific choices of the parameter C where $\varphi \in [0, 2\pi C)$. For example, if the axis $\theta = \pi$ between the black holes is taken to be regular, then there is a conical singularity on $\theta = 0$ which corresponds to a cosmic string under tension. This has deficit angle

$$\delta_0 = \frac{8\pi \, \alpha \, m}{1 + 2\alpha m + \alpha^2(a^2 + e^2) + \frac{1}{3}\Lambda a^2}.$$

Provided $\mathcal{Q}$ has three positive real roots, the metric (14.43) with (14.45) represents a charged and rotating black hole that is caused to accelerate in a de Sitter or anti-de Sitter background, under the action of a cosmic string which is represented by a conical singularity. It describes the space-time from the singularity through the inner and outer black hole horizons and out to and beyond the acceleration horizon. However, these coordinates do not cover the complete analytic extension either inside the black hole horizon or beyond the acceleration horizon. For both of these, Kruskal–Szekeres-like coordinates are required. As for the simpler cases, these would show that the maximally extended space-time contains an infinite sequence of pairs of causally separated charged and rotating black holes, which accelerate away from each other in opposite spatial directions.

To elucidate the character of the coordinates introduced in the metric (14.43) with (14.45), it is appropriate to consider the limit in which m, a and e are reduced to zero while α and Λ remain arbitrary. In this case, the black holes are replaced by test particles that are located on worldlines on which $r = 0$. The resulting metric is

$$ds^2 = \frac{1}{(1 - \alpha \, r \cos \theta)^2} \left[- \left(1 - (\alpha^2 + \tfrac{1}{3}\Lambda)r^2\right) dt^2 \right.$$
$$\left. + \frac{dr^2}{1 - (\alpha^2 + \tfrac{1}{3}\Lambda)r^2} + r^2 \left(d\theta^2 + \sin^2 \theta \, d\varphi^2\right) \right], \tag{14.46}$$

which, in fact, is conformal to the standard form of the de Sitter metric (4.9) with a different value of Λ, to which it also reduces when $\alpha = 0$.

When $\Lambda = 0$, the metric (14.46) is the weak-field limit of the C-metric given in (14.23). As shown in Section 14.1.4, this describes a flat space-time in terms of a metric that is adapted to uniformly accelerated test particles. Similarly, when $\Lambda \neq 0$, the metric (14.46) describes a de Sitter or anti-de Sitter universe but now expressed in analogous "accelerated co-ordinates". These space-times can be represented as the four-dimensional hyperboloid

$$-Z_0{}^2 + Z_1{}^2 + Z_2{}^2 + Z_3{}^2 + \varepsilon Z_4{}^2 = 3/\Lambda \,, \tag{14.47}$$

in the flat five-dimensional space

$$\mathrm{d}s^2 = -\mathrm{d}Z_0{}^2 + \mathrm{d}Z_1{}^2 + \mathrm{d}Z_2{}^2 + \mathrm{d}Z_3{}^2 + \varepsilon \mathrm{d}Z_4{}^2 \,, \tag{14.48}$$

where $\varepsilon = \operatorname{sign}\Lambda$ (see (4.1), (4.2) and (5.1), (5.2)). Provided $\alpha^2 + \frac{1}{3}\Lambda > 0$, the metric (14.46) can be expressed in this notation by the parametrisation

$$
\begin{aligned}
Z_0 &= \frac{\sqrt{(\alpha^2 + \frac{1}{3}\Lambda)^{-1} - r^2}}{1 - \alpha\, r \cos\theta}\, \sinh(\sqrt{\alpha^2 + \tfrac{1}{3}\Lambda}\, t)\,, \\[2mm]
Z_1 &= \pm\frac{\sqrt{(\alpha^2 + \frac{1}{3}\Lambda)^{-1} - r^2}}{1 - \alpha\, r \cos\theta}\, \cosh(\sqrt{\alpha^2 + \tfrac{1}{3}\Lambda}\, t)\,, \\[2mm]
Z_2 &= \frac{r \sin\theta \sin\varphi}{1 - \alpha\, r \cos\theta}\,, \\[2mm]
Z_3 &= \frac{r \sin\theta \cos\varphi}{1 - \alpha\, r \cos\theta}\,, \\[2mm]
Z_4 &= \frac{\alpha - (\alpha^2 + \frac{1}{3}\Lambda)\, r \cos\theta}{\sqrt{\frac{1}{3}|\Lambda|}\,\sqrt{\alpha^2 + \frac{1}{3}\Lambda}\,(1 - \alpha\, r \cos\theta)}\,.
\end{aligned}
\tag{14.49}
$$

In particular, the trajectory of a test particle located at $r = 0$ is given by

$$
\begin{aligned}
Z_0 &= (\alpha^2 + \tfrac{1}{3}\Lambda)^{-1/2}\, \sinh(\sqrt{\alpha^2 + \tfrac{1}{3}\Lambda}\, t)\,, \\[2mm]
Z_1 &= \pm(\alpha^2 + \tfrac{1}{3}\Lambda)^{-1/2}\, \cosh(\sqrt{\alpha^2 + \tfrac{1}{3}\Lambda}\, t)\,,
\end{aligned}
\tag{14.50}
$$

$Z_2 = Z_3 = 0$, $Z_4 = \text{const}$. This represents the trajectories of a pair of uniformly accelerated particles in a de Sitter or anti-de Sitter space-time (see Podolský and Griffiths (2001c), Dias and Lemos (2003a,b), Bičák and Krtouš (2005) and Krtouš (2005)).

In the alternative case of a test particle with small acceleration in an anti-de Sitter universe, for which $\alpha^2 + \frac{1}{3}\Lambda < 0$, the metric (14.46) corresponds to the parametrisation

$$Z_0 = \frac{\sqrt{r^2 - (\alpha^2 + \frac{1}{3}\Lambda)^{-1}}}{1 - \alpha r \cos\theta} \sin(\sqrt{-(\alpha^2 + \frac{1}{3}\Lambda)}\, t),$$

$$Z_4 = \frac{\sqrt{r^2 - (\alpha^2 + \frac{1}{3}\Lambda)^{-1}}}{1 - \alpha r \cos\theta} \cos(\sqrt{-(\alpha^2 + \frac{1}{3}\Lambda)}\, t), \qquad (14.51)$$

$$Z_1 = \frac{\alpha - (\alpha^2 + \frac{1}{3}\Lambda)\, r \cos\theta}{\sqrt{\frac{1}{3}|\Lambda|}\, \sqrt{-(\alpha^2 + \frac{1}{3}\Lambda)}\, (1 - \alpha r \cos\theta)},$$

with Z_2, Z_3 as in (14.49). In this case, the trajectory $r = 0$ represents the motion of a single uniformly accelerated test particle in an anti-de Sitter universe. For details see Podolský (2002a), Dias and Lemos (2003a) and Krtouš (2005).

In fact, *any* world-line $x^\mu(\tau) = (t(\tau), r_0, \theta_0, \phi_0)$ in the space-time (14.46), where r_0, θ_0, ϕ_0 are constants and τ is the proper time, represents the motion of a uniformly accelerated test particle. Its four-velocity is $u^\mu = (\dot{t}, 0, 0, 0)$, $\dot{t} = 1/\sqrt{-g_{tt}(r_0, \theta_0)} = \text{const.}$, and the four-acceleration has constant components

$$a^\mu \equiv u^\mu_{\;;\nu} u^\nu = \left(0, \frac{g_{tt,r}}{2g_{tt}g_{rr}}, \frac{g_{tt,\theta}}{2g_{tt}g_{\theta\theta}}, 0\right).$$

Since $a^\mu u_\mu = 0$, this is a spatial vector in the instantaneous rest frame orthogonal to the four-velocity. Its magnitude is a constant given by

$$A^2 \equiv -a^\mu a_\mu = (\alpha^2 + \tfrac{1}{3}\Lambda)\frac{(1 - \alpha r_0 \cos\theta_0)^2}{1 - (\alpha^2 + \frac{1}{3}\Lambda)\, r_0^2} - \tfrac{1}{3}\Lambda. \qquad (14.52)$$

In particular, the uniform acceleration of a test particle located at the origin $r = 0$ is given exactly by $A = \alpha$, independently of θ_0, ϕ_0 or Λ. For this reason, the line element (14.43) with (14.45) appears to use the most convenient coordinates for an accelerating particle in a Minkowski or (anti-)de Sitter background (14.46). (Of course, the acceleration is only that of a real physical particle when m is non-zero and there exists a physical cause that can be modelled by conical singularities on the axis of symmetry.)

Further properties of the non-rotating, but charged C-metric in a de Sitter or anti-de Sitter background have been analysed by Podolský and Griffiths (2001c), Podolský (2002a), Dias and Lemos (2003a,b), Krtouš and Podolský (2003), Podolský, Ortaggio and Krtouš (2003) and Krtouš (2005).

14.5 Some related topics

As shown in Section 14.1.7, the static region of the C-metric is also a member of Weyl's class of solutions that were introduced in Chapter 10, and it has a corresponding combination of potentials as illustrated in Figure 14.12. It is possible to superpose such potentials to construct solutions for the static regions of any number of colinear, possibly accelerating, black holes. For such constructions, see Israel and Khan (1964) and Dowker and Thambyahpillai (2003).

Several authors have used various extensions of the vacuum C-metric to describe the possible creation of a pair of black holes by the breaking of a cosmic string. For this purpose, Eardley *et al.* (1995) have used a charged C-metric, while Hawking and Ross (1995) have included a cosmological constant, Emparan (1995) has used the Ernst solution, and Dowker *et al.* (1994) have also included a dilaton extension. In each case, the additional field is considered to supply the potential energy that is necessary for such a pair-creation process.

This process, with a cosmological constant included in the traditional way, has been further analysed by Mann and Ross (1995), Mann (1997), and Booth and Mann (1998, 1999). However, as shown by Podolský and Griffiths (2001c), the physical interpretation of this process would be clearer if Λ is inserted as described here. This has been investigated by Dias and Lemos (2004) and Dias (2004).

A particular case of the C-metric with $\Lambda < 0$ has also been used for the construction of a solution describing a black hole on the brane by Emparan, Horowitz and Myers (2000a,b).

15

Further solutions for uniformly accelerating particles

The possible motion of particles with negative mass was considered in the context of general relativity by Bondi (1957a). He found that exact solutions exist in which the interaction between two particles, one with positive and the other with negative mass, is such that they are both induced to accelerate in the same direction. He showed that such solutions could be described in terms of a metric that has boost and rotation symmetries, and that this includes "static" regions that can be described by the Weyl metric. Explicit solutions representing such situations were subsequently found by Israel and Khan (1964) and Bonnor and Swaminarayan (1964).

Israel and Khan (1964) started with a static solution in Weyl coordinates that represents three colinear Schwarzschild-like bodies, the outer two being standard black holes and the inner being a negative-mass naked singularity. They then replaced one of the black holes by its limit as a uniformly accelerated reference frame, in which its event horizon becomes an acceleration horizon. As already seen, such situations have a reflection symmetry. The resulting extended space-time then represents two pairs of sources that accelerate in opposite directions. Each pair consists of a black hole and a naked singularity of negative mass. This situation is similar to that of the C-metric that was described in the previous chapter, except that the conical singularities that were considered to cause the acceleration are now replaced by the presence of bodies with negative mass.

Bonnor and Swaminarayan (1964) also found a number of space-times of a similar kind using sources of Curzon-Chazy-type. Some of these contain Curzon-Chazy particles with positive mass only, although conical singularities then exist on parts of the axis of symmetry. Such space-times can also be obtained as limits of the Israel–Khan solutions. Further examples of this type were subsequently found by Bičák, Hoenselaers and Schmidt (1983a,b). These are all reviewed in the present chapter.

291

15.1 Boost-rotation symmetric space-times

The general metric with boost and rotation symmetries can be expressed in the form (14.44) which was presented in the previous chapter. This describes the fields of uniformly accelerating and possibly rotating sources. Solutions of algebraic type D, for which the sources are black holes, have been described in Chapter 14 and will be considered further in Chapter 16. However, alternative accelerating and rotating sources are also possible, and some properties of such algebraically general space-times have been described by Pravdová and Pravda (2002) and in references cited therein.

In this chapter, only non-rotating sources will be considered, so it is appropriate to start with the metric in the form (14.32) in which the Killing vectors are hypersurface orthogonal. In this case, the metric can be re-expressed as

$$ds^2 = \frac{1}{Z^2 - T^2} \left[-e^\mu (Z dT - T dZ)^2 + e^\lambda (Z dZ - T dT)^2 \right] + e^\lambda d\rho^2 + e^{-\mu} \rho^2 d\phi^2,$$
(15.1)

where μ and λ are functions of ρ and $Z^2 - T^2$ only, and it is assumed that $\phi \in [0, 2\pi)$. For the full range of the coordinates Z and T, this metric covers two static regions in which $Z^2 > T^2$, and two time-dependent regions in which $Z^2 < T^2$, as is now familiar for accelerated coordinates (see Section 3.5 and Chapter 14). The static and time-dependent regions are separated by an acceleration horizon where $Z^2 - T^2 = 0$.

Under certain natural conditions, the boost-rotation symmetric space-times with the metric (15.1) are the only non-rotating vacuum space-times with two isometries that are compatible with asymptotic flatness and also admit radiation (see Ashtekar and Xanthopoulos, 1978, Ashtekar and Schmidt, 1980, and Bičák and Schmidt, 1984, 1989b). This result has been generalised by Bičák and Pravdová (1998) to include both rotation and an electromagnetic field.

In either of the static regions, in which $Z^2 > T^2$, the coordinate transformation

$$\left. \begin{array}{l} \bar{z} = \frac{1}{2}(Z^2 - T^2 - \rho^2), \\[2mm] \bar{\rho} = \rho \sqrt{Z^2 - T^2}, \\[2mm] t = \operatorname{arctanh}(T/Z), \end{array} \right\} \Leftrightarrow \left\{ \begin{array}{l} \rho = \sqrt{\sqrt{\bar{\rho}^2 + \bar{z}^2} - \bar{z}}, \\[2mm] Z = \sqrt{\sqrt{\bar{\rho}^2 + \bar{z}^2} + \bar{z}} \cosh t, \\[2mm] T = \sqrt{\sqrt{\bar{\rho}^2 + \bar{z}^2} + \bar{z}} \sinh t, \end{array} \right.$$
(15.2)

takes the metric (15.1) to the Weyl form (10.2), namely

$$ds^2 = -e^{2U} dt^2 + e^{-2U} \left[e^{2\gamma} (d\bar{\rho}^2 + d\bar{z}^2) + \bar{\rho}^2 d\phi^2 \right],$$
(15.3)

where

$$e^{2U} = (Z^2 - T^2)\, e^{\mu}, \qquad e^{2\gamma} = \frac{Z^2 - T^2}{\rho^2 + Z^2 - T^2}\, e^{\mu + \lambda},$$

where U and γ must now be expressed in terms of the coordinates $\bar{\rho}$ and $\bar{z}$. Notice that the transformation (15.2) differs slightly from that given in (14.34) particularly because the solutions described in this chapter do not generally include a unique acceleration parameter α. Although the omission of α in (15.2) has an unfortunate effect on the dimensions of the coordinates, it avoids introducing an unnecessary parameter.

As described in Chapter 10, exact solutions can be generated using Weyl coordinates and interpreted here in terms of accelerating sources. A number of special cases are particularly relevant in this context.

15.2 Bonnor–Swaminarayan solutions

In 1964, Bonnor and Swaminarayan obtained a family of solutions using the line element (15.1) which can be explicitly expressed in the form

$$\mu = -\frac{2m_1}{\alpha_1 R_1} - \frac{2m_2}{\alpha_2 R_2} + 4m_1 \alpha_1 + 4m_2 \alpha_2 + B,$$

$$\lambda = 8m_1 m_2 \frac{\alpha_1^3 \alpha_2^3 (R_1 - R_2)^2}{(\alpha_2^2 - \alpha_1^2)^2 R_1 R_2} - \frac{2m_1 m_2}{\alpha_1 \alpha_2 R_1 R_2} \tag{15.4}$$

$$- \left(\frac{m_1^2}{\alpha_1^2 R_1^4} + \frac{m_2^2}{\alpha_2^2 R_2^4} \right) \rho^2 (Z^2 - T^2)$$

$$+ 2 \left(\frac{m_1 \alpha_1}{R_1} + \frac{m_2 \alpha_2}{R_2} \right) (\rho^2 + Z^2 - T^2) + B,$$

where

$$R_i = \frac{1}{2} \sqrt{ \left(\rho^2 + Z^2 - T^2 - \frac{1}{\alpha_i^2} \right)^2 + \frac{4}{\alpha_i^2} \rho^2 }, \qquad (i = 1, 2) \tag{15.5}$$

and m_1, m_2, α_1, α_2 and B are arbitrary constants (see also Bonnor, 1966b, and Bonnor, Griffiths and MacCallum, 1994).[1]

This solution was initially constructed from the Weyl form (15.3) of the metric in which the potential function is given by

$$U = -\frac{m_1}{\alpha_1 R_1} - \frac{m_2}{\alpha_2 R_2} + 2m_1 \alpha_1 + 2m_2 \alpha_2 + \frac{B}{2} + \frac{1}{2} \log(\sqrt{\bar{\rho}^2 + \bar{z}^2} + \bar{z}),$$

---

[1] If the boost-rotation coordinates ρ, Z, T are each considered to have dimension L, then m_i, α_i and R_i have dimensions L, L^{-1} and L^2, respectively, while B is dimensionless.

where

$$R_i = \sqrt{\bar{\rho}^2 + \left(\bar{z} - \frac{1}{2\alpha_i^2}\right)^2}, \qquad (i = 1, 2).$$

Apart from the constant terms, this potential includes that of two Curzon–Chazy particles at specific points on the axis and the boost potential for a uniform acceleration as given in (10.5).

This family of solutions has curvature singularities at $R_1 = 0$ and at $R_2 = 0$, which both have timelike (hyperbolic) worldlines in the $T\text{-}Z$ plane given by

$$\rho = 0, \quad Z^2 - T^2 = \frac{1}{\alpha_i^2}. \qquad (15.6)$$

These correspond to the trajectories of two singular points of "mass" m_i in each of the distinct stationary regions of the complete space-time, each point having uniform acceleration $\pm\alpha_i$ in the positive or negative Z directions with respect to a Minkowski background. Thus, the space-time appears to represent the uniformly accelerated motion of four point sources or "particles", with two pairs located symmetrically in the two static regions on either side of the Z-axis, as illustrated schematically in Figure 15.1. However, these point masses are not black holes, as they are not surrounded by horizons, but are "Curzon–Chazy" particles as described in Section 10.5.

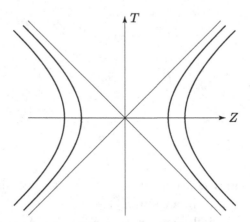

Fig. 15.1 A space-time diagram of the Bonnor–Swaminarayan solutions for the section $\rho = 0$. Worldlines are shown of two causally separated pairs of Curzon–Chazy particles which uniformly accelerate in opposite spatial directions.

It may be noted that the minimum distance of each particle from $Z = 0$ is given by $\pm\alpha_i^{-1}$. Here we will choose the labels such that $\alpha_2 > \alpha_1 > 0$. Also, the parameters adopted in (15.4) differ from those that are sometimes used. The reason for the form presented here is that the "mass" and "acceleration" parameters of each "particle" are identified in such a way that limits can be considered in which $m_i \to 0$ or $\alpha_i \to \infty$.

As pointed out originally by Bonnor and Swaminarayan (1964), this solution can be expressed formally in terms of a Newtonian potential which is the average of the retarded and advanced potentials corresponding to the particles. The radiative properties of these solutions have been extensively studied by Bičák (1968, 1971, 1985, 1987).

The space-time represented by the metric (15.1) generally contains conical singularities, representing strings or struts, along parts of the axis of symmetry $\rho = 0$. It is assumed here that $\phi \in [0, 2\pi)$, and the deficit angle is partly determined by the parameter B. With this range of ϕ, the deficit angle on the axis of symmetry is given in general by the expression

$$\delta = \pi(\mu + \lambda)\Big|_{\rho=0},$$

and hence the condition for a regular section of the axis is that

$$\mu + \lambda \to 0 \qquad \text{as} \qquad \rho \to 0 \tag{15.7}$$

(see Bičák and Schmidt, 1989b). In fact, it is possible to choose the constants m_i, α_i and B such that this regularity condition is satisfied on appropriate sections of the axis.

The cases that are of particular interest are now described.

15.2.1 Four particles and a regular axis

To avoid conical singularities everywhere, it is possible to make the choices

$$m_1 = \frac{(\alpha_2^2 - \alpha_1^2)^2}{4\alpha_1^3 \alpha_2^2}, \qquad m_2 = -\frac{(\alpha_2^2 - \alpha_1^2)^2}{4\alpha_1^2 \alpha_2^3}, \qquad B = 0, \tag{15.8}$$

satisfying (15.7). With this, the axis is regular everywhere except at the locations of the point particles. This situation is illustrated in Figure 15.2.

This case had previously been described (in the static region) by Bondi (1957a). For this choice of constants, the outer particles have positive mass, and the inner particles negative mass. It is then only the interactions between the particles in each pair that causes them both to accelerate towards infinity.

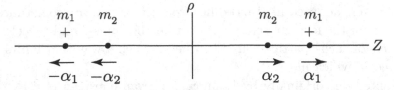

Fig. 15.2 Two pairs of Curzon–Chazy particles of positive and negative mass uniformly accelerate under the action of their mutual gravitational fields.

15.2.2 Four particles with semi-infinite strings

The conditions for the above case can be relaxed by taking

$$m_1 = \frac{(\alpha_2^2 - \alpha_1^2)^2}{4\alpha_1^3 \alpha_2^2}, \qquad m_2 \text{ arbitrary}, \qquad B = 0. \tag{15.9}$$

In this case, in view of (15.7), the axis is only regular between the particles of each pair and between the two pairs of particles. However, there are conical singularities on the parts of the axis that connect the outer particles to infinity. These may be considered to represent two semi-infinite strings. This situation is illustrated in Figure 15.3.

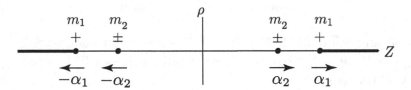

Fig. 15.3 Two Curzon–Chazy particles accelerate in opposite spatial directions under the actions of semi-infinite strings. Two other particles uniformly accelerate under the action of the resulting gravitational field.

In this case, the outer particles must have positive mass and are "pulled" towards infinity by the tension in the string. The inner particles have arbitrary mass which may be positive or negative.

15.2.3 Two particles and semi-infinite strings

Of particular interest is the special case in which $m_2 = 0$. For this choice, the inner particles disappear, and the two remaining particles are pulled towards infinity by semi-infinite strings. The solution representing this situation has been described by Bičák, Hoenselaers and Schmidt (1983a), and is illustrated in Figure 15.4.

In this particular subcase, however, the restriction that the outer particles

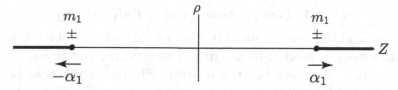

Fig. 15.4 Two Curzon–Chazy particles of positive or negative mass accelerate in opposite spatial directions under the actions of semi-infinite strings.

must have positive mass no longer occurs. It is only required that

$$m_1 \text{ arbitrary}, \qquad m_2 = 0, \qquad B = 0, \tag{15.10}$$

so that the remaining particles may also have negative mass. In this case the conical singularity corresponds to a strut rather than a string. (It may be noticed that this subcase can also be obtained in the limit $\alpha_2 \to \alpha_1$, in which case $m_1 \to 0$ and the masses at the end of the conical singularity are replaced by the original inner masses m_2 which may be arbitrary.)

15.2.4 Four particles and a finite string

A further case can be obtained by imposing the conditions that the axis is regular only between the particles of each pair and between each outer particle and spatial infinity. In this case there remains a conical singularity with a deficit angle between the two inner particles. These two particles must have negative mass. However, the outer particles may have arbitrary (positive or negative) mass. The parameters are given by

$$m_1 \text{ arbitrary}, \qquad m_2 = -\frac{(\alpha_2^2 - \alpha_1^2)^2}{4\alpha_1^2 \alpha_2^3}, \qquad B = -4m_1\alpha_1 - 4m_2\alpha_2. \tag{15.11}$$

This situation is illustrated in Figure 15.5.

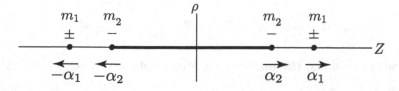

Fig. 15.5 Two pairs of Curzon–Chazy particles uniformly accelerate in opposite spatial directions. In this case, the inner particles are of negative mass and accelerate under the action of the tension in a string which exists between them.

15.2.5 Two particles and a finite string

A special case of this arises if $m_1 = 0$. This would then describe two particles of negative mass connected by a string. However, the restriction on m_2 no longer occurs and it may be chosen arbitrarily (as in the case described in Subsection 15.2.3). This has been described by Bičák, Hoenselaers and Schmidt (1983a). Alternatively, the same situation may be obtained by taking the limit as $\alpha_2 \to \alpha_1$ in (15.11). This subcase is thus given by

$$m_1 \text{ arbitrary}, \qquad m_2 = 0, \qquad B = -4m_1\alpha_1. \qquad (15.12)$$

This is illustrated in Figure 15.6.

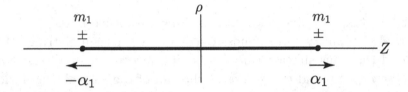

Fig. 15.6 Two Curzon–Chazy particles of positive or negative mass accelerate in opposite spatial directions under the actions of a strut or string which exists between them.

Notice that for the special cases described here and in Subsection 15.2.3, which contain only two accelerated particles, the metric functions (15.4) can be written in a simple form (see Bičák, 1990)

$$\mu = \mu_0 = -\frac{2m}{\alpha R} + 4m\alpha + B,$$

$$\lambda = \lambda_0 = -\frac{m^2}{\alpha^2 R^4}\rho^2(Z^2 - T^2) + \frac{2m\alpha}{R}(\rho^2 + Z^2 - T^2) + B, \qquad (15.13)$$

where $B = 0$ when the conical singularities connect the particles to infinity, as described in Subsection 15.2.3, or $B = -4m\alpha$ when the axis between the particles is singular as described here, and

$$R = \frac{1}{2}\sqrt{\left(\rho^2 + Z^2 - T^2 - \frac{1}{\alpha^2}\right)^2 + \frac{4}{\alpha^2}\rho^2}. \qquad (15.14)$$

(In these expressions, the unnecessary subscripts have been omitted.)

15.2.6 Four particles and two finite strings

A further case has been described in detail by Bičák, Hoenselaers and Schmidt (1983b) for which the axis is regular everywhere except between the two pairs of particles. The parameters here are given by

$$m_1 \text{ arbitrary}, \qquad m_2 = -\frac{\alpha_1}{\alpha_2}m_1, \qquad B = 0. \qquad (15.15)$$

In this case the masses of the two particles in each pair must be of opposite sign. If the outer particles have positive mass, the conical singularity has an excess angle corresponding to a strut. Contrarily, if the outer particles have negative mass the conical singularity corresponds to a string. This is illustrated in Figure 15.7.

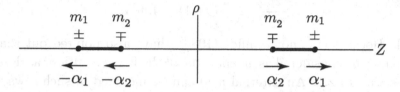

Fig. 15.7 Two pairs of Curzon–Chazy particles, whose masses are of opposite signs, accelerate in opposite spatial directions. A string exists between the particles in each pair.

15.3 Bičák–Hoenselaers–Schmidt solutions

Two particular limits of the above solutions will now be described, together with a generalised version of the second.

15.3.1 Two particles in an external field

A particularly interesting limit of the case described in Subsection 15.2.2 occurs when $\alpha_1 \to 0$. In this case, the outer particles m_1, and the strings attached to them, are scaled out to infinity. This situation has been described by Bičák, Hoenselaers and Schmidt (1983a) and is illustrated in Figure 15.8.

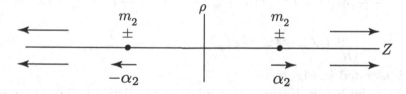

Fig. 15.8 Two Curzon–Chazy particles accelerate in opposite spatial directions under the action of a gravitational field whose sources are at infinity.

The significance of this solution is that the remaining particles move freely and are not connected to conical singularities. However, they move under the action of an exterior gravitational field for which the source is at infinity. Consequently, the space-time cannot be asymptotically flat.

It may be noted that, although $m_1 \sim \alpha_1^{-3}$ in this limit, $R_1 \sim \alpha_1^{-2}$ and μ is bounded. The resulting solution is given by

$$\mu = -\frac{2m_2}{\alpha_2 R_2} + 4m_2\alpha_2 + \alpha_2^2 \left(\rho^2 - Z^2 + T^2\right),$$

$$\lambda = -\left(\frac{m_2^2}{\alpha_2^2 R_2^4} + \alpha_2^4\right)\rho^2(Z^2 - T^2) + \alpha_2^2(\rho^2 + Z^2 - T^2) \quad (15.16)$$

$$+\frac{2m_2}{\alpha_2 R_2}(2\alpha_2^2\rho^2 + 1) - 4m_2\alpha_2 .$$

Bičák, Hoenselaers and Schmidt (1983a) have also pointed out that this solution may be considered as an analogue of the Ernst solution, as described in Subsection 14.2.1. An external field can be included in such a way as to remove the conical singularities either in the C-metric or here in the Bonnor–Swaminarayan solutions.

15.3.2 Monopole-dipole particles

For the case described in Subsection 15.2.6, which contains two pairs of particles with a conical singularity between those of each pair, Bičák, Hoenselaers and Schmidt (1983b) have also considered the limit $\alpha_2 \to \alpha_1$, in which the two particles in each pair coalesce. In this limit, it is convenient to introduce the parameter $M_{01} = 2(\frac{m_1}{\alpha_1} + \frac{m_2}{\alpha_2})$ which, with the constraint (15.15), becomes $M_{01} = \frac{2m_1}{\alpha_1\alpha_2^2}(\alpha_2^2 - \alpha_1^2)$. In this limit, Minkowski space-time is obtained unless the parameter m_1 is rescaled in such a way that M_{01} is kept constant.[2] This particular solution is then given by

$$\mu = -\frac{M_{01}}{R} + \frac{M_{01}}{4\alpha^2 R^3}\left(\rho^2 - Z^2 + T^2 + \frac{1}{\alpha^2}\right),$$

$$\lambda = -\frac{M_{01}^2}{64R^8}\rho^2(Z^2 - T^2)\left[\left((\rho^2 + Z^2 - T^2)^2 - \frac{1}{\alpha^4}\right)^2 - \frac{2}{\alpha^4}\rho^2(Z^2 - T^2)\right]$$

$$-\frac{M_{01}}{4R^3}\left(\rho^2 + Z^2 - T^2\right)\left(\rho^2 - Z^2 + T^2 + \frac{1}{\alpha^2}\right), \quad (15.17)$$

and is illustrated in Figure 15.9.

As shown by Bičák, Hoenselaers and Schmidt (1983b), the source of this space-time corresponds to a combination of monopole and dipole terms (explicit expressions are given in the following subsection), so that the resulting

---

[2] It may be noticed that an analogous limit of the case described in Subsection 15.2.1 does not exist as there is no freedom to rescale the mass m_1 (see (15.8)) and Minkowski space is obtained as the two particles coalesce.

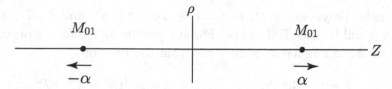

Fig. 15.9 Two particles with monopole and dipole structures accelerate in opposite spatial directions under the action of their own dipole fields.

particle has been referred to as a Curzon–Chazy (01)-pole particle. (For this reason the parameter is denoted by M_{01}.) The complete space-time contains two causally separated accelerating particles of this type. This situation looks superficially similar to that described in Subsection 15.3.1. Interestingly, in this case there is no external field and the particles are accelerated by their internal dipole structure.

15.3.3 Multipole particles

Bičák, Hoenselaers and Schmidt (1983b) have also found other boost-rotationally symmetric space-times that generalise the Bonnor–Swaminarayan solution (15.1)–(15.4) and represent the fields of two accelerating particles with *arbitrary multipole structure*. These are generally attached to conical singularities, as in the cases described in Subsections 15.2.3 and 15.2.5, so that the particles accelerate away from each other in opposite directions and are causally separated. Moreover, for a special choice of parameters, the space-times may be both free of conical singularities, as in the cases described in Subsections 15.3.2 and 15.3.1, and without an external field.

This class of solutions was initially presented using prolate spheroidal coordinates and the Weyl form of the metric in the static regions. In terms of the metric (15.1), the explicit solution for such a general class of multipole solutions can be expressed in the form

$$\mu = 2 \sum_{n=0}^{\infty} M_n \frac{P_n}{(x-y)^{n+1}} + C,$$

$$\lambda = -2 \sum_{k,l=0}^{\infty} M_k M_l \frac{(k+1)(l+1)}{(k+l+2)} \frac{(P_k P_l - P_{k+1} P_{l+1})}{(x-y)^{k+l+2}} \qquad (15.18)$$

$$- \left(\frac{x+y}{x-y} \right) \sum_{n=0}^{\infty} \frac{M_n}{2^n} \sum_{l=0}^{n} \left(\frac{2}{x-y} \right)^l P_l + D,$$

where the constants M_n represent the multipole moments, the argument of the Legendre polynomials P_n is $z = (1 - xy)/(x - y)$, and C, D are constants that will be specified below. For the prolate spheroidal coordinates x and y, and for z, one has to substitute from the relations

$$x - y = 4\alpha^2 R, \qquad x + y = 2\alpha^2(\rho^2 + Z^2 - T^2),$$
$$z = \frac{1}{2R}\left(\rho^2 - Z^2 + T^2 + \frac{1}{\alpha^2}\right), \tag{15.19}$$

where R is given by (15.14). These imply the useful identities

$$(x^2 - 1)(1 - y^2) = 16\alpha^4\rho^2(Z^2 - T^2), \qquad 1 + xy = 2\alpha^2(Z^2 - T^2 - \rho^2).$$

It may be noted that the expression for λ in (15.18) is simpler than that presented by Bičák, Hoenselaers and Schmidt (1983b). The derivation of the above expression, and the identity concerning Legendre polynomials that gives rise to it, have been given by Podolský and Griffiths (2001a). In addition, using the scaling property of the Weyl metric, the acceleration parameter α has been included here explicitly.

It now remains to set the values of the constants C and D. To ensure that the space-time is regular on the acceleration horizon where $Z^2 - T^2 = 0$, it is necessary that $\mu = \lambda$ at $\rho = 0 = Z^2 - T^2$ (see Bičák and Schmidt, 1989b). This yields explicitly that $D = C + \sum_{n=0}^{\infty} 2^{-n}M_n$. Moreover, the metric is regular on the axis provided $\mu + \lambda = 0$ at $\rho = 0$, see (15.7).

The regularity of the axis *between* the two particles (analogous to the case described in Subsection 15.2.3, see Figure 15.4) requires that

$$C = -\sum_{n=0}^{\infty} \frac{M_n}{2^n}, \qquad D = 0. \tag{15.20}$$

In this case there is generally a string connecting the particles to infinity. The alternative situation (analogous to the case described in Subsection 15.2.5, see Figure 15.6) in which there is a string between the particles, and the axis is regular *outside*, is given by

$$C = 0, \qquad D = \sum_{n=0}^{\infty} \frac{M_n}{2^n}. \tag{15.21}$$

Interestingly, if the combination of multipole moments satisfies the condition $\sum_{n=0}^{\infty} 2^{-n}M_n = 0$, so that $C = 0 = D$, then the axis is *regular everywhere*, except at the two particles. (This was pointed out by Bičák, Hoenselaers and Schmidt, 1983b.)

For the *monopole* case in which only the coefficient M_0 is non-zero, the previous formula (15.13) is recovered for accelerating Curzon–Chazy particles as described in Subsections 15.2.3 and 15.2.5, with the identification $M_0 = -4m\alpha$.

It is also straightforward to write an explicit solution representing accelerated *dipole* particles with the single moment M_1:

$$\mu_1 = \frac{M_1}{16\alpha^4 R^3} \left(\rho^2 - Z^2 + T^2 + \frac{1}{\alpha^2} \right) + C,$$

$$\lambda_1 = \frac{M_1^2}{512\,\alpha^8 R^4} \left(9z^4 - 10z^2 + 1 \right) \qquad (15.22)$$

$$- \frac{M_1}{4R} \left[1 + \frac{1}{4\alpha^2 R^2} \left(\rho^2 - Z^2 + T^2 + \frac{1}{\alpha^2} \right) \right] \left(\rho^2 + Z^2 - T^2 \right) + D.$$

With these two observations, it may also be seen that the case given by (15.17) is just a special case of the general class of solutions (15.18). It is a *combination* of both monopole and dipole terms with the identification $2M_0 = -M_1 = -4\alpha^2 M_{01}$, so that the constraint $M_0 + M_1/2 = 0$, which guarantees the regularity of the axis, is automatically satisfied. This is fully consistent with the interpretation of the solution (15.17) described by Bičák, Hoenselaers and Schmidt (1983b). Indeed, the metric functions μ and λ given by (15.17) can be written as $\mu = \mu_0 + \mu_1$ and $\lambda = \lambda_0 + \lambda_1 + \lambda_{01}$, where μ_0 and λ_0 are given by (15.13), μ_1 and λ_1 are given by (15.22), and $\lambda_{01} = M_{01}^2 z(1 - z^2)/4\alpha^2 R^3$.

16

Plebański–Demiański solutions

A number of the previous chapters have described important black hole space-times that are of algebraic type D – specifically Chapters 8, 9, 11, 12 and 14. In fact these are all members of a larger family of solutions that can be expressed in a common form. The purpose of this chapter is to present this wider class, particularly showing the relation between these and related space-times, and to indicate their further generalisations.

A general family of type D space-times with an aligned non-null electromagnetic field and a possibly non-zero cosmological constant can be represented by a metric that was given originally by Debever (1971), and in a more convenient form by Plebański and Demiański (1976). These solutions are characterised by two related quartic functions, each of a single coordinate, whose coefficients are determined by seven arbitrary parameters which include Λ and both electric and magnetic charges. Together with cases that can be derived from it by certain transformations and limiting procedures, this gives the complete family of such solutions.

Non-accelerating solutions of this class were obtained by Carter (1968b). For the vacuum case with no cosmological constant, they include all the particular solutions identified by Kinnersley (1969a). Metrics with an expanding repeated principal null congruence were analysed further by Debever (1969) and Weir and Kerr (1977), where the relations between the different forms of the line element can be deduced. They have also been studied by Debever and Kamran (1980) and Ishikawa and Miyashita (1982). The most general metric form which covers both expanding and non-expanding cases was given by Debever, Kamran and McLenaghan (1984) and García D. (1984).

Unfortunately, the familiar forms of some well-known type D space-times are not included explicitly in the original Plebański–Demiański line element. They can only be obtained from it by using certain degenerate transformations. Moreover, the parameters that were introduced in the original papers are not the most useful ones for their interpretation. In a recent review (Griffiths and Podolský, 2006b), an alternative form of the Plebański–Demiański metric was presented in which the parameters are given a more direct physical meaning and from which the most important special cases can be obtained by explicit reduction. The present chapter is based on that review.

The physical interpretations of many particular subcases of these spacetimes have already been described in previous chapters, while others have no known significance. The emphasis here is on those members of this family of solutions which possess natural foliations by spacelike 2-surfaces of positive curvature. These are the most important physically as they relate to the fields around generalised forms of black holes. The family of solutions whose repeated principal null directions are non-expanding is also briefly described.

16.1 The Plebański–Demiański metric

We consider here the general family of type D solutions of the Einstein–Maxwell equations including a generally non-zero cosmological constant. These may be vacuum or include a non-null electromagnetic field such that the two repeated principal null directions of the Weyl tensor are aligned with the two principal null directions of the electromagnetic field.[1]

This family of solutions was first obtained by Debever (1971), but is best known in the metric form given by Plebański and Demiański (1976) (see §21.1.2 of Stephani *et al.*, 2003), namely

$$
ds^2 = \frac{1}{(1 - \hat{p}\hat{r})^2} \left[-\frac{\widehat{Q}(d\hat{\tau} - \hat{p}^2 d\hat{\sigma})^2}{\hat{r}^2 + \hat{p}^2} + \frac{\widehat{P}(d\hat{\tau} + \hat{r}^2 d\hat{\sigma})^2}{\hat{r}^2 + \hat{p}^2} \right.
$$
$$
\left. + \frac{\hat{r}^2 + \hat{p}^2}{\widehat{P}} d\hat{p}^2 + \frac{\hat{r}^2 + \hat{p}^2}{\widehat{Q}} d\hat{r}^2 \right]. \tag{16.1}
$$

This contains two generally quartic functions

$$
\widehat{P}(\hat{p}) = \hat{k} + 2\hat{n}\hat{p} - \hat{\epsilon}\hat{p}^2 + 2\hat{m}\hat{p}^3 - (\hat{k} + \hat{e}^2 + \hat{g}^2 + \Lambda/3)\hat{p}^4 ,
$$
$$
\widehat{Q}(\hat{r}) = (\hat{k} + \hat{e}^2 + \hat{g}^2) - 2\hat{m}\hat{r} + \hat{\epsilon}\hat{r}^2 - 2\hat{n}\hat{r}^3 - (\hat{k} + \Lambda/3)\hat{r}^4 , \tag{16.2}
$$

in which $\hat{m}$, $\hat{n}$, $\hat{e}$, $\hat{g}$, $\hat{\epsilon}$, $\hat{k}$ and Λ are arbitrary real parameters. It is often assumed that $\hat{m}$ and $\hat{n}$ are the mass and NUT parameters, respectively, but

---

[1] Other non-null type D electrovacuum solutions exist in which only one of the principal null congruences of the electromagnetic field is aligned with a repeated principal null congruence of the Weyl tensor (see for example Kowalczyński and Plebański, 1977).

this is not generally the case. The parameters $\hat{e}$ and $\hat{g}$ represent electric and magnetic charges.[2]

As is required for type D space-times of this type, the general family of solutions represented by (16.1) admits (at least) two commuting Killing vectors $\partial_{\hat{\sigma}}$ and $\partial_{\hat{\tau}}$, whose group orbits are spacelike in regions with $\widehat{Q} > 0$ and timelike when $\widehat{Q} < 0$. Type D solutions also exist in which the group orbits are null, but these are not considered here.[3]

For purposes of interpreting the Plebański–Demiański metric (16.1), it is convenient (see Griffiths and Podolský, 2005) to introduce the rescaling

$$\hat{p} = \sqrt{\alpha\omega}\, p, \qquad \hat{r} = \sqrt{\frac{\alpha}{\omega}}\, r, \qquad \hat{\sigma} = \sqrt{\frac{\omega}{\alpha^3}}\, \sigma, \qquad \hat{\tau} = \sqrt{\frac{\omega}{\alpha}}\, \tau, \qquad (16.3)$$

with the relabelling of parameters

$$\hat{m} + i\hat{n} = \left(\frac{\alpha}{\omega}\right)^{3/2}(m + in), \qquad \hat{e} + i\hat{g} = \frac{\alpha}{\omega}(e + ig), \qquad \hat{\epsilon} = \frac{\alpha}{\omega}\epsilon, \qquad \hat{k} = \alpha^2 k. \qquad (16.4)$$

This introduces two additional parameters α and ω. With these changes, the metric becomes

$$ds^2 = \frac{1}{(1 - \alpha p r)^2}\left[-\frac{\mathcal{Q}}{r^2 + \omega^2 p^2}(\mathrm{d}\tau - \omega p^2 \mathrm{d}\sigma)^2 + \frac{r^2 + \omega^2 p^2}{\mathcal{Q}}\,\mathrm{d}r^2\right.$$

$$\left. + \frac{\mathcal{P}}{r^2 + \omega^2 p^2}(\omega \mathrm{d}\tau + r^2 \mathrm{d}\sigma)^2 + \frac{r^2 + \omega^2 p^2}{\mathcal{P}}\,\mathrm{d}p^2\right], \quad (16.5)$$

where

$$\mathcal{P} = \mathcal{P}(p) = k + 2\omega^{-1}np - \epsilon p^2 + 2\alpha m p^3 - [\alpha^2(\omega^2 k + e^2 + g^2) + \omega^2 \Lambda/3]p^4$$

$$\mathcal{Q} = \mathcal{Q}(r) = (\omega^2 k + e^2 + g^2) - 2mr + \epsilon r^2 - 2\alpha\omega^{-1}nr^3 - (\alpha^2 k + \Lambda/3)r^4, \qquad (16.6)$$

and m, n, e, g, Λ, ϵ, k, α and ω are arbitrary real parameters of which two can be chosen for convenience. It should be emphasised that, apart from Λ,

---

[2] The parameter γ that is used by Plebański and Demiański (1976) and Stephani *et al.* (2003) is obtained by putting $\hat{k} = \gamma - \hat{g}^2 - \Lambda/6$. However, as shown by Podolský and Griffiths (2001c), it is more convenient for physical interpretation to include the cosmological constant in the form given in (16.2).

[3] Such metrics can be obtained by using a degenerate coordinate transformation. This is described in detail by García Díaz and Plebański (1982) (see also page 322 of Stephani *et al.*, 2003) who confirm that a particular case of Leroy (1978) is included in this general family. All such type D space-times with null group orbits have been given by García Díaz and Salazar (1983). A generalised form of the metric (16.1) which includes all known cases of this type has been given by Debever, Kamran and McLenaghan (1983, 1984), although this form is not well suited for the interpretation of the solutions. A form of the metric which covers the cases of null and non-null orbits simultaneously was used by García D. (1984) (see also García D. and Macias, 1998).

e and g, the parameters included in this metric do not necessarily have their traditional physical interpretation. They only acquire their usual specific well-identified meanings in certain special sub-cases.

Adopting the null tetrad

$$k = \frac{1 - \alpha pr}{\sqrt{2(r^2 + \omega^2 p^2)}} \left[\frac{1}{\sqrt{Q}} \left(r^2 \partial_\tau - \omega \partial_\sigma \right) - \sqrt{Q}\, \partial_r \right],$$

$$l = \frac{1 - \alpha pr}{\sqrt{2(r^2 + \omega^2 p^2)}} \left[\frac{1}{\sqrt{Q}} \left(r^2 \partial_\tau - \omega \partial_\sigma \right) + \sqrt{Q}\, \partial_r \right], \qquad (16.7)$$

$$m = \frac{1 - \alpha pr}{\sqrt{2(r^2 + \omega^2 p^2)}} \left[-\frac{1}{\sqrt{P}} \left(\omega p^2 \partial_\tau + \partial_\sigma \right) + \mathrm{i}\,\sqrt{P}\, \partial_p \right],$$

the spin coefficients (2.15) are given by

$$\kappa = \sigma = \lambda = \nu = 0, \qquad \rho = \mu = \sqrt{\frac{Q}{2(r^2 + \omega^2 p^2)}}\, \frac{(1 + \mathrm{i}\,\alpha \omega p^2)}{(r + \mathrm{i}\,\omega p)}, \qquad (16.8)$$

with $\tau = \pi$, $\epsilon = \gamma$ and $\alpha = \beta$ non-zero. These indicate that the congruences tangent to k and l are both geodesic and shear-free but have *non-zero expansion*. It can also be seen that the *twist* of both congruences is proportional to ω. In some particular cases ω is directly related to both the angular velocity of sources and the effects of the NUT parameter, as will be shown below.

Using the tetrad (16.7), the only non-zero Weyl tensor component is

$$\Psi_2 = -(m + \mathrm{i}\,n) \left(\frac{1 - \alpha pr}{r + \mathrm{i}\,\omega p} \right)^3 + (e^2 + g^2) \left(\frac{1 - \alpha pr}{r + \mathrm{i}\,\omega p} \right)^3 \frac{1 + \alpha pr}{r - \mathrm{i}\,\omega p}. \qquad (16.9)$$

This confirms that these space-times are of algebraic type D, and that the above tetrad vectors k and l are aligned with the repeated principal null directions of the Weyl tensor. Apart from Λ, the only non-zero component of the Ricci tensor is

$$\Phi_{11} = \frac{1}{2} (e^2 + g^2) \frac{(1 - \alpha pr)^4}{(r^2 + \omega^2 p^2)^2}. \qquad (16.10)$$

Together, these indicate the presence of a curvature singularity at $p = 0 = r$ which, if contained within the space-time, may be considered as the source of the gravitational field. The line element (16.5) is flat if $m = 0 = n$, $e = 0 = g$ and $\Lambda = 0$, but the remaining parameters ϵ, k, α and ω may be non-zero in this flat limit.

Having introduced two useful continuous parameters α and ω, the rescaling (16.3) with (16.4) can be used to scale the parameters ϵ and k to any specific value (without changing their signs). For example, they could be set to the values $+1$, 0 or -1, but it will generally be more convenient to scale them to some other appropriate values. It is clear that e and g are the electric and magnetic charges of the sources and Λ is the cosmological constant. For certain choices of the other parameters, it is found that m is related to the mass of the source and n is related to the NUT parameter (although it should not be identified with it in general).

To retain a Lorentzian signature in (16.5), it is necessary that $\mathcal{P} > 0$. And, since $\mathcal{P}(p)$ is generally a quartic function, the coordinate p must be restricted to a particular range between appropriate roots, and this places a restriction on the possible signs of the parameters ϵ and k. In fact, various particular cases can be identified and classified according to the number and types of the roots of $\mathcal{P}$ and the range of p that is adopted (see e.g. Ishikawa and Miyashita, 1982). Points at which $\mathcal{P} = 0$ generally correspond to poles of the coordinates. By contrast, surfaces on which $\mathcal{Q} = 0$ are horizons through which coordinates can be extended. It is also significant that

$$\mathcal{Q}(r) = -\alpha^2 r^4 \mathcal{P}\left(\frac{1}{\alpha r}\right) - \frac{\Lambda}{3}\left(\frac{\omega^2}{\alpha^2} + r^4\right).$$

When $\Lambda = 0$, this relates the number and character of the roots of $\mathcal{P}$ and $\mathcal{Q}$. However, when $\Lambda \neq 0$, this correspondence is obscured.

16.2 A general metric for expanding solutions

Since the character of these solutions depends on the root structure of the quartic $\mathcal{P}$, it is most convenient to introduce a shift and rescaling in p that will place appropriate roots at specifically chosen locations. In fact, this procedure is essential to obtain the familiar metric for rotating (and also accelerating) black holes and for introducing the NUT parameter.

It is thus appropriate to introduce the coordinate transformation

$$p = \frac{l}{\omega} + \frac{a}{\omega}\tilde{p}, \qquad \tau = t - \frac{(l+a)^2}{a}\varphi, \qquad \sigma = -\frac{\omega}{a}\varphi, \qquad (16.11)$$

where a and l are new arbitrary parameters which, in some cases, will be shown below to have significant physical meanings. With this transformation, the metric (16.5), (16.6) becomes

$$ds^2 = \frac{1}{\Omega^2}\left\{ -\frac{\mathcal{Q}}{\varrho^2}\left[dt - \left(a(1-\tilde{p}^2) + 2l(1-\tilde{p})\right)d\varphi\right]^2 + \frac{\varrho^2}{\mathcal{Q}}\,dr^2 \right. \qquad (16.12)$$

$$\left. + \frac{\varrho^2}{\tilde{\mathcal{P}}}\,d\tilde{p}^2 + \frac{\tilde{\mathcal{P}}}{\varrho^2}\left[a\,dt - \left(r^2 + (l+a)^2\right)d\varphi\right]^2 \right\},$$

where

$$\Omega = 1 - \frac{\alpha}{\omega}(l + a\tilde{p})r,$$

$$\varrho^2 = r^2 + (l + a\tilde{p})^2,$$

$$\widetilde{\mathcal{P}} = a_0 + a_1\tilde{p} + a_2\tilde{p}^2 + a_3\tilde{p}^3 + a_4\tilde{p}^4,$$

$$\mathcal{Q} = (\omega^2 k + e^2 + g^2) - 2mr + \epsilon r^2 - 2\alpha\frac{n}{\omega}r^3 - \left(\alpha^2 k + \frac{\Lambda}{3}\right)r^4,$$

in which

$$a_0 = \frac{1}{a^2}\left(\omega^2 k + 2nl - \epsilon l^2 + 2\alpha\frac{l^3}{\omega}m - \left[\frac{\alpha^2}{\omega^2}(\omega^2 k + e^2 + g^2) + \frac{\Lambda}{3}\right]l^4\right),$$

$$a_1 = \frac{2}{a}\left(n - \epsilon l + 3\alpha\frac{l^2}{\omega}m - 2\left[\frac{\alpha^2}{\omega^2}(\omega^2 k + e^2 + g^2) + \frac{\Lambda}{3}\right]l^3\right),$$

$$a_2 = -\epsilon + 6\alpha\frac{l}{\omega}m - 6\left[\frac{\alpha^2}{\omega^2}(\omega^2 k + e^2 + g^2) + \frac{\Lambda}{3}\right]l^2, \qquad (16.13)$$

$$a_3 = 2\alpha\frac{a}{\omega}m - 4\left[\frac{\alpha^2}{\omega^2}(\omega^2 k + e^2 + g^2) + \frac{\Lambda}{3}\right]al,$$

$$a_4 = -\left[\frac{\alpha^2}{\omega^2}(\omega^2 k + e^2 + g^2) + \frac{\Lambda}{3}\right]a^2.$$

When $\widetilde{\mathcal{P}}$ has no roots, this function can only be positive and $\tilde{p} \in (-\infty, \infty)$. Solutions with this property are generalisations of the *AII*-metric described in Section 9.1.1 which, apart from an interpretation as the field of a tachyon, is of little physical significance.

For the cases in which $\widetilde{\mathcal{P}}$ has at least one root, without loss of generality, the parameters can be chosen so that such a root occurs at $\tilde{p} = 1$. This corresponds to a coordinate pole on an axis, and it is then appropriate to take φ as a periodic coordinate.

When another distinct root of $\widetilde{\mathcal{P}}$ exists, it is always possible to use another freedom to set the second root at $\tilde{p} = -1$. The metric component $a(1 - \tilde{p}^2)$ in (16.12) is then regular at this second pole while the component $2l(1 - \tilde{p})$ is not. Thus, the complete metric is regular at $\tilde{p} = 1$, but a singularity of some kind occurs at $\tilde{p} = -1$. (In fact, unless $l = 0$, the region near $\tilde{p} = -1$ contains closed timelike curves.) With this choice, and for the case of positive curvature 2-spaces, a corresponds to a Kerr-like rotation parameter for which the corresponding metric components are regular on the entire axis, while l corresponds to a NUT-like parameter for which the corresponding components are only regular on the half-axis $\tilde{p} = 1$.

With these choices, $\widetilde{\mathcal{P}}$ takes the form

$$\widetilde{\mathcal{P}} = (1 - \tilde{p}^2)(a_0 - a_3\tilde{p} - a_4\tilde{p}^2),$$

and the above coefficients must satisfy the constraints

$$a_1 + a_3 = 0, \qquad a_0 + a_2 + a_4 = 0. \tag{16.14}$$

Substituting from (16.13), these provide two equations, which are linear in ϵ and n, and can be solved to express these parameters in terms of those remaining as

$$\epsilon = \frac{\omega^2 k}{a^2 - l^2} + 4\alpha\frac{l}{\omega}\,m - (a^2 + 3l^2)\left[\frac{\alpha^2}{\omega^2}(\omega^2 k + e^2 + g^2) + \frac{\Lambda}{3}\right], \tag{16.15}$$

$$n = \frac{\omega^2 k\,l}{a^2 - l^2} - \alpha\frac{(a^2 - l^2)}{\omega}\,m + (a^2 - l^2)l\left[\frac{\alpha^2}{\omega^2}(\omega^2 k + e^2 + g^2) + \frac{\Lambda}{3}\right]. \tag{16.16}$$

Putting these expressions into that for a_0 implies that

$$\left(\frac{\omega^2}{a^2 - l^2} + 3\alpha^2 l^2\right)k = a_0 + 2\alpha\frac{l}{\omega}\,m - 3\alpha^2\frac{l^2}{\omega^2}(e^2 + g^2) - l^2\Lambda. \tag{16.17}$$

The character of the solution then partly depends on whether a_0 is positive, negative or zero. If it is non-zero, the scaling freedom can then be used to set $a_0 = \pm 1$, and (16.17) then effectively defines the parameter k.

According to the above construction, these solutions are generally considered to be determined by seven essential parameters m, e, g, a, l, α and Λ. The three parameters ϵ, n and k are given by the expressions (16.15), (16.16) and (16.17). One remaining (scaling) freedom of the type (16.3) is therefore still available which may be used to set ω to any convenient value (assuming a and l do not both vanish).

Of course, the complete family of Plebański–Demiański solutions includes those in which $\widetilde{\mathcal{P}}$ has up to four real roots. And within each of these cases, different ranges of p exist for which $\widetilde{\mathcal{P}} > 0$. For the vacuum case with $\Lambda = 0$, these possibilities lead to the classification given by Kinnersley (1969a). In particular, the expanding but non-twisting case ($\omega = 0$) is identified with the type D family of Robinson–Trautman space-times. For the general expanding case, the most physically significant solutions of this class are those for which $\widetilde{\mathcal{P}}$ has at least two roots between which it is positive. Such solutions contain spacelike 2-surfaces with positive curvature and usually include black hole-like structures. It is just this subfamily of solutions that will now be described in detail.[4]

---

[4] Space-times of this family in which $\widetilde{\mathcal{P}}$ has different root structures may also be significant. For

16.3 Generalised black holes

This section concentrates on the physically most relevant case of the line element (16.12) for which $\widetilde{\mathcal{P}}$ has at least two distinct roots and $a_0 > 0$, so that it is possible to set $a_0 = 1$.

For such a case, $\tilde{p}$ is taken to cover the range between the roots $\tilde{p} = \pm 1$, and it is natural to put $\tilde{p} = \cos\theta$, where $\theta \in [0, \pi]$. In this case, the metric (16.12) becomes

$$\mathrm{d}s^2 = \frac{1}{\Omega^2}\left\{ -\frac{\mathcal{Q}}{\varrho^2}\left[\mathrm{d}t - \left(a\sin^2\theta + 4l\sin^2\tfrac{1}{2}\theta\right)\mathrm{d}\varphi\right]^2 + \frac{\varrho^2}{\mathcal{Q}}\,\mathrm{d}r^2 \right. \tag{16.18}$$

$$\left. +\frac{\varrho^2}{P}\,\mathrm{d}\theta^2 + \frac{P}{\varrho^2}\sin^2\theta\left[a\mathrm{d}t - \left(r^2 + (a+l)^2\right)\mathrm{d}\varphi\right]^2 \right\},$$

where (having put $\widetilde{\mathcal{P}} = \sin^2\theta\, P$)

$$
\begin{aligned}
&\Omega = 1 - \frac{\alpha}{\omega}(l + a\cos\theta)\,r, \\
&\varrho^2 = r^2 + (l + a\cos\theta)^2, \\
&P = 1 - a_3\cos\theta - a_4\cos^2\theta, \\
&\mathcal{Q} = (\omega^2 k + e^2 + g^2) - 2mr + \epsilon r^2 - 2\alpha\frac{n}{\omega}r^3 - \left(\alpha^2 k + \frac{\Lambda}{3}\right)r^4,
\end{aligned}
\tag{16.19}
$$

and

$$
\begin{aligned}
a_3 &= 2\alpha\frac{a}{\omega}m - 4\alpha^2\frac{al}{\omega^2}(\omega^2 k + e^2 + g^2) - 4\frac{\Lambda}{3}al, \\
a_4 &= -\alpha^2\frac{a^2}{\omega^2}(\omega^2 k + e^2 + g^2) - \frac{\Lambda}{3}a^2,
\end{aligned}
\tag{16.20}
$$

with ϵ, n and k given by (16.15), (16.16) and (16.17) with $a_0 = 1$. It is also assumed that $|a_3|$ and $|a_4|$ are sufficiently small that P has no roots with $\theta \in [0, \pi]$.

This solution contains eight arbitrary constants: the mass parameter m of the source, its electric and magnetic charges e and g, its Kerr-like rotation parameter a, a NUT parameter l, its acceleration α, the cosmological constant Λ. In addition, there is the parameter ω that can be set to any convenient value if a or l are not both zero; otherwise $\omega \equiv 0$.

These space-times thus represent accelerating and rotating charged black holes with a generally non-zero NUT parameter. An arbitrary cosmological constant is also included, so that the "background" is either Minkowski, de Sitter or anti-de Sitter space-time. For the vacuum case in which e, g

example, some C-metrics with $\Lambda < 0$ have recently been interpreted in terms of black funnels and droplets: see Hubeny, Marolf and Rangamani (2010) and Caldarelli *et al.* (2011).

and Λ vanish, the general structure of this family of solutions is given in Figure 16.1. Most special cases have been described in previous chapters.

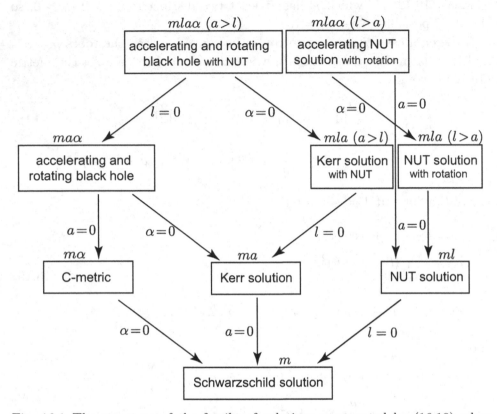

Fig. 16.1 The structure of the family of solutions represented by (16.18) when $\Lambda = 0$, $e = 0 = g$ and $m \neq 0$. This family has four parameters m, l, a and α. An accelerating Kerr solution with a small NUT parameter has been distinguished from an accelerating NUT solution with a small rotation, as their singularity structures differ significantly even though their metric forms are identical. For the same reason, the Kerr–NUT solution has been similarly divided. An accelerating NUT solution without rotation has not been identified. All the special cases indicated have obvious charged versions and versions with a non-zero cosmological constant.

The non-zero components of the curvature tensor are given by (16.9) and (16.10), in which ωp is replaced by $l + a \cos\theta$. It can be seen that the metric (16.18) has a *curvature singularity* when $\varrho^2 = 0$. If $|l| \leq |a|$, this occurs when $r = 0$ and $\cos\theta = -l/a$. On the other hand, if $|l| > |a|$, ϱ^2 cannot be zero and the metric is non-singular. These two cases have to be considered separately, as they clearly have very different global and singularity structures. This distinction is indicated in Figure 16.1 and will be clarified below.

When $a^2 \geq l^2$, the metric (16.18) has a ring-like curvature singularity at

$r = 0$, $\cos\theta = -l/a$. The space-time is asymptotically flat with conformal infinity $\mathcal{I}$ located at $\Omega = 0$, where $r = \omega/\alpha(l + a\cos\theta)$ for $\cos\theta > -l/a$. But note that these coordinates do not extend as far as conformal infinity for $\cos\theta < -l/a$.

In the alternative case in which $a^2 < l^2$, there is no curvature singularity, and the range of r can also take negative values as in the NUT solution.

Conical singularities generally occur on the axis of symmetry. However, by specifying the range of φ appropriately, the singularity on one half of the axis can be removed. For example, that on $\theta = \pi$ is removed by taking $\varphi \in [0, 2\pi C)$, where $C = (1 + a_3 - a_4)^{-1}$. In this case, the acceleration of the "source" is achieved by a "string" of deficit angle

$$\delta_0 = \frac{4\pi a_3}{1 + a_3 - a_4} \qquad (16.21)$$

connecting it to infinity. Alternatively, the singularity on $\theta = 0$ could be removed by setting $C = (1 - a_3 - a_4)^{-1}$, and the acceleration would then be achieved by a "strut" between the sources in which the excess angle is

$$-\delta_\pi = \frac{4\pi a_3}{1 - a_3 - a_4}. \qquad (16.22)$$

The expressions (16.21) or (16.22) are closely related to the tension or stress in the string or strut, respectively, and these should be equal for any given acceleration α, at least according to Newtonian theory. In fact, these deficit/excess angles are the same fractions of the range of the periodic coordinate in each case. This, presumably, corresponds to an equality of different types of forces which produce the same acceleration.

It should also be recalled that, when $l \neq 0$, the metric (16.18) has an additional singularity on $\theta = \pi$ which here corresponds to the "axis" between two causally separated "sources". However, this can be switched to the other half-axis by the transformation $t' = t - 4l\varphi$. It can thus be seen that the topological singularity on the axis which causes the acceleration, and the singularity on the axis associated with the NUT parameter and the existence of closed timelike curves (see e.g. Bonnor, 1969a), are mathematically independent. They may each be set on whatever parts of the axis may be considered to be most physically significant.

Intriguingly, there exists a possibility that the conical singularity may vanish on both halves of the axis simultaneously. This occurs when $a_3 = 0$: i.e. when

$$2\alpha^2 l(\omega^2 k + e^2 + g^2) - \tfrac{2}{3}\omega^2 l\Lambda = \alpha\omega m.$$

However, this only occurs when the NUT parameter l is non-zero and, in this case, one of the "half-axes" is singular in a different way and is surrounded by a region containing closed timelike curves. Thus, the presence of charges or a cosmological constant is not sufficient to cause the sources to accelerate uniformly: some string-like structure is required (at least within this family of solutions).

Some special cases will now be considered in more detail. Namely, the non-accelerating solutions and those with no NUT parameter. The general case with $\Lambda = 0$ has been described in detail elsewhere (Griffiths and Podolský, 2005). For the case with $a = 0$, α is a redundant parameter, indicating that there are no accelerating NUT solutions without rotation within this family of type D solutions.[5]

16.3.1 The Kerr–Newman–NUT–(anti-)de Sitter solution

When $\alpha = 0$, these solutions are contained within those found by Carter (1968b). In the above notation, (16.17) becomes $w^2 k = (1 - l^2\Lambda)(a^2 - l^2)$ and with this (16.15) and (16.16) imply that $\epsilon = 1 - (\frac{1}{3}a^2 + 2l^2)\Lambda$ and $n = l + \frac{1}{3}(a^2 - 4l^2)l\Lambda$, respectively. Setting $\varphi = \phi \in [0, 2\pi)$, the metric is then given by

$$ds^2 = -\frac{Q}{\varrho^2} \left[dt - \left(a \sin^2\theta + 4l \sin^2 \tfrac{1}{2}\theta \right) d\phi \right]^2 + \frac{\varrho^2}{Q} dr^2 \qquad (16.23)$$

$$+ \frac{\varrho^2}{P} d\theta^2 + \frac{P}{\varrho^2} \sin^2\theta \left[adt - \left(r^2 + (a+l)^2 \right) d\phi \right]^2,$$

with

$\varrho^2 = r^2 + (l + a\cos\theta)^2$,

$P = 1 + \frac{4}{3}\Lambda al \cos\theta + \frac{1}{3}\Lambda a^2 \cos^2\theta$,

$Q = (a^2 - l^2 + e^2 + g^2) - 2mr + r^2 - \Lambda \left[(a^2 - l^2)l^2 + (\frac{1}{3}a^2 + 2l^2)r^2 + \frac{1}{3}r^4 \right]$.

This is exactly the Kerr–Newman–NUT–(anti-)de Sitter solution in the form that is regular on the half-axis $\theta = 0$. It represents a non-accelerating black hole with mass m, electric and magnetic charges e and g, a rotation parameter a and a NUT parameter l, in a de Sitter or anti-de Sitter background. It reduces to the standard forms (11.10) and (12.19) when combinations of l, a or Λ are set to zero. The uncharged case was obtained by Frolov (1974b) using the Newman–Penrose formalism.

---

[5] Chng, Mann and Stelea (2006) have used a solution-generating technique to derive an apparently accelerating NUT solution. This includes the C-metric and the NUT metric as special cases, but is algebraically general.

For these solutions, it is important to distinguish between the two cases in which $|a|$ is greater or less than $|l|$. When $a^2 \geq l^2$, $k \geq 0$, the metric has a Kerr-like ring singularity at $r = 0$, $\cos\theta = -l/a$. This case represents a Kerr–Newman–(anti-)de Sitter solution (a charged black hole) with a small NUT parameter. Alternatively, when $a^2 < l^2$, $k < 0$, the metric is singularity-free, and the range of r includes negative values. This case is best described as a charged NUT–(anti-)de Sitter solution with a small Kerr-like rotation. Although these two cases have identical metric forms, their singularity and global structures differ substantially.

16.3.2 Accelerating Kerr–Newman–(anti-)de Sitter black holes

In the physically most significant case in which α is arbitrary but $l = 0$, (16.17) implies that $\omega^2 k = a^2$. It is then convenient to use the remaining scaling freedom to put $\omega = a$, and hence

$$\epsilon = 1 - \alpha^2(a^2 + e^2 + g^2) - \tfrac{1}{3}\Lambda a^2, \qquad k = 1, \qquad n = -\alpha a m.$$

For this case, it can be seen explicitly that the Plebański–Demiański parameter n is non-zero, while the NUT parameter l vanishes. The metric (16.18) now takes the form

$$ds^2 = \frac{1}{\Omega^2}\left\{ -\frac{Q}{\varrho^2}\left(\mathrm{d}t - a\sin^2\theta\,\mathrm{d}\varphi\right)^2 + \frac{\varrho^2}{Q}\,\mathrm{d}r^2 \right.$$
$$\left. + \frac{\varrho^2}{P}\,\mathrm{d}\theta^2 + \frac{P}{\varrho^2}\sin^2\theta\left[a\mathrm{d}t - (r^2 + a^2)\mathrm{d}\varphi\right]^2 \right\}, \tag{16.24}$$

where

$$\Omega = 1 - \alpha r\cos\theta,$$
$$\varrho^2 = r^2 + a^2\cos^2\theta,$$
$$P = 1 - 2\alpha m\cos\theta + \left[\alpha^2(a^2 + e^2 + g^2) + \tfrac{1}{3}\Lambda a^2\right]\cos^2\theta,$$
$$Q = \left((a^2 + e^2 + g^2) - 2mr + r^2\right)(1 - \alpha^2 r^2) - \tfrac{1}{3}\Lambda(a^2 + r^2)r^2.$$

Apart from Λ, the only non-zero components of the curvature tensor are

$$\Psi_2 = \left(-m(1 - \mathrm{i}\,\alpha a) + (e^2 + g^2)\frac{1 + \alpha r\cos\theta}{r - \mathrm{i}\,a\cos\theta} \right)\left(\frac{1 - \alpha r\cos\theta}{r + \mathrm{i}\,a\cos\theta} \right)^3,$$
$$\Phi_{11} = \frac{1}{2}(e^2 + g^2)\frac{(1 - \alpha r\cos\theta)^4}{(r^2 + a^2\cos^2\theta)^2}. \tag{16.25}$$

These indicate the presence of a Kerr-like ring singularity at $r = 0$, $\theta = \frac{\pi}{2}$. Further properties have been described in Podolský and Griffiths (2006).

This is the family of solutions whose special cases have been described in detail in Chapter 14. It represents an accelerating and rotating black hole without any NUT-like behaviour, in which the acceleration is characterised by α. It may be observed that

$$a_3 = 2\alpha m, \qquad a_4 = -\alpha^2(a^2 + e^2 + g^2) - \tfrac{1}{3}\Lambda a^2.$$

Hence the expressions (16.21) or (16.22) indicate the presence of a deficit or excess angle which represents a string or strut that causes the black holes to accelerate. It may also be seen that there are no closed timelike curves near the axis, confirming that this is the appropriate metric for describing a pair of accelerating and rotating black holes, as argued by Hong and Teo (2005) for the case when $\Lambda = 0$.

It should be emphasised that, when $e = g = 0$ and $\Lambda = 0$, the solution (16.24) is different from the one that is obtained by putting $n = 0$ in (16.5) and which is usually called the "spinning C-metric". This case, whose properties have been described by Farhoosh and Zimmerman (1980b,c), Bičák and Pravda (1999), Letelier and Oliveira (2001) and Pravda and Pravdová (2002), still retains NUT-like properties such as the existence of closed timelike curves near one half of the axis.

16.4 Non-expanding solutions

The above sections have only considered cases of the Plebański–Demiański family in which the repeated principal null congruences have non-zero expansion. The alternative situation will now be described in which these congruences are non-expanding. And, since solutions with a non-expanding shear-free null geodesic congruence must be twist-free, the twist must also vanish in this case. Such solutions are necessarily type D solutions of Kundt's class (see Chapter 18). It follows from (16.8) that these are not covered by the metric (16.5). Interestingly, however, such solutions can be generated from it using a specific limiting procedure.

Starting with the metric (16.5) with $\omega = 1$,[6] the general family of non-expanding type D solutions can be obtained by applying the transformation

$$r = \gamma + \kappa q, \qquad \sigma = \kappa^{-1} t, \qquad \tau = \psi - \gamma^2 \kappa^{-1} t, \tag{16.26}$$

where γ and κ are arbitrary parameters, and taking the limit in which $\kappa \to 0$. As for the case in which $a = 0$ above, it is again found that the parameter

---

[6] Another distinct non-expanding space-time can be obtained from the metric (16.5) with $\omega = 0$. This will be described in Section 16.5.

α is redundant and can be removed by the coordinate transformation.

$$p = \frac{\tilde{p} - \alpha\gamma^3\mu}{\mu + \alpha\gamma\tilde{p}}, \qquad \psi = \frac{\tilde{\psi}}{\mu}, \qquad \text{where} \qquad \mu^2 = \frac{1}{1 + \alpha^2\gamma^4},$$

and a relabelling of parameters. The resulting metric can then be written in the form

$$\mathrm{d}s^2 = \varrho^2\left(-\tilde{\mathcal{Q}}\,\mathrm{d}t^2 + \frac{1}{\tilde{\mathcal{Q}}}\,\mathrm{d}q^2\right) + \frac{\mathcal{P}}{\varrho^2}\left(\mathrm{d}\psi + 2\gamma q\,\mathrm{d}t\right)^2 + \frac{\varrho^2}{\mathcal{P}}\,\mathrm{d}p^2, \qquad (16.27)$$

where

$$\varrho^2 = \gamma^2 + p^2,$$
$$\tilde{\mathcal{Q}} = \epsilon_0 - \epsilon_2 q^2,$$
$$\mathcal{P} = \left(-\epsilon_2\gamma^2 - (e^2 + g^2) + \Lambda\gamma^4\right) + 2np + (\epsilon_2 - 2\Lambda\gamma^2)p^2 - \tfrac{1}{3}\Lambda p^4.$$

It is seen that the first terms in (16.27) are conformal to a (timelike) surface of constant curvature whose sign is that of ϵ_2 (see Appendix A). This family of solutions is characterised by two discrete parameters ϵ_2 and ϵ_0 that can be set to the values $1, -1, 0$, and five continuous parameters γ, n, e, g and Λ. It may also be noted that the parameter γ is formally the analogue of the NUT parameter in these space-times. Specifically, if $\gamma \neq 0$, these solutions have no curvature singularities.

This family of solutions was given by Carter (1968b) as his form $[\tilde{B}(-)]$. It was also discussed by Plebański (1975) who referred to these as anti-NUT metrics since the transformation (16.26) is analogous to (16.11), which gives rise to generalised NUT metrics. The different space-times that correspond to the distinct canonical forms of $\tilde{\mathcal{Q}}$ have been discussed by Griffiths and Podolský (2006b). For the vacuum case with $\Lambda = 0$, these solutions are of Kinnersley's class IV. An alternative form of the metric for this case was given by Kinnersley (1975).

The cosmological constant is Λ, and e and g are electric and magnetic charge parameters, but the physical meanings of the remaining parameters m, n and α have not been identified. These solutions represent a large family of distinct space-times which depend on the canonical form of $\tilde{\mathcal{Q}}$, the possible roots of $\mathcal{P}$ and the range of p that is chosen to ensure that $\mathcal{P}(p) > 0$. In particular, when γ, ϵ_2 and Λ vanish, they include the Melvin solution that was described in Section 7.3. Apart from this, however, none of these solutions appear to have any known physically significant interpretation.

Since the repeated principal null congruence is now non-expanding and non-twisting (as well as being geodesic and shear-free), this family of solutions must belong to the type D solutions of Kundt's class which will be described in Section 18.6, where explicit transformations will be given.

16.4.1 The B-metrics

Among the above family of space-times, the simplest vacuum solutions are known as the B-metrics following the classification of Ehlers and Kundt (1962). These occur for the line element (16.27) when m, α, e, g and Λ (and hence γ and k) all vanish with $n \neq 0$. The three cases are then distinguished by the discrete parameters ϵ_0 and $\epsilon_2 = -\epsilon$, so that $\tilde{Q} = \epsilon_0 + \epsilon q^2$ and $\mathcal{P}/\varrho^2 = \frac{2n}{p} - \epsilon$. They represent space-times which admit four Killing vectors.

In these three cases, the only non-zero component of the Weyl tensor is given by $\Psi_2 = -n/p^3$ (relative to a suitable tetrad). This confirms that they are of type D and possess a curvature singularity at $p = 0$. It is therefore convenient to consider only solutions for which p is positive, but the parameter n may have either sign.

The BI-metric occurs when $\epsilon = -1$. Its most general form has $\epsilon_0 = 1$, and the line element then becomes

$$\mathrm{d}s^2 = -p^2(1-q^2)\,\mathrm{d}t^2 + \frac{p^2}{1-q^2}\mathrm{d}q^2 + \left(1 + \frac{2n}{p}\right)\mathrm{d}\psi^2 + \left(1 + \frac{2n}{p}\right)^{-1}\mathrm{d}p^2, \quad (16.28)$$

so that, if $n > 0$, $p \in (0, \infty)$. Alternatively, if $n < 0$, this represents a non-singular space-time with $p \in (2|n|, \infty)$. Both cases are asymptotically flat as $p \to \infty$.

Following an analogy with the A-metrics, it is common practice to put $q = \cos\theta$. However, this is unnecessarily restrictive. The coordinate q may cover the complete range $q \in (-\infty, \infty)$. For the ranges $-\infty < q < -1$ and $1 < q < \infty$, the space-time is time-dependent and q is a timelike coordinate. Horizons exist at $q = \pm 1$, between which the space-time is static.

An alternative form of the metric for parts of this space-time that do not include horizons occurs for the choice $\epsilon_0 = -1$. In this case, putting $q = \sinh\tau$, $t = \varphi$ and replacing p by $-p$, the metric takes the form

$$\mathrm{d}s^2 = p^2\left(-\mathrm{d}\tau^2 + \cosh^2\tau\,\mathrm{d}\varphi^2\right) + \left(1 - \frac{2n}{p}\right)\mathrm{d}\psi^2 + \left(1 - \frac{2n}{p}\right)^{-1}\mathrm{d}p^2.$$

This has been interpreted by Gott (1974) as representing those parts of the space-time containing a tachyon that are not causally related to the source.

For **the BII-metric**, $\epsilon = +1$, it is necessary that $n > 0$, and the general case occurs when $\epsilon_0 = -1$. In this case, the metric takes the form

$$ds^2 = -p^2(q^2 - 1)\, dt^2 + \frac{p^2}{q^2 - 1}\, dq^2 + \left(\frac{2n}{p} - 1\right)d\psi^2 + \left(\frac{2n}{p} - 1\right)^{-1}dp^2, \quad (16.29)$$

with $p \in (0, 2n)$ and $q \in (-\infty, \infty)$. Again, horizons exist at $q = \pm 1$, but now the space-time is time-dependent in the range $-1 < q < 1$, and static elsewhere.

Finally, **the $BIII$-metric** occurs when $\epsilon = 0$. In this case, it is appropriate to use a scaling freedom to set $2n = 1$, and the metric becomes

$$ds^2 = p^2\left(-dt^2 + dq^2\right) + \frac{1}{p}\, d\psi^2 + p\, dp^2. \quad (16.30)$$

This is the Levi-Civita solution (10.9) in the limiting case when $\sigma = \frac{1}{4}$, i.e. (10.14), which is also locally isometric to the asymptotic form of the Melvin solution (7.21). It can also be expressed in the form (10.8) with the Kasner-like parameters $(p_0, p_2, p_3) = (\frac{2}{3}, \frac{2}{3}, -\frac{1}{3})$. However, these space-times are not understood physically.

In fact, almost none of the B-metrics have any known physical significance. Nevertheless, they are a very simple family of type D space-times that have been known for a long time, since they are formally related to the well-known A-metrics by a complex coordinate transformation. If the ψ-coordinate is taken to have a finite range $[0, 2\pi)$, with $\psi = 2\pi$ identified with $\psi = 0$, the static regions of these space-times can be expressed in Weyl form. In this case, the associated Newtonian potentials have been identified by Martins (1996) as semi-infinite line masses. However, the physical interpretation of these space-times clearly requires further investigation.

16.5 An alternative extension

Another family of space-times, which includes some expanding and non-expanding type D and conformally flat solutions, can be obtained from the Plebański–Demiański metric (16.5) in the non-accelerating ($\alpha = 0$) and non-twisting ($\omega = 0$) case. The coordinate transformation

$$p = \beta + \kappa\tilde{p}, \qquad r = \gamma + \kappa\tilde{q}, \qquad \sigma = \kappa^{-1}\tilde{\sigma}, \qquad \tau = b^2\kappa^{-1}\tilde{\tau}, \quad (16.31)$$

where β, γ, κ and b are arbitrary constants, involves shifts in both r and p simultaneously. This leads to the metric

$$ds^2 = -\frac{b^4\tilde{Q}}{(\gamma + \kappa\tilde{q})^2}\, d\tilde{\tau}^2 + (\gamma + \kappa\tilde{q})^2\left(\frac{1}{\tilde{Q}}\, d\tilde{q}^2 + \frac{1}{\tilde{P}}\, d\tilde{p}^2 + \tilde{P}d\tilde{\sigma}^2\right), \quad (16.32)$$

where

$$\widetilde{\mathcal{P}} = a_0 + a_1 \tilde{p} + a_2 \tilde{p}^2,$$

$$\widetilde{\mathcal{Q}} = b_0 + b_1 \tilde{q} + b_2 \tilde{q}^2 - \tfrac{4}{3}\Lambda\gamma\kappa\tilde{q}^3 - \tfrac{1}{3}\Lambda\kappa^2\tilde{q}^4,$$

and

$$b_2 = \epsilon - 2\Lambda\gamma^2, \qquad\qquad a_2 = -\epsilon,$$

$$b_1 = 2\kappa^{-1}(-m + \epsilon\gamma - \tfrac{2}{3}\Lambda\gamma^3), \qquad a_1 = 2\kappa^{-1}(n' - \epsilon\beta),$$

$$b_0 = \kappa^{-2}(e^2 + g^2 - 2m\gamma + \epsilon\gamma^2 - \tfrac{1}{3}\Lambda\gamma^4), \qquad a_0 = \kappa^{-2}(k + 2\beta n' - \epsilon\beta^2).$$

When $\kappa \neq 0$, the metric (16.32) explicitly contains the family of expanding Schwarzschild–Reissner–Nordström–de Sitter space-times. But when $\kappa = 0$, it also includes the non-expanding space-times that are a direct product of two two-dimensional spaces of constant curvature, such as the Bertotti–Robinson solution. For the case in which the cosmological constant vanishes, the metric (16.32) includes that of Ray and Wei (1977) (rediscovered by Halilsoy, 1993) which combines the Schwarzschild, the Reissner–Nordström and the Bertotti–Robinson solutions in a single metric.

The non-expanding solutions are obtained by setting $\kappa = 0$. In this case, rescaling $\widetilde{\mathcal{Q}}$ by b^2/γ^2, the metric becomes

$$\mathrm{d}s^2 = b^2\left(-\widetilde{\mathcal{Q}}\,\mathrm{d}\tilde{\tau}^2 + \frac{1}{\widetilde{\mathcal{Q}}}\,\mathrm{d}\tilde{q}^2\right) + \gamma^2\left(\widetilde{\mathcal{P}}\,\mathrm{d}\tilde{\sigma}^2 + \frac{1}{\widetilde{\mathcal{P}}}\,\mathrm{d}\tilde{p}^2\right), \qquad (16.33)$$

where $\widetilde{\mathcal{Q}}$ and $\widetilde{\mathcal{P}}$ are quadratic functions of $\tilde{q}$ and $\tilde{p}$, respectively, which can be transformed to their standard canonical forms (see Appendix A). This is the family of solutions given by Carter (1968b) as his form [D], which has also been given by Plebański (1975) as his case C.

In fact there are three distinct canonical forms for each 2-space of constant curvature of signatures $(-, +)$ and $(+, +)$. The metric (16.33) thus represents the complete family of direct-product space-times, which includes the Bertotti–Robinson, Nariai and Plebański–Hacyan solutions. Excluding cases with negative energy density, there are six possible geometries, as described in Section 7.2. Allowing for the possibilities of both electric and magnetic charges, this family of space-times essentially depends on only three free dynamical parameters, which may be taken as e, g and Λ.

16.6 The general family of type D space-times

To summarise, the Plebański–Demiański space-times include all type D vacuum and aligned non-null Einstein–Maxwell solutions with a possible cosmological constant. There are distinct metrics for the cases in which the repeated principal null directions are either expanding or non-expanding. For the expanding case, the metric which includes all possibilities is given by (16.12). For the non-expanding case, the general metric is given by (16.27) although, when $\Lambda \neq 0$, other type D solutions are given by (16.33), which also includes certain associated conformally flat solutions.

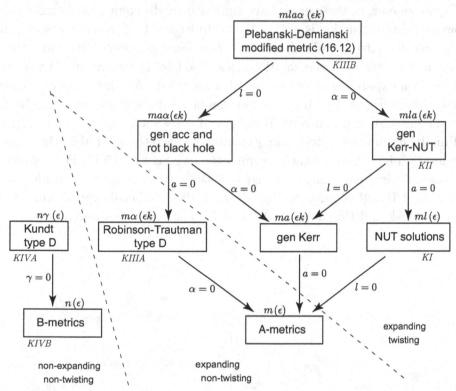

Fig. 16.2 The complete family of vacuum type D solutions with $\Lambda = 0$ and m or n non-zero. In the expanding case, these solutions generally have four continuous parameters m, l, a and α, and two auxiliary parameters ϵ and k that can be set to any convenient specific values. The non-expanding solutions generally have two continuous parameters n and γ, and one auxiliary parameter ϵ. The parameters associated with each solution are shown above each box – those in brackets have discrete values. The abbreviation "gen" indicates that the named solution is generalised to include the associated solutions with other values of the discrete parameters. The Kinnersley class of each sub-family is also indicated below each box. All particular cases have obvious generalisations which include charges and/or a cosmological constant.

The structure of this complete family of solutions for the vacuum case with $e = g = \Lambda = 0$ is illustrated in Figure 16.2. The expanding solutions are characterised by four continuous parameters m, l, a and α, and two auxiliary parameters ϵ and k that can be set to any convenient specific values. The classification of those cases which naturally contain surfaces of positive curvature, and for which $a_0 = +1$, has been illustrated in Figure 16.1. In that case, the family of Robinson–Trautman type D solutions is represented by the C-metric. It was also appropriate there to distinguish the cases in which the Kerr-like rotation parameter is greater than or less than the NUT parameter, as these cases have significantly different global structures. However, this distinction is not continued in Figure 16.2, where the solutions are generalised to include those with 2-surfaces of alternative curvatures (thus also generalising the solutions described by Ishikawa and Miyashita, 1983). Non-expanding solutions are also included. All the particular cases illustrated in Figure 16.2 have obvious generalisations with non-zero charge parameters and/or a non-zero cosmological constant.

Finally it may be noted that expanding space-times of this class with acceleration have boost-rotation symmetry (see Section 15.1). For a vanishing cosmological constant, their asymptotic properties have been analysed in terms of Bondi coordinates by Bičák (1968), Farhoosh and Zimmerman (1979), Bičák and Pravdová (1998) and Lazkoz and Valiente-Kroon (2000).

17

Plane and pp-waves

The previous nine chapters have been mainly devoted to various black hole space-times and other solutions that at least contain stationary regions. The present chapter, and the following four, will concentrate on some of the most important exact solutions which represent gravitational waves in general relativity. It is convenient to start here with the simplest case of non-expanding waves, known as pp-waves, which are also important in the context of higher-dimensional theories, and their subclass of plane waves. General classes of solutions with non-expanding and expanding waves will be described, respectively, in Chapters 18 and 19, and both types of impulsive waves will be reviewed in Chapter 20. The collision and interaction of plane waves will be covered in Chapter 21. These may all be associated with other forms of pure radiation. Cylindrical gravitational waves will be more briefly described in Section 22.3.

The class of pp-waves describes plane-fronted waves with parallel rays. They are defined geometrically by the property that they admit a covariantly constant null vector field k, and may represent gravitational waves, electromagnetic waves, some other forms of matter, or any combination of these.

Using (2.15) and the properties described in Section 2.1.3, it follows from the defining property $k^{\mu}_{;\nu} = 0$ that the vector field k is tangent to an expansion-free, shear-free and twist-free null geodesic congruence. Since this congruence is non-twisting, there exists a family of 2-surfaces orthogonal to k which may be considered as wave surfaces (see Kundt, 1961). It is also implied by the fact that k is covariantly constant that these wave surfaces are indeed planar and that rays orthogonal to them are parallel.

The particular pp-waves for which the curvature tensor components are constant over each wave surface, are known as *plane waves*. These will be treated separately in Section 17.5.

17.1 The class of pp-wave space-times

The family of pp-wave space-times was first discussed by Brinkmann (1925), and interpreted in terms of gravitational waves by Peres (1959). Using a coordinate u to label the null wave surfaces such that $k_\mu = -u_{,\mu}$, and an affine parameter r defined such that $\mathbf{k} = \partial_r$, the metric for any vacuum, aligned null electromagnetic or pure radiation pp-wave space-time can be written in the Brinkmann form

$$ds^2 = -2\,du\,dr - 2H(\zeta, \bar{\zeta}, u)\,du^2 + 2\,d\zeta\,d\bar{\zeta}, \qquad (17.1)$$

where the complex coordinate ζ spans the wave surfaces, and $H(\zeta, \bar{\zeta}, u)$ is an arbitrary real function.[1] With the additional null vectors $\mathbf{l} = \partial_u - H\partial_r$ and $\mathbf{m} = \partial_{\bar{\zeta}}$, the only non-zero components of the curvature tensor are

$$\Psi_4 = H_{,\zeta\zeta}, \qquad \Phi_{22} = H_{,\zeta\bar{\zeta}}. \qquad (17.2)$$

The cosmological constant necessarily vanishes.[2] It follows from these expressions that such space-times are either of algebraic type N, with a quadruply repeated principal null direction $\mathbf{k}$ that is also aligned with any radiation field that may be included, or are conformally flat. The vacuum field equations reduce to the simple two-dimensional Laplace equation

$$H_{,\zeta\bar{\zeta}} = 0. \qquad (17.3)$$

(Terms in H that are linear in ζ, $\bar{\zeta}$ or independent of it can be removed by a simple coordinate transformation.)

For reviews of these space-times see Takeno (1961a,b), Ehlers and Kundt (1962), Zakharov (1973) or Section 24.5 in Stephani *et al.* (2003). In general they only have the single isometry whose generator is $\mathbf{k}$, but may admit up to six Killing vectors.

When H vanishes, the metric (17.1) reduces to Minkowski space-time in the form (3.4). The case with small H can thus be regarded as a "perturbation" of flat space. For a general $H(\zeta, \bar{\zeta}, u)$, the corresponding space-times may represent strong gravitational (or electromagnetic or other) waves with arbitrary profiles in terms of the retarded time u. These space-times are *exact* solutions, which can be used to investigate certain features of gravitational (and other) waves, such as their focusing properties and possible nonlinear interactions, that cannot be determined by only considering approximation schemes.

---

[1] It should be emphasised that the metric (17.1) is not the most general form of pp-wave space-times when other forms of matter are permitted. An example of a such a space-time will be described in Section 18.5.

[2] The analogous family of solutions with a non-zero cosmological constant are members of Kundt's class of solutions that will be described in the following chapter.

The source for the Ricci tensor component Φ_{22}, can be interpreted as any type of null matter or radiation. It may generally be described as a null fluid. However, since Maxwell's equations can be satisfied for a null electromagnetic field, it may also be interpreted in terms of electromagnetic radiation in which the electric and magnetic fields are locally orthogonal, of equal magnitude and orthogonal to the propagation vector $\mathbf{k}$. And since Weyl's equations for a spin $\frac{1}{2}$ field can also be satisfied, the source may alternatively be interpreted in terms of massless neutrino fields.

Components of the Weyl tensor that propagate with the speed of light can be interpreted as *gravitational waves*. In *pp*-wave space-times, such a wave is characterised by the single curvature component $\Psi_4 = |\Psi_4| e^{i\theta}$, which depends on the null coordinate u. Using the analogy between the Weyl tensor component Ψ_4 and the radiative electromagnetic field tensor component Φ_2, the function $|\Psi_4|$ can be regarded as the *amplitude* of the gravitational wave and θ as its *polarisation*. Waves for which θ is constant are said to be "linearly polarised".

17.2 Focusing properties

The main properties of gravitational waves can be seen clearly in the study of the deviation of neighbouring geodesics (Pirani 1957, Bondi, Pirani and Robinson 1959, Szekeres 1965). This describes the influence of the space-time curvature on the motion of a neighbouring set of test particles. For the metric (17.1), it is convenient to consider the orthonormal frame for which the timelike vector $e^\mu_{\hat{0}}$ is the four-velocity $u^\mu = (\dot{u}, \dot{r}, \dot{\zeta}, \dot{\bar{\zeta}})$ of a test parti-cle, where the dot denotes differentiation with respect to its proper time τ. Corresponding (parallelly transported) orthonormal spatial vectors can be chosen as $e^\mu_{\hat{1}} = -\frac{1}{\sqrt{2}}(0, 2\mathcal{R}e\{\dot{\zeta}\}/\dot{u}, 1, 1)$, $e^\mu_{\hat{2}} = -\frac{1}{\sqrt{2}}(0, 2\mathcal{I}m\{\dot{\zeta}\}/\dot{u}, i, -i)$, and $e^\mu_{\hat{3}} = (0, 1/\dot{u}, 0, 0) - e^\mu_{\hat{0}}$. For *vacuum pp*-wave space-times, the invari-ant form of the geodesic deviation equation (2.17) with respect to this frame, along *any* timelike geodesic, is then

$$\ddot{\xi}^{\hat{1}} = -\mathcal{A}_+ \xi^{\hat{1}} + \mathcal{A}_\times \xi^{\hat{2}},$$
$$\ddot{\xi}^{\hat{2}} = \mathcal{A}_+ \xi^{\hat{2}} + \mathcal{A}_\times \xi^{\hat{1}}, \qquad (17.4)$$
$$\ddot{\xi}^{\hat{3}} = 0,$$

where the amplitudes of the two possible modes are given by

$$\mathcal{A}_+ = \mathcal{R}e\{\Psi_4 \dot{u}^2\}, \qquad \mathcal{A}_\times = \mathcal{I}m\{\Psi_4 \dot{u}^2\}, \qquad (17.5)$$

evaluated along the geodesic. It may be noted here that $\dot{u}$ is a constant for any geodesic in a pp-wave space-time. The equations (17.4) express the relative accelerations of neighbouring free test particles in terms of their actual positions. The invariant distances between the test particles are determined by the frame components of the displacement vector $\xi^{\hat{i}} \equiv \xi^\mu e_\mu{}^{\hat{i}}$, while $\ddot{\xi}^{\hat{i}} \equiv (D^2\xi^\mu/d\tau^2)\, e_\mu{}^{\hat{i}} = d^2\xi^{\hat{i}}/d\tau^2$ are their relative accelerations.

It is seen from (17.4) that the particles move freely with respect to each other if $\Psi_4 = 0$, in which case the space-time is flat. When $\Psi_4 \neq 0$, the amplitudes $\mathcal{A}_+$ and $\mathcal{A}_\times$ do not both vanish, but there is still no relative acceleration in the $e_{\hat{3}}$ direction. In fact, this direction coincides with the projection of the quadruply repeated principal null direction k onto the hypersurface orthogonal to the four-velocity u, and thus may be interpreted as the spatial direction of propagation. It is then obvious from (17.4) that the gravitational wave has a *transverse character* as the relative motions of test particles occur only in the plane $(e_{\hat{1}}, e_{\hat{2}})$ perpendicular to $e_{\hat{3}}$.

Moreover, (17.4) shows that test particles are influenced by an exact gravitational pp-wave in just the same way as in the standard linearised theory for gravitational waves in a Minkowski background (see e.g. Misner, Thorne and Wheeler, 1973). In particular, any ring of test particles in the transverse plane is locally deformed into an ellipse after a short period of time. The waves have *two polarisation modes*, "+" and "×", with corresponding amplitudes $\mathcal{A}_+$ and $\mathcal{A}_\times$. Under rotations by ϑ in this plane they transform according to $\mathcal{A}'_+ = \cos 2\vartheta \mathcal{A}_+ - \sin 2\vartheta \mathcal{A}_\times$, $\mathcal{A}'_\times = \sin 2\vartheta \mathcal{A}_+ + \cos 2\vartheta \mathcal{A}_\times$, so that the helicity of the wave is 2, as with linearised waves on a flat background. In the special case of constant polarisation ($\vartheta = \text{const.}$), it is always possible to rotate the transverse frame such that either $\mathcal{A}'_\times = 0$ or $\mathcal{A}'_+ = 0$. In the first case, the frame vectors $e_{\hat{1}}, e_{\hat{2}}$ are aligned with the principal axes of the deformed ellipses.

For the general *non-vacuum case*, the equations (17.4) generalise to

$$
\begin{aligned}
\ddot{\xi}^{\hat{1}} &= -\mathcal{A}_+\, \xi^{\hat{1}} + \mathcal{A}_\times\, \xi^{\hat{2}} - \mathcal{B}\xi^{\hat{1}}, \\
\ddot{\xi}^{\hat{2}} &= \mathcal{A}_+\, \xi^{\hat{2}} + \mathcal{A}_\times\, \xi^{\hat{1}} - \mathcal{B}\xi^{\hat{2}}, \\
\ddot{\xi}^{\hat{3}} &= 0,
\end{aligned}
\tag{17.6}
$$

where $\mathcal{B} = \Phi_{22}\,\dot{u}^2$.

For the case in which $\mathcal{A}_+ = 0 = \mathcal{A}_\times$, these equations show that any ring of test particles in the transverse plane radially converges on the passing of a pure radiation field since $\mathcal{B}$ is necessarily non-negative. The worldlines of particles that are initially at rest relative to each other are *focused*.

By contrast, it can be seen from (17.4) that for purely gravitational waves, for which $\mathcal{B} = 0$, initially parallel worldlines of test particles tend to focus in one direction and diverge in the direction perpendicular to it. This effect may be described as a kind of *astigmatic focusing* and may lead to the formation of caustics after a sufficient period of time.

These focusing effects also apply to null geodesics which are not aligned with $\boldsymbol{k}$. When an initially parallel congruence of null geodesics meets an electromagnetic wave (or some other form of pure radiation), it is focused by it. In particular, as argued by Penrose (1966), the amplitude of the wave is directly related to its focusing power. Similarly, when an initially parallel congruence of null geodesics meets a gravitational wave, it is focused by it astigmatically. In this case, the focusing power of the wave is again a measure of its amplitude, while the polarisation of the wave is determined by the direction of the astigmatism (i.e. the shear).

17.3 Linear superposition and beams of pure radiation

Since the vacuum field equation (17.3) is linear, distinct solutions with different expressions for H may be simply superposed. Indeed, even for the non-vacuum case, H may be taken to be the sum of any number of independent components. It follows that distinct gravitational wave components of this type that propagate in the same direction do not interact directly. This was first pointed out by Peres (1960), Bonnor (1969b) and Aichelburg (1971). In the coordinates of the line element (17.1), both the metric function H and the curvature tensor components Ψ_4 and Φ_{22} satisfy a superposition principle. Waves with different expressions for H add linearly both at the same value of u and for different ranges of u. However, the effects of each component, such as their focusing properties and the formation of caustics that will be described below, do not in general superpose linearly.

Specifically, it is possible in exact solutions of this type for the Ricci tensor component Φ_{22} to be non-zero only over some finite regions of the transverse wave surfaces spanned by ζ. Outside these regions, the metric function H has to satisfy the two-dimensional Laplace equation (17.3) and be C^1 over each wave surface. This generally implies that Ψ_4 must be non-zero over the complete wave surface. A simple example of this was given by Bonnor (1969b) in which

$$H = \begin{cases} c(\zeta\bar{\zeta} - a^2) & \text{for} \quad |\zeta| \leq a, \\ ca^2 \log(\zeta\bar{\zeta}/a^2) & \text{for} \quad |\zeta| > a. \end{cases}$$

This represents a beam of pure radiation ($\Psi_4 = 0$) of constant amplitude $\Phi_{22} = c$ in a region of circular cross-section $|\zeta| \leq a$ on each wave surface with constant u. Outside this beam the solution is a purely gravitational pp-wave such that $\Phi_{22} = 0$ and $\Psi_4 = -ca^2\,\zeta^{-2}$.

Solutions of this type may be simply superposed, and Bonnor (1969b) has presented an explicit solution for two beams of pure radiation with their associated gravitational wave fields given by

$$H = \begin{cases} c_1(\rho_1^2 - a_1^2) + c_2 a_2^2 \log(\rho_2^2/a_2^2) & \text{for} \quad |\zeta - \bar\zeta_1| \le a_1, \\ c_2(\rho_2^2 - a_2^2) + c_1 a_1^2 \log(\rho_1^2/a_1^2) & \text{for} \quad |\zeta - \bar\zeta_2| \le a_2, \\ c_1 a_1^2 \log(\rho_1^2/a_1^2) + c_2 a_2^2 \log(\rho_2^2/a_2^2) & \text{elsewhere}, \end{cases}$$

where $\rho_1^2 = |\zeta - \zeta_1|^2$ and $\rho_2^2 = |\zeta - \zeta_2|^2$. This is illustrated in Figure 17.1. In this case $\Psi_4 = -c_1 a_1^2\,(\zeta - \zeta_1)^{-2} - c_2 a_2^2\,(\zeta - \zeta_2)^{-2}$ outside both beams.

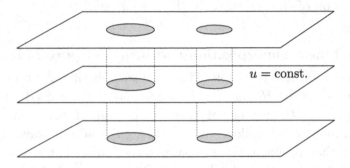

Fig. 17.1 Two beams of pure radiation are represented as two circular regions of radius a_1 and a_2 on each wave surface $u = $ const. that are spanned by the complex coordinate ζ. (This is not a space-time diagram.)

Generally, such space-times may be globally regular and the beams of pure radiation can be regarded as the sources of the gravitational waves. In particular, it may be concluded that parallel beams of this type do not mutually interact. Solutions of the Einstein–Maxwell equations with similar properties have also been given by Bonnor (1970a). These represent gravitational and electromagnetic pp-waves that are generated by null currents (i.e. by beams of charge moving at the speed of light).

17.4 Shock, sandwich and impulsive waves

It may be noticed that the curvature tensor components (17.2) do not depend on the u-derivatives of the function $H(\zeta, \bar\zeta, u)$ in the Brinkmann form of the pp-wave metric (17.1). It is therefore not necessary for this function to be continuous with respect to the retarded time coordinate u. Indeed it may even be distributional.

This lack of any continuity requirement in the function that determines the amplitude and polarisation of the wave may be used to admit the possibilities of shock (or step) waves and sandwich waves that have finite duration. Even impulsive waves, given formally as multiples of the Dirac δ-function, are permitted. However, when attempting to analyse the global properties of such space-times, it is not appropriate to work with such a discontinuous metric. For this, it is necessary to transform the metric to an alternative form which is usually C^1 (or C^0 for an impulse) everywhere and piecewise at least C^2. In general, such transformations are found to depend heavily on the particular form of the function $H(\zeta, \bar{\zeta}, u)$.

A *shock* or step wave can be constructed simply by requiring that H be some multiple of the Heaviside step function $\Theta(u)$. It is then straightforward to extend this to *sandwich waves*. Illustrated in Figure 17.2, these may be represented by a function $H(\zeta, \bar{\zeta}, u)$ which is only non-trivial for a finite period $u_0 \le u \le u_1$ of retarded time. In this case the regions both ahead of the wavefront $u = u_0$ and behind $u = u_1$ are described by flat Minkowski half-spaces $\mathcal{M}^-$ and $\mathcal{M}^+$, and the wave in the region with non-zero curvature tensor propagates along the null hypersurfaces $u = \text{const.}$ Actually, the physical interpretation of such sandwich gravitational waves is not as straightforward as might appear at first sight. Although the regions on either side of the wave are flat, it is not clear a priori how the different Minkowski half-spaces on each side are related to each other.

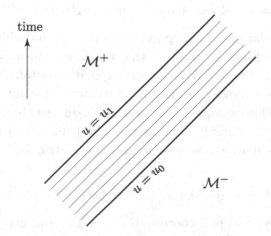

Fig. 17.2 A sandwich wave in which the function $H(\zeta, \bar{\zeta}, u)$ is non-trivial only for $u_0 \le u \le u_1$ is joined before and after to two different regions of Minkowski space.

It is also possible to proceed to the limit of a thin sandwich wave corresponding to an *impulse*. In such a limit, the amplitude function H is modelled by a Dirac δ-function of u, which then (inconveniently) appears explicitly in the Brinkmann form of the metric. In fact, the same limit occurs irrespective of the actual profile of the wave of which it is the limit. Such impulsive waves will be described further in Section 20.2.

17.5 Plane wave space-times

Plane waves are defined as the particular subclass of *pp*-waves for which, on each wave surface, the field components are the same at every point. This is the sense in which they are said to have 'plane symmetry'. This special and physically important class of solutions was first considered by Baldwin and Jeffery (1926). See also Brdička (1951), Bonnor (1957b) and Bondi (1957b).

Using the above notation, the condition for plane waves requires that Ψ_4 and Φ_{22} given by (17.2) are independent of the complex coordinate ζ which spans the wave surfaces. The function $H(\zeta, \bar{\zeta}, u)$ thus has to be at most quadratic in ζ and $\bar{\zeta}$. And since a coordinate transformation can be used to remove terms in H that are at most linear in ζ, the line element for a plane wave can always be written in the form

$$ds^2 = -2\,du\,dr - \Big(A(u)\zeta^2 + \bar{A}(u)\bar{\zeta}^2 + 2B(u)\zeta\bar{\zeta}\Big)du^2 + 2\,d\zeta\,d\bar{\zeta}, \quad (17.7)$$

where $A(u)$ is complex and $B(u)$ is real. The non-zero curvature tensor components are then given by

$$\Psi_4 = A(u), \qquad \Phi_{22} = B(u).$$

For a "linearly polarised" plane gravitational wave, for which $\arg A$ is constant, it is always possible to rotate the coordinates to make A real.

The form of the metric (17.7), which explicitly contains the profile functions of the wave components, is not convenient in many contexts as will be shown below (for example in Section 21.1). In fact, to avoid any discontinuities and also to directly demonstrate the transverse character of these space-times, it is more convenient to use the metric form given by Rosen (1937), namely

$$ds^2 = -2\,du\,dv + g_{ij}(u)\,dx^i\,dx^j, \qquad i,j = 1,2. \quad (17.8)$$

This form employs two null coordinates u and v, and can be derived from (17.7) using the transformation

$$\zeta = c_i\,x^i, \qquad r = v + \tfrac{1}{4}\dot{g}_{ij}\,x^i\,x^j, \qquad g_{ij}(u) = \bar{c}_i\,c_j + c_i\,\bar{c}_j, \quad (17.9)$$

where c_i are complex functions of u which satisfy the constraints

$$\mathcal{R}e\left\{\bar{c}_{(i}\,\ddot{c}_{j)} + A(u)\,c_i\,c_j + B(u)\,\bar{c}_i\,c_j\right\} = 0, \quad (17.10)$$

and a dot denotes a derivative with respect to u.

The line element (17.8) may be explicitly written in the form

$$ds^2 = -2\,du\,dv + \alpha \left(\chi\,dy^2 + \chi^{-1}(dx + \omega\,dy)^2 \right), \tag{17.11}$$

where α, χ and ω are functions of u only. In the case in which the gravitational wave is "linearly polarised" it is always possible to rotate the coordinates to put $\omega = 0$, but the condition for a vacuum is rather complicated. Unfortunately, the metric in the form (17.11) always contains a coordinate singularity where $\alpha = 0$. This is a quasiregular singularity which can be interpreted in terms of the focusing of congruences in directions other than $\boldsymbol{k}$.

Bondi, Pirani and Robinson (1959) have shown that the metric given by (17.11) admits a 5-parameter group of isometries that are generated by the Killing vectors

$$\boldsymbol{\xi}_1 = \partial_x, \qquad \boldsymbol{\xi}_2 = \partial_y, \qquad \boldsymbol{\xi}_3 = \partial_v,$$
$$\boldsymbol{\xi}_4 = x\partial_v + R(u)\,\partial_x + N(u)\,\partial_y, \tag{17.12}$$
$$\boldsymbol{\xi}_5 = y\partial_v + P(u)\,\partial_y + N(u)\,\partial_x,$$

where $P(u) = \int \alpha\chi^{-1}du$, $R(u) = \int \alpha(\chi^2 + \omega^2)\chi^{-1}du$, $N(u) = \int \alpha\omega\chi^{-1}du$. All these commute except for $[\boldsymbol{\xi}_1, \boldsymbol{\xi}_4] = \boldsymbol{\xi}_3$ and $[\boldsymbol{\xi}_2, \boldsymbol{\xi}_5] = \boldsymbol{\xi}_3$, which indicates that the structure constants for plane gravitational waves are the same as those for plane electromagnetic waves in a flat space-time. This analogy with plane electromagnetic waves further justifies their interpretation as gravitational plane waves.

17.5.1 Shock, impulsive and sandwich plane waves

The plane wave metric in the Brinkmann form (17.7) explicitly includes the functions $A(u)$ and $B(u)$ which represent the gravitational wave and the pure radiation components of the curvature tensor, respectively. These functions may include steps or impulses. In these cases, it is often not appropriate to work with such a discontinuous metric but, for plane waves, the alternative continuous Rosen form of the metric (17.11) is available.

As an initial example, consider a purely *electromagnetic shock* plane wave with a step wavefront at $u = 0$ and subsequently constant amplitude a. This can be represented by the metric (17.7) with $A = 0$ and $B = a^2\Theta(u)$, so that the only non-zero component of the curvature tensor is given by $\Phi_{22} = a^2\Theta(u)$. In this case, the transformation (17.9)–(17.10) becomes $\ddot{c}_i + a^2\Theta(u)c_i = 0$, and one can take $c_2 = \mathrm{i}\,c_1 = \frac{1}{\sqrt{2}}\mathrm{i}\cos\left(au\Theta(u)\right)$. Then, putting $\zeta = c_1 x + c_2 y$, the line element takes the Rosen form

$$ds^2 = -2\,du\,dv + \cos^2\left(au\Theta(u)\right)\left(dx^2 + dy^2\right), \qquad (17.13)$$

which is C^1 everywhere. (The collision of waves of this type will be considered later in Section 21.6.)

The metric (17.13) contains a singularity at $u = \pi/2a$. This is obviously a consequence of the choice of coordinates as it does not occur in the Brinkmann form of the metric. In fact, it is a manifestation of the focusing properties of the wave as described in Section 17.2. In this case, it can be clearly seen that null geodesics for which v, x and y are constant are parallel when $u < 0$, but focus on the null hypersurface $u = \pi/2a > 0$.

As another example which represents a pure gravitational wave, consider an *impulsive gravitational* plane wave. This is obtained by taking $A = a\,\delta(u)$ and $B = 0$ in (17.7), so that $\Psi_4 = a\,\delta(u)$. In this case, the transformation to Rosen form is explicitly given in (20.4), and yields

$$ds^2 = -2\,du\,dv + (1 - au\Theta(u))^2 dx^2 + (1 + au\Theta(u))^2 dy^2, \qquad (17.14)$$

which is C^0 on the null hypersurface $u = 0$ where the impulsive gravitational wave occurs. This form of the metric clearly has a coordinate singularity at $u = 1/|a|$. It also nicely illustrates the astigmatic focusing that is induced by a gravitational wave. In this case, null geodesics for which v, x and y are constant (or similar timelike geodesics) are parallel when $u < 0$ but, after passing the wave, contract in the x-direction and expand in the y-direction. Further properties of this solution and the collision of such waves will be described in Section 21.3.

The properties of sandwich plane gravitational waves were investigated in detail by Bondi, Pirani and Robinson (1959). In particular they inferred that, since the passing of a sandwich gravitational wave produces relative accelerations in the motion of free test particles, such waves must transport energy – a conclusion that had been widely questioned until that time. To illustrate this property, they considered the motion of a bead in a suitably aligned rod with some friction.

17.5.2 Focusing properties of sandwich plane waves

One very interesting property of plane waves that also has important implications was pointed out by Penrose (1965a). He considered the structure of the future light cone originating at an event Q in front of a sandwich wave. As this cone expands, part of it crosses the wave and is distorted. Null geodesics on this cone which pass through the wave will be focused by it. For a gravitational wave, they will focus on a line through some future event P. They thus form part of the past light cone through the point P. It must therefore be concluded that the *future* light cone of an event Q in front of the wave includes part of the *past* light cones of a family of events behind the wave. Suppressing one spatial dimension, this is illustrated in Figure 17.3.

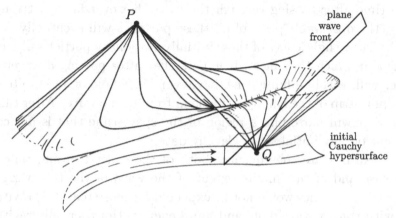

Fig. 17.3 (Penrose, 1965a) The future light cone of the point Q is distorted as it passes through a plane wave and is again focused to another vertex P which may be a point or a line (one spatial dimension has been suppressed).

As Penrose (1965a) deduced, the light-cone structure described above implies that a plane wave space-time cannot contain a global Cauchy hypersurface. In other words, it is not possible to set up initial data for a plane wave on any global spacelike hypersurface. Such a hypersurface must lie entirely in the past of any future null cone from any point on the surface. However, in this case, the future null cone of a point in front of the wave folds down to focus again on some line behind it. The Cauchy hypersurface passing through the initial point must therefore lie entirely below the past null cone of the line. It cannot, therefore, extend to spatial infinity behind the wave. Thus, no complete spacelike hypersurface exists that is adequate for the global specification of Cauchy data. It also follows from this that, although a plane wave space-time is strongly causal, it is *not globally hyperbolic*.

The focusing properties of plane waves were analysed in greater detail by Bondi and Pirani (1989). They considered the effect of a sandwich plane gravitational wave on a set of test particles that are strung out in a fixed direction, which depends on the polarisation of the wave, and which are initially at rest in a flat space-time in front of the wave. They proved, at least for waves that have constant linear polarisation, that all these particles collide after a finite time independent of how far apart they were initially. They described such focusing in terms of caustics. These are associated with the coordinate singularities that occur in the Rosen form of the metric.

Treating one such test particle as an observer who passes through the wave and looks out along past null cones, Bondi and Pirani (1989) also described some further interesting properties of plane waves as seen by an observer. Since the worldlines of a row of initially stationary test particles all meet a finite time after passing through the wave, however far apart they were initially, the observable pasts of all these particles will eventually coincide. Since the future light cones of these initially stationary particles also contain the event at which they meet, it follows that after passing a wavefront an observer will, within a finite time, have seen the whole of an infinite spatial volume in a strip of the half hyperplane in front of the wave. After this time, the observer will move into a region of the space-time that is not causally dependent on any initial set of Cauchy data.

These features are consequences both of the exact focusing that occurs in this case and of the infinite extent of the wave in the transverse directions. Such properties would not be expected for more realistic gravitational waves with convex wavefronts and finite energy. However, all gravitational waves will induce astigmatic focusing, so that caustics and some of the features associated with them may still be found to occur even in less idealised situations.

17.6 Some related topics

As described above for the case of plane waves, geodesics which pass through any gravitational waves are focused astigmatically: they converge in some directions and diverge in others. In fact, test particles propagating through certain types of vacuum pp-waves could end up following significantly different types of trajectory. In particular, they could diverge in a number of distinct directions.

Families of such geodesics in particular vacuum pp-wave space-times have been investigated in detail by Podolský and Veselý (1998c,d). They have demonstrated that, in the polynomial case in which $2H = C(\zeta^n + \bar{\zeta}^n)$ for $n = 3, 4, 5 \ldots$, geodesic motion is *chaotic*. The geodesic motions have n distinct "outcomes" representing n distinct "channels to infinity" at $\zeta = \infty$, and the boundaries between geodesics that have different outcomes have fractal structure. There is thus a highly complicated, indeed chaotic, dependence of the outcome on the initial data.

In a subsequent paper, Podolský and Veselý (1999) investigated geodesic motions in the space-times of corresponding sandwich *pp*-waves. They found that, as the duration of the wave decreases, the fractal structure of the motions is cut at an ever decreasing level. Thus, although the motions are no longer strictly "chaotic" in the complete sense, the apparently chaotic structure disappears slowly as the duration of the wave decreases, and fully disappears only in the limit of an impulsive wave. In this sense, the detailed fractal structure "smears" as the "width" of the wave reduces.

18

Kundt solutions

Solutions are said to belong to *Kundt's class* if they admit a null geodesic congruence, generated by a vector field $\boldsymbol{k}$, which is shear-free, twist-free and expansion-free. They include the *pp*-waves and plane wave space-times that have been described in the previous chapter. The whole class, initially investigated by Kundt (1961, 1962) in the case of vacuum or with an aligned pure radiation field, is however much wider. Because the null vector field $\boldsymbol{k}$ is not in general covariantly constant, the rays of the corresponding non-expanding waves are not necessarily parallel, as in the case of *pp*-waves, and the wave surfaces need not be planar. This greater freedom permits, for example, the presence of a cosmological constant Λ or aligned electromagnetic fields, both null and non-null.

For vacuum or some specific matter content, generalised Goldberg–Sachs theorems imply that the Kundt space-times must be algebraically special, that is of Petrov type II, D, III or N (or conformally flat), with $\boldsymbol{k}$ being a repeated principal null direction of the Weyl tensor. Any Einstein–Maxwell and pure radiation fields must also be aligned: that is, $\boldsymbol{k}$ is the common eigendirection of the Weyl and Ricci tensor.

A number of physically interesting space-times of this class are explicitly known. These will be briefly described in this chapter, with emphasis given to those that describe exact non-expanding gravitational waves of various kinds.

18.1 The general class of Kundt space-times

As shown by Kundt (1961), see also Kundt and Trümper (1962), Ehlers and Kundt (1962) or Section 31.2 in Stephani *et al.* (2003), the general line element which admits a geodesic, shear-free, twist-free and non-expanding null congruence can be expressed in the form

$$ds^2 = -2\,du\,(dr + H du + W d\zeta + \bar{W} d\bar{\zeta}) + 2P^{-2}d\zeta d\bar{\zeta}. \qquad (18.1)$$

Here $P(\zeta, \bar{\zeta}, u)$ and $H(\zeta, \bar{\zeta}, u, r)$ are real functions, while $W(\zeta, \bar{\zeta}, u, r)$ is a complex function, which are to be determined by field equations. The coordinate u labels null surfaces (it represents a retarded time for which $k_\mu = -u_{,\mu}$), r is an affine parameter along the repeated principal null congruence such that $\boldsymbol{k} = \partial_r$, and ζ is a complex coordinate which spans the transverse spatial 2-space.

For any line element of the form (18.1), the surfaces given by constant u (at any r) are spacelike with the metric $ds^2 = 2P^{-2}d\zeta\,d\bar{\zeta}$. The vector fields $P\partial_\zeta$ and $P\partial_{\bar{\zeta}}$ are tangent to them and are orthogonal to the principal null direction $\boldsymbol{k}$. Such surfaces can thus be regarded as "wave surfaces" associated with radiation propagating along $\boldsymbol{k}$. The Gaussian curvature $K(\zeta, \bar{\zeta}, u)$ of these wave surfaces is given by $K = 2P^2(\log P)_{,\zeta\bar{\zeta}}$. It is independent of the coordinate r, thus explicitly demonstrating the non-expanding character of the Kundt space-times.

Some special properties of this general family of space-times have been investigated by Coley *et al.* (2009). In particular, they have shown that these include the only space-times that cannot be uniquely determined by their scalar polynomial curvature invariants constructed from the Riemann tensor and its covariant derivatives (see also Coley, Hervik and Pelavas, 2009).

18.2 Vacuum and pure radiation Kundt space-times

The simplest solutions of the extensive Kundt's class are those which are empty. In the original papers of Kundt (1961, 1962), all such space-times of algebraic types N and III were given explicitly. In particular, it was demonstrated that the subclass of type N solutions contains not only the previously known *pp*-waves but also a specific family of the *Kundt waves* for which the principal null vector field $\boldsymbol{k}$ is not covariantly constant. All type D vacuum solutions of Kundt's class were later found by Kinnersley (1969a) and Plebański and Demiański (1976) (see Section 16.4). However, only some empty Kundt space-times of type II are known explicitly (see Stephani *et al.*, 2003).

Natural generalisations to a non-zero cosmological constant Λ are also possible. According to the value of Λ, the conformally flat vacuum space-times are just Minkowski, de Sitter and anti-de Sitter spaces, see Chapters 3, 4 and 5. A particular subclass of non-expanding type N vacuum space-times with $\Lambda \neq 0$ was found by García Díaz and Plebański (1981), and a geometrically different solution with $\Lambda < 0$ was discovered by Siklos (1985). The complete family was described and classified by Ozsváth, Robinson and Rózga (1985). Mutual relations between these solutions and some of their physical properties were clarified by Bičák and Podolský (1999a,b), Podolský (1998a) and Podolský and Ortaggio (2003). All type III solutions with a cosmological constant have been obtained by Griffiths, Docherty and Podolský (2004), while the type D Kundt space-times with Λ are contained in the Plebański–Demiański family of solutions.

In most of these cases, a field of pure radiation aligned with the principal null direction $\boldsymbol{k}$ can be included – in particular in the type III and type N space-times, but also in Kundt space-times of more general algebraic types,[1] see e.g. Wils and Van den Bergh (1990), Podolský and Ortaggio (2003). Special attention has been paid to conformally flat pure radiation space-times of Kundt's class, see for example Wils (1989), Van den Bergh, Gunzig and Nardone (1990), Koutras and McIntosh (1996), Edgar and Ludwig (1997a,b), Skea (1997), Griffiths and Podolský (1998) or Barnes (2001).

These specific solutions will now be described in more detail, with the emphasis being given to their physical relevance. Some other references will also be mentioned.

18.3 Type III, N and O solutions

The complete class of Kundt space-times of types III, N or conformally flat, with a possibly non-vanishing cosmological constant Λ and pure radiation, can be written in the metric form (18.1) in which

$$W = \frac{2\bar{\tau}}{P}\,r + W^\circ, \qquad H = -\left(\tau\bar{\tau} + \tfrac{1}{6}\Lambda\right)r^2 + 2G^\circ r + H^\circ. \qquad (18.2)$$

Here the functions W°, G°, H°, P and τ are independent of r, so that W is (at most) linear and H quadratic in r. The field equations also imply that $P^2(\log P)_{,\zeta\bar\zeta} = \tfrac{1}{6}\Lambda$, which ensures that the Gaussian curvature of the wave surfaces is constant, namely $K = \tfrac{1}{3}\Lambda$. Thus, a coordinate freedom can be used (see Appendix A) to put P into the canonical form

$$P = 1 + \tfrac{1}{6}\Lambda\zeta\bar\zeta. \qquad (18.3)$$

When $\Lambda = 0$, the wave surfaces are planes (as in the case of pp-waves described in the previous chapter). For $\Lambda > 0$ they are 2-spheres, while for

---

[1] All type D pure radiation fields of this type, including a cosmological constant, have recently been found by De Groote, Van den Bergh and Wylleman (2010).

$\Lambda < 0$ the wave surfaces have constant negative curvature (they represent hyperboloidal "Lobatchevski planes").

The function τ in (18.2) is, in fact, the spin coefficient $\tau \equiv -k_{\alpha;\beta}\,m^\alpha l^\beta$ for the natural null tetrad $\boldsymbol{k} = \partial_r$, $\boldsymbol{l} = \partial_u - H\partial_r$, $\boldsymbol{m} = P\partial_{\bar\zeta} - P\bar{W}\partial_r$ that is adapted to the repeated principal null direction. Thus τ can be interpreted as a measure of the rotation of the principal null congruence about a spacelike direction (see Section 2.1.3). With the choice (18.3), the field equations lead to the following general form of the spin coefficient τ

$$\tau = \frac{-b + \tfrac{1}{3}\Lambda a\,\zeta + \tfrac{1}{6}\Lambda\bar{b}\,\zeta^2}{(1 - \tfrac{1}{6}\Lambda\zeta\bar\zeta)\,a + \bar{b}\,\zeta + b\,\bar\zeta}\,, \qquad (18.4)$$

where $a(u)$ and $b(u)$ are arbitrary real and complex functions of u, respectively. It is also convenient to consider the expression

$$k \equiv \tfrac{1}{6}\Lambda\,a^2 + b\bar{b}\,, \qquad (18.5)$$

which was first identified by Ozsváth, Robinson and Rózga (1985) as a quantity whose sign is invariant under transformations that preserve the metric. It can therefore be used to assist in the classification of Kundt's family of solutions. For each subfamily that is invariantly defined by the sign of k, the function τ can be expressed in an appropriate canonical form.

When $\Lambda = 0$:
In this case, $\tau = -b/(a + \bar{b}\zeta + b\,\bar\zeta)$ and $k = b\bar{b}$. There are obviously two geometrically distinct types of solutions, namely the cases $k = 0$ and $k > 0$, which correspond to vanishing and non-vanishing τ. Using the remaining coordinate freedom, these subclasses can be put into the canonical forms with $a = 1$, $b = 0$ and $a = 0$, $b = 1$, respectively. This identifies two types of solutions:

$$k = 0: \qquad \text{generalised } pp\text{-waves} \qquad \tau = 0,$$

$$k > 0: \qquad \text{generalised Kundt waves} \qquad \tau = -\frac{1}{\zeta + \bar\zeta}.$$

These are generally type III solutions, for which the principal null vector field $\boldsymbol{k}$ is not covariantly constant. However, for type N solutions, the subclass $k = 0$ gives exactly the pp-waves described in Chapter 17 (in which case the functions W° and G° vanish).

When $\Lambda > 0$:

Since $k = \frac{1}{6}\Lambda a^2 + b\bar{b}$ is now always positive, there exists just one canonical case. Interestingly, it is possible to use a coordinate transformation to put either $a = 1$, $b = 0$ or $a = 0$, $b = 1$. Thus, the function τ can always be transformed to either of the following two canonical forms which, for this case, are completely equivalent:

$$k > 0: \qquad \text{generalised } pp \text{ or Kundt waves} \qquad \tau = \frac{\frac{1}{3}\Lambda\zeta}{1 - \frac{1}{6}\Lambda\zeta\bar{\zeta}}$$

$$\text{or} \qquad \tau = -\frac{1 - \frac{1}{6}\Lambda\zeta^2}{\zeta + \bar{\zeta}}.$$

These forms of τ, respectively, reduce to the above two cases when $\Lambda = 0$. This family of solutions may therefore be considered to be a generalisation of either the pp-waves or the Kundt waves, in the sense that they reduce to either of these forms for type N solutions in the appropriate limit, depending on the particular coordinates adopted.

When $\Lambda < 0$:

In this case there are three distinct possibilities which are identified by the sign of k. If $k < 0$, it is always possible to put $a = 1$, $b = 0$ to obtain generalised pp-waves. Alternatively, if $k > 0$, it is possible to put $a = 0$, $b = 1$ and hence to obtain generalised Kundt waves. However, another interesting case arises here when $k = 0$. Due to (18.5), this occurs when $b = \sqrt{-\frac{1}{6}\Lambda}\, a\, e^{i\theta}$ for an arbitrary function $\theta(u)$. Such solutions generalise the type N spacetimes that have been described in detail by Siklos (1985) using a different coordinate system. It can thus be concluded that for a negative cosmological constant there exist the following three canonical subfamilies of vacuum Kundt's solutions:

$$k < 0: \quad \text{generalised } pp\text{-waves} \qquad \tau = \frac{\frac{1}{3}\Lambda\zeta}{1 - \frac{1}{6}\Lambda\zeta\bar{\zeta}},$$

$$k > 0: \quad \text{generalised Kundt waves} \quad \tau = -\frac{1 - \frac{1}{6}\Lambda\zeta^2}{\zeta + \bar{\zeta}},$$

$$k = 0: \quad \text{generalised Siklos waves} \quad \tau = -\sqrt{-\frac{1}{6}\Lambda}\left(\frac{1 + \sqrt{-\frac{1}{6}\Lambda}\,\zeta\, e^{-i\theta}}{1 + \sqrt{-\frac{1}{6}\Lambda}\,\bar{\zeta}\, e^{i\theta}}\right)e^{i\theta}.$$

It may finally be mentioned that, since $a(u)$ and $b(u)$ in (18.4) are arbitrary functions, it is possible also to construct composite space-times in which these functions are non-zero for different ranges of u.

18.3.1 Type III solutions

The complete family of Kundt space-times of algebraic type III with a cosmological constant Λ, which are empty or contain pure radiation field, can thus be described by the line element (18.1), in which the metric functions W, H and P are given by (18.2) and (18.3). The remaining field equations and coordinate freedoms then imply that it is possible to take $W^\circ(u, \zeta)$ as an arbitrary holomorphic function of ζ (in the special case $\Lambda = 0 = k$, it can also depend on $\bar{\zeta}$). With this, $H^\circ(\zeta, \bar{\zeta}, u)$ is an arbitrary real function, and $G^\circ(\zeta, \bar{\zeta}, u)$ is given by

$$G^\circ = -\tfrac{1}{2}P(\tau W^\circ + \bar{\tau}\bar{W}^\circ) - \tfrac{1}{6}\Lambda \int W^\circ \mathrm{d}\zeta.$$

The function τ takes one of the canonical forms listed above (in the case $\Lambda < 0$, $k = 0$ it is possible to set $\theta = 0$ without loss of generality).

The only non-vanishing components of the Weyl and Ricci tensors are

$$\Psi_3 = -\tau(P^2W^\circ),_\zeta - \tfrac{1}{3}\Lambda PW^\circ,$$

$$
\begin{aligned}
\Psi_4 = &\left[-P\tau(P^2W^\circ),_{\zeta\zeta} + 2(\tau\bar{\tau} - \tau P,_\zeta - \tfrac{1}{3}\Lambda)(P^2W^\circ),_\zeta + \Lambda P\bar{\tau}W^\circ\right] r \\
&+ P^2H^\circ,_{\zeta\zeta} + 2P(\bar{\tau} + P,_\zeta)H^\circ,_\zeta + 2\bar{\tau}^2H^\circ - (P^2W^\circ),_{u\zeta} \\
&+ \left[3P\tau W^\circ + P\bar{\tau}\bar{W}^\circ + \tfrac{1}{3}\Lambda(f + \bar{f})\right](P^2W^\circ),_\zeta \\
&+ 2P^2\bar{\tau}^2W^\circ\bar{W}^\circ + \Lambda P^2W^{\circ 2},
\end{aligned}
\tag{18.6}
$$

$$
\begin{aligned}
\Phi_{22} = &P^2H^\circ,_{\zeta\bar{\zeta}} + P\tau H^\circ,_\zeta + P\bar{\tau}H^\circ,_{\bar{\zeta}} + 2(\tau\bar{\tau} + \tfrac{1}{3}\Lambda)H^\circ \\
&+ P\tau\bar{W}^\circ(P^2W^\circ),_\zeta + P\bar{\tau}W^\circ(P^2\bar{W}^\circ),_{\bar{\zeta}} + P^2(2\tau\bar{\tau} + \Lambda)W^\circ\bar{W}^\circ,
\end{aligned}
$$

in which a potential function $f(u, \zeta)$ has been introduced such that $W^\circ = f,_\zeta$.

In the exceptional case $\Lambda = 0 = \tau$, it is not possible as above to set W° independent of $\bar{\zeta}$. In this case, however, the coordinate freedom can alternatively be used to set $W^\circ = W^\circ(\bar{\zeta}, u)$, and the curvature tensor components take the form

$$
\begin{aligned}
\Psi_3 &= \tfrac{1}{2}\bar{W}^\circ,_{\zeta\zeta}, \\
\Psi_4 &= \tfrac{1}{2}\bar{W}^\circ,_{\zeta\zeta} r + H^\circ,_{\zeta\zeta} - W^\circ\bar{W}^\circ,_{\zeta\zeta}, \\
\Phi_{22} &= H^\circ,_{\zeta\bar{\zeta}} - \tfrac{1}{2}(W^\circ W^\circ,_{\bar{\zeta}\bar{\zeta}} + \bar{W}^\circ\bar{W}^\circ,_{\zeta\zeta}) \\
&\quad - \tfrac{1}{2}(W^\circ,_{u\bar{\zeta}} + \bar{W}^\circ,_{u\zeta}) - \tfrac{1}{2}(W^\circ,_{\bar{\zeta}}^2 + \bar{W}^\circ,_\zeta^2).
\end{aligned}
\tag{18.7}
$$

Further details are given in Griffiths, Docherty and Podolský (2004).

18.3.2 Type N solutions

When $W^\circ = 0$, the Weyl tensor component Ψ_3 vanishes and the above solutions of Kundt's class are of algebraic type N. In this case, it is convenient to perform the transformation $r = (Q^2/P^2)\, v$ where

$$P = 1 + \tfrac{1}{6}\Lambda\zeta\bar\zeta\,, \qquad Q = (1 - \tfrac{1}{6}\Lambda\zeta\bar\zeta)\, a + \bar b\,\zeta + b\,\bar\zeta\,, \tag{18.8}$$

with $a(u)$, $b(u)$ being functions of u, as in (18.4). This puts the metric (18.1), (18.2) with $G^\circ = -\tfrac{1}{2}(\log Q)_{,u}$ and $H^\circ = (Q/2P)\,\mathcal{H}$ to the explicit form

$$\mathrm{d}s^2 = -2\,\frac{Q^2}{P^2}\,\mathrm{d}u\,\mathrm{d}v + \left(2k\,\frac{Q^2}{P^2}\,v^2 - \frac{(Q^2)_{,u}}{P^2}\,v - \frac{Q}{P}\,\mathcal{H}\right)\mathrm{d}u^2 + \frac{2\,\mathrm{d}\zeta\,\mathrm{d}\bar\zeta}{P^2}\,. \tag{18.9}$$

Here $k = \tfrac{1}{6}\Lambda\, a^2 + b\bar b$, see (18.5), and $\mathcal{H}(\zeta, \bar\zeta, u)$ is an arbitrary function.

This metric was first presented by Ozsváth, Robinson and Rózga (1985). It represents all type N Kundt space-times with a cosmological constant which are either vacuum or contain pure radiation. Indeed, with the null tetrad $\boldsymbol{k} = \partial_v$, $\boldsymbol{l} = (P^2/Q^2)\,\partial_u + (P^4/2Q^4)F\,\partial_v$, $\boldsymbol{m} = P\partial_{\bar\zeta}$ where $F = g_{uu}$, the only non-zero curvature tensor components are given by

$$\Psi_4 = \tfrac{1}{2}(P\,\mathcal{H})_{,\zeta\zeta}\,\frac{P^4}{Q^3}\,,$$

$$\Phi_{22} = \tfrac{1}{2}(P^2\mathcal{H}_{,\zeta\bar\zeta} + \tfrac{1}{3}\Lambda\,\mathcal{H})\frac{P^3}{Q^3}\,. \tag{18.10}$$

The space-times are *vacuum* when $P^2\mathcal{H}_{,\zeta\bar\zeta} + \tfrac{1}{3}\Lambda\,\mathcal{H} = 0$, which has a general solution

$$\mathcal{H}(\zeta, \bar\zeta, u) = (f_{,\zeta} + \bar f_{,\bar\zeta}) - \frac{\Lambda}{3P}(\bar\zeta f + \zeta\bar f)\,, \tag{18.11}$$

where $f(\zeta, u)$ is an arbitrary function of ζ and u, holomorphic in ζ. Moreover, these vacuum space-times are conformally flat if, and only if, the function $\mathcal{H}$ is given by

$$P\,\mathcal{H} = \mathcal{A}(u)(1 - \tfrac{1}{6}\Lambda\zeta\bar\zeta) + \bar{\mathcal{B}}(u)\zeta + \mathcal{B}(u)\bar\zeta\,, \tag{18.12}$$

with $\mathcal{A}(u)$ and $\mathcal{B}(u)$ arbitrary real and complex functions, respectively. Since this form of $\mathcal{H}$ corresponds to (18.11) with f quadratic in ζ, it follows that the above vacuum solutions with $f = c_0(u) + c_1(u)\zeta + c_2(u)\zeta^2$, where $c_i(u)$ are complex functions of u related to $\mathcal{A}$ and $\mathcal{B}$, are isometric to Minkowski (if $\Lambda = 0$), de Sitter (if $\Lambda > 0$) and anti-de Sitter space-time (if $\Lambda < 0$). For all other choices of the function $f(\zeta, u)$, the vacuum Kundt type N space-times describe non-expanding gravitational waves propagating in these maximally symmetric background spaces of constant curvature.

As described above, various geometrically distinct subclasses of such gravitational waves exist. For the different possible signs of the function k defined in (18.5), each subclass can be represented by the corresponding canonical form of the function τ, i.e. by the most natural choice of the parameters a and b, which also directly occur in the function Q introduced in (18.8).

In particular, it follows that for $\Lambda = 0$, i.e. $P = 1$, there exist two different types of Kundt type N gravitational waves propagating in flat space, namely the pp-waves (for $k = 0$) and the Kundt waves (for $k > 0$). In the first case, the canonical choice is $a = 1$, $b = 0$ so that $Q = 1$, and the metric (18.9) reduces to the standard form (17.1). For the Kundt waves, the canonical choice is $a = 0$, $b = 1$, so that $Q = \zeta + \bar{\zeta}$, see Kundt (1961).

For a positive cosmological constant $\Lambda > 0$, there is only the subclass $k > 0$ of possible solutions which generalises both pp-waves and the Kundt waves. Taking their canonical representation as $a = 0$, $b = 1$, the metric for these space-times can be written as

$$ds^2 = \frac{1}{(1 + \frac{1}{6}\Lambda\zeta\bar{\zeta})^2}\left[-2(\zeta + \bar{\zeta})^2 du\,dv + \left(2(\zeta + \bar{\zeta})^2 v^2 - \tilde{H}\right)du^2 + 2\,d\zeta\,d\bar{\zeta}\right],$$
(18.13)

where $\tilde{H}(\zeta, \bar{\zeta}, u) = (\zeta + \bar{\zeta})(1 + \frac{1}{6}\Lambda\zeta\bar{\zeta})\,\mathcal{H}$. This family was first identified by García Díaz and Plebański (1981). For vanishing cosmological constant, this immediately gives the metric for Kundt waves in Minkowski space in the form (18.9). Moreover, for any $\Lambda \neq 0$, the line element (18.13) is conformal to this form for the Kundt waves. In fact, the solutions given by the metric (18.13) represent the only non-trivial class of vacuum space-times which are conformal to Kundt waves with $\Lambda = 0$ (Bičák and Podolský, 1999a).

Interestingly, for $\Lambda < 0$, the structure of possible space-times is richer. Apart from the above solution (18.13) for which $k > 0$, there exists another family of space-times with $k < 0$ which can be described by the canonical choice $a = 1$, $b = 0$, i.e. by the functions $P = 1 + \frac{1}{6}\Lambda\zeta\bar{\zeta}$ and $Q = 1 - \frac{1}{6}\Lambda\zeta\bar{\zeta}$ in (18.9). For this case, $k = \frac{1}{6}\Lambda$, so that for vanishing cosmological constant this reduces to standard pp-waves in flat space. In addition, there also exists an exceptional subclass of solutions with $\Lambda < 0$ such that $k = 0$. In this case, the canonical choice of a and b in Q is $a = 1$, $b = \sqrt{-\frac{1}{6}\Lambda}\,e^{i\theta}$ with an arbitrary function $\theta(u)$. In the particular subcase when θ is independent of u, the principal null direction $\boldsymbol{k} = \partial_v$ is a Killing vector (indeed, the metric functions in (18.9) are then manifestly independent of v) and, using the remaining coordinate freedom, the function Q can be expressed in the factorised form $Q = \left(1 + \sqrt{-\frac{1}{6}\Lambda}\,\xi\right)\left(1 + \sqrt{-\frac{1}{6}\Lambda}\,\bar{\xi}\right)$. Such solutions are identical to the family of type N space-times which were found and investigated by Siklos (1985) using different coordinates. The Siklos metric

$$ds^2 = \frac{3}{(-\Lambda)\, x^2} \left(-2\, du\, dr + \tilde{H}\, du^2 + dx^2 + dy^2 \right), \qquad (18.14)$$

where $\tilde{H}(x, y, u) = -\frac{1}{6}\Lambda\, x\, \mathcal{H}$, is obtained by applying the coordinate transformation $\zeta = -\sqrt{\frac{6}{-\Lambda}}\, (x + \frac{1}{2} + i y)/(x - \frac{1}{2} + i y)$, $v = \frac{12}{-\Lambda}\, r$. The metric (18.14) is obviously conformal to pp-waves described by the standard line element (17.1). In fact, Siklos (1985) proved that this is the only non-trivial class of vacuum space-times with such a property.

Further physical and geometrical properties of type N vacuum and pure radiation Kundt space-times have been described by Salazar, García and Plebański (1983), Ozsváth, Robinson and Rózga (1985), Podolský (1998a), Bičák and Podolský (1999a,b), Bičák and Pravda (1998), Pravda *et al.* (2002), Podolský and Kofroň (2007), and Milson and Pelavas (2008).

18.3.3 Conformally flat solutions

The complete family of type O Kundt space-times with pure radiation are obtained by setting $\Psi_4 = 0 = \Psi_3$. These solutions can simply be written in terms of the Ozsváth–Robinson–Rózga metric (18.9), (18.8) in which, in view of (18.10), the function $\mathcal{H}$ takes the particular form

$$\mathcal{H} = \frac{\mathcal{A}(u) + \bar{\mathcal{B}}(u)\zeta + \mathcal{B}(u)\bar{\zeta} + \mathcal{C}(u)\zeta\bar{\zeta}}{1 + \frac{1}{6}\Lambda\zeta\bar{\zeta}}, \qquad (18.15)$$

where $\mathcal{A}(u)$, $\mathcal{B}(u)$ and $\mathcal{C}(u)$ are arbitrary functions of u. The pure radiation component is then

$$\Phi_{22} = \frac{1}{2} \left(\mathcal{C} + \frac{1}{6}\Lambda\, \mathcal{A} \right) \frac{P^3}{Q^3}. \qquad (18.16)$$

Thus, for vacuum space-times, $\mathcal{C} = -\frac{1}{6}\Lambda\mathcal{A}$, which leads to (18.12).

As explained in Section 18.3, several distinct invariant subclasses can be distinguished whose canonical representations are given by specific choices of the parameters a and b. For vanishing cosmological constant Λ, there are just two possibilities, namely conformally flat pp-waves ($P = 1 = Q$),[2] see Chapter 17, and conformally flat Kundt waves ($P = 1$, $Q = \zeta + \bar{\zeta}$) with pure radiation. For $\Lambda \neq 0$, there exist conformally flat generalised pp-waves, Kundt waves and Siklos waves.

In particular, conformally flat generalised Kundt waves are given by the

---

[2] These are necessarily *plane* waves which satisfy the Einstein–Maxwell field equations, see e.g. McLenaghan, Tariq and Tupper (1975).

metric (18.13) with $\tilde{H}(\zeta,\bar{\zeta},u) = (\zeta+\bar{\zeta})(\mathcal{A}+\bar{\mathcal{B}}\zeta+\mathcal{B}\bar{\zeta}+\mathcal{C}\zeta\bar{\zeta})$. In the coordinates of (18.1), which are obtained by the transformation $r = (Q^2/P^2)\,v$, this corresponds to the metric functions

$$W = -\frac{2\left(1-\frac{1}{6}\Lambda\,\bar{\zeta}^2\right)}{(\zeta+\bar{\zeta})\left(1+\frac{1}{6}\Lambda\,\zeta\bar{\zeta}\right)}\,r\,,\tag{18.17}$$

$$H = \frac{(\zeta+\bar{\zeta})\left(A(u)+\bar{B}(u)\zeta+B(u)\bar{\zeta}+C(u)\zeta\bar{\zeta}\right)}{2\left(1+\frac{1}{6}\Lambda\,\zeta\bar{\zeta}\right)^2} - \left(\frac{1+\frac{1}{6}\Lambda\,\zeta\bar{\zeta}}{\zeta+\bar{\zeta}}\right)^2 r^2\,.$$

An alternative form of this metric has been presented by Van den Bergh, Gunzig and Nardone (1990). Introducing real spatial coordinates x and y via the standard relation $\zeta = \frac{1}{\sqrt{2}}(x+\mathrm{i}\,y)$, the metric for $\Lambda = 0$ reduces to

$$\mathrm{d}s^2 = -2\,\mathrm{d}u\,\mathrm{d}r + \left(\frac{r^2}{x^2} - 2H^\circ\right)\mathrm{d}u^2 + 4\frac{r}{x}\,\mathrm{d}u\,\mathrm{d}x + \mathrm{d}x^2 + \mathrm{d}y^2\,,\tag{18.18}$$

where $2H^\circ = x\Big(\alpha(u)+\beta(u)\,x+\gamma(u)\,y+\delta(u)(x^2+y^2)\Big)$, in which $\alpha,\beta,\gamma,\delta$ are arbitrary functions of u, of which $\beta(u)$ can be removed by a suitable coordinate transformation. This metric was explicitly presented by Edgar and Ludwig (1997a,b). It contains special subcases that were previously found by Wils (1989). In fact, the solution given by (18.18) covers the complete class of conformally flat pure radiation Kundt metrics with $\Lambda = 0$ that are not pp-waves. And since the Bianchi identities imply that conformally flat pure radiation metrics must be expansion-free (and therefore necessarily belong to Kundt's class), it follows that the metrics described in this section represent the *entire* class of conformally flat space-times with pure radiation.

Some interesting properties of these space-times have been identified. In general, they contain no scalar invariants or Killing or homothetic vectors, see Koutras and McIntosh (1996) or Barnes (2001). They thus provide an exceptional case for the invariant classification of exact solutions. It was demonstrated by Skea (1997) that, to distinguish the Wils (1989) metric within the most general solution (18.18), it is necessary to go as far as, but no further than, the fourth covariant derivative of the curvature tensor. Moreover, it has been demonstrated by Koutras and McIntosh (1996), Bičák and Pravda (1998) and Pravda *et al.* (2002) that, for conformally flat pure radiation (and some other type N, III and O) Kundt space-times, *all* scalar curvature invariants constructed from the Riemann tensor and its covariant derivatives of *all orders* identically vanish.

18.4 Geometry of the wave surfaces

The solutions of Kundt's class presented above are radiative space-times in which the rays are non-expanding (as well has having zero shear and twist). The wave surfaces have constant curvature proportional to the cosmological constant. However, the fact that τ is generally non-zero indicates that subsequent wave surfaces are rotated relative to each other about a spacelike direction, see Urbantke (1979). These properties will be illustrated below.

The metric functions H and W given by (18.2), and any non-zero components of the curvature tensor (18.6), generally diverge if τ diverges. For example, type III, N or conformally flat (generalised) Kundt waves, for which $\tau = -(1 - \frac{1}{6}\Lambda\zeta^2)/(\zeta + \bar{\zeta})$, contain a curvature singularity at $\zeta + \bar{\zeta} = 0$. This is not a scalar polynomial curvature singularity. Nevertheless, some tetrad components of the Riemann tensor diverge as this singularity is approached, so that an observer approaching it would experience unbounded tidal forces. These particular non-scalar polynomial curvature singularities will be interpreted below in terms of envelopes of wave surfaces.

Apart from the quantities Λ, τ and $P = 1 + \frac{1}{6}\Lambda\zeta\bar{\zeta}$, solutions of Kundt's class (18.1), (18.2) also depend on the functions W°, G° and H°. In the weak-field limit, these functions become arbitrarily small. This identifies the background in which weak gravitational waves of this type propagate. In view of (18.6), such a background is conformally flat and vacuum, and thus is Minkowski, de Sitter or anti-de Sitter space according as the cosmological constant Λ is, respectively, zero, positive or negative. The background metric is thus given by (18.1) with

$$H = -(\tau\bar{\tau} + \tfrac{1}{6}\Lambda)\,r^2, \qquad W = \frac{2\bar{\tau}}{1 + \frac{1}{6}\Lambda\zeta\bar{\zeta}}\,r\,, \qquad (18.19)$$

where τ takes one of the canonical forms described in Section 18.3. In these background space-times, it is possible to explicitly investigate the geometry of the waves surfaces $u = u_0 = $ const. and the way in which they foliate the space-time. This will also determine the location and character of the singularities of the solutions in this weak-field limit. This approach is also relevant to the study of exact sandwich waves as it explicitly determines the geometries of the shock fronts and backs of these waves.

For this purpose, it is convenient to put the background metrics (18.19), of constant curvature $R = 4\Lambda$, into the Ozsváth–Robinson–Rózga form (18.9) with $\mathcal{H} = 0 = Q_{,u}$ by the transformation $r = (Q^2/P^2)\,v$. This yields

$$ds^2 = -2\frac{Q^2}{P^2}\,\mathrm{d}u\,\mathrm{d}v + 2k\frac{Q^2}{P^2}\,v^2\mathrm{d}u^2 + \frac{2\,\mathrm{d}\zeta\mathrm{d}\bar{\zeta}}{P^2}\,, \qquad (18.20)$$

where

$$P = 1 + \tfrac{1}{6}\Lambda\zeta\bar\zeta, \quad Q = (1 - \tfrac{1}{6}\Lambda\zeta\bar\zeta)\,a + \bar b\,\zeta + b\bar\zeta, \quad k = \tfrac{1}{6}\Lambda\,a^2 + b\bar b, \quad (18.21)$$

and a, b are constant parameters that identify the canonical subclasses.

18.4.1 Kundt waves in a Minkowski background

First consider the case in which $\Lambda = 0$. Here, space-times for which $\tau = 0$ are the pp-waves ($a = 1$, $b = 0$) that have been described in Chapter 17 and need not be further discussed here. It is therefore only necessary to consider the case of Kundt waves ($a = 0$, $b = 1$) for which $\tau = -(\zeta + \bar\zeta)^{-1} \neq 0$. Putting $\zeta = \frac{1}{\sqrt{2}}(x + \mathrm{i}\,y)$, the line element (18.20) takes the form

$$ds^2 = -4\,x^2 \mathrm{d}u\,\mathrm{d}v + 4\,x^2 v^2 \mathrm{d}u^2 + \mathrm{d}x^2 + \mathrm{d}y^2. \quad (18.22)$$

This is related to the Cartesian form of Minkowski space (3.1),

$$ds^2 = -\mathrm{d}T^2 + \mathrm{d}X^2 + \mathrm{d}Y^2 + \mathrm{d}Z^2,$$

by the transformation

$$\left.\begin{aligned} T &= x\,[v + u(1 + uv)], \\ Z &= x\,[v - u(1 + uv)], \\ X &= x(1 + 2uv), \\ Y &= y, \end{aligned}\right\} \Leftrightarrow \left\{\begin{aligned} u &= \frac{X \mp \sqrt{X^2 + Z^2 - T^2}}{T + Z}, \\ v &= \pm\frac{T + Z}{2\sqrt{X^2 + Z^2 - T^2}}, \\ x &= \pm\sqrt{X^2 + Z^2 - T^2}, \\ y &= Y. \end{aligned}\right. \quad (18.23)$$

It can be seen from (18.23) that the null wave surfaces $u = u_0 = \text{const.}$ in this flat background space are given by $(1 + u_0^2)\,T - 2u_0\,X - (1 - u_0^2)\,Z = 0$, which describes a family of null hyperplanes whose orientation varies for different values of u_0. Putting $u_0 = \tan(\alpha/2)$, where α is a constant on any of these surfaces, these can be written in the form

$$\sin\alpha\,X + \cos\alpha\,Z = T. \quad (18.24)$$

This clearly demonstrates that, at any time, successive wave surfaces $u = u_0$ are rotated about the Y-axis as u_0 increases from $-\infty$ to $+\infty$ (that is as α goes from $-\pi$ to $+\pi$). The rotation of these planes for different values of u_0 is consistent with the non-zero value of τ for these metrics.

For this family of solutions, the curvature tensor is unbounded when

$\zeta + \bar{\zeta} = 0$ (or $x = 0$ in the above notation). In the background space-time, this singularity occurs on the hypersurface

$$X^2 + Z^2 = T^2, \qquad Y \text{ arbitrary}, \qquad (18.25)$$

which is a *cylinder* centred on the Y-axis, whose radius expands with the speed of light. It may be noticed that each wave surface (18.24) touches this cylinder on the line $X = T \sin\alpha$, $Z = T \cos\alpha$. Thus, the singularity at $x = 0$ on the expanding cylinder (18.25) can be interpreted as the *caustic* formed from the envelope of the family of wave surfaces, rotated one with respect to the other along the cylinder. This is illustrated in Figure 18.1.

Obviously, two tangent plane wave surfaces (18.24) pass through each point outside the cylinder (18.25). However, this repetition must be excluded since the tangent vector to the wave surface at any point must be unique. The complete family of wave surfaces therefore has to be taken as the family of *half-planes* for which $x \geq 0$ (Griffiths and Podolský, 1998).

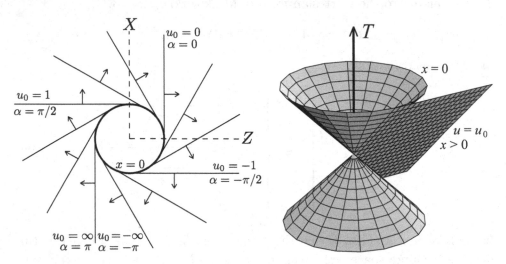

Fig. 18.1 The geometry of the Kundt wave surfaces in the Minkowski background. For constant T and Y (left), the wave surfaces $u = u_0 = $ const. are a family of half-lines at a perpendicular distance T from an origin. The envelope of the lines forms a circle corresponding to the coordinate singularity at $x = 0$. As T increases, the circle expands and each wave surface propagates perpendicular to the tangent to that surface. The half-plane wave surfaces are tangent to a two-dimensional null cone on which $x = 0$ (right). The picture illustrates the expanding singularity and a single null wave surface in a section with constant Y.

No wave surfaces pass through points that are inside the expanding cylinder (18.25). In fact, the coordinates used in (18.22) do not cover this part of the space-time. However this is not necessary as, in the general Kundt curved space-time, the expanding cylinder is a (non-scalar) curvature "envelope" singularity through which the space-time cannot be physically extended. For further details see Podolský and Belán (2004).

It may also be observed from Figure 18.1 that any observer outside the envelope singularity would intersect a sequence of wave surfaces whose direction of propagation will rotate until it is reached by the singularity itself.

Finally, it should be pointed out that the complete solution is time-symmetric, so that the envelope of null wave surfaces is a cylinder whose radius decreases to zero at the speed of light and then increases.

18.4.2 Kundt waves in an (anti-)de Sitter background

These results can be extended to include a non-zero cosmological constant Λ. In this case, the line element of generalised Kundt waves has the canonical form with $a = 0$, $b = 1$. The corresponding de Sitter or anti-de Sitter background metric is given by (18.20) with $P = 1 + \frac{1}{6}\Lambda\zeta\bar{\zeta}$, $Q = \zeta + \bar{\zeta}$, and $k = 1$. In real spatial coordinates, this reads

$$ds^2 = \frac{4x^2}{P^2}\left(-\,du\,dv + v^2\,du^2\right) + \frac{1}{P^2}\left(dx^2 + dy^2\right), \tag{18.26}$$

where $P = 1 + \frac{1}{12}\Lambda(x^2 + y^2)$. As explained in Chapters 4 and 5, the (anti-) de Sitter space can be represented as a four-dimensional hyperboloid

$$-Z_0{}^2 + Z_1{}^2 + Z_2{}^2 + Z_3{}^2 + \varepsilon Z_4{}^2 = \varepsilon a_\Lambda^2, \tag{18.27}$$

embedded in a five-dimensional flat space

$$ds^2 = -dZ_0{}^2 + dZ_1{}^2 + dZ_2{}^2 + dZ_3{}^2 + \varepsilon dZ_4{}^2, \tag{18.28}$$

where $a_\Lambda = \sqrt{3/|\Lambda|}$, $\varepsilon = 1$ for a de Sitter background ($\Lambda > 0$), and $\varepsilon = -1$ for an anti-de Sitter background ($\Lambda < 0$). The two forms of the metric (18.26) and (18.28) are related by the transformation

$$\left.\begin{aligned}
Z_0 &= \frac{x}{P}\left[v + u(1 + uv)\right],\\[4pt]
Z_1 &= \frac{x}{P}\left[v - u(1 + uv)\right],\\[4pt]
Z_2 &= \frac{x}{P}(1 + 2uv),\\[4pt]
Z_3 &= \frac{y}{P},\\[4pt]
Z_4 &= a_\Lambda\frac{2 - P}{P},
\end{aligned}\right\} \Leftrightarrow \left\{\begin{aligned}
u &= \frac{Z_2 \mp \sqrt{Z_1{}^2 + Z_2{}^2 - Z_0{}^2}}{Z_0 + Z_1},\\[4pt]
v &= \pm\frac{Z_0 + Z_1}{2\sqrt{Z_1{}^2 + Z_2{}^2 - Z_0{}^2}},\\[4pt]
x &= \pm\frac{2a_\Lambda\sqrt{Z_1{}^2 + Z_2{}^2 - Z_0{}^2}}{a_\Lambda + Z_4},\\[4pt]
y &= \frac{2a_\Lambda Z_3}{a_\Lambda + Z_4}.
\end{aligned}\right. \tag{18.29}$$

Obviously, the singularity which occurs at $\zeta + \bar{\zeta} = 0$ (or $x = 0$) is located on

$$Z_1{}^2 + Z_2{}^2 = Z_0{}^2 \qquad \text{and} \qquad Z_3{}^2 + \varepsilon Z_4{}^2 = \varepsilon a_\Lambda^2 . \tag{18.30}$$

For the de Sitter background (for which $\varepsilon = 1$), this is an *expanding torus*. The sections in the Z_1, Z_2 plane are circles which are expanding at the speed of light with the time coordinate Z_0, and sections in the Z_3, Z_4 plane are circles of constant circumference $2\pi a$ that may be considered to be the circumference of the closed de Sitter universe.

For the anti-de Sitter background ($\varepsilon = -1$), however, the singularity is located on an *expanding hyperboloid* in which the sections in the Z_1, Z_2 plane are circles which are expanding at the speed of light with the time coordinate Z_0, and sections in the Z_3, Z_4 plane are hyperbolae.

For both of these cases, it can be seen that the null wave surfaces $u = u_0$ are given in this five-dimensional representation by the intersection of the hyperboloid (18.27) with the plane $(1 + u_0^2)Z_0 - (1 - u_0^2)Z_1 - 2\,u_0\,Z_2 = 0$. Putting $\sin\alpha = \frac{2\,u_0}{1 + u_0^2}$, $\cos\alpha = \frac{1 - u_0^2}{1 + u_0^2}$, this becomes

$$\cos\alpha\, Z_1 + \sin\alpha\, Z_2 = Z_0 . \tag{18.31}$$

This describes a family of planes which rotate relative to each other in the Z_1, Z_2 plane. They cut the four-dimensional hyperboloid at

$$(\sin\alpha\, Z_1 - \cos\alpha\, Z_2)^2 + Z_3{}^2 + \varepsilon Z_4{}^2 = \varepsilon a_\Lambda^2 . \tag{18.32}$$

For the de Sitter background ($\varepsilon = 1$), these intersections representing the null wave surfaces are a family of spheres with constant area $4\pi a_\Lambda^2$. Moreover, the plane cuts are clearly tangent to the expanding torus, so that the singularity can again be interpreted as a caustic formed from the envelope of wave surfaces. Also, since two spheres pass through each point within the region covered by these coordinates, it is appropriate to restrict the family of wave surfaces to the hemispheres on which $x \geq 0$ whose boundary is located on the expanding torus. (The situation is obviously the positive curvature equivalent of the half-plane wave surfaces for the Kundt waves in a Minkowski background.) This is illustrated in Figure 18.2. (Notice that the view along the axis corresponds to the first picture in Figure 18.1.)

For the anti-de Sitter background ($\varepsilon = -1$), the null wave surfaces are hyperboloidal. They are tangent to the singularity which is an expanding hyperboloid that can again be interpreted as an envelope of wave surfaces. Again, since it is only possible for one wave surface to pass through any point, it is appropriate to take the wave surfaces as the family of semi-infinite hyperboloids on which $x \geq 0$. For this case with $\Lambda < 0$, an additional apparent

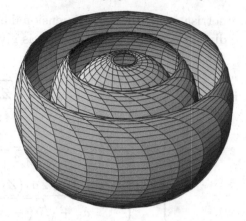

Fig. 18.2 Portions of the de Sitter universe covered by the coordinates of (18.26) for three values of the time coordinate Z_0. Using Z_1, Z_2, Z_3 coordinates (Z_3 being the symmetry axis) with $Z_4 = 0$, the expanding 3-sphere of the universe reduces to an expanding 2-sphere. The null wave surfaces $u = u_0$ are represented by semicircles of constant radius a which are all tangent to the two expanding circles at $Z_3 = \pm a_\Lambda$, which are sections of the expanding torus corresponding to the singularity $x = 0$. The wave surfaces cover a decreasing portion of each successive 2-sphere.

singularity occurs when $P = 0$, i.e. when $x^2 + y^2 = 12/|\Lambda|$. However, it can easily be seen that this simply corresponds to anti-de Sitter infinity.

Just as for the Kundt waves in a Minkowski background, it may finally be observed that the above wave surfaces in de Sitter and anti-de Sitter backgrounds only exist outside the expanding singular torus or hyperboloid, respectively. They therefore foliate a decreasing portion of the complete background for increasing time coordinate Z_0.

18.4.3 "pp-waves" in an (anti-)de Sitter background

In the case of generalised *pp*-waves, the canonical choice is $a = 1$, $b = 0$, so that $\tau = \frac{1}{3}\Lambda\zeta/(1 - \frac{1}{6}\Lambda\zeta\bar{\zeta})$. For $\Lambda = 0$, these reduce to the *pp*-waves with vanishing τ. The metric of the corresponding (anti-)de Sitter background takes the form (18.20) with $P = 1 + \frac{1}{6}\Lambda\zeta\bar{\zeta}$, $Q = 1 - \frac{1}{6}\Lambda\zeta\bar{\zeta}$, $k = \frac{1}{6}\Lambda$, i.e.

$$ds^2 = 2\frac{Q^2}{P^2}\left(-\,du\,dv + \tfrac{1}{6}\Lambda\,v^2\,du^2\right) + \frac{2}{P^2}\,d\zeta\,d\bar{\zeta}. \qquad (18.33)$$

The appropriate parametrisation of the four-dimensional hyperboloid (18.27) embedded in the five-dimensional flat space (18.28) is given by

$$
\left.
\begin{aligned}
Z_0 &= \frac{Q}{\sqrt{2}\,P}(v + u + \tfrac{1}{6}\Lambda u^2 v),\\[4pt]
Z_1 &= \frac{Q}{\sqrt{2}\,P}(v - u - \tfrac{1}{6}\Lambda u^2 v),\\[4pt]
Z_2 + \mathrm{i}Z_3 &= \sqrt{2}\,\frac{\zeta}{P},\\[4pt]
Z_4 &= a_\Lambda \frac{Q}{P}(1 + \tfrac{1}{3}\Lambda uv),
\end{aligned}
\right\}
\Leftrightarrow
\left\{
\begin{aligned}
\frac{u}{\sqrt{2}} &= a_\Lambda \frac{Z_4 - \sqrt{Z_4{}^2 + \varepsilon(Z_1{}^2 - Z_0{}^2)}}{\varepsilon(Z_0 + Z_1)},\\[4pt]
\sqrt{2}\,v &= a_\Lambda \frac{Z_0 + Z_1}{\sqrt{Z_4{}^2 + \varepsilon(Z_1{}^2 - Z_0{}^2)}},\\[4pt]
\zeta &= \frac{\sqrt{2}\,a_\Lambda(Z_2 + \mathrm{i}Z_3)}{a_\Lambda + \sqrt{a_\Lambda^2 - \varepsilon(Z_2{}^2 + Z_3{}^2)}}.
\end{aligned}
\right.
$$

$$(18.34)$$

When $\Lambda > 0$ ($\varepsilon = 1$), the null wave surfaces $u = u_0$, are identical to those for the Kundt waves with $\Lambda > 0$, as given in (18.29) but with the roles of Z_2 and Z_4 interchanged. (For this de Sitter background, Z_2 and Z_4 are both spacelike coordinates and their interchange is trivial.) This result is consistent with the fact that these two cases are equivalent for a positive cosmological constant. The wave surfaces are a family of hemispheres of common constant area $4\pi a_\Lambda^2$, as illustrated in Figure 18.2. These hemispheres are tangent to the expanding torus $Z_2{}^2 + Z_3{}^2 = a_\Lambda^2$ and $Z_1{}^2 + Z_4{}^2 = Z_0{}^2$, which here corresponds to the coordinate singularity $1 - \tfrac{1}{6}\Lambda\zeta\bar{\zeta} = 0$. This is the weak-field limit of the curvature singularity that forms the boundary of the general space-time.

For the case in which $\Lambda < 0$, the null wave surfaces $u = u_0$ are the intersections of the hyperboloid (18.27) with $\varepsilon = -1$ with the hyperplane $(1 - (u_0^2/2a_\Lambda^2))\,Z_0 - (1 + (u_0^2/2a_\Lambda^2))\,Z_1 - (\sqrt{2}\,u_0/a_\Lambda)\,Z_4 = 0$. Now, putting $u_0 = \sqrt{2}\,a_\Lambda \tan(\alpha/2)$, this is given by

$$Z_1 - \cos\alpha\,Z_0 + \sin\alpha\,Z_4 = 0. \tag{18.35}$$

This corresponds to a family of planes which are rotated relative to each other in the section of the timelike coordinates (Z_0, Z_4). In this case, there is no singularity as the term $a_\Lambda^2 + Z_2{}^2 + Z_3{}^2$ cannot be zero. The apparent coordinate singularity that occurs in the metric when $P = 1 + \tfrac{1}{6}\Lambda\zeta\bar{\zeta} = 0$ again corresponds to anti-de Sitter infinity. These space-times are generalisations of the *pp*-waves and, for $\Lambda < 0$, the background space-time is open and the wave surfaces are hyperboloids which foliate the entire anti-de Sitter universe.

18.4.4 Siklos waves in an anti-de Sitter background

Finally, the special solutions for which $\Lambda < 0$ and $k = 0$ have the canonical form $Q = \left(1 + \sqrt{-\frac{1}{6}\Lambda}\,\zeta\right)\left(1 + \sqrt{-\frac{1}{6}\Lambda}\,\bar{\zeta}\right)$. The corresponding anti-de Sitter background space (18.20) is

$$\mathrm{d}s^2 = -2\frac{Q^2}{P^2}\,\mathrm{d}u\,\mathrm{d}v + \frac{2\,\mathrm{d}\zeta\,\mathrm{d}\bar{\zeta}}{P^2}. \tag{18.36}$$

In fact, this metric can be represented as the hyperboloid (18.27) with $\varepsilon = -1$ embedded in a five-dimensional flat space-time (18.28) using the parametrisation

$$
\left.
\begin{aligned}
Z_0 &= \frac{(v+u)}{\sqrt{2}}\frac{Q}{P}, \\
Z_1 &= \frac{(v-u)}{\sqrt{2}}\frac{Q}{P}, \\
Z_2 &= \frac{uv}{a_\Lambda}\frac{Q}{P} + \frac{(\zeta+\bar{\zeta})}{\sqrt{2}\,P}, \\
Z_3 &= -\frac{\mathrm{i}(\zeta-\bar{\zeta})}{\sqrt{2}\,P}, \\
Z_4 &= \left(a_\Lambda - \frac{uv}{a_\Lambda}\right)\frac{Q}{P} - \frac{(\zeta+\bar{\zeta})}{\sqrt{2}\,P},
\end{aligned}
\right\}
\Leftrightarrow
\left\{
\begin{aligned}
u &= \frac{a_\Lambda}{\sqrt{2}}\frac{(Z_0 - Z_1)}{(Z_2 + Z_4)}, \\
v &= \frac{a_\Lambda}{\sqrt{2}}\frac{(Z_0 + Z_1)}{(Z_2 + Z_4)}, \\
\zeta &= \sqrt{2}\,a_\Lambda \frac{(Z_2 + \mathrm{i}Z_3 + Z_4)^2 - a_\Lambda^2}{(a_\Lambda + Z_2 + Z_4)^2 + Z_3^2}.
\end{aligned}
\right.
\tag{18.37}
$$

The null wave surfaces $u = u_0 = $ const. are located on the intersection of the hyperboloid with the null hyperplane

$$Z_0 - Z_1 = \frac{\sqrt{2}\,u_0}{a_\Lambda}(Z_2 + Z_4). \tag{18.38}$$

The complete family of these hyperboloidal wave surfaces foliate the entire background space-time and, again, the apparent singularity at $P = 0$ simply corresponds to anti-de Sitter infinity (see Podolský, 2001).

Note also that an equivalent unified form of the five-dimensional parametrisations (18.29), (18.34) and (18.37) of the (anti-)de Sitter space by the general Ozsváth–Robinson–Rózga metric (18.20) has been given by Tran, as quoted in Barrabès and Hogan (2007).

18.5 "Spinning" null matter and "gyratons"

A physically interesting example of a type III Kundt solution which contains some kind of "spinning" null fluid was given by Bonnor (1970b). In many ways, this is like the beam of pure radiation that was described at the end of Section 17.3, but in this case the energy-momentum tensor has an extra non-diagonal term.

Taking the metric (18.1) with (18.2) and $\zeta = \frac{1}{\sqrt{2}}\rho e^{i\phi}$, one particularly interesting and simple example[3] of this type is given by the axially symmetric solution in which $\Lambda = 0$, $\tau = 0$, $W^\circ = -\frac{i}{\sqrt{2}}J(u,\rho)e^{-i\phi}/\rho$, $G^\circ = 0$, and $H^\circ = H(u,\rho)$. In this case, the metric takes the form

$$ds^2 = -2\,du\,dr - 2H(u,\rho)\,du^2 - 2J(u,\rho)\,du\,d\phi + d\rho^2 + \rho^2\,d\phi^2. \quad (18.39)$$

Using the null tetrad $\boldsymbol{k} = \partial_r$, $\boldsymbol{l} = \partial_u - (H + \frac{1}{2}J^2/\rho^2)\partial_r + (J/\rho^2)\partial_\phi$ and $\boldsymbol{m} = \frac{1}{\sqrt{2}}(\partial_\rho + (i/\rho)\partial_\phi)$, the non-zero curvature tensor components for this metric are given by

$$\Psi_4 = \frac{1}{2}\left(H_{,\rho\rho} - \frac{1}{\rho}H_{,\rho}\right) + \frac{i}{2\rho}\left(J_{,u\rho} - \frac{2}{\rho}J_{,u}\right) + \frac{J}{2\rho^2}\left(J_{,\rho\rho} - \frac{1}{\rho}J_{,\rho}\right),$$

$$\Psi_3 = \frac{i}{4\sqrt{2}\,\rho}\left(J_{,\rho\rho} - \frac{1}{\rho}J_{,\rho}\right),$$

$$\Phi_{22} = \frac{1}{2}\left(H_{,\rho\rho} + \frac{1}{\rho}H_{,\rho}\right) + \frac{J}{2\rho^2}\left(J_{,\rho\rho} - \frac{1}{\rho}J_{,\rho}\right) + \frac{1}{4\rho^2}J_{,\rho}{}^2,$$

$$\Phi_{12} = \frac{i}{4\sqrt{2}\,\rho}\left(J_{,\rho\rho} - \frac{1}{\rho}J_{,\rho}\right).$$

For this family of solutions, the repeated principal null direction $\boldsymbol{k}$ is covariantly constant. They therefore belong to the class of pp-wave spacetimes, even though their metric cannot be written in the simple form (17.1). It thus follows that the null congruence aligned with $\boldsymbol{k}$ is geodesic, non-expanding, shear-free and twist-free everywhere. The geodesic congruence tangent to $\boldsymbol{k}$ is therefore *not rotating*.

The interpretation of this solution as representing some kind of spinning source arises purely from the non-diagonal structure of the energy-momentum tensor (arising from the non-zero Φ_{12} component), as the integrated angular momentum evaluated from the energy-momentum tensor is non-zero. This was fully emphasised by Bonnor (1970b). However, it is not clear what kind of null matter is represented by this solution, or in what sense it can be considered as "spinning". It cannot represent a purely electromagnetic field but, in some cases, the source can be interpreted as a massless neutrino field (Griffiths, 1972). This would indicate that it may, in these cases, be considered to possess some kind of intrinsic spin.

Note that Φ_{12} and Ψ_3 *both* vanish when $J = \kappa(u)\rho^2 + \lambda(u)$, where κ and λ

---

[3] It may be noted that all the different subclasses of Kundt wave solutions, summarised in Section 18.3, can be generalised in a very similar way, although only this simple case will be described here.

are arbitrary functions of retarded time. With this choice, the solutions are standard pp-waves, and it is possible to use the transformation $r = \bar{r} - \lambda\bar{\phi}$, $\phi = \bar{\phi} + \int \kappa\,du$ to put the metric into the canonical form (17.1) in which H is replaced by $\bar{H} = H + \frac{1}{2}\kappa^2\rho^2 + \kappa\lambda - \lambda_{,u}\bar{\phi}$. Clearly, this transformation can always be used to remove the component $\kappa(u)\rho^2$ from J. However, although it is also possible to remove the component $\lambda(u)$ *locally*, it is not always appropriate to do this – the introduced $\bar{\phi}$-dependence of the function $\bar{H}$ when λ is not a constant would lead to a discontinuity in the metric which arises from the identification of $\bar{\phi} = 2\pi$ with $\bar{\phi} = 0$. Retaining this term, the remaining vacuum field equation in this case is

$$H_{,\rho\rho} + \frac{1}{\rho}H_{,\rho} = 0.$$

Hence, the general solution in any vacuum region may be given in the form

$$J = \lambda(u), \qquad H = 2\mu(u)\log\rho + \nu(u), \qquad (18.40)$$

where μ and ν are additional arbitrary functions, of which $\mu(u)$ represents the profile of a pure gravitational pp-wave and $\nu(u)$ has no physical significance. In this region

$$\Psi_4 = -\frac{2\mu + \mathrm{i}\,\lambda_{,u}}{\rho^2}. \qquad (18.41)$$

It is now appropriate to investigate the properties of space-times of this type in which the null matter is contained within some cylindrical beam ($\rho \leq a$ say), with the remaining space being vacuum. Among many possibilities, one simple case that is regular everywhere has been given by Bonnor (1970b) in which

$$\rho \leq a : \quad \begin{cases} J = \lambda\rho^2\,(3a - 2\rho)/a^3, \\ H = \mu\,\rho^2/a^2, \end{cases}$$

$$\rho > a : \quad \begin{cases} J = \lambda, \\ H = \mu(1 + 2\log(\rho/a)), \end{cases}$$

where λ and μ are arbitrary functions of u. Within the beam, the non-zero components of the curvature tensor are

$$\Phi_{22} = 2\mu/a^2 + 3\lambda^2(5\rho^2 - 9\rho a + 3a^2)/a^6,$$

$$\Phi_{12} = \Psi_3 = -3\mathrm{i}\lambda/2\sqrt{2}a^3,$$

$$\Psi_4 = -\mathrm{i}\lambda_{,u}\rho/a^3 + 3\lambda^2\rho(2\rho - 3a)/a^6.$$

It is possible to satisfy the condition that $\Phi_{22} \geq 0$ everywhere by choosing $\mu > 63\lambda^2/40a^2$, but this particular example does not admit a neutrino interpretation. In the external region for which $\rho > a$, the only non-zero curvature tensor component is given by (18.41).

Of course, there also exists a similar class of solutions for which $J = 0$ outside the beam. In addition, for any exterior solution, an interior massless neutrino solution always exists.

Since the u-dependence of the metric functions in solutions of this type is arbitrary, it is possible to consider these to be non-zero only for some finite interval of retarded time. For example, Bonnor (1970b) suggested that the profile functions λ and μ above could be proportional to $(b^2 - u^2)^4$ for $|u| \leq b$ and zero otherwise. The resulting solution thus describes the interior and exterior field of a finite region containing a finite pulse of some kind of null matter that Bonnor called a "spinning nullicon".

The exterior field in such cases is a sandwich wave which is locally isomorphic to a pp-wave. And it is important to emphasise that these are locally indistinguishable from type N sandwich pp-waves except within the interior non-vacuum region. The exterior field contains no information about the "rotating" character of the interior matter.

Recently, these solutions have been investigated in greater detail by Frolov and his collaborators who have called them "gyratons". By considering solutions in the linear approximation and their electromagnetic analogue, Frolov and Fursaev (2005) have demonstrated that they represent circularly polarised radiation, and that the interior region contains matter with energy and internal angular momentum. They have further considered the limit of these solutions in which the size of the source becomes infinitesimally small, and have permitted the two profile functions $\mu(u)$ and $\lambda(u)$ to be independent. They have also investigated the geodesic motion of test particles in the exterior field and have shown that, if the beam passes through the centre of a ring of test particles, then those particles appear to rotate about the centre during the period of the pulse. However, a cloud of test particles must rotate about the beam in a way that is locally irrotational, and this explains much of the dynamics described. Frolov, Israel and Zelnikov (2005) and Frolov and Zelnikov (2006) have presented further details and generalisations of these gyraton solutions. The inclusion of a cosmological constant has been given by Frolov and Zelnikov (2005).

It can be seen from (18.41) that, when $\lambda(u)$ contains a step, the gravitational wave component Ψ_4 contains an impulse. Impulsive limits of the profile functions $\mu(u)$ and $\lambda(u)$ can also be obtained. These are special cases of the family of non-expanding impulsive waves that will be described in more detail in Chapter 20.

18.6 Type D (electro)vacuum solutions

All type D vacuum solutions of Kundt's class were found and classified by Kinnersley (1969a). More general space-times which also admit a non-vanishing cosmological constant Λ and an aligned electromagnetic field were given by Carter (1968b) and Plebański (1975).

As described in Section 16.4, and summarised in Figure 16.2, this family of type D non-expanding space-times of Kundt's class can be obtained from the metric of Plebański and Demiański (1976) (see Griffiths and Podolský, 2006b, for more details). Specifically, the metric (16.27) can be explicitly written in the standard form of the Kundt solutions using the transformation

$$z = p, \qquad y = \psi + 2\gamma \int \frac{q}{\widetilde{Q}}\,\mathrm{d}q\,, \qquad u = t - \int \frac{\mathrm{d}q}{\widetilde{Q}}\,, \qquad r = \varrho^2 q\,.$$

Putting

$$\sqrt{2}\,\zeta = x + \mathrm{i}\,y\,, \qquad x = \int P^2(z)\,\mathrm{d}z\,, \qquad \text{where} \quad P^2 = \frac{\varrho^2}{\mathcal{P}}\,, \tag{18.42}$$

this leads to the metric (18.1),

$$\mathrm{d}s^2 = -2\mathrm{d}u\Big(\mathrm{d}r + H\,\mathrm{d}u + W\,\mathrm{d}\zeta + \bar{W}\,\mathrm{d}\bar{\zeta}\Big) + 2\,P^{-2}\mathrm{d}\zeta\mathrm{d}\bar{\zeta}\,,$$

with

$$P^2 = \frac{z^2 + \gamma^2}{-(e^2 + g^2 + \epsilon_2\gamma^2 - \Lambda\gamma^4) + 2nz + (\epsilon_2 - 2\Lambda\gamma^2)z^2 - \frac{1}{3}\Lambda z^4}\,,$$

$$W = -\frac{\sqrt{2}}{(z + \mathrm{i}\,\gamma)\,P^2}\,r\,, \tag{18.43}$$

$$H = \tfrac{1}{2}\epsilon_0(z^2 + \gamma^2) - \left[\frac{\epsilon_2}{2(z^2 + \gamma^2)} + \frac{2\gamma^2}{(z^2 + \gamma^2)^2\,P^2}\right]r^2\,,$$

in which z is to be expressed in terms of $\zeta + \bar{\zeta}$ using (18.42). This is a generalisation of the solution given as equations (31.41) and (31.58) in Stephani *et al.* (2003) which now includes a cosmological constant Λ.

This family of solutions is characterised by a discrete parameter $\epsilon_2 = \pm 1, 0$ and six continuous parameters. The cosmological constant is Λ, while e and g are electric and magnetic charge parameters. The physical meanings of the remaining parameters γ, n and ϵ_0 have not yet been identified. However, it can be noted that the parameter γ is the analogue of the NUT parameter in these space-times. Specifically, if $\gamma \neq 0$, such solutions have no curvature singularities.

Among the large family of space-times described above, the simplest vacuum solutions are known as the *B-metrics* following the classification of

Ehlers and Kundt (1962). These occur for the line element (18.43) when γ, e, g and Λ all vanish with $n \neq 0$, and ϵ_0, ϵ_2 are chosen appropriately to be ± 1 or 0. Such metrics have been described in Section 16.4.1.

When γ, ϵ_2 and Λ vanish, the parameter ϵ_0 can be removed by a coordinate transformation $r \to r - \frac{1}{2}\epsilon_0 z^2 u$. The metric (18.43) then reduces to the Melvin solution (7.28) that was described in Section 7.3, in which $B^2 = e^2 + g^2 = 2n$.

Interestingly, some Kundt space-times of type D can also be written in a simple form which is a slight generalisation of the Ozsváth–Robinson–Rózga metric (18.9), namely

$$ds^2 = -2\frac{Q^2}{P^2}\,du\,dv + \left(D\frac{Q^2}{P^2}v^2 - \frac{(Q^2)_{,u}}{P^2}v - \frac{Q}{P}\mathcal{H}\right)du^2 + \frac{2\,d\zeta d\bar\zeta}{P^2}\,, \quad (18.44)$$

with

$$P = 1 + \alpha\,\zeta\bar\zeta\,,$$
$$Q = (1 + \beta\,\zeta\bar\zeta)\,\varepsilon + C\,\zeta + \bar{C}\,\bar\zeta\,, \quad (18.45)$$

and

$$\mathcal{H} = \frac{\mathcal{A}(u) + \bar{\mathcal{B}}(u)\zeta + \mathcal{B}(u)\bar\zeta + \mathcal{C}(u)\zeta\bar\zeta}{1 + \alpha\,\zeta\bar\zeta}\,. \quad (18.46)$$

Here α, β and ε are real constants (without loss of generality it can be assumed that $\varepsilon = 0$ or $\varepsilon = 1$), $C(u)$ and $D(u)$ are arbitrary complex and real functions of the coordinate u, respectively, and $\mathcal{A}(u)$, $\mathcal{B}(u)$ and $\mathcal{C}(u)$ are also arbitrary functions of u ($\mathcal{A}$ and $\mathcal{C}$ are real, $\mathcal{B}$ is complex). Using the null tetrad introduced in Section 18.3.2, the only non-zero component of the Weyl tensor and the Ricci scalar are

$$\Psi_2 = -\tfrac{1}{6}\left(D + 2\varepsilon\beta - 2C\bar{C}\right)\frac{P^2}{Q^2}\,, \quad (18.47)$$

$$R = 24\alpha - 12\varepsilon(\alpha + \beta)\frac{P}{Q} + 2\left(D + 2\varepsilon\beta - 2C\bar{C}\right)\frac{P^2}{Q^2}\,, \quad (18.48)$$

so that the space-times are indeed of algebraic type D (or conformally flat). The Ricci tensor components Φ_{11} and Φ_{22} are in general non-trivial.

For the special choice $\alpha = \beta$, $\varepsilon = 1$, $C = 0$, $\mathcal{H} = 0$ and $D = 2(\Lambda - \alpha)$, the metric (18.44) reduces to

$$ds^2 = -2\,du\,dv + 2(\Lambda - \alpha)\,v^2 du^2 + \frac{2\,d\zeta d\bar\zeta}{(1 + \alpha\,\zeta\bar\zeta)^2}\,. \quad (18.49)$$

In this case $\Psi_2 = -\tfrac{1}{3}\Lambda$, $R = 4\Lambda$, $\Phi_{11} = \alpha - \tfrac{1}{2}\Lambda$ and $\Phi_{22} = 0$. These are exactly the (electro)vacuum space-times with the geometry of a direct product of two constant-curvature 2-spaces described in Section 7.2. In particular, the Nariai or anti-Nariai vacuum universe arises when $\alpha = \frac{1}{2}\Lambda$, while the exceptional Plebański–Hacyan electrovacuum space-time occurs when $\alpha = 0$. For further details and references, see Sections 7.2.1, 7.2.2 or Ortaggio and Podolský (2002).

There also exist more general solutions which include pure radiation in the Kundt type D space-times, such as those described by Wils and Van den Bergh (1990) or Podolský and Ortaggio (2003). However, these pure radiation solutions cannot be Einstein–Maxwell null fields, neutrino or massless scalar fields.

18.7 Type II solutions

Some particular solutions of Kundt's class which are of algebraic type II are known. The physically most interesting are those which represent non-expanding gravitational waves which propagate on the specific type D backgrounds described at the end of the previous section. Indeed, by considering the metric (18.44), (18.45) with a completely *general* function $\mathcal{H}(u, \zeta, \bar{\zeta})$, exact space-times of this type are obtained. The only non-vanishing components of the curvature tensor are then

$$\Psi_2 = -\tfrac{1}{6}\left(D + 2\varepsilon\beta - 2C\bar{C}\right)\frac{P^2}{Q^2},$$

$$\Psi_4 = \tfrac{1}{2}\left(P\mathcal{H}\right)_{,\zeta\zeta}\frac{P^4}{Q^3},$$

$$R = 24\alpha - 12\varepsilon(\alpha + \beta)\frac{P}{Q} + 2\left(D + 2\varepsilon\beta - 2C\bar{C}\right)\frac{P^2}{Q^2}, \qquad (18.50)$$

$$\Phi_{11} = \tfrac{1}{2}\varepsilon(\alpha + \beta)\frac{P}{Q} - \tfrac{1}{4}\left(D + 2\varepsilon\beta - 2C\bar{C}\right)\frac{P^2}{Q^2},$$

$$\Phi_{22} = \tfrac{1}{2}\left\{\left[\varepsilon(\alpha - \beta)(1 - \alpha\,\zeta\bar{\zeta}) + 2\alpha\left(C\,\zeta + \bar{C}\,\bar{\zeta}\right)\right]\mathcal{H} + P^2 Q\,\mathcal{H}_{,\zeta\bar{\zeta}}\right.$$
$$\left. - 2\varepsilon(\alpha + \beta)\,v\,(\dot{C}\,\zeta + \dot{\bar{C}}\,\bar{\zeta})\right\}\frac{P^3}{Q^4}.$$

If $\mathcal{H} = 0$ (more generally, if $\mathcal{H}$ is of the form (18.46)) then $\Psi_4 = 0$, and the space-time is of type D, for example the (anti-)Nariai vacuum universe or the Plebański–Hacyan electrovacuum universe. For non-trivial $\mathcal{H}(u, \zeta, \bar{\zeta})$, the solution (18.44), (18.45) describes type II gravitational (and possibly pure radiation or electromagnetic) waves of arbitrary profiles in these background universes. The wave surfaces $u = $ const. are 2-spaces of a constant curvature equal to 2α.

If, in addition, the function D takes the special form $D = -2\varepsilon\beta + 2C\bar{C}$, the scalar Ψ_2 vanishes and type N waves propagating in conformally flat space-times are obtained. In particular, for $\alpha = -\beta = \frac{1}{6}\Lambda$, the pure radiation or vacuum waves described in Section 18.3.2 are recovered yielding the Ozsváth–Robinson–Rózga metric (18.9), (18.8) with the identification $a = \varepsilon$, $b = \bar{C}$, $2k = D$.

More details concerning these Kundt gravitational waves of type II, including the classes of solutions that were investigated by Lewandowski (1992) and García D. and Alvarez (1984), can be found in Podolský and Ortaggio (2003). In fact, all such metrics representing pure gravitational waves in electromagnetic universes are specific subcases of a solution presented by Khlebnikov (1986).

19

Robinson–Trautman solutions

Another large and physically important family of solutions was introduced and initially investigated by Robinson and Trautman (1960, 1962). These are defined geometrically by the property that they admit a geodesic, shear-free, twist-free but expanding null congruence. They thus differ from those of Kundt's class, that were described in the previous chapter, by having non-zero expansion.

According to the Goldberg–Sachs theorem and its generalisations, the existence of a shear-free null geodesic congruence for vacuum, aligned pure radiation, or an aligned non-null or null electromagnetic field, implies that the Robinson–Trautman space-times have to be algebraically special, that is of Petrov types II, D, III, N, or conformally flat. In all these cases a cosmological constant Λ can easily be included.

Apart from various classes of radiative space-times, this family also includes the Schwarzschild and Reissner–Nordström black holes, the C-metric which represents accelerating black holes, the Vaidya solution with pure radiation, photon rockets and their non-rotating generalisations.

19.1 Robinson–Trautman vacuum space-times

As shown in the original works of Robinson and Trautman (1960, 1962) (see also Newman and Tamburino, 1962, and Stephani *et al.*, 2003), the general metric for a vacuum space-time with the geometric properties defined above can be written in the form

$$\mathrm{d}s^2 = -2\,\mathrm{d}u\,\mathrm{d}r - 2\,H\,\mathrm{d}u^2 + 2\,\frac{r^2}{P^2}\,\mathrm{d}\zeta\,\mathrm{d}\bar\zeta, \qquad (19.1)$$

in which

$$2H \equiv \Delta \log P - 2r(\log P)_{,u} - \frac{2m}{r} - \frac{\Lambda}{3}r^2, \qquad (19.2)$$

where $\Delta \equiv 2P^2 \partial_\zeta \partial_{\bar\zeta}$ and Λ is the cosmological constant. This metric contains two functions, $P = P(\zeta, \bar\zeta, u)$ and $m = m(u)$ which, for vacuum solutions, are required to satisfy the fourth-order nonlinear equation

$$\Delta\Delta(\log P) + 12m(\log P)_{,u} - 4m_{,u} = 0. \tag{19.3}$$

This is referred to as the Robinson–Trautman equation.[1] It can be explicitly written as $P^4 P_{,\zeta\zeta\bar\zeta\bar\zeta} - P^3 P_{,\zeta\zeta} P_{,\bar\zeta\bar\zeta} + 3m P_{,u} - P m_{,u} = 0$.

The coordinates employed in the line element (19.1) are adapted to the assumed geometry which admits a geodesic, shear-free, twist-free and expanding null congruence generated by $\boldsymbol{k} = \partial_r$. Specifically, r is an affine parameter along the rays of this principal null congruence, u is a retarded time coordinate, and ζ is a complex spatial stereographic-type coordinate. They admit the coordinate freedoms

$$u' = U(u), \quad r' = \frac{r}{U_{,u}}, \quad \zeta' = F(\zeta), \quad P' = P\frac{|F_{,\zeta}|}{U_{,u}}, \quad m' = \frac{m}{U_{,u}^3}. \tag{19.4}$$

The Gaussian curvature of the 2-surfaces spanned by ζ, on which u is any constant and $r = 1$, is given by

$$K(\zeta, \bar\zeta, u) \equiv \Delta \log P. \tag{19.5}$$

When this is everywhere positive, these surfaces may be thought of as distorted spheres (if they are closed), so that such Robinson–Trautman solutions have been traditionally referred to as describing a kind of expanding gravitational radiation with "spherical-like" wave surfaces. For general fixed values of r and u, the Gaussian curvature of these 2-spaces is $K(\zeta, \bar\zeta, u)/r^2$ so that, as $r \to \infty$, they (locally) become planar.

Using the null tetrad $\boldsymbol{k} = \partial_r$, $\boldsymbol{l} = \partial_u - H\partial_r$, $\boldsymbol{m} = (P/r)\partial_{\bar\zeta}$, the non-zero components of the Weyl tensor for the metric (19.1) are

$$\Psi_2 = -\frac{m}{r^3},$$

$$\Psi_3 = -\frac{P}{2r^2}K_{,\zeta}, \tag{19.6}$$

$$\Psi_4 = \frac{1}{2r^2}\left(P^2 K_{,\zeta}\right)_{,\zeta} - \frac{1}{r}\left[P^2(\log P)_{,u\zeta}\right]_{,\zeta}.$$

When $m \neq 0$, there is obviously a scalar polynomial curvature singularity at $r = 0$, since the invariant I defined in (2.13) is here given by $3\Psi_2^2$. These space-times are of algebraic types II or D, and m can (in some cases) be related to the mass of a source. (In particular, the type D solutions, which

---

[1] For the axially symmetric case, this equation can be written in the form $U_{,t} + U^2 U_{,xxxx} = 0$. However, this is not integrable, see Kramer (1988) and Hoenselaers and Perjés (1993).

include the Schwarzschild solution and the C-metric, are non-twisting members of the Plebański–Demiański family of expanding solutions given in Section 16.2.) Many of these space-times include gravitational radiation. However, they also generally contain various types of singularities, and no solutions in this family are known that *globally* describe an isolated radiating source. A curvature singularity always occurs at $r = 0$, and other topological or curvature singularities usually occur on the wave surfaces.

When $m = 0$, the solutions are of algebraic types III or N and represent pure gravitational waves. A second-order invariant for type N space-times with expansion has been found by Bičák and Pravda (1998) whose square root in this case is proportional to

$$\frac{|\Psi_4|^2}{r^4}.$$ (19.7)

Subsequently, Pravda (1999) has shown that expanding type III space-times have an analogous invariant

$$\frac{|\Psi_3|^2}{r^2},$$

confirming that both type N and III Robinson–Trautman vacuum solutions also have a curvature singularity at $r = 0$.

To complete a general analysis of the global structure of these space-times, it is necessary to investigate the conformal infinity $\mathcal{I}$. This is given by $r = \infty$ where, as can be seen from (19.6), the space-times become conformally flat; i.e. they are asymptotically Minkowski, de Sitter or anti-de Sitter. Indeed, introducing an inverse radial coordinate $l = r^{-1}$, and taking the conformal factor to be $\Omega = l$, the Robinson–Trautman metric becomes

$$\mathrm{d}s^2 = \frac{1}{\Omega^2}\left[2\mathrm{d}u\,\mathrm{d}l + \left(\tfrac{1}{3}\Lambda + 2l(\log P)_{,u} - l^2\Delta\log P + 2ml^3\right)\mathrm{d}u^2 + \frac{2}{P^2}\,\mathrm{d}\zeta\,\mathrm{d}\bar\zeta\right].$$ (19.8)

This is clearly conformal to the metric in square brackets that, for smooth $P(\zeta, \bar\zeta, u)$ is regular at conformal infinity $\mathcal{I}$, located at $l = \Omega = 0$. In addition, $\mathcal{I}$ is null, spacelike or timelike according to the sign of the cosmological constant, i.e. whether $\Lambda = 0$, $\Lambda > 0$, or $\Lambda < 0$, respectively. The character of the conformal infinity of vacuum Robinson–Trautman space-times thus has (at least locally) one of the forms described in Sections 3.3, 4.2 or 5.2.

19.2 Type N solutions

For solutions of type N, the condition $\Psi_2 = 0$ implies that $m = 0$, and $\Psi_3 = 0$ implies that $K_{,\bar\zeta} = 0 = K_{,\zeta}$. Each wave surface $u = \text{const.}$, on a fixed r, thus has Gaussian curvature proportional to $K(u)$, that is independent of $\zeta, \bar{\zeta}$. The Robinson–Trautman field equation (19.3) is identically satisfied. Moreover, it is possible to use the freedom (19.4) in the labelling of the coordinate u to normalise K to the constant

$$\epsilon = +1, 0 \text{ or } -1.$$

All wave surfaces (at a given r) have the same curvature, and the metric is

$$ds^2 = -2du\,dr - \left(\epsilon - 2r(\log P)_{,u} - \tfrac{1}{3}\Lambda r^2\right)du^2 + 2\frac{r^2}{P^2}\,d\zeta\,d\bar{\zeta}. \qquad (19.9)$$

The function $P(\zeta, \bar{\zeta}, u)$ has to satisfy the equation $2P^2(\log P)_{,\zeta\bar{\zeta}} = \epsilon$, which has the general solution (Foster and Newman, 1967)

$$P = \left(1 + \tfrac{1}{2}\epsilon F\bar{F}\right)\left(F_{,\zeta}\bar{F}_{,\bar{\zeta}}\right)^{-1/2}, \qquad (19.10)$$

where $F = F(\zeta, u)$ is an arbitrary complex function of u and ζ, holomorphic in ζ. With these conditions, the remaining component of the Weyl tensor in (19.6) can be expressed as

$$\Psi_4 = -\frac{1}{r}\left[P^2(\log P)_{,u\zeta}\right]_{,\zeta} = \frac{P^2}{2r}F_{,\zeta}\left[\frac{1}{F_{,\zeta}}(\log F_{,\zeta})_{,u\zeta}\right]_{,\zeta}. \qquad (19.11)$$

Since Ψ_4 is the only non-vanishing component of the Weyl tensor, all the scalar polynomial invariants vanish. However, as mentioned above, the invariant (19.7) shows that there is a curvature singularity at $r = 0$. In addition, these solutions always contain singular points for some ζ on each wave surface, and these combine to form singular lines which are generally not well understood.

This family of type N solutions can alternatively be expressed in terms of the coordinates of García Díaz and Plebański (1981). Their metric also involves an arbitrary complex function $f(\xi, u)$ which is holomorphic in the complex spatial coordinate ξ, and is given by

$$ds^2 = -2\psi\,du\,dv + 2v^2 d\xi d\bar{\xi} - 2v\bar{A}\,d\xi\,du - 2vA\,d\bar{\xi}\,du + 2(A\bar{A} + \psi B)du^2, \qquad (19.12)$$

where

$$A = \tfrac{1}{2}\epsilon\,\xi + vf,$$
$$B = -\tfrac{1}{2}\epsilon - \tfrac{1}{2}v(f_{,\xi} + \bar{f}_{,\bar{\xi}}) + \tfrac{1}{6}\Lambda v^2\psi,$$
$$\psi = 1 + \tfrac{1}{2}\epsilon\,\xi\bar{\xi}.$$

The complete family of solutions described by this metric can be denoted as $RTN(\Lambda,\epsilon)[f]$, in which the free parameters Λ and ϵ and the arbitrary function $f(\xi,u)$ are identified explicitly.

The transformation which relates the above two forms of the Robinson–Trautman solutions is explicitly (see Bičák and Podolský, 1999a)

$$\xi = F(\zeta,u) = \int f(\xi(\zeta,u),u)\mathrm{d}u\,, \qquad v = \frac{r}{1+\frac{1}{2}\epsilon F\bar{F}}\,. \qquad (19.13)$$

(If f is independent of ξ, it is necessary to put $\xi = F = \zeta + \int f(u)\mathrm{d}u$.) The Weyl tensor component (using an adapted tetrad) is now given simply by

$$\Psi_4 = -\frac{1}{2r}\,f_{,\xi\xi\xi}\,. \qquad (19.14)$$

The line element (19.12) thus represents a conformally flat Minkowski or (anti-)de Sitter space when f is no more than quadratic in ξ, otherwise it describes a radiative vacuum space-time of type N. Using the equation of geodesic deviation, it can be shown that the effect of these exact gravitational waves on freely moving nearby test particles is transverse. In addition, the real and imaginary parts of Ψ_4 correspond to the two "+" and "×" polarisation modes (see Bičák and Podolský, 1999b).

In order to investigate the physical interpretation of these solutions, it is instructive to consider the properties of sandwich waves (see Griffiths, Podolský and Docherty, 2002) in which the space-time is conformally flat in the background regions in front of and behind the wave. In view of (19.11), such background space-times of the form (19.9), (19.10) occur when F is independent of u (they also represent the weak-field limits of this general family of solutions). These are just Minkowski, de Sitter or anti-de Sitter space according to the value of Λ. Although there are other possibilities to achieve $\Psi_4 = 0$, they are the simplest conformally flat limits of these solutions and will be considered separately in the following sections. They correspond to the choice $f(\xi,u) = 0$ in the García–Plebański coordinates of (19.12).

19.2.1 Minkowski background coordinates

Starting with the Minkowski metric in the form

$$\mathrm{d}s^2 = -2\,\mathrm{d}\mathcal{U}\,\mathrm{d}\mathcal{V} + 2\,\mathrm{d}\eta\,\mathrm{d}\bar{\eta}, \qquad (19.15)$$

in which the null coordinates are related to the usual Cartesian coordinates by $\mathcal{U} = \frac{1}{\sqrt{2}}(t-z)$, $\mathcal{V} = \frac{1}{\sqrt{2}}(t+z)$ and $\eta = \frac{1}{\sqrt{2}}(x+\mathrm{i}y)$ as in (3.4), the transformation

$$\mathcal{U} = u + \frac{r\,F\bar{F}}{1 + \frac{1}{2}\epsilon F\bar{F}}, \qquad \mathcal{V} = \tfrac{1}{2}\epsilon u + \frac{r}{1 + \frac{1}{2}\epsilon F\bar{F}}, \qquad \eta = \frac{r\,F}{1 + \frac{1}{2}\epsilon F\bar{F}}, \quad (19.16)$$

where $F = F(\zeta)$, puts the Minkowski metric (19.15) into the form

$$ds^2 = -2\,\mathrm{d}u\,\mathrm{d}r - \epsilon\,\mathrm{d}u^2 + 2\frac{r^2}{P^2}\,\mathrm{d}\zeta\,\mathrm{d}\bar{\zeta},$$

where P is given by (19.10). This is exactly the Minkowski limit of the Robinson–Trautman type N family of solutions (19.9) in the case when $\Lambda = 0$ and F is independent of u. The inverse transformation to (19.16) is

$$
\begin{aligned}
u &= \mathcal{U} - \frac{1}{\epsilon}\left[\sqrt{(\mathcal{V} - \tfrac{1}{2}\epsilon\mathcal{U})^2 + 2\epsilon\eta\bar{\eta}} - (\mathcal{V} - \tfrac{1}{2}\epsilon\mathcal{U})\right], \\
F(\zeta) &= \frac{1}{\epsilon\bar{\eta}}\left[\sqrt{(\mathcal{V} - \tfrac{1}{2}\epsilon\mathcal{U})^2 + 2\epsilon\eta\bar{\eta}} - (\mathcal{V} - \tfrac{1}{2}\epsilon\mathcal{U})\right], \qquad (19.17) \\
r &= \sqrt{(\mathcal{V} - \tfrac{1}{2}\epsilon\mathcal{U})^2 + 2\epsilon\eta\bar{\eta}},
\end{aligned}
$$

which reduces to $u = \mathcal{U} - \eta\bar{\eta}/\mathcal{V}$, $F = \eta/\mathcal{V}$, $r = \mathcal{V}$ when $\epsilon = 0$.

Using (19.17) and rescaled null coordinates (corresponding to a simple Lorentz boost) such that $\mathcal{U} = t - z$ and $\mathcal{V} = \frac{1}{2}(t + z)$, the null wave surfaces $u = u_0 = $ const. in this background are given by

$$-\left(t - \tfrac{1}{2}(1 + \epsilon)u_0\right)^2 + x^2 + y^2 + \left(z + \tfrac{1}{2}(1 - \epsilon)u_0\right)^2 = 0. \qquad (19.18)$$

This equation describes a *family of null cones*. The future null cones relative to each vertex, located at $t = \frac{1}{2}(1 + \epsilon)u_0$, $z = -\frac{1}{2}(1 - \epsilon)u_0$, $x = 0 = y$, are clearly expanding spheres for all values of ϵ and u_0. Successive cones are each displaced in specific t and z directions, as indicated in the space-time diagrams given in Figure 19.1.

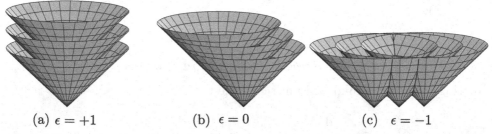

    (a) $\epsilon = +1$ (b) $\epsilon = 0$ (c) $\epsilon = -1$

Fig. 19.1 Families of null cones given by $u = u_0$ foliate Minkowski space-time in different ways in the three cases for which $\epsilon = +1, 0, -1$. One spatial dimension (x or y) is suppressed.

When $\epsilon = +1$, the vertices of the cones are located on a *timelike* line. The future null cones naturally fit inside each other and foliate the entire space-time. These are null hypersurfaces representing concentric spheres which expand at the speed of light. It follows from (19.17) that the origin of the Robinson–Trautman coordinate $r = 0$ corresponds to $x = y = z = 0$, with t arbitrary. This represents the weak-field "remnant" of the curvature singularity as a timelike line on the t-axis.

When $\epsilon = 0$, the cones all have a common *null* line $t = -z$, $x = 0 = y$. The future ones, i.e. the expanding spheres, only foliate the half of the Minkowski space for which $t > -z$. (The other half $t < -z$, is covered by the past null cones representing contracting spheres.) In this case, the origin of the Robinson–Trautman coordinate $r = 0$ corresponds to the boundary plane $t = -z$, with x and y arbitrary.

Finally, when $\epsilon = -1$, the vertices of the null cones given by (19.18) are located on a *spacelike* line at $t = 0$, and the null cones all intersect each other. These may be considered as the surfaces of null waves emitted by a tachyon propagating across Minkowski space at this instant. (In fact, this is identical to the limit of the AII-metric in which $m \to 0$ that was initially described in Section 9.1.1.) In this case, the origin $r = 0$ corresponds to $t^2 = x^2 + y^2$, with z arbitrary. This clearly represents the envelope of the above family of cones, and is a cylinder that expands at the speed of light. The null cones which are expanding spheres doubly foliate the inner region of the cylinder for which $t \geq \sqrt{x^2 + y^2}$. (Similarly, the null cones representing contracting spheres would doubly foliate the region $t \leq -\sqrt{x^2 + y^2}$.) Points for which $|t| < \sqrt{x^2 + y^2}$ are excluded in these foliations. However, for the weak-field limit of the Robinson–Trautman solutions, an overlapping of the wave surfaces is not permitted. Each wave surface must therefore be restricted to a family of half null cones, such as those for which $z \geq -u_0$, as shown in Figure 19.2. This restriction is necessary to maintain a consistent and unambiguous foliation of the space-time.

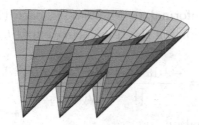

Fig. 19.2 Three typical members of the family of null cones $u = $ const. which foliate part of Minkowski space in the case when $\epsilon = -1$. One spatial dimension (x or y) has been suppressed.

For the case $\epsilon = -1$, there appears to be another singularity when $P = 0$, i.e. when $|F|^2 = 2$. It follows from (19.17) that this occurs for $t^2 = x^2 + y^2$, with z arbitrary. However, this exactly coincides with $r = 0$, which is the *envelope* of the null cones discussed above. It can thus be seen that the parts of null cones $t \geq 0$, $z \geq -u_0$, which are the wave surfaces for this solution, are spanned by the above coordinates with $|F|^2 \leq 2$. (An alternative set of half-cone wave surfaces for which $z \leq -u_0$ would be spanned by $|F|^2 \geq 2$.)

19.2.2 (Anti-)de Sitter background coordinates

The above analysis of the character of the Robinson–Trautman coordinates applies to the Minkowski background. However, a similar structure occurs in the de Sitter and anti-de Sitter backgrounds too, but the global structure is different.

As already described in Chapters 4 and 5, (anti-)de Sitter space can be represented as a four-dimensional hyperboloid

$$-Z_0{}^2 + Z_1{}^2 + Z_2{}^2 + Z_3{}^2 + \varepsilon Z_4{}^2 = \varepsilon a^2, \tag{19.19}$$

see (4.1) and (5.1), embedded in a flat five-dimensional space-time

$$ds^2 = -dZ_0{}^2 + dZ_1{}^2 + dZ_2{}^2 + dZ_3{}^2 + \varepsilon dZ_4{}^2,$$

where $a = \sqrt{3/|\Lambda|}$, $\varepsilon = 1$ for a de Sitter background ($\Lambda > 0$), and $\varepsilon = -1$ for an anti-de Sitter background ($\Lambda < 0$). As in (4.18), the parametrisation

$$\mathcal{U} = \sqrt{2}\,a\,\frac{Z_0 - Z_1}{Z_4 + a}, \qquad \mathcal{V} = \sqrt{2}\,a\,\frac{Z_0 + Z_1}{Z_4 + a}, \qquad \eta = \sqrt{2}\,a\,\frac{Z_2 + \mathrm{i}\,Z_3}{Z_4 + a}, \tag{19.20}$$

takes this metric to the form (4.19), namely

$$ds^2 = \frac{-2\,d\mathcal{U}\,d\mathcal{V} + 2\,d\eta\,d\bar{\eta}}{\left[1 - \frac{1}{6}\Lambda(\mathcal{U}\mathcal{V} - \eta\bar{\eta})\right]^2}, \tag{19.21}$$

which is explicitly conformal to Minkowski space (19.15) to which it reduces when $\Lambda = 0$. Applying the transformation (19.16), the metric (19.21) becomes

$$ds^2 = \frac{-2\,du\,dr - \epsilon\,du^2 + 2r^2 P^{-2}\,d\zeta\,d\bar{\zeta}}{\left[1 - \frac{1}{6}\Lambda u(r + \frac{1}{2}\epsilon u)\right]^2}.$$

The further transformation

$$\tilde{r} = \frac{r}{1 - \frac{1}{6}\Lambda u(r + \frac{1}{2}\epsilon u)}, \qquad \tilde{u} = \int \frac{du}{1 - \epsilon\frac{1}{12}\Lambda u^2},$$

yields

$$ds^2 = -2\,d\tilde{u}\,d\tilde{r} - \left(\epsilon - \tfrac{1}{3}\Lambda\,\tilde{r}^2\right)d\tilde{u}^2 + 2\frac{\tilde{r}^2}{P^2}\,d\zeta\,d\bar{\zeta}. \tag{19.22}$$

This is the (anti-)de Sitter background expressed in the Robinson–Trautman form (19.9) given by (19.10) with F independent of u.

Now consider the null wave surfaces $\tilde{u} = \tilde{u}_0 = \text{const.}$ Clearly, a constant $\tilde{u}$ implies that u is also a constant. Substituting $u = u_0$ into (19.17) and then applying the parametrisation (19.20) but with the rescaled (boosted) null coordinates $\mathcal{U} \to \sqrt{2}\mathcal{U}$, $\mathcal{V} \to \frac{1}{\sqrt{2}}\mathcal{V}$, gives

$$(1+\epsilon)\tfrac{u_0}{2a}\,Z_0 + (1-\epsilon)\tfrac{u_0}{2a}\,Z_1 - \left(\varepsilon + \epsilon\tfrac{u_0{}^2}{4a^2}\right)Z_4 + \left(\varepsilon - \epsilon\tfrac{u_0{}^2}{4a^2}\right)a = 0. \tag{19.23}$$

This is linear in Z_i, and thus represents a family of planes in the five-dimensional flat space-time. Their sections through the four-dimensional hyperboloid (19.19) give the family of null wave surfaces $\tilde{u} = \text{const.}$ in the (anti-)de Sitter universe. In fact, these planes are tangent to the hyperboloid. In the following, it will be demonstrated that all the sections are *null cones* relative to a vertex which is the point at which the plane touches the hyperboloid. For this discussion, it is convenient to introduce a dimensionless parameter α such that $\tan(\alpha/2) = \frac{u_0}{2a}$, or

$$\sin\alpha = \frac{\frac{u_0}{a}}{1 + \frac{u_0^2}{4a^2}}, \qquad \cos\alpha = \frac{1 - \frac{u_0^2}{4a^2}}{1 + \frac{u_0^2}{4a^2}}, \tag{19.24}$$

which parametrises the family of wave surfaces $u = u_0 = \text{const.}$

The de Sitter background

For the case in which $\Lambda > 0$ (i.e. $\varepsilon = 1$), there are three cases to consider, according to the sign of ϵ.

When $\epsilon = +1$, the null wave surfaces $u = u_0$ given by (19.23) can be re-expressed using (19.24) as the planes $Z_0 \sin\alpha - Z_4 + a\cos\alpha = 0$ whose intersections with the hyperboloid (19.19) are

$$(Z_0 \cos\alpha - a\sin\alpha)^2 = Z_1^2 + Z_2^2 + Z_3^2. \tag{19.25}$$

This is a family of null cones with vertices on the timelike hyperbola given by $Z_0 = a\tan\alpha$, $Z_1 = Z_2 = Z_3 = 0$, $Z_4 = a\sec\alpha$, which represents the origin of the Robinson–Trautman coordinates $\tilde{r} = 0$. Since $\alpha \in (-\frac{\pi}{2}, \frac{\pi}{2})$, all the vertices are located on an infinite *timelike hyperbola* with $Z_0 \in (-\infty, \infty)$. The future null cones from these vertices naturally foliate the half of the de Sitter space for which $Z_0 + Z_4 \geq 0$. These are illustrated in Figure 19.3(a).

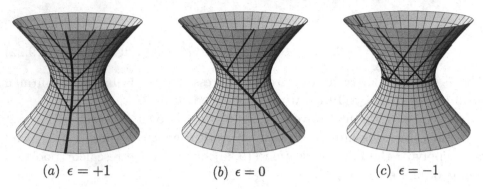

(a) $\epsilon = +1$ (b) $\epsilon = 0$ (c) $\epsilon = -1$

Fig. 19.3 Families of future null cones in de Sitter space given by $u = u_0 = $ const. are sections of a hyperboloid in a five-dimensional Minkowski space. With two dimensions (Z_2 and Z_3) suppressed, they appear as straight (null) lines. These foliate parts of the hyperboloid in different ways for the three cases in which $\epsilon = +1, 0, -1$. The vertices, which correspond to $\tilde{r} = 0$, are located, respectively, along timelike, null and spacelike lines.

When $\epsilon = 0$, the null wave surfaces are the planes $Z_4 - a = \frac{u_0}{2a}(Z_0 + Z_1)$. In this case, their intersections with the hyperboloid are given by

$$\left(Z_0 - \tfrac{u_0}{4a}(Z_4 + a)\right)^2 = \left(Z_1 + \tfrac{u_0}{4a}(Z_4 + a)\right)^2 + Z_2^2 + Z_3^2. \qquad (19.26)$$

Again, this is a family of null cones but with vertices now located along one common *straight null line* $Z_0 = -Z_1 = \frac{u_0}{2}$, $Z_2 = Z_3 = 0$, $Z_4 = a$. The origin of the Robinson–Trautman coordinates $\tilde{r} = 0$, which is the weak-field remnant of the curvature singularity, occurs on the null hypersurface $Z_0 + Z_1 = 0$, with $Z_2^2 + Z_3^2 + Z_4^2 = a^2$, which is a sphere with radius a equal to that of the universe at the instant $Z_0 = 0$. For this case, $u_0 = \tilde{u}_0 \in (-\infty, \infty)$. The future null cones from the vertices foliate the half of the de Sitter space for which $Z_0 + Z_1 \geq 0$. This is illustrated in Figure 19.3(b).

Finally, when $\epsilon = -1$, and again using (19.23) and (19.24), the hypersurfaces $u = u_0$ are given by the planes $Z_1 \sin\alpha - Z_4 \cos\alpha + a = 0$. Their intersections with the hyperboloid are

$$Z_0^2 = (Z_1 \cos\alpha + Z_4 \sin\alpha)^2 + Z_2^2 + Z_3^2, \qquad (19.27)$$

which is a family of null cones with vertices on the *spacelike circle* given by $Z_0 = Z_2 = Z_3 = 0$, $Z_1 = -a \sin\alpha$, $Z_4 = a \cos\alpha$, as illustrated in Figure 19.3(c). For this case, $\alpha = \frac{\tilde{u}_0}{a} \in (-\pi, \pi]$. The vertices are located on a closed circle (a possible path of a tachyon) around the de Sitter universe with $Z_0 = Z_2 = Z_3 = 0$. The future null cones from these vertices cover the future half ($Z_0 \geq 0$) of the de Sitter space twice. As for the Minkowski case with $\epsilon = -1$, it is again only appropriate to consider the family of half null cones with $Z_1 \geq -Z_4 \tan \alpha$. Locally these are similar to the wave surfaces illustrated in Figure 19.2 but, in this case, they wrap around the entire universe. Moreover, $\tilde{r} = 0$ now corresponds to $Z_0^2 = Z_2^2 + Z_3^2$ with $Z_1^2 + Z_4^2 = a^2$. When $Z_0 = 0$, this is the closed circle around the de Sitter universe containing all the vertices. For $Z_0 > 0$, the origin of the Robinson–Trautman coordinates is a torus whose radius expands at the speed of light in the de Sitter background. The apparent singularity which occurs when $P = 0$ coincides exactly with this expanding toroidal origin $\tilde{r} = 0$.

The anti-de Sitter background

For the complementary case in which $\Lambda < 0$ (i.e. $\varepsilon = -1$), there are again three geometrically distinct cases to consider.

When $\epsilon = +1$, the null wave surfaces $u = u_0$ of (19.23) are given by the planes $Z_0 \sin \alpha + Z_4 \cos \alpha - a = 0$. Their intersections with (19.19) are

$$(Z_0 \cos \alpha - Z_4 \sin \alpha)^2 = Z_1^2 + Z_2^2 + Z_3^2, \qquad (19.28)$$

which are null cones with vertices on the *(closed) timelike line* $Z_0 = a \sin \alpha$, $Z_1 = Z_2 = Z_3 = 0$, $Z_4 = a \cos \alpha$ in the anti-de Sitter universe. This corresponds to the origin $\tilde{r} = 0$. The future null cones from these vertices, illustrated in Figure 19.4(a), now cover the complete anti-de Sitter space.

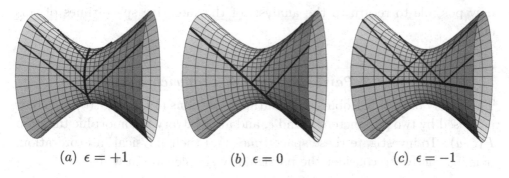

(a) $\epsilon = +1$ (b) $\epsilon = 0$ (c) $\epsilon = -1$

Fig. 19.4 Families of null cones $u = u_0$ in anti-de Sitter space are sections of a hyperboloid in a five-dimensional flat space with two temporal coordinates. With the spatial dimensions Z_2 and Z_3 suppressed, they are straight (null) lines. These foliate parts of the hyperboloid in different ways for the three cases in which $\epsilon = +1, 0, -1$, with vertices located, respectively, along timelike, null and spacelike lines.

When $\epsilon = 0$, the wave surfaces are the planes $Z_4 - a = -\frac{u_0}{2a}(Z_0 + Z_1)$. In this case, their intersections with the hyperboloid are the null cones

$$\left(Z_0 - \tfrac{u_0}{4a}(Z_4 + a)\right)^2 = \left(Z_1 + \tfrac{u_0}{4a}(Z_4 + a)\right)^2 + Z_2^2 + Z_3^2, \qquad (19.29)$$

with vertices located on one *common null line* $Z_0 = -Z_1 = \frac{u_0}{2}$, $Z_2 = Z_3 = 0$, $Z_4 = a$. The origin $\tilde{r} = 0$ occurs on the null hypersurface $Z_0 + Z_1 = 0$ with $Z_4^2 = a^2 + Z_2^2 + Z_3^2$, which is a hyperboloidal surface. With $u_0 \in (-\infty, \infty)$, the future null cones foliate the half of the anti-de Sitter space for which $Z_0 + Z_1 \geq 0$, as illustrated in Figure 19.4(*b*).

When $\epsilon = -1$, the wave surfaces are the planes $Z_1 \sin \alpha + Z_4 - a \cos \alpha = 0$, which intersect the hyperboloid on the null cones

$$Z_0^2 = (Z_1 \cos \alpha + a \sin \alpha)^2 + Z_2^2 + Z_3^2, \tag{19.30}$$

as in Figure 19.4(*c*). Their vertices are located on the *spacelike hyperbola* $Z_0 = Z_2 = Z_3 = 0$, $Z_1 = -a \tan \alpha$, $Z_4 = a \sec \alpha$, which could represent the path of a tachyon. As for the previous cases in which $\epsilon = -1$, an unambiguous foliation is obtained only for the family of half null cones with $Z_1 \geq -a \tan \alpha$, which are analogous to the wave surfaces illustrated in Figure 19.2. Moreover, the origin $\tilde{r} = 0$ of the Robinson–Trautman coordinates corresponds to $Z_0^2 = Z_2^2 + Z_3^2$ with $Z_4^2 = a^2 + Z_1^2$. When $Z_0 = 0$, this is a hyperbola across the anti-de Sitter universe which contains all the vertices of the null cones. For $Z_0 > 0$, it is an expanding cylindrical-type surface centred on this hyperbola. Again, the singularity $P = 0$ coincides with $\tilde{r} = 0$.

With the above geometrical understanding of the character of the Robinson–Trautman coordinates in these conformally flat background cases, it is now possible to return to the analysis of the radiative space-times of this class.

19.2.3 Particular type N sandwich waves

The family of type N Robinson–Trautman solutions (19.9), (19.10) are characterised by two parameters Λ and ϵ, and an arbitrary holomorphic function $F(\zeta, u)$. To investigate these space-times and their physical interpretation, it is instructive to consider the particularly simple case in which

$$F(\zeta, u) = \zeta^{g(u)}, \tag{19.31}$$

where $g(u)$ is an arbitrary positive function of retarded time. For this choice, the expressions involving P given by (19.10), which appear in the metric (19.9), take the forms

$$\frac{1}{P^2} = \frac{g^2 (\zeta \bar{\zeta})^{g-1}}{\left[1 + \frac{1}{2} \epsilon (\zeta \bar{\zeta})^g \right]^2}, \quad (\log P)_{,u} = -\frac{g_{,u}}{g} \left(1 + \frac{1}{2} \log (\zeta \bar{\zeta})^g \left[\frac{1 - \frac{1}{2} \epsilon (\zeta \bar{\zeta})^g}{1 + \frac{1}{2} \epsilon (\zeta \bar{\zeta})^g} \right] \right),$$
$$\tag{19.32}$$

and the explicit expression for Ψ_4 given by (19.11) is

$$\Psi_4 = -\frac{g_{,u}\bar{\zeta}}{2g\,\zeta\,r}\left(\frac{1}{|\zeta|^g} + \frac{\epsilon|\zeta|^g}{2}\right)^2.$$

(19.33)

These solutions are obviously conformally flat (i.e. the Minkowski, de Sitter or anti-de Sitter backgrounds) if, and only if, g is a constant. For a general g, they represent exact gravitational waves with arbitrary amplitudes. They reduce to weak radiation fields if g is approximately constant, i.e. when $g_{,u}/g$ is small. Interestingly, the metric component $(\log P)_{,u}$ and the Weyl tensor component Ψ_4 are both proportional to the same wave profile $g_{,u}/g$. In view of the invariant (19.7), provided $g_{,u} \neq 0$, these space-times possess a curvature singularity at $r = 0$, and also when $\zeta = 0$ or ∞.

Sandwich type N waves

A particularly simple sandwich Robinson–Trautman wave of this type has been described by Griffiths and Docherty (2002) for the case $\Lambda = 0$ and $\epsilon = +1$.[2] This has the form (19.31) with the continuous function

$$g(u) = \begin{cases} 1 - A\,u_1 & \text{for } u < 0, \\ 1 - A\,(u_1 - u) & \text{for } 0 \leq u \leq u_1, \\ 1 & \text{for } u > u_1, \end{cases}$$

(19.34)

where A and u_1 are positive constants. It represents a Robinson–Trautman wave confined to the region $0 \leq u \leq u_1$, in which $g_{,u} = A$. It has an expanding spherical wavefront $u = 0$, and the wave continues until $u = u_1$, which is also a concentric expanding sphere. Ahead of and behind this wave zone, the space-time is the flat Minkowski background, see (19.33). However, taking $\arg\zeta \in [0, 2\pi)$, the Minkowski region ahead of the wave contains a topological defect at $\zeta = 0$ and $\zeta = \infty$. This can be interpreted as a cosmic string with the deficit angle $2\pi A u_1$. By contrast, the Minkowski region behind the wave contains no such defect. This solution has thus been interpreted as representing a breaking cosmic string in a flat background in which the tension (deficit angle) of the string reduces uniformly to zero over a finite interval of retarded time. This decay of the cosmic string may be considered to generate the gravitational wave.

The above solution has been generalised to arbitrary functions $g(u)$, and

---

[2] A different step-like solution has been given by Nutku (1991) by choosing $\sqrt{2}\,F(\zeta, u) = (\zeta + u\Theta(u))^{-1/2}$. However, a global solution of this kind cannot be obtained in view of the non-analytic nature of the stereographic-type coordinate $F(\zeta, u)$ over the complete complex plane.

to all possible values of Λ and ϵ by Griffiths, Podolský and Docherty (2002). Obviously, $g(u)$ specifies the wave profile, Λ determines the background and ϵ characterises the geometrical properties of the wave surfaces.

On any hypersurface $u = $ const., the complex function $F = \zeta^g$ represents a stereographic-type coordinate. Introducing the parametrisations,

$$\zeta^g = \begin{cases} \sqrt{2}\tan(\theta/2)\,\mathrm{e}^{\mathrm{i}\varphi} & \text{for } \epsilon = +1, \\ \frac{1}{\sqrt{2}}\rho\,\mathrm{e}^{\mathrm{i}\varphi} & \text{for } \epsilon = 0, \\ \sqrt{2}\tanh(R/2)\,\mathrm{e}^{\mathrm{i}\varphi} & \text{for } \epsilon = -1, \end{cases} \tag{19.35}$$

(see Appendix A) each wave surface takes the standard form

$$\frac{2}{P^2}\,\mathrm{d}\zeta\,\mathrm{d}\bar{\zeta} = \begin{cases} \mathrm{d}\theta^2 + \sin^2\theta\,\mathrm{d}\varphi^2 & \text{for } \epsilon = +1, \\ \mathrm{d}\rho^2 + \rho^2\,\mathrm{d}\varphi^2 & \text{for } \epsilon = 0, \\ \mathrm{d}R^2 + \sinh^2 R\,\mathrm{d}\varphi^2 & \text{for } \epsilon = -1. \end{cases} \tag{19.36}$$

This is simply the metric on a 2-sphere, 2-plane, and 2-hyperboloid, respectively. (For the case $\epsilon = -1$, ζ is restricted to the range $|\zeta| \le 1$ to cover a single sheet of the hyperboloid.) However, if it is assumed that the argument of ζ covers the full range $[0, 2\pi)$, then $\varphi \in [0, 2\pi g)$ and these surfaces thus in general include a *deficit angle* $2\pi(1 - g)$ around $\theta = 0$ or π if $\epsilon = +1$, around $\rho = 0$ if $\epsilon = 0$ and around $R = 0$ if $\epsilon = -1$.

Any region in which $g = $ const. has to be one of the Minkowski, de Sitter or anti-de Sitter backgrounds according to the value of Λ. If $g < 1$, these regions contain a constant deficit angle on all sections $u = $ const. Together, these are interpreted as cosmic strings with constant tension. (If $g > 1$, there is an excess angle corresponding to a strut under constant compression.)

In general, any region in which $g(u)$ is not constant is a Robinson–Trautman type N solution, and the deficit angle on each wave surface will vary. It can be seen from (19.33) that the poles about which there is a deficit angle actually correspond to curvature singularities in the Weyl tensor.

By combining the above two possibilities, solutions can be constructed in which g is non-constant only over a *finite* range of u. Such solutions clearly represent *sandwich Robinson–Trautman waves*. The situation in which $g(u)$ is constant (< 1) in front of the wave and then increases continuously to 1 behind the wave can be interpreted as a disintegrating string (the deficit angle reduces continuously to zero). One particular case of this for $\epsilon = +1$, is represented by (19.34) and illustrated in Figure 19.5(a). The analogous solutions for alternative values of ϵ are also illustrated in Figure 19.5.

When $\epsilon = +1$, the wave surfaces $u = $ const. at any time are a family of concentric spheres. For the above example, the conformally flat background

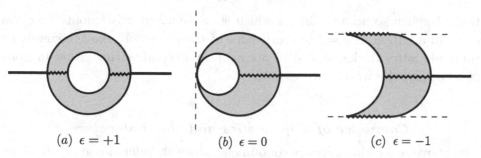

Fig. 19.5 The shaded regions represent axially symmetric Robinson–Trautman sandwich waves at some fixed time for different values of ϵ. The expanding spherical (hemispherical for $\epsilon = -1$) wave surfaces are given by $u = $ const. The region behind the wave is Minkowski or (anti-)de Sitter, while the region ahead of the wave is Minkowski or (anti-)de Sitter with a deficit angle representing a cosmic string. The dashed lines denote the boundaries of the coordinate system adopted.

space ahead of the wave contains strings of equal tension at opposite sides of the expanding spherical wavefront at $\theta = 0$ and $\theta = \pi$.

In the case when $\epsilon = 0$, there is only a single pole in $\rho = 0$. The background region thus contains a single string, but only part of the complete space-time is now covered by the coordinate system (that to the right of the dashed line in Figure 19.5(b)). At any time, the spherical wave surfaces contain a common point, opposite to the pole, as may be observed from the family of null cones in Figure 19.1(b).

When $\epsilon = -1$, since there is the restriction $|\zeta| \leq 1$ to span only hemispherical surfaces, there is only a single pole $R = 0$ on each expanding hemisphere. This may again be attached to a cosmic string in the background region ahead of the wavefront. As explained in previous sections, the envelope of these surfaces $r = 0$ is a physical (curvature) singularity within the wave zone. In the background space-time, it is a cylindrical surface of finite length whose radius is expanding at the speed of light. Its section is denoted by the outer zigzag lines in Figure 19.5(c). The boundary of the coordinate system is illustrated in Figure 19.5(c) by the dashed lines.

For the particular choice of the sandwich wave (19.34), the function $g(u)$ is linear within the wave zone. This gives rise to discontinuities in $\Psi_4 \sim g_{,u}$ at $u = 0$ and $u = u_1$, see (19.33), which correspond to shocks on the boundary wave surfaces indicated in Figure 19.5. However, for this example, the $(\log P)_{,u}$ term contained in the metric is also discontinuous on these shock fronts, see (19.32). Of course, more general families of sandwich waves without discontinuities in the metric and Ψ_4 can easily be constructed by permitting g to vary in a suitably smooth way over a finite range of the retarded time. Explicit solutions exist for which Ψ_4 is an arbitrarily smooth function of u. In addition, if $g = 1$ on both sides of the sandwich, the Minkowski or (anti-)de Sitter background does not contain a cosmic string either in front of or behind the wave.

Character of singularities and global structure

The above space-times always contain curvature singularities at $r = 0$ which may naturally be considered as the sources of the Robinson–Trautman solutions. The character of the coordinate origin $r = 0$ in the background space-times for different values of ϵ and Λ has already been discussed in Sections 19.2.1 and 19.2.2. For sandwich Robinson–Trautman waves, this curvature singularity occurs only within the wave zones, and does not extend to the regions in front of and behind the waves where $g_{,u} = 0$. The location of this singularity for sections of constant ζ and different values of ϵ is illustrated in the space-time pictures in Figure 19.6.

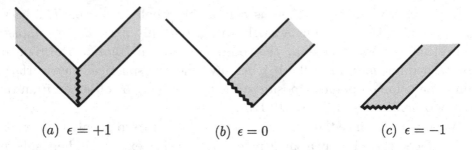

(a) $\epsilon = +1$ (b) $\epsilon = 0$ (c) $\epsilon = -1$

Fig. 19.6 Schematic space-time pictures for constant ζ for Robinson–Trautman sandwich waves with different values of ϵ. The shaded areas are the wave regions. The zigzag lines represent the singularities at $r = 0$. If the particular section corresponds to $\zeta = 0$, the region external to the wave also contains a cosmic string but this becomes a string-like curvature singularity through the wave zone.

It should be noted that Figure 19.5 corresponds to a spacelike section through the space-time pictures in Figure 19.6 with the addition of one spatial dimension. The particular section in Figure 19.5 is at a sufficiently late time to contain the complete sandwich. Earlier spacelike sections than those shown in Figure 19.5 could intersect the naked curvature singularity at $r = 0$, which corresponds to the source of the wave.

When $\epsilon = +1$, the solution can be interpreted as representing the disintegration of a cosmic string in which the deficit angle of the string reduces to zero through a region in which it generates the gravitational wave.

A similar interpretation can be given for the case when $\epsilon = 0$. However, there is now a boundary of the coordinate systems which is represented as a dashed line in Figure 19.5(b). This is a *null plane* in the background and, as such, it is possible to locally extend the space-time to include an external conformally flat background region. Such an extension is non-unique as an arbitrary impulsive gravitational wave may occur on this null hypersurface. In the absence of such an impulsive wave, it is necessary to have the same backgrounds which contain strings in both directions. For example, the region to the left of the dashed line in Figure 19.5(b) could be taken to be the same as the region external to the wavefront, having a string with the same deficit angle. In this case, one end of the string is a clean break, while the deficit angle at the other end reduces to zero over a finite range.

The situation when $\epsilon = -1$ is even more complicated. However, a model for a snapping and disintegrating cosmic string can again be obtained by considering two separate sandwich waves of the type described above and illustrated in Figures 19.5(c) and 19.6(c). These can be combined such that two hemispherical waves propagate in opposite directions, as illustrated in Figure 19.7.

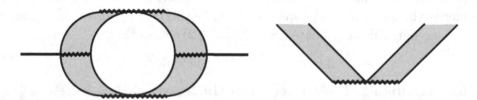

Fig. 19.7 A disintegrating cosmic string in the case $\epsilon = -1$. The left picture represents the situation at a fixed time (one spatial direction is suppressed). The right picture is a space-time diagram with two spatial directions suppressed. The shaded areas are the wave regions. The zigzag lines represent the expanding cylindrical singularity where $r = 0$ and (left only) the string-like singularity at $\zeta = 0$.

The schematic pictures in Figures 19.5 and 19.6 also remain valid if the cosmological constant were non-zero. Interestingly, this Robinson–Trautman family of type N solutions imposes no restrictions at all on the character and decay of string-like structures of the type described above. Any form of snapping or decaying string could be accommodated within this framework.

19.3 Type D solutions

Returning now to the general family of vacuum Robinson–Trautman space-times (19.1), (19.2) introduced in Section 19.1, it follows from the curvature tensor components (19.6) that these solutions are of algebraic types II or D when $m \neq 0$. In this case, the freedom (19.4) can be used to set m equal to a constant. In particular, the space-times are of type D if $3\Psi_2\Psi_4 = 2\Psi_3{}^2$ (see Section 2.1.2). This is equivalent to the conditions

$$\left(P^2 K_{,\zeta}\right)_{,\zeta} = 0, \qquad P^2 \left(K_{,\zeta}\right)^2 = 6m \left[P^2(\log P)_{,u\zeta}\right]_{,\zeta},$$

where $K = \Delta \log P$ is the Gaussian curvature (19.5) of the 2-surfaces spanned by the complex spatial coordinate ζ. Together with the vacuum field equation (19.3), these imply that

$$P^2 K_{,\bar{\zeta}} = h(\zeta),$$

where $h(\zeta)$ is an arbitrary analytic function (see Stephani *et al.*, 2003).

If $h = 0$, the Gaussian curvature K on each wave surface is constant, and P can be expressed in the standard form $P = 1 + \frac{1}{2}K(u)\zeta\bar{\zeta}$ (see Appendix A). The field equation (19.3) implies that K is a constant, and a simple rescaling can thus be used to put K to $\epsilon = -1, 0, +1$ so that

$$P = 1 + \tfrac{1}{2}\epsilon\,\zeta\bar{\zeta}.$$

In the case in which $K > 0$, i.e. $\epsilon = +1$, the wavefronts $u = $ const. are spheres which expand at the speed of light. In this case, ζ may be interpreted as a stereographic coordinate $\zeta = \sqrt{2}\,\tan(\theta/2)\,e^{i\phi}$, so that

$$2\,P^{-2}\mathrm{d}\zeta\,\mathrm{d}\bar{\zeta} = \mathrm{d}\theta^2 + \sin^2\theta\,\mathrm{d}\phi^2.$$

Also, substituting $P = 1 + \frac{1}{2}\zeta\bar{\zeta}$, into (19.2) gives $2H = 1 - \frac{2m}{r} - \frac{\Lambda}{3}r^2$. This leads to the Schwarzschild–(anti-)de Sitter space-time (Section 9.4) expressed in the form

$$\mathrm{d}s^2 = -2\mathrm{d}u\,\mathrm{d}r - \left(1 - \frac{2m}{r} - \frac{\Lambda}{3}r^2\right)\mathrm{d}u^2 + r^2(\mathrm{d}\theta^2 + \sin^2\theta\,\mathrm{d}\phi^2), \quad (19.37)$$

which clearly reduces to the Schwarzschild solution (8.8) when the cosmological constant Λ vanishes. (For cases with other values of ϵ see Section 9.1.)

If $h \neq 0$, it is possible to relate the corresponding Robinson–Trautman metrics to the family of C-metric vacuum solutions which are described in Chapter 14. Indeed, the transformation (see Krtouš and Podolský, 2003)

$$r = \frac{1}{\alpha(x+y)},$$

$$u = \frac{1}{\alpha}\left(\tau + \int \frac{\mathrm{d}y}{F}\right), \qquad (19.38)$$

$$\zeta = \frac{1}{\sqrt{2}}\left(\tau + \int \frac{\mathrm{d}y}{F} - \int \frac{\mathrm{d}x}{G} + i\,\varphi\right),$$

where[3]

$$G(x) = (1 - x^2)(1 + 2\alpha m x), \qquad F(y) = -\frac{\Lambda}{3\alpha^2} - (1 - y^2)(1 - 2\alpha m y),$$

puts the Robinson–Trautman metric (19.1), (19.2) into the form (14.3), namely

$$ds^2 = \frac{1}{\alpha^2(x+y)^2} \left(-F \, d\tau^2 + \frac{dy^2}{F} + \frac{dx^2}{G} + G \, d\varphi^2 \right).$$

In this relation, $2H = \alpha^2 r^2(F + G)$ and the expression for the function P in (19.1) for this particular case is

$$P(\zeta, \bar{\zeta}, u) = G^{-1/2}\Big(x(\zeta, \bar{\zeta}, u)\Big),$$

where the function $x(\zeta, \bar{\zeta}, u)$ is obtained from (19.38) as

$$\int \frac{dx}{G(x)} = \alpha u - \tfrac{1}{\sqrt{2}}(\zeta + \bar{\zeta}).$$

The Robinson–Trautman vacuum solutions of type D with $h \neq 0$ thus include those which describe uniformly accelerating black holes.

Note also that an alternative form of the C-metric is obtained by keeping the original Robinson–Trautman coordinates r, u and introducing only the coordinates x, φ by the transformation (19.38). This leads to the line element

$$ds^2 = -2 \, du \, dr - (2H - \alpha^2 r^2 \, G) \, du^2 - 2 \, \alpha \, r^2 \, du \, dx$$
$$+ r^2 \left(\frac{dx^2}{G} + G \, d\varphi^2 \right), \tag{19.39}$$

which, for $\Lambda = 0$, was studied by Kinnersley and Walker (1970) and Ishikawa and Miyashita (1983), and for a non-vanishing cosmological constant by Mann (1997, 1998).

19.4 Type II solutions

The class of vacuum Robinson–Trautman space-times (19.1), (19.2) of algebraic type II describe exact expanding gravitational waves. Indeed, with a constant $m \neq 0$ and P depending on the retarded time u, their gravitation field contains the radiative component Ψ_4 of (19.6) which decays as r^{-1}.

An interesting feature of these space-times is that many solutions of this

---

[3] In this particular form generalising (14.4) to include a cosmological constant, the functions $G(x)$ and $F(y)$ have three real roots that have been located at convenient points using the available coordinate freedom. These forms lead to the C-metric which contains 2-surfaces of positive curvature. Other forms of these cubic functions, possibly with fewer roots, lead to the other solutions in this family as described in Section 16.2.

family with $\Lambda = 0$ decay to the Schwarzschild space-time asymptotically as $u \to \infty$. Since, as will be shown below, this boundary corresponds to the event horizon, such solutions may be considered as possible radiating exterior fields for a spherically symmetric Schwarzschild black hole. This property was first demonstrated by Foster and Newman (1967) using a perturbative approach. Subsequently, exact treatments of the existence and asymptotic behaviour of the solutions to the Robinson–Trautman field equation (19.3) were given by Lukács *et al.* (1984), Vandyck (1987), Schmidt (1988), Rendall (1988), Tod (1989), Singleton (1990), Frittelli and Moreschi (1992), Hoenselaers and Perjés (1993) and others. A detailed analysis has been given by Chruściel (1991, 1992) and Chruściel and Singleton (1992).

These rigorous and systematic studies investigated the development of the space-time from initial data prescribed on an initial surface $u = u_0$. Specifically, the metric function $P(\zeta, \bar{\zeta}, u)$ on this surface is taken to be an arbitrarily smooth function of ζ and $\bar{\zeta}$. (The kind of singularities that were considered in the previous sections are here excluded.) For such initial data, it was demonstrated that Robinson–Trautman type II vacuum space-times (19.1), (19.2) *exist globally* for all values $u \geq u_0$. In addition, they converge asymptotically to the Schwarzschild metric with the corresponding mass m as $u \to \infty$. Specifically, introducing a function $f(\zeta, \bar{\zeta}, u)$ by

$$P \equiv f(\zeta, \bar{\zeta}, u)\,(1 + \tfrac{1}{2}\,\zeta\bar{\zeta})\,, \tag{19.40}$$

and for which the initial data $f(\zeta, \bar{\zeta}, u_0)$ is sufficiently smooth, a solution of the Robinson–Trautman equation (19.3) always exists. For $u \to +\infty$, it is asymptotically given as

$$f = \sum_{i,j \geq 0} f_{i,j} u^j \mathrm{e}^{-2iu/m} = 1 + f_{1,0}\,\mathrm{e}^{-2u/m} + f_{2,0}\,\mathrm{e}^{-4u/m} + \cdots + f_{14,0}\,\mathrm{e}^{-28u/m}$$

$$+ f_{15,1}\,u\,\mathrm{e}^{-30u/m} + f_{15,0}\,\mathrm{e}^{-30u/m} + \cdots\,, \tag{19.41}$$

where $f_{i,j}$ are *smooth* functions of the spatial coordinates $\zeta, \bar{\zeta}$. Thus, for large times u, the function P exponentially approaches $1 + \tfrac{1}{2}\zeta\bar{\zeta}$ which describes the spherically symmetric Schwarzschild solution (19.37), see Figure 19.8.

It can be shown, however, that the extension of the Robinson–Trautman metrics across the Schwarzschild-like event horizon $\mathcal{H}^+$ of the resulting black hole, located at $u = +\infty$, is *not analytic*. It only possesses a finite degree of smoothness. In fact, the metric is C^5 through $\mathcal{H}^+$ in general, and can be of class C^{117}. Moreover, there exist an infinite number of alternative extensions through $\mathcal{H}^+$, which are obtained by gluing the initial Robinson–Trautman space-time to any similar Robinson–Trautman space-time with the same constant value of m.

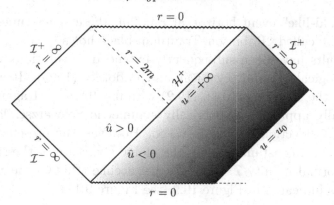

Fig. 19.8 For arbitrary, smooth initial data for the Robinson–Trautman equation at $u = u_0 = $ const., the Robinson–Trautman type II vacuum metrics (in the region $\hat{u} < 0$) approach the Schwarzschild metric exponentially as $u \to +\infty$. It is possible to extend this by attaching part of the Schwarzschild space-time (for $\hat{u} > 0$). The junction occurs on the null hypersurface $\mathcal{H}^+$ at $\hat{u} = 0$, which bounds the Robinson–Trautman region and coincides with the event horizon of the Schwarzschild region. However, such an extension has only a finite degree of smoothness.

To demonstrate this explicitly, it is convenient to introduce an advanced time coordinate $v = u + 2r + 4m \log(r/2m - 1)$, and then the null Kruskal–Szekeres coordinates $\hat{u}, \hat{v}$ by (see Sections 8.2 and 8.3)

$$\hat{u} = -\exp(-u/4m), \qquad \hat{v} = \exp(v/4m). \qquad (19.42)$$

The null surface $\mathcal{H}^+$ at $u = +\infty$ now becomes a boundary given by $\hat{u} = 0$, and the metric (19.1) takes the form

$$ds^2 = -\frac{32m^3}{r} e^{-r/2m} \, d\hat{u} \, d\hat{v} - 32m^2 \hat{H} \, d\hat{u}^2 + 2\frac{r^2}{P^2} \, d\zeta \, d\bar{\zeta}, \qquad (19.43)$$

where

$$2\hat{H} = e^{u/2m} \left(K - 1 + \frac{r}{6m} \Delta K \right). \qquad (19.44)$$

In terms of $\hat{u}$, the expansion (19.41) becomes

$$f = 1 + f_{1,0}\,\hat{u}^8 + \cdots + f_{14,0}\,\hat{u}^{112} - 4m f_{15,1} (\log|\hat{u}|)(\hat{u})^{120} + f_{15,1}\hat{u}^{120} + \cdots. \qquad (19.45)$$

Due to the presence of the $\log|\hat{u}|$ terms, the function f is not smooth at $\hat{u} = 0$. Indeed, it is at most C^{119} if $f_{15,1} \neq 0$. The full metric (19.43) is then C^{117}, since $\hat{H}$ contains the additional factor $e^{u/2m} \approx 1/\hat{u}^2$. In general, f is C^7 and the metric C^5 if $f_{1,0} \neq 0$. In principle, the differences between the C^5, C^{117} or smooth metrics are observable. Also, since they have a

"Schwarzschild-like" event horizon at $r = 2m$, these space-times could appropriately be called "Robinson–Trautman black holes".

These results have been subsequently extended to solutions with a positive cosmological constant by Bičák and Podolský (1995). It was demonstrated that the solutions (19.1), (19.2) with $0 < 9\Lambda m^2 < 1$ again exist and asymptotically approach a spherically symmetric Schwarzschild–de Sitter space-time (see Section 9.4) as $u \to +\infty$. However, the presence of a positive cosmological constant changes the global structure of the space-times. Future conformal infinity $\mathcal{I}^+$ now has a spacelike (that is "de Sitter-like") character, as indicated in Figure 19.9 (see Figure 9.11).

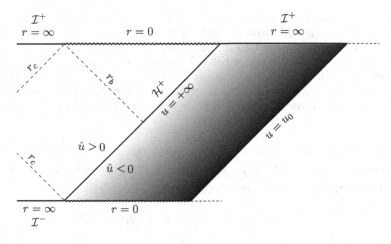

Fig. 19.9 For smooth initial data prescribed at $u = u_0$, the Robinson–Trautman metrics with $0 < 9\Lambda m^2 < 1$ always exist and converge to the spherically symmetric Schwarzschild–de Sitter metric with the same values of Λ and m as $u \to +\infty$. The metric at this black hole horizon $\mathcal{H}^+$ has only a finite degree of smoothness, but this can be higher than in the case $\Lambda = 0$. Future and past conformal infinities $\mathcal{I}^+$ and $\mathcal{I}^-$ have a spacelike character.

These Robinson–Trautman solutions serve as explicit models which satisfy the *cosmic no-hair conjecture*. (See e.g. Gibbons and Hawking, 1977, Wald, 1983, Starobinskii, 1983, Barrow and Götz, 1989 or Maeda, 1989 for more details and other references.) Close to $\mathcal{I}^+$ they *locally* asymptotically approach the de Sitter space-time, as can be shown by performing a suitable coordinate transformation.

The presence of $\Lambda > 0$ also influences the degree of smoothness of an extension across the black hole horizon $\mathcal{H}^+$ located at $u = +\infty$. The function f defined by (19.40), which expresses the way the solutions approach the Schwarzschild–de Sitter metric given by $f = 1$, is again characterised by the expansion (19.41). However, for $0 < 9\Lambda m^2 < 1$, it is necessary to employ an alternative transformation to Kruskal–Szekeres null coordinates. This is similar to (19.42), but the constant $4m$ is replaced by $2\delta_+$, where the parameter δ_+ is related to the location of the black hole horizon in the Schwarzschild–de Sitter space-time (see Section 9.4) by $\delta_+ = r_b/(1 - \Lambda r_b^2)$. This monotonically increases from $2m$ to ∞, as Λm^2 increases from 0 to its maximal value of $1/9$. Using this, the term $\hat{u}^8$ in (19.45) is replaced by $\hat{u}^{4\delta_+/m}$, and the term $\hat{u}^{120}$ is replaced by $\hat{u}^{60\delta_+/m}$. For $\Lambda > 0$, the extension across $\mathcal{H}^+$ can thus be *smoother* than in the $\Lambda = 0$ case since $\delta_+ > 2m$. In fact, the horizon may become "arbitrarily smooth" in the limit $\Lambda m^2 \to 1/9$.

This interesting effect motivated Bičák and Podolský (1997) to analyse in detail the properties of the Robinson–Trautman type II vacuum solutions in the *extreme* case in which $9\Lambda m^2 = 1$. They demonstrated that, in this case, the asymptotic expansion (19.41), in appropriate null coordinates, takes the form

$$f = 1 + f_{1,0}\, e^{-(2\delta/m)\cot\hat{u}} + \cdots + f_{14,0}\, e^{-(28\delta/m)\cot\hat{u}}$$

$$+ \delta f_{15,1}\, \cot\hat{u}\, e^{-(30\delta/m)\cot\hat{u}} + \cdots , \qquad (19.46)$$

where $-\delta/m = 3 - 2\log 2 > 0$. From this expansion, it follows that these Robinson–Trautman space-times can be extended across the horizon $\mathcal{H}^+$ located at $\hat{u} = 0$ in a *smooth* C^∞ way. They can be smoothly matched, for example, to the solutions representing for $\hat{u} > 0$ an extreme Schwarzschild–de Sitter space-time with the same values of Λ and m, for which the black hole horizon r_b and the cosmological horizon r_c coincide at $r = 3m$ (see Section 9.4). A schematic representation of the resulting conformal structure is given in Figure 19.10. Such an extension is smooth, but *not analytic* because it is not unique. This somewhat surprising result has been mentioned by Chruściel (1996) as an argument against a "natural" assumption of analyticity, which is usually considered in proofs of the fundamental "rigidity" theorem, according to which stationary analytic electrovacuum black holes are necessarily either static or axially symmetric.

When $9\Lambda m^2 > 1$, the solutions represent the formation of a naked singularity in the de Sitter universe. Solutions with $\Lambda < 0$ describe the formation of a black hole in the anti-de Sitter universe. With decreasing negative Λ, the smoothness of the extension across the black hole horizon decreases.

It can thus be concluded that vacuum radiative Robinson–Trautman space-times of algebraic type II with $m > 0$ settle down to spherically symmetric Schwarzschild (if $\Lambda = 0$), or Schwarzschild–(anti-)de Sitter (if $\Lambda \neq 0$) type D solutions at large retarded times u. This is true for "arbitrary strong" smooth initial data at u_0 in the Robinson–Trautman class of metrics. For

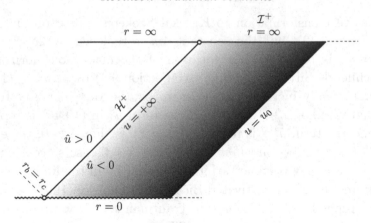

Fig. 19.10 Global structure of extreme Robinson–Trautman metrics of type II with $9\Lambda m^2 = 1$. For smooth initial data at $u = u_0$, the solution exists and converges to the extreme Schwarzschild–de Sitter metric with the same Λ and m as $u \to +\infty$. The continuation across the horizon $\mathcal{H}^+$ is smooth but non-analytic. Both conformal infinity $\mathcal{I}^+$ at $r = \infty$ and the singularity at $r = 0$ have a spacelike character.

$\Lambda > 0$, these models demonstrate the cosmic no-hair conjecture in the presence of gravitational waves. In fact, they are exact examples of black hole formation in non-spherical space-times that are not asymptotically flat. The asymmetries are radiated away, and the space-times approach a spherically symmetric solution. However, as shown by Bičák (1989), such space-times contain not only outgoing but also incoming radiation.

Finally, it should be emphasised that the properties of solutions of the Robinson–Trautman equation described above apply only when smooth initial data at u_0 are prescribed. In view of the possible presence of various curvature singularities, conical singularities and other topological defects in many of these space-times, the relevance of such smooth initial data may be too restrictive. In fact, Hoenselaers and Perjés (1993) found a class of the Robinson–Trautman space-times that asymptotically decay to the C-metric which represents a uniformly accelerating pair of black holes, as opposed to the spherically symmetric and static Schwarzschild black hole.

19.5 Robinson–Trautman space-times with pure radiation

All the space-times described in this chapter so far have been vacuum. However, the shear-free, twist-free but expanding Robinson–Trautman metric (19.1) with (19.2) generally admits a Ricci tensor component

$$\Phi_{22} = \frac{1}{4r^2}\Big[\Delta\Delta\log P + 12m(\log P)_{,u} - 4m_{,u}\Big], \qquad (19.47)$$

in addition to the Weyl tensor components (19.6) and the cosmological constant. This corresponds to the presence of an aligned pure radiation field (that is, a flow of matter of zero rest-mass, propagating in the repeated principal null direction) with energy-momentum tensor of the form $T_{\mu\nu} = \rho\, k_\mu k_\nu$, where $\boldsymbol{k} = \partial_r$. The radiation density is given by

$$\rho = \frac{n^2(\zeta, \bar\zeta, u)}{r^2}, \tag{19.48}$$

where the function n^2 is determined by the equation

$$\Delta\Delta(\log P) + 12m(\log P)_{,u} - 4m_{,u} = 16\pi\, n^2, \tag{19.49}$$

while the function $m(u)$ is arbitrary.

There are no type N or conformally flat solutions of this type. This can be seen from (19.6), since the conditions that Ψ_2 and Ψ_3 vanish imply that $m = 0$ and $\Delta \log P = K(u)$ which, in turn, imply from (19.47) that $\Phi_{22} = 0$.

19.5.1 Type D solutions

As in the vacuum case discussed in Section 19.3, the type D Robinson–Trautman space-times with pure radiation also have to satisfy the equation $P^2 K_{,\bar\zeta} = h(\zeta, u)$, where $h(\zeta, u)$ is an arbitrary function.

The simplest situation occurs when $h = 0$. The Gaussian curvature K then depends only on u, and there exist solutions of the form

$$P = A(u) + B(u)\, \zeta + \bar B(u)\, \bar\zeta + C(u)\, \zeta\bar\zeta, \qquad K = 2(AC - B\bar B), \tag{19.50}$$

where A, B, C are arbitrary functions of u (B is complex). In particular, when these are constants, the function P can be put into the canonical form

$$P = 1 + \tfrac{1}{2}\epsilon\,\zeta\bar\zeta,$$

where $\epsilon = +1, 0, -1$. In the case $\epsilon = +1$, the wavefronts $u = $ const. are exact spheres, and ζ is a stereographic coordinate such that $\zeta = \sqrt{2}\,\tan(\theta/2)\,e^{i\phi}$. Consequently, the Robinson–Trautman metric (19.1), (19.2) takes the form

$$ds^2 = -2\,du\,dr - \left(1 - \frac{2m(u)}{r} - \frac{\Lambda}{3}r^2\right)du^2 + r^2(d\theta^2 + \sin^2\theta\,d\phi^2), \tag{19.51}$$

which can be recognised as the Vaidya–(anti-)de Sitter space-time with $n^2 = -\frac{1}{4\pi}m_{,u}$. When the cosmological constant vanishes, this reduces to the standard radiating Vaidya metric (9.32) whose properties have been thoroughly described in Section 9.5. (Cases with $\epsilon = 0$ and $\epsilon = -1$ are analogous.)

Other more general solutions with $h = 0$ are possible. Those which are given by (19.50) are called Kinnersley rockets, and will be described in the next section. In addition, there also exists a large family of solutions for which $h \neq 0$. These are not described here, but are considered for a larger family of sources in Stephani *et al.* (2003).

19.5.2 Kinnersley's rocket

An interesting axially symmetric type D solution of the Robinson–Trautman family (19.1), (19.2), for $h = 0$ and $K = +1$, is that given by the function

$$P = \cosh \left(\int \alpha(u)\,\mathrm{d}u - \tfrac{1}{\sqrt{2}}(\zeta + \bar{\zeta}) \right), \tag{19.52}$$

where $\alpha(u)$ is an arbitrary function of u.[4] In this case, it is convenient to introduce new coordinates θ, ϕ by

$$\cos\theta = \tanh \left(\int \alpha(u)\,\mathrm{d}u - \tfrac{1}{\sqrt{2}}(\zeta + \bar{\zeta}) \right), \qquad \phi = \tfrac{-i}{\sqrt{2}}(\zeta - \bar{\zeta}), \tag{19.53}$$

so that $1/P = \sin\theta$, $\Delta \log P = 1$, and $(\log P)_{,u} = \alpha \cos\theta$. The line element thus takes the form

$$\mathrm{d}s^2 = -2\,\mathrm{d}u\,\mathrm{d}r - \left(1 - \frac{2\,m(u)}{r} - \frac{\Lambda}{3}r^2 - 2\alpha(u)\,r\cos\theta - \alpha^2(u)\,r^2\sin^2\theta \right)\mathrm{d}u^2$$
$$+ 2\alpha(u)\,r^2\sin\theta\,\mathrm{d}u\,\mathrm{d}\theta + r^2\left(\mathrm{d}\theta^2 + \sin^2\theta\,\mathrm{d}\phi^2 \right), \tag{19.54}$$

in which $m(u)$ and $\alpha(u)$ are arbitrary functions of the null coordinate u.

For a pure radiation field $T_{\mu\nu} = \rho\,k_\mu k_\nu$, where ρ is given by (19.48), the field equation (19.49) yields

$$3\,\alpha\,m\,\cos\theta - m_{,u} = 4\pi\,n^2(\theta, u). \tag{19.55}$$

For any given functions $\alpha(u)$ and $m(u)$, the radiation field profile $n^2(\theta, u)$ is thus fully determined. However, for this to be real for all θ, it is necessary that $m_{,u} < -3\,|\alpha\,m|$, which implies that $m(u)$ must be strictly decreasing.

If $\alpha = 0$, the metric (19.54) reduces to the Vaidya–(anti-)de Sitter spacetime expressed in the form (19.51). This describes a non-accelerating and spherically symmetric source with a varying mass determined by $m(u)$ for which the radiation field is given by $n^2(u) = -\tfrac{1}{4\pi}\,m_{,u}$. It is therefore generally assumed that m is a positive and non-increasing function.

When $\Lambda = 0$, the solution (19.54) is known as the *Kinnersley rocket* (see Kinnersley, 1969b). It represents a singular "point-like" source at $r = 0$, of

---

[4] This is obtained from (19.50) by the transformation $\zeta = \exp(\sqrt{2}\,\tilde{\zeta})$ with the identification $A(u) = \tfrac{1}{\sqrt{2}}\exp(\int\alpha(u)\,\mathrm{d}u)$, $B = 0$, $C(u) = \tfrac{1}{\sqrt{2}}\exp(-\int\alpha(u)\,\mathrm{d}u)$, and dropping the tilde.

mass determined by $m(u)$, which emits pure radiation and accelerates (along a straight line which is the axis of symmetry) due to the corresponding net back-reaction. It therefore serves as a simple exact model of a kind of rocket that is propelled by the anisotropic emission of photons.

In the limit in which $m = 0$, the above space-time is a conformally flat vacuum solution, namely Minkowski, de Sitter or anti-de Sitter space according to the sign of the cosmological constant Λ. The function $\alpha(u)$ then determines the acceleration of a test particle located at the origin $r = 0$ which moves along the timelike trajectory $x^\mu = (u(\tau), r = 0, \theta_0, \phi_0)$, where τ is its proper time, and θ_0 and ϕ_0 are constants. Normalisation requires that $u(\tau) = \tau$, so that the four-velocity is $u^\mu = (1, 0, 0, 0)$. For the corresponding four-acceleration $a^\mu = \Gamma^\mu{}_{uu}$, it follows that $u_\mu a^\mu = 0$ and $a_\mu a^\mu = \alpha^2(u)$. This proves that the *instantaneous acceleration* of the test particle is given by $\alpha(u) \equiv \alpha(\tau)$. And, since the function $\alpha(u)$ in the metric (19.54) can be chosen arbitrarily, it is possible to prescribe any specific acceleration for the source at $r = 0$. In particular, it is possible to construct a rocket that has a constant acceleration α.

The solution (19.54) describes an accelerating source which emits pure (null) radiation. Note, however, that it is not directly related to the C-metric, which describes uniformly accelerating black holes. The main difference is that the Kinnersley rocket contains pure radiation whereas the C-metric is a vacuum solution. When $n = 0$ in the field equation (19.55), the space-time is vacuum and the only possibilities are given by $\alpha = 0$, $m = $ const., which yields the Schwarzschild–(anti-)de Sitter space-time, or $m = 0$, which corresponds to the Minkowski or (anti-)de Sitter universe. Also, the complete axis given by $\theta = 0, \pi$ in (19.54) is regular, without strings or struts, while the C-metric space-time must always include a topological singularity of some kind on at least one half of the axis. This is the alternative "physical cause" of the acceleration of the black holes in this case. Nevertheless, these two distinct types of solutions are similar. Both of these classes of accelerating space-times can be written in the common form (19.39). The Kinnersley rocket (19.54) is obtained when $x = \cos\theta$, $G = 1 - x^2$ and $2H = 1 - 2\,m(u)/r - \frac{\Lambda}{3}r^2 - 2\alpha(u)\,r\,x$.

In fact, by considering the limit $m \to 0$, the Kinnersley solution (19.54) with $\Lambda = 0$ reduces to the flat-space metric whose origin is accelerating, as introduced already by Newman and Unti (1963). Using the transformation to standard Minkowski coordinates, Bonnor (1994) found, quite surprisingly, that with respect to the natural flat background frame there is no energy loss corresponding to gravitational radiation emitted by the accelerating rocket. This *absence of gravitational radiation* in the Kinnersley space-time has been subsequently confirmed and clarified by Damour (1995) in the context of the associated problem using the post-Minkowskian perturbation formalism. He argued that, although the Kinnersley rocket emits null fluid anisotropically, it does not produce (linearised) gravitational waves because the matter emission in it has no quadrupole moment. In fact, the radiation is the sum of waves generated by the point-like rocket and of those generated by the distribution of the photon fluid. To the highest order, these two distinct contributions cancel each other.

Further studies of this effect have been undertaken by Dain, Moreschi and Gleiser (1996) who employed general properties of exact Robinson–Trautman geometries with pure radiation to distinguish, in particular, gravitational and matter-energy radiation. By transforming to Bondi–Sachs coordinates, von der Gönna and Kramer (1998) have demonstrated that the Kinnersley photon rocket family is the only axisymmetric and asymptotically flat Robinson–Trautman solution with pure radiation that does not contain gravitational radiation. This has been confirmed by Cornish (2000) even without the assumption of axial symmetry (see also Section 19.5.4).

19.5.3 Type II solutions

As summarised in Section 19.4, when smooth data are prescribed on the initial surface $u = u_0$, the Robinson–Trautman vacuum space-times of algebraic type II exist and evolve towards the spherically symmetric Schwarzschild or Schwarzschild–(anti-)de Sitter space-time as $u \to +\infty$. It was shown by Bičák and Perjés (1987) that analogous behaviour occurs also when pure radiation is present. Specifically, it has been demonstrated that type II Robinson–Trautman space-times with pure radiation approach asymptotically the spherically symmetric Vaidya metric (see Section 9.5).

These results have been further generalised by Podolský and Svítek (2005) to include also a non-vanishing cosmological constant Λ. Indeed, space-times from the Robinson–Trautman family of exact solutions with pure radiation approach, for smooth initial data at u_0, the Vaidya–(anti-)de Sitter space-time (19.51) as $u \to +\infty$. Extensions of these metrics across the horizon are again possible, but their order of smoothness is in general only finite.

These models can be used, for example, to describe the evaporation of a white hole in a Minkowski or (anti-)de Sitter universe. Or, using the "advanced" rather than the "retarded" form of these space-times, a non-spherical generalisation of the gravitational collapse of a shell of null dust forming a naked singularity can be constructed and investigated.

19.5.4 Bonnor's rocket

Physically interesting type II space-times with pure radiation include those which generalise the Kinnersley photon rocket (see Section 19.5.2). Restricting attention to axially symmetric Robinson–Trautman solutions, it is convenient to introduce the polar-type coordinates x, ϕ such that

$$\zeta = \frac{1}{\sqrt{2}} \left(-\int \frac{\mathrm{d}x}{G(x,u)} + \int \alpha(u)\,\mathrm{d}u + i\,\phi \right), \qquad (19.56)$$

where $G(x, u)$ is an arbitrary function of x and u, while $\alpha(u)$ is any function of u. With the identification

$$P(\zeta, \bar\zeta, u) = G^{-1/2}\left(x(\zeta, \bar\zeta, u), u \right), \qquad (19.57)$$

in which the function $x(\zeta, \bar\zeta, u)$ is obtained by inverting (19.56) as

$$\int \frac{\mathrm{d}x}{G(x,u)} = \int \alpha(u)\,\mathrm{d}u - \tfrac{1}{\sqrt{2}}(\zeta + \bar\zeta), \qquad (19.58)$$

the standard Robinson–Trautman metric (19.1) takes the form

$$\mathrm{d}s^2 = -2\,\mathrm{d}u\,\mathrm{d}r - \left(-\frac{1}{2}G_{,xx} - \frac{2m(u)}{r} - \frac{\Lambda}{3}r^2 - r\,(b\,G)_{,x} - b^2\,G\,r^2 \right) \mathrm{d}u^2$$
$$+ 2\,b\,r^2\,\mathrm{d}u\,\mathrm{d}x + r^2 \left(\frac{\mathrm{d}x^2}{G} + G\,\mathrm{d}\phi^2 \right), \qquad (19.59)$$

where

$$b(x, u) = -\alpha(u) - \int \frac{G_{,u}(x, u)}{G^2(x, u)}\,\mathrm{d}x.$$

This is the generalisation of the metric (19.39) in which $G(x)$ is independent of u and thus $b(u) = -\alpha(u)$.

In particular, the Kinnersley rocket is obtained by taking $G(x) = 1 - x^2$. Then, introducing an angular coordinate by putting $x = \cos\theta$, the metric (19.54) is immediately recovered, and the general expressions (19.58) and (19.57) reduce exactly to (19.53) and (19.52), respectively.

A generalisation of the type D Kinnersley rocket was presented by Bonnor (1996). This is contained within the above family of axially symmetric Robinson–Trautman type II solutions (19.59) with $\Lambda = 0$. This exact family of photon rockets, which accelerate due to the anisotropic emission of null fluid, is given by the function

$$G(x, u) = (1 - x^2)\left[1 + (1 - x^2)\,h(x, u) \right], \qquad (19.60)$$

where $h(x, u)$ is an arbitrary smooth bounded function (greater than -1). Since G must be positive, it is appropriate to put $x = \cos\theta$, and the metric can be seen to be regular on the axis $x = \pm 1$ (i.e. $\theta = 0, \pi$). The function b in (19.59) now reads

$$b(x, u) = -\alpha(u) - \int \frac{h_{,u}(x, u)}{[1 + (1 - x^2)\, h(x, u)]^2} \, \mathrm{d}x \,,$$

and the pure radiation field, see (19.48) and (19.49), is

$$4\pi\, n^2(x, u) = -\tfrac{1}{8}(GG_{,xxx})_{,x} + \tfrac{3}{2}m(b\, G)_{,x} - m_{,u} \,, \tag{19.61}$$

which generalises (19.55). The angular distribution of photons thus now has a more complicated character that can be prescribed almost arbitrarily by a specific choice of the smooth function $h(x, u)$.

Despite the fact that the photon flux in general contains quadrupole terms, Bonnor (1996) has observed that such photon rockets with $\Lambda = 0$ lose no energy as gravitational radiation. However, this is in disagreement with later results obtained by von der Gönna and Kramer (1998) that such general type II axisymmetric and asymptotically flat Robinson–Trautman solutions with a pure radiation field admit gravitational radiation. The energy balance at future null infinity shows that the mass loss is due to a superposition of both the pure and gravitational radiation parts. The only exception is the Kinnersley photon rocket (see Section 19.5.2) which is of type D, and for which there is no gravitational radiation. This result has been subsequently extended by Cornish (2000) who demonstrated that the class of pure radiation Robinson–Trautman metrics for which the news function vanishes is the same as the class of all (not necessarily axially symmetric) Kinnersley rocket metrics (19.50).

Further generalisations of these rocket solutions have been given by Ivanov (2005) in the class of (possibly twisting) Kerr–Schild fields without gravitational radiation, and by Podolský (2008) to include a cosmological constant. The case with a charged null radiation and an electromagnetic field is included in the solutions of Khlebnikov (1978).

19.6 Comments on further solutions

Apart from the physically most important type N, D and II vacuum and pure radiation space-times described in this chapter, there exist other Robinson–Trautman solutions. For example, some type III solutions are known for the vacuum and pure radiation case, see Stephani *et al.* (2003). However, these do not seem to have any known obvious physical relevance.

Also, there are Robinson–Trautman solutions of the Einstein–Maxwell equations. This large family of expanding but non-twisting electrovac-uum space-times contains, for example, the Reissner–Nordström black holes (see Section 9.2) or the charged version of the C-metric (see Section 14.2). They are also included in the non-twisting subfamily of type D Plebański–Demiański solutions (see Chapter 16).

All such Robinson–Trautman solutions with an electromagnetic field are summarised in Stephani *et al.* (2003), where references to original works can be found. Some physical properties of these type II solutions have been studied by Kozameh, Newman and Silva-Ortigoza (2006). Note also that it has been demonstrated by Kozameh, Kreiss and Reula (2008) that the corresponding Robinson–Trautman–Maxwell equations are unstable against linear perturbations. In fact, they do not constitute a well-posed initial value problem. Any smooth perturbation of an exact solution would produce an arbitrarily large solution when the initial data are evolved by these field equations.

20

Impulsive waves

Physical situations exist in which different regions of space-time have different matter contents. These can be modelled by compound space-times. (For example, in Subsection 9.5.2, a space-time was discussed in which a region represented by the Vaidya metric is sandwiched between a Minkowski and a Schwarzschild region.) In such cases, we have followed the approach of Lichnerowicz (1955) and used a global metric form[1] that is at least C^3 everywhere except on junctions, represented by hypersurfaces $\mathcal{N}$, on which the metric is only C^1. Since the curvature of the space-time involves second derivatives of the metric, such situations give rise to discontinuities in the curvature across $\mathcal{N}$. When $\mathcal{N}$ is null, these may represent various forms of *shock waves* which propagate with the speed of light.

More extreme situations may also be considered for which the metric is still (at least) C^3 almost everywhere, but merely C^0 on some hypersurface $\mathcal{N}$. In such situations, some components of the curvature of the space-time will formally contain a δ-function. When $\mathcal{N}$ is null, these may be interpreted as *impulsive waves*. They are regarded as impulsive gravitational waves when the δ-function components occur in the Weyl tensor, or impulsive components of some kind of null matter when they occur in the Ricci tensor.

The geometry of impulsive waves in flat space was first described in detail by Penrose (1972). However, some particular examples of exact solutions which include impulsive gravitational waves or thin sheets of null matter were known before then. Many further examples have subsequently been obtained, and involve a variety of backgrounds. These have been reviewed from different points of view, and with different emphases, by Podolský (2002b) and Barrabès and Hogan (2003a).

---

[1] Darmois (1927) has expressed junction conditions across a non-null hypersurface $\mathcal{N}$ by requiring equality of the first and second fundamental forms evaluated on either side. This actually guarantees the existence of coordinates in which the metric is continuous across $\mathcal{N}$.

In the present chapter, we will concentrate mainly on impulsive waves that occur in Minkowski, de Sitter or anti-de Sitter backgrounds, as these are the simplest cases that illustrate their generic properties. In these backgrounds, such waves belong to two distinct families according to whether the corresponding null congruences tangent to the propagation directions have zero or non-zero expansion.

20.1 Methods of construction

In constant-curvature background space-times, for both non-expanding and expanding cases, particular exact solutions can be constructed by three basic and distinct methods:

1. As a distributional limit of a class of radiative space-times in which the radiation component initially has an arbitrary profile. In this case, the limit is considered in which the profile function approaches a Dirac distribution with support on a single null hypersurface. This is the most natural approach to impulsive waves, as impulses are intuitively regarded as mathematical idealisations of short pulses of radiation (sandwich waves). In this case, the non-expanding and expanding waves arise as limits of the *pp*- or other Kundt waves and the Robinson–Trautman type N space-times, respectively. However, technical problems may arise in the limit when considering the continuity of a space-time across the impulse.

2. By a geometrical construction in which two different space-time regions are joined together across a null hypersurface $\mathcal{N}$. For a Minkowski background, a specific method was introduced by Penrose (1968a,b, 1972) which involves cutting the space-time along a null hypersurface and re-attaching the two pieces with a suitable warp. Both non-expanding and expanding impulsive waves can be constructed in this way according to whether the null hypersurface is a plane or a cone. In both cases, the construction involves the introduction of an arbitrary function that can be chosen to represent particular physical situations. In this approach, it is important to identify a coordinate system that is continuous across the impulse, although this may be transformed to one in which the metric explicitly contains an impulsive component.

3. As a limit of a suitable solution in which either the velocity of the source approaches that of light or its acceleration becomes unbounded (with the other physical parameters being scaled appropriately). This approach was initially pioneered by Aichelburg and Sexl (1971) in boosting the

Schwarzschild space-time, and has subsequently been developed and generalised by many other researchers. Non-expanding and expanding impulsive waves are obtained, respectively, as limits of solutions with boosted or accelerating sources.

Each of the above approaches needs to be modified when the cosmological constant is non-zero. The first two methods of construction lead to complete families of solutions, while the third leads to particular, but physically significant, cases only.

The following sections generally start by using the second approach. It will then be shown how this is related to the first. Particular solutions that have been derived using the third method will also be commented on.

20.2 Non-expanding impulsive waves in Minkowski space

It is convenient to start here with the "cut and paste" method for constructing impulsive wave space-times that was introduced by Penrose (1968a,b, 1972). This involves cutting Minkowski space along a plane null hypersurface $\mathcal{N}$ and then re-attaching the two halves $\mathcal{M}^-$ and $\mathcal{M}^+$ with an arbitrary "warp". This process is illustrated in Figure 20.1.

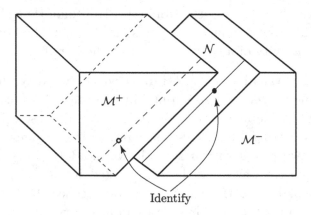

Fig. 20.1 Minkowski space is cut into two parts $\mathcal{M}^-$ and $\mathcal{M}^+$ along a plane null hypersurface $\mathcal{N}$. These parts are then re-attached with an arbitrary "warp" in which points are shunted along the null generators of the cut and then identified.

Technically, it is appropriate to start with Minkowski space in the form (3.4) using complex spatial coordinates, namely

$$\mathrm{d}s^2 = -2\,\mathrm{d}u\,\mathrm{d}v + 2\,\mathrm{d}\zeta\,\mathrm{d}\bar{\zeta}. \tag{20.1}$$

Applying the coordinate transformation $u = U$, $v = V - H + UH_{,Z}H_{,\bar{Z}}$,

$\zeta = Z - U H_{,\bar{Z}}$, where $H = H(Z, \bar{Z})$ is an arbitrary real function of the spatial coordinates, leads to the line element

$$ds^2 = -2\,dU\,dV + 2\left|dZ - U(H_{,Z\bar{Z}}dZ + H_{,\bar{Z}\bar{Z}}d\bar{Z})\right|^2 . \qquad (20.2)$$

Clearly the hypersurface $u = 0$, (or, equivalently $U = 0$) is a common null hyperplane $\mathcal{N}$ in the two representations of the same space-time. Following the "cut and paste" method, one of these metric forms can be taken on each side of the null hypersurface. Specifically, for $\mathcal{M}^-$, (20.1) is taken with the trivial re-parametrisation $u = U$, $v = V$ and $\zeta = Z$ for $U < 0$. This may be combined with $\mathcal{M}^+$ given by (20.2) for $U > 0$. The resulting line element is

$$ds^2 = -2\,dU\,dV + 2\left|dZ - U\Theta(U)(H_{,Z\bar{Z}}dZ + H_{,\bar{Z}\bar{Z}}d\bar{Z})\right|^2 , \qquad (20.3)$$

where $\Theta(U)$ is the Heaviside step function. This is evidently continuous across the entire null hypersurface $U = 0$ (except at any poles in the derivatives of H). However, the discontinuity in the derivatives of the metric across this hypersurface produces impulsive components in the curvature tensor as will be specified below. Elsewhere, for $U \neq 0$, the space-time is flat. The metric (20.3) therefore represents a planar impulsive wave in a Minkowski background.

In fact, the combination of the above transformations, applied to both regions of the space-time simultaneously, can be expressed in the explicit form

$$u = U, \qquad v = V - H\,\Theta(U) + U\,\Theta(U)\,H_{,Z}H_{,\bar{Z}}, \qquad \zeta = Z - U\,\Theta(U)\,H_{,\bar{Z}}, \qquad (20.4)$$

which is obviously discontinuous at $u = 0$ in such a way that

$$(u = 0, v, \zeta, \bar{\zeta})_{\mathcal{M}^-} = (u = 0, v + H(\zeta, \bar{\zeta}), \zeta, \bar{\zeta})_{\mathcal{M}^+} . \qquad (20.5)$$

This is the general *Penrose junction condition* for re-attaching the two halves of the space-time $\mathcal{M}^-(u < 0)$ and $\mathcal{M}^+(u > 0)$ with a "warp" that is determined by the arbitrary real function $H(Z, \bar{Z})$. It can also be seen that this warp function represents an arbitrary shunt along the null generators ∂_v of the cut, as illustrated in Figure 20.1.

Clearly, the transformation (20.4) relates both $\mathcal{M}^-$ and $\mathcal{M}^+$ to the initial metric (20.1). However, and most significantly, when applied (formally) to the combined continuous form of the impulsive wave metric (20.3), it leads to the metric

$$ds^2 = -2\,du\,dv - 2H(\zeta, \bar{\zeta})\,\delta(u)\,du^2 + 2\,d\zeta\,d\bar{\zeta}. \qquad (20.6)$$

By comparing this with (20.1), it can be seen that an impulsive component proportional to the Dirac δ-function and located on the wavefront $u = 0$ is explicitly included in this metric.

It may now be observed that the space-time (20.6) is included in the general class of pp-waves expressed in the standard Brinkmann form (17.1). In this case the wave profile, which could be an arbitrary function of the null coordinate u, is simply taken to be proportional to the Dirac δ-function distribution. This may be considered physically as a natural limit of a "sandwich" wave that initially has compact support on a particular range of u. (This approach has been described as method 1 in Section 20.1.) As in any impulse, the u-dependent profile in the sandwich may be arbitrary, but the limit is taken in which its duration becomes arbitrarily small while its amplitude correspondingly increases. (For some explicit examples of such formal limits, see Podolský and Veselý, 1998a.) In fact, the same solutions (20.6) also arise (Podolský, 1998b) as impulsive limits of the more general class of Kundt waves (18.1) with $P = 1$ (since $\Lambda = 0$ in this case).

It also follows from (17.2), that the only non-zero components of the curvature tensor for the metric (20.6) are given by

$$\Psi_4 = H_{,\zeta\zeta}\, \delta(u), \qquad \Phi_{22} = H_{,\zeta\bar\zeta}\, \delta(u). \qquad (20.7)$$

It is therefore clear that the space-time (20.6) indeed generally represents an arbitrary impulsive wave, having a plane surface which propagates with the speed of light along a null hyperplane $u = 0$ in a Minkowski background.

The mathematical problems associated with the discontinuity in the transformation (20.4) and the correct distributional limit of sandwich waves have been investigated by Steinbauer (1998) and Kunzinger and Steinbauer (1999a,b). By considering the impulsive solution as limits of sandwich waves in the two coordinate systems, they have given a precise meaning to the "physical equivalence" of the two metrics (20.3) and (20.6). They have also shown that the transformation can be defined rigorously within Colombeau's theory of generalised functions.

Among the above family of impulsive pp-waves, the case in which $H(\zeta, \bar\zeta)$ is quadratic in ζ and $\bar\zeta$ is of particular interest. (Terms that are independent of, or are linear in, ζ and $\bar\zeta$ may be removed by a coordinate transformation.) This represents the familiar impulsive *plane* wave that is contained within the wider family of plane wave space-times given by (17.7). A simple and important case of this will be described in more detail in Section 21.3.

Another case that is of particular interest is that given by the line element

$$\mathrm{d}s^2 = -2\,\mathrm{d}u\,\mathrm{d}v - 4\mu \log(\zeta\bar\zeta)\,\delta(u)\,\mathrm{d}u^2 + 2\,\mathrm{d}\zeta\,\mathrm{d}\bar\zeta. \qquad (20.8)$$

Derived initially by Aichelburg and Sexl (1971) using the third method outlined in Section 20.1, it describes the field of a single particle or black hole boosted up to the speed of light. This solution has been obtained by considering a black hole of mass m, represented by the Schwarzschild metric, moving with speed v in the limit as $v \to 1$ and $m \to 0$ such that the quantity $\mu = m(1 - v^2)^{-1/2}$ is held constant. By considering the Ricci tensor component of the resulting metric (20.8), it can be seen that the energy-momentum tensor for the space-time in this limit has the form

$$T_{\alpha\beta} = \mu\, \delta(u)\, \delta(\zeta)\, k_\alpha\, k_\beta,$$

where $\boldsymbol{k}$ is a null covector such that $k_\alpha = -u_{,\alpha}$. This indicates that the solution represents the field of a point-like monopole particle at $\zeta = 0$ which moves along the null hypersurface $u = 0$. It is of interest to note that there is no event horizon in this limiting case as the rest-mass of the source has been scaled to zero.

Aichelburg and Balasin (1998, 2000) have shown that the Bondi and ADM four-momenta for this space-time are well defined, equal and null. For this case, a continuous metric has been given by D'Eath (1978) in the form[2]

$$\mathrm{d}s^2 = -2\,\mathrm{d}U\,\mathrm{d}V + \left(1 + \frac{4\mu}{\rho^2}\, U\,\Theta(U)\right)^2 \mathrm{d}\rho^2 + \left(1 - \frac{4\mu}{\rho^2}\, U\,\Theta(U)\right)^2 \rho^2 \mathrm{d}\phi^2.$$

This line element and (20.8) correspond, respectively, to (20.3) and (20.6) with the particular choices $H(Z, \bar{Z}) = 2\mu \log(Z\bar{Z})$ and $Z = \frac{1}{\sqrt{2}}\rho\, \mathrm{e}^{\mathrm{i}\phi}$.

Using the fact that the field equations for the metric (20.6) reduce to the two-dimensional Poisson equation

$$\Delta H = 2\,H_{,\zeta\bar{\zeta}} = 8\pi\, T_{uu},$$

Griffiths and Podolský (1997) have explicitly constructed exact solutions for which the sources are null particles with arbitrary multipole structures.

Since the vacuum field equation (17.3) for pp-waves is linear, distinct solutions of the above type may be superposed, provided only that the individual particles are all propagating in the same direction. Since the above solution is that of a null particle, it may alternatively be interpreted as the field of a photon. Hence solutions describing beams of photons may be constructed (Bonnor, 1969b).

By boosting the Kerr solution to the speed of light along the axis of symmetry using a similar limiting process, Ferrari and Pendenza (1990) have obtained an impulsive gravitational pp-wave for which the source corresponds to a ring of null particles (see also Balasin and Nachbagauer, 1995, 1996, and

---

[2] A continuous form of the metric for a more general family of impulsive pp-wave space-times has been given by Podolský and Veselý (1998b).

Barrabès and Hogan, 2003b). Further particular impulsive *pp*-wave space-times can also be obtained by boosting some other well-known space-times to the ultrarelativistic limit. For example, some interesting examples have been obtained by boosting the Kerr–Newman family (Loustó and Sánchez, 1992, 1995), and the Weyl solutions for fields of null particles with arbitrary multipole structure (Podolský and Griffiths, 1998b). Other limits of boosts of the Lewis–Papapetrou family of solutions have been summarised by Aichelburg (1991).

20.3 Non-expanding impulsive waves in (anti-)de Sitter space

Non-expanding impulsive waves propagating in a de Sitter (or anti-de Sitter) universe can be obtained in exactly analogous ways to those described above. They can also all be derived using Penrose's "cut and paste" method (see Podolský and Griffiths, 1999b). In this case, the initial metric of the constant-curvature space-time is taken in the form (4.19), namely

$$\mathrm{d}s^2 = \frac{-2\,\mathrm{d}\mathcal{U}\,\mathrm{d}\mathcal{V} + 2\,\mathrm{d}\xi\,\mathrm{d}\bar{\xi}}{[\,1 - \frac{1}{6}\Lambda(\mathcal{U}\mathcal{V} - \xi\bar{\xi})\,]^2}. \tag{20.9}$$

It is now possible to proceed exactly as above, applying different transformations either side of the null hypersurface $u = 0$ and then re-attaching them. In fact, exactly the same transformations may be adopted as those given in the combined form (20.4) with $\mathcal{U}$, $\mathcal{V}$ and ξ replacing u, v and ζ, respectively, provided the Penrose junction condition (20.5) for attaching the two parts is also applied. The resulting line element is then given by

$$\mathrm{d}s^2 = \frac{-2\,\mathrm{d}U\,\mathrm{d}V + 2\left|\mathrm{d}Z - U\Theta(U)(H_{,Z\bar{Z}}\mathrm{d}Z + H_{,\bar{Z}\bar{Z}}\mathrm{d}\bar{Z})\right|^2}{[\,1 - \frac{1}{6}\Lambda(UV - Z\bar{Z} - U\Theta(U)G)\,]^2}, \tag{20.10}$$

where $G = H - ZH_{,Z} - \bar{Z}H_{,\bar{Z}}$. This clearly reduces to (20.3) when $\Lambda = 0$. In an exact analogue of the previous case, the transformation (20.4) relates the metric (20.10), not to (20.9), but to the form

$$\mathrm{d}s^2 = \frac{-2\,\mathrm{d}\mathcal{U}\,\mathrm{d}\mathcal{V} - 2H(\xi,\bar{\xi})\,\delta(\mathcal{U})\,\mathrm{d}\mathcal{U}^2 + 2\,\mathrm{d}\xi\,\mathrm{d}\bar{\xi}}{[\,1 - \frac{1}{6}\Lambda(\mathcal{U}\mathcal{V} - \xi\bar{\xi})\,]^2}, \tag{20.11}$$

which explicitly represents an impulsive wave in any background space-time of constant curvature. The metric (20.11) is clearly conformal to an impulsive *pp*-wave.

This family of space-times can also be obtained as the impulsive limit of the family of Kundt waves with a non-zero cosmological constant. Full details of this have been given by Podolský (1998b). In particular, it may be noticed that the transformation

$$\mathcal{U} = (\zeta + \bar\zeta)(1 + uv)u, \qquad \mathcal{V} = (\zeta + \bar\zeta)\,v, \qquad \xi = \zeta + (\zeta + \bar\zeta)\,uv,$$

takes the line element (20.11) to the form

$$ds^2 = \frac{(\zeta + \bar\zeta)^2}{(1 + \frac{1}{6}\Lambda\zeta\bar\zeta)^2}\left(-2\,\mathrm{d}u\,\mathrm{d}v + 2v^2\mathrm{d}u^2 + 2H(\zeta,\bar\zeta)\,\delta(u)\,\mathrm{d}u^2\right) + \frac{2\,\mathrm{d}\zeta\,\mathrm{d}\bar\zeta}{(1 + \frac{1}{6}\Lambda\zeta\bar\zeta)^2},$$

which is clearly a special case of (18.13). Then, denoting

$$P = 1 + \tfrac{1}{6}\Lambda\zeta\bar\zeta, \qquad \tau = -\frac{1 - \frac{1}{6}\Lambda\zeta^2}{\zeta + \bar\zeta},$$

as described in Section 18.3, and putting

$$v = \frac{P^2}{(\zeta + \bar\zeta)^2}\,r,$$

the metric can be rewritten as

$$ds^2 = -2\mathrm{d}u\left(\mathrm{d}r + \tilde{H}\,\mathrm{d}u + \frac{2\bar\tau}{P}r\,\mathrm{d}\zeta + \frac{2\tau}{P}r\,\mathrm{d}\bar\zeta\right) + \frac{2\,\mathrm{d}\zeta\,\mathrm{d}\bar\zeta}{P^2}, \tag{20.12}$$

where

$$\tilde{H} = -(\tau\bar\tau + \tfrac{1}{6}\Lambda)r^2 - \frac{(\zeta + \bar\zeta)^2}{P^2}\,H(\zeta,\bar\zeta)\,\delta(u).$$

This form corresponds exactly to an impulsive wave of Kundt's class (18.1) in which H and G are defined by (18.2) with $H^\circ = -(\zeta + \bar\zeta)^2 P^{-2}H(\zeta,\bar\zeta)\delta(u)$ and $G^\circ = 0 = W^\circ$. Moreover, it is expressed in the form that is referred to in Section 18.3 as a generalised Kundt wave, and for which the background is described in Section 18.4.2. However, the identical form (20.12) can also be achieved from (20.11) for impulsive limits of generalised pp-waves and generalised Siklos waves, which have different canonical forms of τ.

Except on the impulse itself, this space-time is either de Sitter or anti-de Sitter space, and this can be conveniently represented as the four-dimensional hyperboloid (18.27) in the five-dimensional flat space (18.28) (see also Chapters 4 and 5). The parametrisation of the hyperboloid in terms of the coordinates of the metric (20.9) have already been given in (4.17). In fact, as shown by Podolský and Griffiths (1999b), the metric (20.11) corresponds to the parametrisation of the class of nonexpanding impulsive wave solutions that can be described in a five-dimensional formalism as

$$ds^2 = -\mathrm{d}Z_0{}^2 + \mathrm{d}Z_1{}^2 + \mathrm{d}Z_2{}^2 + \mathrm{d}Z_3{}^2 + \varepsilon\mathrm{d}Z_4{}^2$$
$$+ \tilde{H}(Z_2, Z_3, Z_4)\delta(Z_0 - Z_1)(\mathrm{d}Z_0 - \mathrm{d}Z_1)^2,$$

where $\varepsilon = \operatorname{sign} \Lambda$ and

$$\tilde{H}(\xi, \bar{\xi}) = -\sqrt{2} \left(1 + \tfrac{1}{6} \Lambda \xi \bar{\xi}\right)^{-1} H(\xi, \bar{\xi}). \tag{20.13}$$

In this relation, the parametrisation (4.17), restricted to the impulsive wave surface $\mathcal{U} = 0$ is used to express the arguments of $\tilde{H}$ in terms of ξ and $\bar{\xi}$ only. Moreover, the impulsive wave surface itself corresponds to the null section of the hyperboloid on which $Z_0 - Z_1 = 0$, and thus from (18.27), it satisfies

$$Z_2{}^2 + Z_3{}^2 + \varepsilon Z_4{}^2 = \varepsilon a^2,$$

where $a = \sqrt{3/|\Lambda|}$. The location of these sections for both de Sitter and anti-de Sitter spaces is illustrated in Figure 20.2.

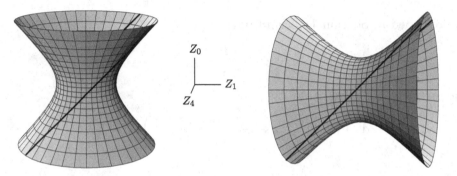

Fig. 20.2 Non-expanding impulsive waves can be pictured as sections of the four-dimensional hyperboloids representing de Sitter and anti-de Sitter spaces on which $Z_0 - Z_1 = 0$. The coordinates Z_2 and Z_3 have been suppressed, so that the impulsive waves are represented as pairs of null lines.

The geometrical and global properties of the impulsive wave surface are clearly different in these two cases. For the de Sitter space in which $\Lambda > 0$ and $\varepsilon = 1$, the wave surface is given by $Z_2{}^2 + Z_3{}^2 + Z_4{}^2 = a^2$. This is clearly *spherical* and has constant area $4\pi a^2 = 12\pi/\Lambda$, which confirms that the waves are non-expanding in this family of solutions. Moreover, the impulsive wave surface is identified with a cosmological horizon of the de Sitter space. The null line on the left hyperboloid in Figure 20.2, together with the parallel line on the opposite side, may be considered as the trajectories of opposite poles of the same spherical wave surface on which $Z_2 = 0$, $Z_3 = 0$ and $Z_4 = \pm a$.

Alternatively, in an anti-de Sitter background for which $\Lambda < 0$ and $\varepsilon = -1$, the null wave surface is the *hyperboloid* $Z_2{}^2 + Z_3{}^2 - Z_4{}^2 = -a^2$. The null line on the face of the right hyperboloid in Figure 20.2, on which $Z_4 = a$, represents the trajectory of the point on a wave surface on which $Z_2 = 0$ and $Z_3 = 0$. This propagates from one side of the universe to the other.

The null line on the opposite side of this hyperboloid on which $Z_4 = -a$ represents the trajectory of the same point when it propagates back in the opposite direction. (Further representations of these wave surfaces, for both cases and in various coordinate systems, have been given by Podolský and Griffiths, 1997.)

Using a natural null tetrad, the non-zero components of the Weyl and Ricci tensors for the metric (20.11) are given by

$$\Psi_4 = \left(1 + \tfrac{1}{6}\Lambda\,\xi\,\bar\xi\right)^2 H_{,\xi\xi}\,\delta(\mathcal{U}),$$

$$\Phi_{22} = \left(1 + \tfrac{1}{6}\Lambda\,\xi\,\bar\xi\right)\left[\left(1 + \tfrac{1}{6}\Lambda\,\xi\,\bar\xi\right)H_{,\xi\bar\xi} + \tfrac{1}{6}\Lambda\left(H - \xi H_{,\xi} - \bar\xi H_{,\bar\xi}\right)\right]\delta(\mathcal{U}).$$

This generalises the expressions (20.7) for impulsive waves in a Minkowski background. In terms of the function $\tilde{H}$, defined by (20.13), the vacuum field equations $\Phi_{22} = 0$ on the impulse can be expressed in the form

$$\left(1 + \tfrac{1}{6}\Lambda\,\xi\,\bar\xi\right)^2 \tilde{H}_{\xi\bar\xi} + \tfrac{1}{3}\Lambda\,\tilde{H} = 0, \tag{20.14}$$

which is simply $(\Delta + \tfrac{2}{3}\Lambda)\tilde{H} = 0$, where Δ is the Laplacian operator on the two-dimensional impulsive wave surface (see Sfetsos, 1995, and Horowitz and Itzhaki, 1999). Podolský and Griffiths (1998a, 1999b) have obtained explicit solutions of this equation which represent non-expanding impulsive waves generated by null particles of arbitrary multipole structure.

Just as impulsive wave space-times in a Minkowski background can be constructed by taking limits of boosted stationary solutions, impulsive waves in de Sitter or anti-de Sitter backgrounds can also be obtained by a similar process. However, in these cases, the limiting procedure must be undertaken carefully using elements of the (anti-)de Sitter group of isometries. This was successfully achieved first by Hotta and Tanaka (1993), who boosted the Schwarzschild–(anti-)de Sitter space-time (9.26) to the ultrarelativistic limit as the mass parameter is also scaled to zero (see also Podolský and Griffiths, 1997). Using this approach, the particular case of the above family of solutions corresponding to a null monopole particle is obtained. This is the analogue of the Aichelburg–Sexl solution (20.8). In the de Sitter background, there are two such particles situated at opposite poles of the non-expanding spherical impulsive wave that is located on the cosmological horizon. They are thus causally separated from each other.

20.4 Expanding impulsive waves in Minkowski space

In his classic article, Penrose (1972) also showed that an impulsive *spherical* gravitational wave could be obtained by cutting Minkowski space-time along a null cone $\mathcal{N}$ and then re-attaching the two pieces with a suitable "warp". This procedure is illustrated in Figure 20.3. It is exactly analogous to the case for non-expanding impulsive waves. However, there is one important difference. Although these solutions may also contain null sources on the wavefront $\mathcal{N}$, additional sources necessarily occur in the background regions $\mathcal{M}^+$ and/or $\mathcal{M}^-$ in this case due to the global structure of the space-times. For the non-expanding case, the Minkowski space on either side of the impulse is the entire space-time cut along a planar null hypersurface. For spherical waves, on the other hand, the complete set of points on one side of the null cone does not match to the entire null cone on the other side. This necessarily results in an incompleteness or excess that can be interpreted in terms of deficit angles, cosmic strings or other forms of topological structure in the Minkowski space on at least one side of the wave surface.

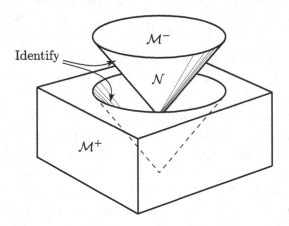

Fig. 20.3 Minkowski space is cut into two parts $\mathcal{M}^-$ and $\mathcal{M}^+$ along a null cone $\mathcal{N}$. These parts are then re-attached with an arbitrary "warp" in which points on each null cone are identified.   ·

To construct these solutions explicitly, it is again convenient to start with the line element of Minkowski space in the form (20.1), but this time to apply the transformation

$$u = U + V Z \bar{Z}, \qquad v = V, \qquad \zeta = V Z, \qquad (20.15)$$

giving

$$\mathrm{d}s^2 = -2\,\mathrm{d}U\,\mathrm{d}V + 2\,V^2\,\mathrm{d}Z\,\mathrm{d}\bar{Z}. \qquad (20.16)$$

The (Robinson–Trautman) coordinates used in this metric are related to the familiar Cartesian coordinates of Minkowski space (3.1) by

$$U = \frac{1}{\sqrt{2}} \left(\frac{t^2 - x^2 - y^2 - z^2}{t - z} \right), \qquad V = \frac{t - z}{\sqrt{2}}, \qquad Z = \frac{x + iy}{t - z}.$$

It is then obvious that the hypersurfaces $U = $ const. are a family of null cones (see the $\epsilon = 0$ case of Section 19.2.1 and Figure 19.1). In particular, the hypersurface on which $U = 0$ is the cone

$$x^2 + y^2 + z^2 = t^2,$$

which represents a sphere that contracts (for $t < 0$) and then expands (for $t > 0$) at the speed of light.

Introducing an arbitrary complex holomorphic "warp" function $h(Z)$, one can now consider the alternative transformation:[3]

$$
\begin{aligned}
u &= \frac{|h|^2}{|h'|} \left(V + \left| \frac{h'}{h} - \frac{h''}{2h'} \right|^2 U \right), \\[2mm]
v &= \frac{1}{|h'|} \left(V + \left| \frac{h''}{2h'} \right|^2 U \right), \\[2mm]
\zeta &= \frac{h}{|h'|} \left(V - \left(\frac{h'}{h} - \frac{h''}{2h'} \right) \frac{\bar{h}''}{2\bar{h}'} U \right),
\end{aligned}
\qquad (20.17)
$$

where $h' = h_{,Z}$ etc. This takes (20.1) to the following form of flat space

$$ds^2 = -2dU\,dV + 2 \left| V dZ - U \bar{H}\,d\bar{Z} \right|^2, \qquad (20.18)$$

where

$$H(Z) = \frac{h'''}{2\,h'} - \frac{3\,h''^2}{4\,h'^2}. \qquad (20.19)$$

For the particular choice $h(Z) = Z$, (20.17) reduces exactly to (20.15).

According to Penrose's method, a null cut $\mathcal{N}$ is taken to be the null cone $U = 0$. It is then possible to take $\mathcal{M}^-$ to be given by (20.16) for $U < 0$, and $\mathcal{M}^+$ to be given by (20.18) for $U > 0$. The combined line element is then

$$ds^2 = -2dU\,dV + 2 \left| V dZ - U\Theta(U)\bar{H}\,d\bar{Z} \right|^2, \qquad (20.20)$$

which is continuous across the junction $U = 0$ (except at possible poles of H). This is the metric for an expanding impulsive spherical wave on the null cone $U = 0$ in a Minkowski background as considered by Nutku and Penrose

---

[3] A more general transformation which includes all three of the distinct (Robinson–Trautman) foliations of null cones distinguished by the parameter ϵ was given by Hogan (1994). An alternative extension to the case when $\Lambda \neq 0$ was presented by Hogan (1992). This will be described in the following section. The combined transformation covering all possible cases has been given by Podolský and Griffiths (1999a, 2000).

(1992), Hogan (1993, 1994), Podolský and Griffiths (2000) and Aliev and Nutku (2001). The only non-zero components of the curvature tensor turn out to be

$$\Psi_4 = \frac{H}{V}\,\delta(U), \qquad \Phi_{22} = -\frac{H\bar{H}}{V^2}\,U\,\delta(U). \tag{20.21}$$

This indicates the presence of an impulsive gravitational wave component with amplitude proportional to the function H, which is given by (20.19). It also confirms that the space-time is vacuum everywhere except for particular points on the wave surface $U = 0$. These correspond either to the singularity at $V = 0$ which may be considered as the source of the wave, or to singular points of $H(Z)$.

By comparing the transformations (20.15) with (20.17), the identifications across the impulse are given by

$$\left(U = 0, V, Z, \bar{Z}\right)_{\mathcal{M}^-} = \left(U = 0, \frac{V}{|h'|}, h(Z), \bar{h}(\bar{Z})\right)_{\mathcal{M}^+}. \tag{20.22}$$

These are the junction conditions given by Penrose (1972).

Significantly, the complex coordinate Z can be interpreted nicely as a *stereographic coordinate*. It is well known that any point P on a unit sphere can be represented by a unique point on its equatorial plane by projecting a line from the North pole (N) through P to the plane. By taking the Cartesian coordinates of the point in the plane as the real and imaginary parts of a complex number, that complex number can be considered to represent uniquely the point P on the sphere.

The above space-time represents an impulsive spherical wave surface that is expanding at the speed of light and, according to (20.15), the complex number $Z = \frac{\zeta}{v} = \frac{x+\mathrm{i}y}{t-z}$ identifies stereographically a unique point on this expanding sphere. Thus any point P_- on the spherical boundary $\mathcal{N}$ of $\mathcal{M}^-$ can be identified with a point Z on a complex plane. Similarly, a point P_+ on the spherical boundary $\mathcal{N}$ of $\mathcal{M}^+$ can be identified with the equivalent point $\frac{\zeta}{v}$ on $U = 0$ that, in view of (20.17), is now given by $h(Z)$. Thus the "warp" mapping of $Z \to h(Z)$ in the complex plane corresponds to identifying the points P_- on the "inside" of the wave surface and P_+ on the "outside", as illustrated in Figure 20.4 (see Podolský and Griffiths, 2000).

It is normally assumed here that the "inside" represented by Z covers the complete sphere, but the function $h(Z)$ will not generally cover the entire sphere on the "outside". In the cases in which the range of $h(Z)$ is the same as that of Z, the function H will be zero, and the mapping $Z \to h(Z)$ simply corresponds to a Lorentz transformation (see Penrose and Rindler, 1984). In the more general case, the restrictions on the range of the function $h(Z)$,

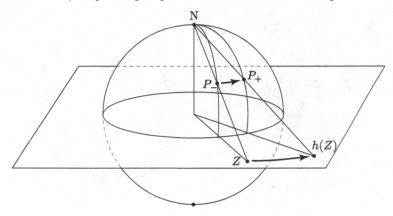

Fig. 20.4 The stereographic correspondence between the Riemann sphere and the complex plane enables a geometrical description of the Penrose junction conditions. Mapping in the complex plane $Z \to h(Z)$ is equivalent to identifying the points P_- inside the impulsive spherical surface with the corresponding points P_+ outside.

together with its specific character, will correspond to particular physical situations.

For non-trivial choices of $h(Z)$, i.e. choices such that $H(Z) \neq 0$, the complete sections of Minkowski space on either side of the wave surface $U = 0$ do not cover each other exactly. This necessarily indicates the presence of topological singularities such as cosmic strings in the region exterior to the spherical wave where $U > 0$. (Of course, it is always possible to reverse the above choice so that topological singularities occur inside the wave.) The simple example in which a single cosmic string, as described in Section 3.4, occurs in the exterior region is obtained by the choice $h = Z^{1-\delta}$. By putting $Z = |Z|e^{i\phi}$ in this case, where $\phi \in [0, 2\pi)$, it can be seen that $h(Z)$ covers the plane minus a wedge, specifically $\arg h(Z) \in [0, (1-\delta)2\pi)$. Thus the string is demonstrated to have a deficit angle $2\pi\delta$, as shown in Figure 20.5.

As in the nonexpanding case, it is possible to combine the two transformations (20.15) and (20.17) either side of the wavefront (for details see Podolský and Griffiths, 1999a). Then, putting $F(\xi) = h(Z) - Z$ at $U = 0$, the combined transformation formally relates the continuous form of the line element (20.20), not to (20.16), but to the form

$$\begin{aligned} ds^2 = &-2du\,dv + 2v^2 d\xi\,d\bar{\xi} - 2v^2 \bar{F}\delta(u)d\xi\,du - 2v^2 F\delta(u)d\bar{\xi}\,du \\ &+ \left[2v^2 F\bar{F}\delta(u) - v(F_{,\xi} + \bar{F}_{,\bar{\xi}})\right]\delta(u)\,du^2, \end{aligned} \tag{20.23}$$

which includes impulsive components explicitly located on the spherical wavefront $u = 0$. This can immediately be seen to be identical to the impul-

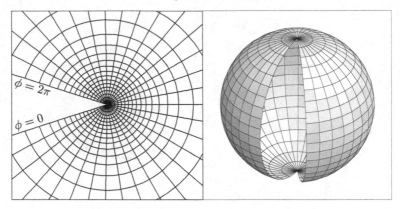

Fig. 20.5 The complex mapping $Z \to h(Z) = Z^{1-\delta}$ (left) corresponds to the cut through the Riemann sphere (right). This represents Minkowski space with a deficit angle $2\pi\delta$ outside the expanding spherical impulse, i.e. a "snapping cosmic string" along the z-axis.

sive limit of the Robinson–Trautman type N solution expressed in García–Plebański coordinates in the form (19.12) with $\Lambda = 0$ and $\epsilon = 0$, namely

$$
\begin{aligned}
\mathrm{d}s^2 = &-2\mathrm{d}u\,\mathrm{d}v + 2v^2\mathrm{d}\xi\,\mathrm{d}\bar{\xi} - 2v^2\bar{f}\mathrm{d}\xi\,\mathrm{d}u - 2v^2 f\mathrm{d}\bar{\xi}\,\mathrm{d}u \\
&+ \left[2v^2 f\bar{f} - v(f_{,\xi} + \bar{f}_{,\bar{\xi}})\right]\mathrm{d}u^2,
\end{aligned} \tag{20.24}
$$

where $f = f(\xi, u)$ is an arbitrary complex function holomorphic in ξ, and the above impulsive limit is obtained by putting $f(\xi, u) = F(\xi)\,\delta(u)$.

It must however be emphasised that, in this impulsive limit, the metric (20.23) includes a square of the Dirac δ-function, and so this form should be treated with considerable caution. Nevertheless, this at least demonstrates that the spherical impulsive waves obtained using the above "cut and paste" method are equivalent to the impulsive limit of Robinson–Trautman type N solutions. This can be seen explicitly by considering the limit of the sandwich wave described in Section 19.2.3 as its "width" becomes arbitrarily small.

As pointed out above, the family of null cones on which $U = 0$ correspond to the $\epsilon = 0$ case which is illustrated in Figure 19.1. These have a common null line and cover only half of the complete Minkowski background. Thus, the metric (20.20) strictly covers only half of a complete space-time which contains an expanding impulsive wave. An equivalent (time-reversed) region needs to be attached in order for the entire space-time to represent an impulsive wave that contracts to a point and then expands. A form of the metric which explicitly covers both the incoming and outgoing spherical wave has been given by Hogan (1995). This uses the coordinates for the case $\epsilon = +1$

(see Figure 19.1), and leads to impulsive limits of the Robinson–Trautman sandwich waves considered in Sections 19.2 and 19.2.3.

Interestingly, Bičák and Schmidt (1989a) obtained a particular case of the above class as a limit of a solution with boost-rotation symmetry (see Chapter 15, specifically the cases described in Subsections 15.2.3 and 15.2.5). In this limit, in which the acceleration parameter becomes unbounded ($\alpha \to \infty$), two null particles recede from a common point generating an impulsive spherical gravitational wave, and there are either cosmic strings attaching each particle to infinity, or there is an expanding strut along the axis of symmetry separating the two particles. This solution can therefore be interpreted as representing the spherical impulsive wave generated by a snapping cosmic string. It was given independently by Gleiser and Pullin (1989) and discussed further by Bičák (1990), who pointed out that it does not strictly describe a snapping cosmic string, but rather two semi-infinite cosmic strings which initially approach each other at the speed of light and separate again at the instant at which they collide. The same solution was also explicitly obtained as null limits of the C-metric, the Bonnor–Swaminarayan solutions and the Bičák–Hoenselaers–Schmidt solutions (Podolský and Griffiths, 2001a,b).

Nutku and Penrose (1992) have also outlined an interesting situation in which two cosmic strings collide and both snap at their point of intersection generating a spherical impulsive gravitational wave. The explicit solution representing this case has been constructed by Podolský and Griffiths (2000) using geometrical techniques based on the use of stereographic projections, illustrated in Figure 20.4.

20.5 Expanding impulsive waves in (anti-)de Sitter space

Just as in the previous sections, it can be shown that there are again three possible approaches to the construction of expanding impulsive waves in de Sitter and anti-de Sitter backgrounds.

Using Penrose's "cut and paste" procedure, Hogan (1992) has constructed a solution for a spherical impulsive gravitational wave in the vacuum de Sitter or anti-de Sitter universe. Starting with the metric for these backgrounds in the conformally flat form (4.19), or (20.9), it is convenient to make the transformation $\mathcal{U} = U + VZ\bar{Z}$, $\mathcal{V} = V$, $\xi = VZ$, which transforms the metric to the form

$$ds^2 = \frac{-2\,dU\,dV + 2V^2\,dZ\,d\bar{Z}}{(1 - \frac{1}{6}\Lambda UV)^2},$$

(20.25)

in which the null hypersurface $U = 0$ is an expanding null cone in the (anti-) de Sitter background.

It is also possible to consider the alternative transformation (20.17) with $\mathcal{U}$, $\mathcal{V}$ and ξ replacing u, v and ζ, respectively. For the particular choice $h(Z) = Z$, this is exactly the above transformation which leads to (20.25). However, for arbitrary $h(Z)$, it gives a form of the metric that is conformal to (20.18). As in the previous section, the metric for an impulsive wave can be constructed by taking $h(Z) = Z$ for $U < 0$, and letting $h(Z)$ remain arbitrary for $U > 0$. This leads to the continuous metric, generalising (20.20)

$$ds^2 = \frac{-2\,dU\,dV + 2\left|V dZ - U\,\Theta(U)\,\bar{H}\,d\bar{Z}\right|^2}{(1 - \frac{1}{6}\Lambda UV)^2}, \tag{20.26}$$

where $H(Z)$ is given by (20.19). The non-zero components of the Weyl and Ricci tensor remain exactly as in (20.21). A stereographic representation of this construction can again be given as in Figure 20.4.

The metric (20.26) describes an arbitrary spherical impulsive gravitational wave which expands in a de Sitter or anti-de Sitter universe. In this case, the spherical wavefront $U = 0$ corresponds to $\mathcal{U}\mathcal{V} - \xi\bar{\xi} = 0$ in the coordinates of (4.19) or (20.9). In the representation of the (anti-)de Sitter space as a four-dimensional hyperboloid in a five-dimensional flat space, the parametrisation (4.17) indicates that the impulse occurs on the section on which

$$Z_4 = a, \qquad Z_1{}^2 + Z_2{}^2 + Z_3{}^2 = Z_0{}^2.$$

This clearly represents a contracting and expanding spherical wave surface. It is indicated by a pair of straight null lines in Figure 20.6, in which the coordinates Z_2 and Z_3 are suppressed.

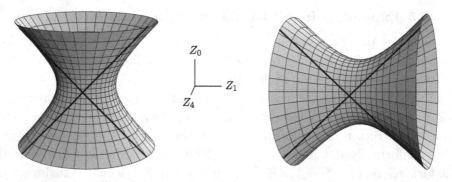

Fig. 20.6 The location of an expanding impulsive wave can be pictured as a section $Z_4 = a$ of the four-dimensional hyperboloids representing de Sitter and anti-de Sitter spaces. The lines shown on the hyperboloids above represent trajectories of opposite poles of an expanding spherical wave for which $Z_2 = 0 = Z_3$.

These solutions typically model the impulsive waves that are generated by snapping cosmic strings or similar structures in a de Sitter or anti-de Sitter space-time. Topological singularities must occur on at least one side of the impulsive wave surface. Particular solutions can be constructed by taking specific expressions for the arbitrary function $h(Z)$, and hence $H(Z)$. For example, as in Minkowski space, a single string of deficit angle $2\pi\delta$ arises from the choice $h = Z^{1-\delta}$ (see Podolský and Griffiths, 2004). When $\Lambda < 0$, this would represent the wave generated by the snapping of an infinite cosmic string in an anti-de Sitter background. However, when $\Lambda > 0$, a single string goes around the entire closed de Sitter universe and snaps at one point generating the impulsive spherical wave.

As is the case with $\Lambda = 0$, these solutions may also be considered as impulsive limits of the Robinson–Trautman type N solutions with a non-zero cosmological constant. The continuous metric (20.26) can be explicitly transformed to an impulsive limit of the Robinson–Trautman type N solutions in García–Plebański coordinates. This leads to a generalised form of the metric (20.23) as described by Podolský and Griffiths (1999a). The alternative parameters $\epsilon = +1$ or -1 can again also be considered.

In view of the third possible approach described in previous sections, it may be anticipated that particular cases of this family of solutions could be obtained as limits of space-times with boost-rotation symmetry and a non-zero cosmological constant. For example, it should be possible to obtain an exact solution for an expanding impulsive spherical wave, generated by a snapping cosmic string, as a limit of the solution for a pair of (causally separated) black holes in a de Sitter or anti-de Sitter universe which are accelerating due to the presence of cosmic strings or struts (see Section 14.4). In such a limit, the acceleration parameter α would become unbounded while the mass is appropriately scaled to zero. However, although this can be achieved in a Minkowski background as mentioned at the end of Section 20.4, the method applied there cannot be extended to the case with a non-zero cosmological constant as coordinates that are analogous to the Weyl coordinates do not exist when $\Lambda \neq 0$.

A different method for obtaining such limits, and which is applicable in the present case, has nevertheless been developed by Podolský and Griffiths (2004). This is based on the original Robinson–Trautman coordinates of (19.1). However, a discontinuity in the derivatives of the metric functions is necessary in this case. The need for this is evident from the geometry of the foliations of the null cones that are illustrated in Figure 20.7. These future-oriented null cones cover half of the complete space-time (above the diagonal in the left diagram) which is symmetric about the vertical axis. In

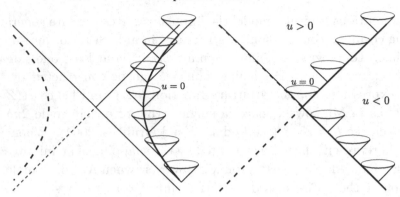

Fig. 20.7 In Robinson–Trautman coordinates for the solution representing an accelerating black hole (left), the surfaces $u = $ const. define a family of future-oriented null cones centred on the worldline of the source. The right figure shows, schematically, the limit as the acceleration becomes unbounded, in which the foliation of the null cones becomes necessarily discontinuous at $u = 0$.

the limit in which the acceleration becomes unbounded, the "source" at the vertex of these cones approaches the null asymptotes as in the diagram on the right. The null surfaces still foliate the same half of the space-time, but are stacked in different ways in the two halves $u < 0$ and $u > 0$. There is an obvious discontinuity in the foliation across the impulsive wave at $u = 0$. These schematic pictures apply in Minkowski, de Sitter and anti-de Sitter backgrounds.

20.6 Other impulsive wave space-times

In previous sections in this chapter, impulsive waves have been considered that occur in Minkowski, de Sitter and anti-de Sitter space-times. However, exact solutions involving impulsive waves can also be considered in other simple background space-times. In fact, early work on impulsive and shock waves in general relativity was related to establishing the appropriate junction conditions across null hypersurfaces. Such studies of the non-null case by Darmois (1927) were followed by the classical works of O'Brien and Synge (1952), Lichnerowicz (1955) and Israel (1966a) (see also Bonnor and Vickers, 1981). The analysis of junction conditions on null hypersurfaces was considered by Stellmacher (1938), Pirani (1965), Penrose (1972), Robson (1973), Taub (1980), Clarke and Dray (1987), Barrabès (1989) and Barrabès and Israel (1991) among others. To conclude the present chapter, it is appropriate to include a few observations on some particular results which involve explicit impulsive waves.

The earliest work on impulsive waves tended to concentrate on the properties of thin shells of null matter. Thus, Synge (1957) presented a model for an expanding spherical impulse of radiation from a white hole. Specifically, he took the space-time to be Schwarzschild outside the shell and Minkowski inside. Thus, he effectively took an impulsive limit of the Vaidya solution as discussed in Section 9.5. His solution can be pictured as the time reverse of Figure 9.19 in the limit in which the duration of the Vaidya part of the space-time is reduced to zero.

Dray and 't Hooft (1985) have obtained necessary and sufficient conditions for a non-expanding impulsive gravitational wave in a large class of vacuum solutions of Einstein's equations. The sources of these waves are massless particles moving along a null hypersurface, such as a horizon of a Schwarzschild black hole. These solutions generally involve an arbitrary "shift function" $h(Z)$ which is equivalent to the "warp" function introduced in the "cut and paste" method described above. This approach has been generalised by Loustó and Sánchez (1989) and Sfetsos (1995) to cover cases in which non-vanishing matter fields and a cosmological constant are also present.

The approaches described in the previous sections have also been applied by Ortaggio (2002) to construct non-expanding impulsive waves in the Nariai universe, and subsequently generalised to Bertotti–Robinson and other direct-product space-times (Ortaggio and Podolský, 2002).

Further examples of impulsive waves are included in the extensive review by Barrabès and Hogan (2003a), together with a detailed discussion of the associated theory.

21

Colliding plane waves

Since Einstein's equations are essentially nonlinear, gravitational fields and waves cannot be simply superposed. In general relativity, even electromagnetic waves experience a nonlinear interaction through the gravitational equations, in spite of the fact that Maxwell's equations remain linear. The physical phenomena that arise as a result of this nonlinearity need to be understood. And the simplest situation for which this can be modelled exactly is in the collision and subsequent interaction of plane waves in a flat Minkowski background. In fact, many explicit solutions are now known which describe situations of this type. Thorough reviews of early work on this topic can be found in the book by Griffiths (1991) and also in Chapter 25 of Stephani *et al.* (2003). The purpose of the present chapter is to use an up-to-date approach to review the basic results that have been found, with a particular emphasis on the physically significant features that arise. It will also be shown how certain solutions that have been studied in previous chapters reappear in this context.

Clearly, the collision of plane waves, is a highly idealised situation. Realistic waves have convex wavefronts, and only become approximately planar at a large distance from their source. Moreover, exact plane waves are infinite in transverse spatial directions and therefore have unbounded energy. These features may have unfortunate consequences in exact colliding plane wave space-times. Thus, when seeking to interpret these solutions, care has to be taken to distinguish properties that apply to general wave interactions from those that arise as a consequence of the idealised assumptions.

The importance of this topic also derives from the fact that it is one of the few situations in which exact solutions representing wave interactions can be obtained explicitly. In fact, they demonstrate a number of interesting properties. For example, the mutual focusing of interacting waves generally leads to singularities, but the structure of these may be of different types.

21.1 Initial data

Plane waves have already been described in Section 17.5. In considering the collision of shock waves of this type, it is natural to start with a background Minkowski region. Two distinct shock waves that propagate in different directions can then be introduced. Before their collision, these will have null wave surfaces that can be expressed in terms of two different null coordinates as $u = $ const. and $v = $ const., with shock wavefronts at $u = 0$ and $v = 0$, respectively. It is always possible to adopt a frame of reference in which the waves approach from exactly opposite spatial directions. With this choice, u and v can be adopted as double-null coordinates globally, and two other spatial coordinates x and y may be chosen to span the (initially parallel) wave surfaces.

The general scheme for considering such a collision of two shock plane waves is illustrated in Figure 21.1. The initial background Minkowski region between the waves, in which $u < 0$ and $v < 0$, is denoted as I. The approaching waves occur in the two separate regions labeled as II ($u > 0$, $v < 0$) and III ($u < 0$, $v > 0$). The wavefronts collide on the spacelike two-dimensional plane given by $u = 0$, $v = 0$, and the waves subsequently continue to interact throughout the region, denoted as IV, in which $u > 0$ and $v > 0$.

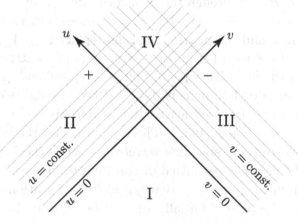

Fig. 21.1 An initial set up for considering the collision of two plane waves. Region I is a flat background. Regions II and III contain the approaching waves with wavefronts $u = 0$ and $v = 0$. Region IV is the interaction region following the collision. Each point represents a plane that is spanned by the coordinates x and y.

When set up in this way, the colliding plane wave situation forms a well-posed characteristic initial value problem. For given arbitrary approaching waves in regions II and III, the metric is explicitly known up to the null characteristics $v = 0$, $u \geq 0$ and $u = 0$, $v \geq 0$. Functions that determine the initial solution in region II will be denoted with a subscript $+$, and those in region III with a subscript $-$. With this initial data, a unique solution of the field equations exists in the interaction region IV (at least for a finite time).

The approaching wave in region II is a plane wave which could be described in the Brinkmann form (17.7), namely

$$ds^2 = -2\,du\,dr - \left(A_+\zeta^2 + \bar{A}_+\bar{\zeta}^2 + 2B_+\zeta\bar{\zeta}\right) du^2 + 2\,d\zeta\,d\bar{\zeta}. \qquad (21.1)$$

In this form, the metric functions explicitly represent the amplitude and phase of the initial gravitational wave component $\Psi_4 = A_+(u)$ and pure radiation component $\Phi_{22} = B_+(u)$. It may be noted that the metric (21.1) may be considered to cover both regions I and II, if it is assumed that A_+ and B_+ are both zero for $u < 0$. However, this form of the metric is not convenient for a discussion of the colliding plane wave problem. The Rosen form (17.11) which employs double-null coordinates is to be preferred. The line element (21.1) should therefore be transformed to the form

$$ds^2 = -2\,du\,dv + \alpha_+ \left(\chi_+ dy^2 + \chi_+^{-1}(dx + \omega_+ dy)^2\right), \qquad (21.2)$$

where the new metric functions $\alpha_+(u)$, $\chi_+(u)$ and $\omega_+(u)$ are determined from $A_+(u)$ and $B_+(u)$ through the transformation (17.9).

The other initial wave in region III can be expressed in a similar form, but with v replacing u and the metric functions replaced by $A_-(v)$ and $B_-(v)$ or, in a metric of the form (21.2), by $\alpha_-(v)$, $\chi_-(v)$ and $\omega_-(v)$.

The background region I is part of Minkowski space, and can be expressed in terms of the metric (3.3), namely $ds^2 = -2\,du\,dv + dx^2 + dy^2$. The junction conditions to be imposed across the wavefronts $u = 0$ and $v = 0$ are those of O'Brien and Synge (1952). These require that the metric functions $\alpha_\pm$ are C^1 across the corresponding wavefronts, while the functions $\chi_\pm$ and $\omega_\pm$ need only be C^0. The permitted discontinuities in the derivatives of $\chi_\pm$ and $\omega_\pm$ arise from the possibility that impulsive gravitational wave components may either be present initially, or may be generated by the collision, while impulsive sheets of matter are excluded. Accordingly, it is required that

$$\begin{aligned} \alpha_+(0) = 1, \quad & \alpha_{+,u}(0) = 0, \quad & \chi_+(0) = 1, \quad & \omega_+(0) = 0, \\ \alpha_-(0) = 1, \quad & \alpha_{-,v}(0) = 0, \quad & \chi_-(0) = 1, \quad & \omega_-(0) = 0. \end{aligned}$$

If the functions $\alpha_\pm$, $\chi_\pm$ and $\omega_\pm$ have been derived from expressions for $A_+(u)$, $B_+(u)$ and $A_-(v)$, $B_-(v)$ that have been appropriately chosen, these conditions are automatically satisfied in view of the conditions (17.10).

It also follows from the transformation (17.9) with (17.10) that, for any initial combination of gravitational and electromagnetic waves and pure radiation fields with positive energy density, $\alpha_+(u)$ and $\alpha_-(v)$ are necessarily decreasing functions. Thus, coordinate singularities necessarily occur in regions II and III when $\alpha_+(u) = 0$ and $\alpha_-(v) = 0$, respectively. These may be considered to arise as a consequence of the focusing properties of the fields. Since they do not occur in the Brinkmann form of the metric (21.1), these would appear to be merely coordinate singularities. However, in the context of colliding plane waves, they generically become curvature singularities as will be described below.

21.2 The interaction region

For the collision of gravitational waves, electromagnetic waves or pure radiation, or any combinations of these, it is found that the metric in the interaction region IV can always be expressed in the form

$$ds^2 = -2\, f \, du \, dv + \alpha \left(\chi \, dy^2 + \chi^{-1}(dx + \omega \, dy)^2 \right), \qquad (21.3)$$

where α, χ, ω and f are functions of both u and v satisfying the initial data:

$$\alpha(u,0) = \alpha_+(u), \qquad \chi(u,0) = \chi_+(u), \qquad \omega(u,0) = \omega_+(u), \qquad f(u,0) = 1,$$
$$\alpha(0,v) = \alpha_-(v), \qquad \chi(0,v) = \chi_-(v), \qquad \omega(0,v) = \omega_-(v), \qquad f(0,v) = 1. \tag{21.4}$$

The space-time in the interaction region retains the two isometries ∂_x and ∂_y that are common to the initial regions, see (17.12).

The metric function $\alpha(u,v)$, which is the area measure of the orbits of the isometry group, must satisfy the field equation

$$\alpha_{,uv} = 0. \tag{21.5}$$

Solutions of this can always be expressed in the form

$$\alpha = \tfrac{1}{2}\left(\xi(u) - \eta(v) \right).$$

In terms of the initial data, the functions ξ and η are given by

$$\xi(u) = 2\alpha_+(u) - 1, \qquad \eta(v) = 1 - 2\alpha_-(v),$$

which imply that α remains a decreasing function of both u and v throughout the interaction region. This can be interpreted as a manifestation of the mutual focusing of the two waves. It also follows that a singularity will necessarily occur on the spacelike surface given by $\alpha(u,v) = 0$. Generically, this is a curvature singularity but exceptions occur as will be discussed below.

For the case of colliding plane gravitational waves in which both waves have constant polarisation and both polarisations are aligned, it is always possible to choose coordinates such that $\omega = 0$ everywhere. In this case, the metric has the simpler form

$$\mathrm{d}s^2 = -2 f \, \mathrm{d}u \, \mathrm{d}v + \alpha \left(\chi^{-1} \mathrm{d}x^2 + \chi \, \mathrm{d}y^2 \right), \tag{21.6}$$

and the remaining principal field equation, expressed in terms of ξ and η, is linear in $\log \chi$. The characteristic initial value problem can then be solved, in principle, using Riemann's method (Szekeres, 1972; Yurtsever, 1988c), but this is not a procedure that can be adopted in practice. Nevertheless, an alternative method for constructing exact solutions from initial data has been found by Hauser and Ernst (1989a,b). This employs the use of Abel transforms and has been implemented by Griffiths and Santano-Roco (2002). However, for many cases with physically significant initial data, it is found that the corresponding exact solutions in the interaction region cannot be expressed in terms of elementary functions.

For the more general case in which the approaching gravitational waves have different or non-constant polarisations, the metric (21.3) representing the interaction region is essentially non-diagonal and the corresponding field equations are nonlinear. The field equations are also nonlinear for the collision of electromagnetic waves, whether or not they are also coupled to gravitational waves. Nevertheless, the space-time retains its two isometries and the field equations are equivalent to the hyperbolic form of the Ernst equations (Ernst, 1968a,b).[1] These equations are known to be integrable and various solution-generating techniques are known for solving them. However, none of the standard techniques are suitable for generating solutions from the initial data for colliding plane waves. To cover this case, Hauser and Ernst (1989b, 1990, 1991) have reformulated this problem as a homogeneous Hilbert problem with corresponding matrix linear integral equations. Using this, they have been able to investigate various global aspects of these space-times including the global existence and uniqueness of solutions and a detailed proof (Hauser and Ernst, 2001) of the hyperbolic form of the Geroch conjecture that all solutions can be generated in principle. However, this approach does not provide an effective method for explicitly constructing solutions for given characteristic initial data.

A different method which does enable exact solutions for the interaction

---

[1] For the vacuum case, this form is given in Equation (22.5). In this hyperbolic case, the Ernst potential is replaced by metric functions in the combination $E = \chi + \mathrm{i}\,\omega$.

region to be generated from initial data has been obtained by Alekseev and Griffiths (2001, 2004). This arises from a different approach to the structure of integrable reductions of the Einstein and Einstein–Maxwell field equations that was developed by Alekseev (1985, 1988). In this approach, every solution can be characterised by a set of functions of an auxiliary (spectral) parameter that are interpreted as the monodromy data on the spectral plane of the fundamental solution of an associated linear system. These monodromy data functions can be determined (at least in principle) from the given initial data. The solution of the initial value problem can then be found by first solving some linear singular integral equations whose scalar kernel is constructed using the monodromy data functions. Since solutions of the associated linear equations for colliding plane waves are nonanalytic at the point at which the waves collide, the original form of the "monodromy transform" approach has to be modified by introducing some "dressing" or "scattering" matrices (see Alekseev, 2001). This construction has the effect that the monodromy data become evolving functions of the null coordinates rather than being conserved quantities. However, although exact solutions can be generated from initial data by this method, as for the colinear case, for many cases with physically significant initial data, the consequent solutions are not expressible in terms of elementary functions.

Colliding plane wave problems are essentially characteristic initial value problems, but effective tools for dealing with these have only been developed very recently. In fact, almost all the known exact solutions representing such situations were obtained previously by an inverse method. Since the metric in the interaction region is known to be expressible in the form (21.3), the field equation in this region can be written explicitly. And since these equations are integrable, various families of exact solutions can be explicitly obtained. Although many of these have to be rejected in this context as they cannot be joined to appropriate initial waves, a large number of solutions of the colliding plane wave problem were constructed in this way. A few of these will be described below.

In seeking to apply this inverse method for finding explicit solutions, it may first be noted that it is always possible to adopt the functions $\xi(u)$ and $\eta(v)$ as null coordinates throughout the interaction region. Moreover, since the form of the metric (21.3) is invariant under the transformations $u \to \tilde{u}(u)$, $v \to \tilde{v}(v)$, the functions $\xi(u)$ and $\eta(v)$ do not need to be expressed in terms of initial data $\alpha_+(u)$ and $\alpha_-(v)$ at this stage. Solutions of the main (integrable) field equations involving any number of arbitrary parameters can then be generated. The remaining field equations then depend on the forms of $\xi(u)$, $\eta(v)$ and their derivatives. The so-called "colliding plane wave conditions" correspond to the integrability conditions for these equations and their compatibility with junction conditions on the wavefronts. They normally provide constraints on some of the parameters introduced. It may be noted, however, that solutions derived by this procedure are not necessarily expressed in terms of the same u, v-coordinates as are employed above. In particular, in view of the freedom in the choice of the null coordinates, a removable metric function $f_+(u)$ may appear in the line element (21.2) in region II, and a similar function $f_-(v)$ may appear in the metric for region III such that

$$f(u,0) = f_+(u), \qquad f(0,v) = f_-(v), \qquad f(0,0) = f_+(0) = f_-(0) = 1.$$
$$(21.7)$$

If $f_+(u)$ and $f_-(v)$ are not constants, the parameters u and v are not affine in regions II and III, respectively. The remaining initial data functions are also modified in this approach, so that the initial plane wave components have to be calculated explicitly *after* the solution has been found, rather than being chosen initially for their physical significance. This approach will be described in more detail in Section 21.4.

In the following sections, some particularly significant explicit colliding plane wave solutions will be briefly described.

21.3 The Khan–Penrose solution

One of the first exact solutions to be found and analysed was that of Khan and Penrose (1971). This is not only one of the simplest such solutions, but is also probably the most significant physically. It describes the collision and subsequent interaction of two *impulsive* purely gravitational waves of amplitudes a and b, and with aligned polarisation.[2] Prior to their collision, these are given by

$$\Psi_4 = a\,\delta(u), \quad \text{for} \quad v < 0, \qquad \Psi_0 = b\,\delta(v), \quad \text{for} \quad u < 0.$$

In this case, the metric function $\omega(u, v)$ in (21.3) vanishes everywhere and the line elements in the regions I, II and III take the forms

$$\text{I:} \qquad \mathrm{d}s^2 = -2\,\mathrm{d}u\,\mathrm{d}v + \mathrm{d}x^2 + \mathrm{d}y^2,$$

$$\text{II:} \qquad \mathrm{d}s^2 = -2\,\mathrm{d}u\,\mathrm{d}v + (1-au)^2\mathrm{d}x^2 + (1+au)^2\mathrm{d}y^2, \qquad (21.8)$$

$$\text{III:} \qquad \mathrm{d}s^2 = -2\,\mathrm{d}u\,\mathrm{d}v + (1-bv)^2\mathrm{d}x^2 + (1+bv)^2\mathrm{d}y^2.$$

---

[2] The two waves can always be given the same amplitude, and the null coordinates can always be scaled to put $a = b = 1$, but it is more instructive here to retain individual amplitudes explicitly.

For regions I and II, these metrics have previously been given in a combined form in (17.14). Those for regions I and III are obviously analogous.

The initial data corresponding to the above are thus

$$\alpha_+(u) = 1 - a^2u^2, \qquad \chi_+(u) = \frac{1 + au}{1 - au}, \qquad f_+(u) = 1,$$

$$\alpha_-(v) = 1 - b^2v^2, \qquad \chi_-(v) = \frac{1 + bv}{1 - bv}, \qquad f_-(v) = 1.$$

Notice that the functions $\alpha_\pm$ are decreasing as required, and that coordinate singularities occur in regions II and III at $u = 1/a$ and $v = 1/b$, respectively. These are manifestations of focusing properties, with the focal lengths being inversely proportional to the amplitudes of the waves. Moreover, the astigmatic character of the focusing is also manifest. For example, in region II, lines on which $v = $ const. contract in the x-direction but expand in the y-direction.

The explicit solution in the interaction region is given by the metric (21.6) in which

$$\alpha = 1 - a^2u^2 - b^2v^2,$$

$$\chi = \frac{1 + au\sqrt{1 - b^2v^2} + bv\sqrt{1 - a^2u^2}}{1 - au\sqrt{1 - b^2v^2} - bv\sqrt{1 - a^2u^2}}, \tag{21.9}$$

$$f = \frac{(1 - a^2u^2 - b^2v^2)^{3/2}}{\sqrt{1 - a^2u^2}\sqrt{1 - b^2v^2}\,w^2},$$

where $w = \sqrt{1 - a^2u^2}\sqrt{1 - b^2v^2} + ab\,uv$. In this region, the non-zero components of the Weyl tensor are given by

$$\Psi_4 = \frac{a}{\sqrt{1 - b^2v^2}}\,\delta(u) - \frac{3a^2bv\sqrt{1 - b^2v^2}\,w^3}{\alpha^{7/2}},$$

$$\Psi_2 = \frac{ab\,w^2}{\alpha^{7/2}}\left(w^2 - ab\,uv\,\sqrt{1 - a^2u^2}\sqrt{1 - b^2v^2}\right),$$

$$\Psi_0 = \frac{b}{\sqrt{1 - a^2u^2}}\,\delta(v) - \frac{3b^2au\sqrt{1 - a^2u^2}\,w^3}{\alpha^{7/2}}.$$

These expressions illustrate the general properties that the amplitudes Ψ_4 and Ψ_0 of the impulsive waves are modified after the collision, and that both waves develop tails. In addition, the Coulomb component of the Weyl tensor Ψ_2 appears as an interaction component. This region of the space-time is algebraically general.

By considering the scalar polynomial invariants (2.13), it is clear that this solution contains a *curvature singularity* on the spacelike hypersurface given by $a^2u^2 + b^2v^2 = 1$, where $\alpha = 0$. This may be considered to be analogous to a cosmological time-reversed "big bang", or "big crunch", type of singularity. Well-known singularity theorems indicate that in the presence of matter or gravitational waves, if there exists a spacelike hypersurface on which the space-time is non-contracting (non-expanding), then there must exist a past (future) curvature singularity. In this case, the assumption of Minkowski space as a background prior to the gravitational wave interaction inevitably implies that some kind of future singularity will occur. This singularity, however, has directional properties as the shear as well as the convergence become unbounded as the singularity is approached.

Having established that a curvature singularity occurs in the interaction region, it is now appropriate to reconsider the apparent coordinate singularities that occur at $au = 1$ and $bv = 1$ in regions II and III, respectively. For this solution, the initial impulsive waves occur only on the junctions between regions I and II and between regions I and III. The interiors of regions I, II and III are each parts of Minkowski space, so their geometries are easy to understand. For example, the metric for region II is given in (21.8), and this can be expressed in the standard form $ds^2 = -2\,d\tilde{u}\,d\tilde{v} + d\tilde{x}^2 + d\tilde{y}^2$ using the substitution

$$u = \tilde{u}, \qquad v = \tilde{v} + \frac{\frac{1}{2}a\tilde{x}^2}{1 - a\tilde{u}} - \frac{\frac{1}{2}a\tilde{y}^2}{1 + a\tilde{u}}, \qquad x = \frac{\tilde{x}}{1 - a\tilde{u}}, \qquad y = \frac{\tilde{y}}{1 + a\tilde{u}}.$$

Using the Cartesian-like coordinates $\tilde{u}$, $\tilde{v}$, $\tilde{x}$ and $\tilde{y}$ in region II, the hypersurfaces on which $v = \text{const.} < 0$ appear to be different when considered on either side of the impulsive wave. This is illustrated for the section $y = 0$ in Figure 21.2.

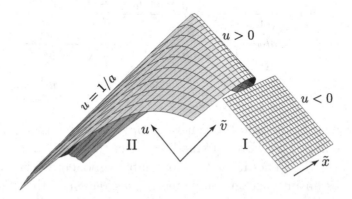

Fig. 21.2 A typical null surface on which $v = \text{const.} < 0$, $y = 0$ in the Khan–Penrose solution. Such a flat null surface in region I is distorted when it is extended into region II. The edges of each surface on which $u = 0_-$ and $u = 0_+$ either side of the impulsive wave at $u = 0$ have to be identified.

From Figure 21.2 it can be seen that a typical null hypersurface on which $v = \text{const.} < 0$ appears to fold down when it is extended through the impulsive wave into region II. Indeed, the lines on this surface on which $u = \text{const.}$ fold down and become degenerate at $u = 1/a$. Moreover, this line $u = 1/a$ is common to the complete family of surfaces with different constant values of v. If there were no opposing wave, this would correspond to a coordinate singularity. However, in the colliding plane wave context, there is a curvature singularity in region IV and the surface (wavefront of the opposing wave) on which $v = 0$ also folds down at $u = 1/a$, so that the fold at $u = 1/a$ in region II is identified with the curvature singularity in region IV. Thus it is not possible to extend the space-time through $u = 1/a$ even for surfaces with $v < 0$. The geometry of these surfaces has been described in detail by Matzner and Tipler (1984) who have introduced the term "fold singularity".

That the singularity at $u = 1/a$ has the character of a curvature singularity can also be seen by considering a null geodesic given in region I by $v = v_0 = \text{const.} < 0$, $y = 0$ and $x = \epsilon$. After passing through the gravitational wave at $u = 0$, it is given in region II by $\tilde{v} = v_0$, $\tilde{y} = 0$ and $\tilde{x} = \epsilon$, namely by

$$ x = \frac{\epsilon}{1 - au}, \qquad y = 0, \qquad v = v_0 + \frac{\frac{1}{2}a\epsilon^2}{1 - au}. $$

From this it is clear that, unless $\epsilon = 0$, v increases indefinitely as $u \to 1/a$ even for arbitrarily small values of ϵ. Thus, although the geodesic with $\epsilon = 0$ approaches the "fold singularity" in region II, an arbitrarily close geodesic that is initially parallel in region I diverges from it and crosses into region IV *before* it reaches the hypersurface $u = 1/a$. This geodesic subsequently terminates in the curvature singularity in region IV.

From the above discussion, it can be concluded that the global structure of this solution is as represented in Figure 21.3. In this case, the surface on which the opposing waves (or congruences) focus each other is a curvature singularity in region IV (or is a fold singularity in regions II and III). The space-time can be extended only up to this surface. As is clear from the previous paragraphs, it is important to regard this figure as a section through the space-time rather than a projection.

The Khan–Penrose solution represents a space-time in which two impulsive plane gravitational waves collide. The non-flat solution in region IV can be considered as the scattered field produced by the collision. And, as this space-time can be regarded as representing a limit of the collision of two sandwich waves, it approximately represents a case in which the initial waves

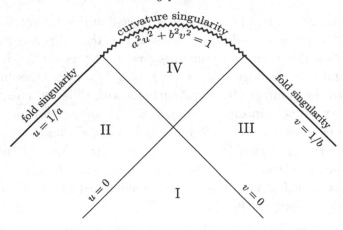

Fig. 21.3 A section through the Khan–Penrose solution showing the general structure of this space-time. This structure is typical of almost all colliding plane wave space-times.

have short finite duration. In fact, the structure illustrated in Figure 21.3 represents that of almost all colliding plane wave space-times.

In the above solution, the polarisations of the two approaching gravitational waves are aligned. An important solution in which this condition is relaxed has been obtained by Nutku and Halil (1977). In this case, the initial waves are still purely impulsive, but their polarisations are distinct. This solution is expressed in terms of the non-diagonal metric (21.3) with $\omega \neq 0$. However, although it is mathematically more complicated, the basic structure of the space-time remains as represented in Figure 21.3. (See also Chandrasekhar and Ferrari, 1984.)

An alternative generalisation of this solution, in which the initial impulsive gravitational waves are also accompanied by electromagnetic shock waves, was later given by Chandrasekhar and Xanthopoulos (1985). This particular space-time will not be discussed further, but other solutions representing colliding plane electromagnetic waves will be described in following sections.

21.4 The degenerate Ferrari–Ibáñez solution

In 1987, Ferrari and Ibáñez obtained a family of exact vacuum colliding plane wave space-times by applying an inverse scattering technique to the field equations within the interaction region and then imposing the colliding plane wave condition. This family of solutions includes one particular degenerate case which is of algebraic type D and appears to have unusual properties.

It was analysed in further detail by Ferrari and Ibáñez (1988) and will be described here.

This special case appears at first sight to be a slight variation of the Khan–Penrose solution (21.9) in which, in the interaction region, the metric function $\chi(u, v)$ is multiplied by the function $\alpha(u, v)$. (Since the main field equation for the metric (21.6) is linear in $\log \chi$, this is a very simple modification.) The resulting metric functions in region IV are given by

$$\alpha = 1 - a^2 u^2 - b^2 v^2,$$

$$\chi = (1 - a^2 u^2 - b^2 v^2)\frac{(1 + au\sqrt{1 - b^2 v^2} + bv\sqrt{1 - a^2 u^2})}{(1 - au\sqrt{1 - b^2 v^2} - bv\sqrt{1 - a^2 u^2})}, \qquad (21.10)$$

$$f = \frac{(1 + au\sqrt{1 - b^2 v^2} + bv\sqrt{1 - a^2 u^2})^2}{\sqrt{1 - a^2 u^2}\sqrt{1 - b^2 v^2}}.$$

With these expressions, the non-zero components of the Weyl tensor in the interaction region are given by

$$\Psi_4 = \frac{a}{(1 + bv)^2\sqrt{1 - b^2 v^2}}\delta(u) - \frac{3a^2}{(1 + au\sqrt{1 - b^2 v^2} + bv\sqrt{1 - a^2 u^2})^3},$$

$$\Psi_2 = \frac{ab}{(1 + au\sqrt{1 - b^2 v^2} + bv\sqrt{1 - a^2 u^2})^3}, \qquad (21.11)$$

$$\Psi_0 = \frac{b}{(1 + au)^2\sqrt{1 - a^2 u^2}}\delta(v) - \frac{3b^2}{(1 + au\sqrt{1 - b^2 v^2} + bv\sqrt{1 - a^2 u^2})^3}.$$

The focusing singularity, which occurs when $\alpha = 0$, appears on the space-like surface $1 - a^2 u^2 - b^2 v^2 = 0$, exactly as in the Khan–Penrose solution. However, it is significant that the Weyl tensor components (21.11) are *regular* on this hypersurface. This therefore does not correspond to a curvature singularity as occurs generically for colliding plane waves. This is, in fact, the simplest of a number of colliding plane wave space-times in which the usual curvature singularity is replaced by some kind of coordinate singularity. The character of this will be considered below.

It may also be noted that the metric function $f \neq 1$ on the junctions with the initial regions II and III where $u = 0$ and $v = 0$, respectively. This solution therefore does not satisfy the initial conditions prescribed in (21.4). Nevertheless, it is possible to replace the coordinates u and v in the metric functions (21.10) by $u\,\Theta(u)$ and $v\,\Theta(v)$, respectively, where Θ is the Heaviside step function. When this is applied to the full range of u and v, all four regions of the space-time are covered, with the metric functions being dependent on u only in region II, dependent on v only in region III and constants in region I. In this case the necessary junction conditions across the wavefronts are satisfied, and the resulting space-time represents a colliding plane wave solution. The expressions for the metric function f in the initial regions are then determined by the conditions (21.7). The fact that these are not equal to unity simply indicates that the null coordinates u and v are not affinely parametrised in these regions. In terms of these coordinates, the approaching waves have forms in which the non-zero Weyl tensor components are given by

$$\text{region II}: \qquad \Psi_4 = a\,\delta(u) - \frac{3a^2}{(1+au)^3}\,\Theta(u),$$

$$\text{region III}: \qquad \Psi_0 = b\,\delta(v) - \frac{3b^2}{(1+bv)^3}\,\Theta(v).$$

Both of the initial approaching waves are thus a combination of impulsive and shock waves.

It can be seen that the scalar polynomial curvature invariants are bounded everywhere within the interaction region IV. However, the points $u = 1/a$, $v = 0$ and $u = 0$, $v = 1/b$ are problematic in view of the unbounded amplitudes of the impulsive gravitational wave components (21.11). No single coordinate system has been found to cover these points, and it has been conjectured (Hayward, 1989) that they correspond to some kind of curvature singularity. For the space-time to be Hausdorff, these points must be removed. It follows that geodesics that reach these points cannot be extended through them, so the appropriate singularity theorem (Tipler, 1980) still applies.

For this particular case, it is sometimes helpful to apply the coordinate transformation

$$au = \sin a\tilde{u}, \qquad bv = \sin b\tilde{v}. \tag{21.12}$$

With this, the metric in the interaction region takes the form

$$ds^2 = -2\Big(1 + \sin(a\tilde{u} + b\tilde{v})\Big)^2 d\tilde{u}\,d\tilde{v} + \frac{1 - \sin(a\tilde{u} + b\tilde{v})}{1 + \sin(a\tilde{u} + b\tilde{v})}\,dx^2$$

$$+ \cos^2(a\tilde{u} - b\tilde{v})\Big(1 + \sin(a\tilde{u} + b\tilde{v})\Big)^2 dy^2.$$

In these coordinates, the junctions to the initial regions II and III occur on the null hypersurfaces $\tilde{v} = 0$, $0 < \tilde{u} < \pi/2a$ and $\tilde{u} = 0$, $0 < \tilde{v} < \pi/2b$, respectively. It can also be seen that $\alpha = \cos(a\tilde{u} + b\tilde{v})\cos(a\tilde{u} - b\tilde{v})$. Thus, the focusing singularity, where $\alpha = 0$, is now given by $a\tilde{u} + b\tilde{v} = \pi/2$, where $0 < \tilde{u} < \pi/2a$ and $0 < \tilde{v} < \pi/2b$. However, although the extension to the initial regions can easily be constructed using the above method, the new null coordinates remain non-affinely parametrised in these regions.

In fact, it is instructive to apply the further coordinate transformation

$$au\sqrt{1 - b^2v^2} + bv\sqrt{1 - a^2u^2} = \sin(a\tilde{u} + b\tilde{v}) = \frac{r}{m} - 1,$$

$$au\sqrt{1 - b^2v^2} - bv\sqrt{1 - a^2u^2} = \sin(a\tilde{u} - b\tilde{v}) = \cos\theta, \qquad (21.13)$$

$$x = t, \qquad\qquad y = m\phi,$$

where $m^2 = 1/2ab$. With this, the line element becomes

$$ds^2 = -\left(1 - \frac{2m}{r}\right)dt^2 + \left(1 - \frac{2m}{r}\right)^{-1}dr^2 + r^2\left(d\theta^2 + \sin^2\theta\,d\phi^2\right), \quad (21.14)$$

which may immediately be recognised as the Schwarzschild metric (8.1).

From the coordinate transformation (21.13), it is clear that $r > m$ in the interaction region. Moreover, the focusing singularity now occurs at $a\tilde{u} + b\tilde{v} = \pi/2$, which is where $r = 2m$. Thus, the interaction region IV of the Ferrari–Ibáñez solution is isomorphic to that part of the Schwarzschild metric for which $m < r < 2m$, which is inside the horizon where r is a time-like coordinate and t is spacelike. This is consistent with the fact that the interaction region for colliding plane waves is time-dependent, not static. Moreover, whereas the ϕ coordinate was taken to be periodic in the spherically symmetric Schwarzschild space-time, a plane symmetry is assumed here in which $t, \phi \in (-\infty, \infty)$. Thus, although the line element in the form (21.14) appears to be very familiar, the geometry of the space-time it represents in this context is very different physically.

In these coordinates, the collision of the two waves occurs at the event $r = m$, $\theta = \pi/2$, and the wave surfaces are spanned by t and ϕ. The boundaries between the interaction region IV and regions II and III occur at $r = m(1 + \cos\theta)$ and $r = m(1 - \cos\theta)$, respectively, while the focusing singularity occurs when $r = 2m$, $0 < \theta < \pi$. The focusing singularity thus corresponds to the horizon of the Schwarzschild space-time. In the colliding plane wave situation, however, this does not have the character of an event horizon. It is clearly still a Killing horizon (associated with the Killing vector ∂_t), which here replaces the curvature singularity of the Khan–Penrose solution. More significantly, it can be seen to be a Cauchy horizon, which occurs because of the mutual focusing of the two waves. The space-time can be extended through this Killing–Cauchy horizon, but the extended region is not uniquely determined by the relevant initial data.[3]

---

[3] It may be recalled that the space-time of a single plane wave is not globally hyperbolic (see Section 17.5). For the collision of two plane waves in a Minkowski background, an initial

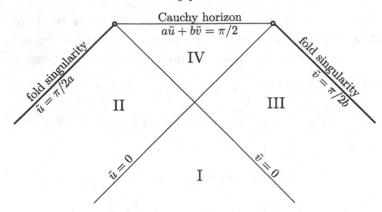

Fig. 21.4 A section through the degenerate Ferrari–Ibáñez solution showing the general structure of this space-time. The points $\tilde{u} = \pi/2a$, $\tilde{v} = 0$ and $\tilde{u} = 0$, $\tilde{v} = \pi/2b$ that are denoted by circles, have to be removed. The space-time can be extended (non-uniquely) through the focusing singularity, which is here a Cauchy horizon as explained in the text.

The general structure of this degenerate Ferrari–Ibáñez solution is illustrated in Figure 21.4. The coordinate singularities at $\tilde{u} = \pi/2a$ in region II and $\tilde{v} = \pi/2b$ in region III retain the character of "fold singularities" as they are identified with the two points that have been removed from the space-time. They correspond to the location of the caustic which forms due to the focusing effect of the two initial waves.

Various possible extensions of the space-time through the Cauchy horizon have been suggested, the simplest of which is just the time reverse of the prior region. A static extension has also been suggested which is equivalent to the Schwarzschild space-time outside the horizon (Hayward, 1989). However, this extended space-time is not easy to visualise.

It may be noted that, since $m^2 = 1/2ab$, stronger approaching gravitational waves correspond to a smaller value of m. This gives rise to a shorter proper time between the collision and the subsequent horizon, and a greater curvature at the focusing singularity (the Killing–Cauchy horizon).

Finally, it is instructive to consider the development of a cluster of test particles which initially follow parallel timelike geodesics in region I with constant values of x and y. As the worldlines of these pass through regions II or III and into region IV, they are focused astigmatically – they diverge in the y-direction but contract in the x-direction, reaching a focal line at

spacelike surface on which to specify Cauchy data can only be extended from region I as far as the fold singularities in regions II and III. The space-time is thus only uniquely determined as far as the fold singularities and the focusing singularity in region IV.

the Cauchy horizon. Each test particle can pass through this horizon, but the orientation of the cluster is reversed once they have passed through it. Their subsequent development is then not uniquely determined.

21.5 Other type D solutions

In view of the unexpected character of its focusing singularity, the degenerate Ferrari–Ibáñez space-time has received considerable attention, as have various generalisations of it. These will now be considered briefly.

21.5.1 The alternative Ferrari–Ibáñez (Schwarzschild) solution

The solution described above is one of the two degenerate cases analysed by Ferrari and Ibáñez (1987, 1988). In the other case, the metric function $\chi(u, v)$ of the Khan–Penrose solution (21.9) is divided by $\alpha(u, v)$ rather than being multiplied by it. In this case, the solution in the interaction region is also isomorphic to part of the Schwarzschild solution inside the horizon, with ϕ being a linear rather than a periodic coordinate. The two waves again collide at the spacelike surface which is equivalent to $r = m$, $\theta = \pi/2$, but in this case the interaction region corresponds to that in which the r-coordinate decreases. The focusing singularity is then a curvature singularity that corresponds to that of the Schwarzschild space-time at $r = 0$, $0 < \theta < \pi$. Although the initial waves that give rise to this solution are a combination of impulsive and shock components, the general structure of this space-time is identical to that of the Khan–Penrose solution, as illustrated in Figure 21.3. It therefore need not be described in further detail here.

21.5.2 The Chandrasekhar–Xanthopoulos (Kerr) solution

Interestingly, a non-diagonal generalisation of the degenerate Ferrari–Ibáñez solution was found by Chandrasekhar and Xanthopoulos (1986) before the simpler diagonal case described above. In this non-diagonal case, the interaction region of the non-colinear colliding plane wave space-time is locally isomorphic to a part of the Kerr space-time in the time-dependent region between the two horizons.

In this case, it is convenient to start with the Kerr metric in the standard Boyer–Lindquist form (11.1), namely

$$ds^2 = -\frac{\Delta_r}{\varrho^2}\left(dt - a\sin^2\theta\,d\phi\right)^2 + \frac{\varrho^2}{\Delta_r}\,dr^2 + \varrho^2\,d\theta^2 + \frac{\sin^2\theta}{\varrho^2}\left(a\,dt - (r^2 + a^2)d\phi\right)^2,$$

where $\varrho^2 = r^2 + a^2 \cos^2 \theta$, and $\Delta_r = r^2 - 2mr + a^2$, and apply the coordinate transformation

$$r = m + \sqrt{m^2 - a^2}\, \eta, \qquad \cos\theta = \mu, \qquad t = x, \qquad \phi = \frac{y}{\sqrt{m^2 - a^2}}.$$

$$(21.15)$$

For $\eta \in [0, 1]$, these coordinates cover part of the time-dependent region of the space-time with η being a timelike coordinate and $\mu \in [-1, 1]$. Introducing the parameters

$$p = \frac{\sqrt{m^2 - a^2}}{m}, \qquad q = \frac{a}{m},$$

so that $p^2 + q^2 = 1$, the line element takes the form

$$ds^2 = m^2 X\left(-\frac{d\eta^2}{1 - \eta^2} + \frac{d\mu^2}{1 - \mu^2}\right) + (1 - \eta^2)(1 - \mu^2)\frac{X}{Y}\, dy^2 + \frac{Y}{X}(dx + \omega\, dy)^2,$$

where

$$X = (1 + p\,\eta)^2 + q^2 \mu^2,$$

$$Y = p^2(1 - \eta^2) + q^2(1 - \mu^2) = 1 - p^2\eta^2 - q^2\mu^2,$$

$$\omega = -\frac{2q}{p}\frac{(1 - \mu^2)(1 + p\,\eta)}{Y}.$$

To interpret this (interior Kerr) solution in terms of colliding plane waves, it is appropriate to make the further coordinate transformation

$$\eta = \sin(a\tilde{u} + b\tilde{v}), \qquad \mu = \sin(a\tilde{u} - b\tilde{v}), \qquad (21.16)$$

where $2ab = 1/m^2$, and to remove the periodicity in ϕ so that $y \in (-\infty, \infty)$. (Note that a is a new parameter that represents the amplitude of one of the approaching waves. It is not the rotation parameter of the Kerr solution that has here been absorbed into q.) The metric is then given by

$$ds^2 = -2 X\, d\tilde{u}\, d\tilde{v} + \alpha^2 \frac{X}{Y}\, dy^2 + \frac{Y}{X}(dx + \omega\, dy)^2,$$

where

$$\alpha = \cos(a\tilde{u} + b\tilde{v})\cos(a\tilde{u} - b\tilde{v}) = 1 - \sin^2 a\tilde{u} - \sin^2 b\tilde{v},$$

$$X = (1 + p\sin(a\tilde{u} + b\tilde{v}))^2 + q^2 \sin^2(a\tilde{u} - b\tilde{v}),$$

$$Y = p^2 \cos^2(a\tilde{u} + b\tilde{v}) + q^2 \cos^2(a\tilde{u} - b\tilde{v}),$$

$$\omega = -\frac{2q}{p}\frac{\cos^2(a\tilde{u} - b\tilde{v})\left(1 + p\sin(a\tilde{u} + b\tilde{v})\right)}{\left(p^2 \cos^2(a\tilde{u} + b\tilde{v}) + q^2 \cos^2(a\tilde{u} - b\tilde{v})\right)},$$

which is of the form (21.3) with $\chi = \alpha X/Y$ and $f = X$. If the approaching plane waves are taken to have wavefronts given by $\tilde{u} = 0$ and $\tilde{v} = 0$, in view of the transformation (21.15) and (21.16), it can be seen that the collision occurs on a spacelike surface $r = m$, $\theta = \pi/2$ which is between the two horizons of the corresponding Kerr solution. The interaction region between the two waves is then the causal future of this surface up to the focusing singularity at $a\tilde{u} + b\tilde{v} = \pi/2$, which corresponds to the outer (event) horizon of the Kerr solution. As for the simpler case of the degenerate Ferrari–Ibáñez solution described in Section 21.4, which corresponds to the colinear case in which $q = 0$, the focusing singularity of this colliding plane wave solution is a coordinate singularity through which the space-time can be extended. This again has the character of a Cauchy horizon, and any extension through it is non-unique. The general structure of this solution is therefore the same as that illustrated in Figure 21.4.

If $q \neq 0$, the approaching waves consist of a combination of impulsive gravitational wave components with different polarisations, and shock waves with variable polarisation. The null coordinates in regions II and III are not affinely parametrised, but there is an obvious symmetry between the two approaching waves.

A different colliding plane wave space-time can be obtained by modifying the transformation (21.15) by alternatively setting $r = m - \sqrt{m^2 - a^2}\,\eta$, with $\eta \in [0, 1]$ and still using (21.16). In this case, the spacelike surface on which the waves collide is still given by $r = m$, $\theta = \pi/2$, but the interaction region now corresponds to that with decreasing r, so that the focusing singularity at $a\tilde{u} + b\tilde{v} = \pi/2$ now corresponds to the inner (Cauchy) horizon of the Kerr solution. These alternative cases have been discussed in detail by Hoenselaers and Ernst (1990).[4] In both, the focusing singularity of the colliding plane wave space-time is not a curvature singularity, but a Cauchy horizon through which it may be non-uniquely extended, as pictured in Figure 21.4. The difference here is that the polarisations of the initial impulsive wave components are not aligned, and the step components have variable polarisation, with different amplitude functions in the two cases.

---

[4] In fact, it was the region which develops toward the inner horizon of the Kerr solution that was initially given a colliding plane wave interpretation by Chandrasekhar and Xanthopoulos (1986). The existence of the alternative solution in which the focusing singularity corresponds to the outer horizon was pointed out by Hoenselaers and Ernst (1990). These authors considered extensions of the space-time through the Cauchy horizon either towards the ring singularity or the asymptotically flat region, respectively. However, although an extension is not uniquely determined, the more natural extension in the colliding plane wave context is simply the time-reversed space-time.

21.5.3 The Ferrari–Ibáñez (Taub–NUT) solution

Ferrari and Ibáñez (1988) have additionally shown that part of the Taub–NUT solution within the time-dependent Taub region is also isomorphic to the interaction region of a specific colliding plane wave solution. In this case, it is convenient to start with the metric in the form (12.1), namely

$$ds^2 = -f(r)(d\bar{t} - 2l\cos\theta\, d\phi)^2 + \frac{dr^2}{f(r)} + (r^2 + l^2)\left(d\theta^2 + \sin^2\theta\, d\phi^2\right),$$

where

$$f(r) = \frac{r^2 - 2mr - l^2}{r^2 + l^2},$$

and apply the coordinate transformation

$$\frac{r - m}{\sqrt{m^2 + l^2}} = \sin(a\tilde{u} + b\tilde{v}), \quad \cos\theta = \sin(a\tilde{u} - b\tilde{v}), \quad \bar{t} = x, \quad \phi = \frac{y}{\sqrt{m^2 + l^2}}.$$

If the constants are chosen such that $m^2 + l^2 = 1/2ab$, the line element takes the form

$$ds^2 = -2F\, d\tilde{u}\, d\tilde{v} + F^{-1}\cos^2(a\tilde{u} + b\tilde{v})\,(dx - 2q\sin(a\tilde{u} - b\tilde{v})\, dy)^2$$
$$+ F\cos^2(a\tilde{u} - b\tilde{v})\, dy^2,$$

where

$$F = 1 + 2p\sin(a\tilde{u} + b\tilde{v}) + \sin^2(a\tilde{u} + b\tilde{v}), \quad p = \frac{m}{\sqrt{m^2 + l^2}}, \quad q = \frac{l}{\sqrt{m^2 + l^2}},$$

so that $p^2 + q^2 = 1$. If the periodicity of ϕ is removed by assuming that $y \in (-\infty, \infty)$, this metric can represent the interaction region of a colliding plane wave space-time with wavefronts at $\tilde{u} = 0$ and $\tilde{v} = 0$. The solutions in the initial regions II and III are obtained by putting $b = 0$ and $a = 0$, respectively. The physical situation represented by this solution is very similar to that of the solutions described in the previous subsection, but the approaching waves in this case have a different non-constant polarisation which is quantified here by the parameter l. The metric function

$$\alpha = 1 - \sin^2 a\tilde{u} - \sin^2 b\tilde{v}$$

decreases throughout the interaction region as required. The focusing singularity at $\alpha = 0$ occurs when $a\tilde{u} + b\tilde{v} = \pi/2$, and this clearly corresponds to one of the horizons of the Taub–NUT solution, namely $r_+ = m + \sqrt{m^2 + l^2}$. (With a slight variation in the transformation above, a colliding plane wave space-time could have been shown to correspond to part of the Taub–NUT solution from a point inside the Taub region to the other horizon r_-.) As above, the focusing singularity is a Cauchy horizon rather than a curvature singularity, as illustrated in Figure 21.4, and any extension beyond this is non-unique.

21.5.4 Further type D solutions

In view of the fact that appropriate non-stationary regions of the Schwarzschild, Kerr and Taub–NUT space-times can be interpreted as interaction regions of colliding plane waves, as shown above, it is natural to ask how many of the other type D space-times can also be interpreted in this way. What are the necessary conditions for such solutions to admit this alternative interpretation?

The first observation to make in answering this question is that colliding plane wave space-times admit two (spacelike) Killing vectors in region IV. Since all type D space-times also admit the two isometries, denoted here by ∂_t and ∂_ϕ, they are locally candidates for this interpretation only in time-dependent regions in which t is a spacelike coordinate, provided ϕ is taken as a linear rather than an angular coordinate.

Secondly, since the two waves mutually focus each other, the non-expanding (non-contracting) type D solutions cannot admit this interpretation. Thus, the only possibilities must be contained in the class of expanding Plebański–Demiański metrics that can be expressed in the form (16.12).

It may also be observed that the initial data for colliding plane waves are given on two null characteristics which extend from the point of collision in both directions, as far as the fold singularities where the opposing waves are focused. These points correspond to coordinate singularities that are identified as roots of the metric function $\tilde{P}$. Since the required region corresponds to that between these two roots, possible solutions are further restricted to the metric form (16.18), which generally represents the family of rotating, charged and accelerating black holes with a NUT parameter, and possibly a cosmological constant.

Appropriate regions of black hole solutions with mass and either a rotation or a NUT parameter have already been shown to admit a colliding plane wave interpretation. So it is appropriate next to consider the possible inclusion of the acceleration parameter. When this is non-zero, however, the characterising metric functions become cubics (or more generally quartics) rather than quadratics. The resulting transformation to double-null coordinates, as is appropriate for a possible colliding plane wave interpretation, therefore necessarily introduces Jacobi elliptic functions, and these have been less widely used in the study of exact solutions. In fact, it has been shown (Griffiths and Halburd, 2007) for the simplest case, that part of the C-metric space-time inside the black hole horizon can be given a colliding plane wave interpretation in which there is an asymmetry between the two approaching waves.

Electric and magnetic charge parameters may also be included. An electromagnetic extension of the case described in Section 21.5.2 has been given by Chandrasekhar and Xanthopoulos (1987b). In this case, the interaction region is isomorphic to part of the time-dependent region of the Kerr–Newman solution between the two horizons. In the colliding plane wave context, this extension represents the collision of a combination of gravitational and electromagnetic waves.[5] (A collision of purely electromagnetic waves will be considered in the following section.)

Finally, it must be noted that the cosmological constant cannot be non-zero in this context. The method of extension to the prior regions II, III and I involves replacing the null coordinates $\tilde{u}$ and $\tilde{v}$ by $\tilde{u}\,\Theta(\tilde{u})$ and $\tilde{v}\,\Theta(\tilde{v})$, respectively, and this inevitably leads to region I being Minkowski space, not de Sitter or anti-de Sitter spaces.

It may therefore be concluded that vacuum and electrovacuum type D colliding plane wave solutions could possibly be given by the metric (16.18) with $\Lambda = 0$ and a range of r which corresponds to the interior of the black hole where $\mathcal{Q} < 0$. This family of solutions has six arbitrary parameters m, a, l, α, e and g. Members of this family with non-zero m and a, or m and l, have been described in detail above, and alternative extensions with non-zero α and e have been referred to. An extension to the five-parameter family of type D (Kerr–Newman–NUT) solutions (with parameters m, a, l, e and g) that can be interpreted as part of a colliding plane wave space-time has been identified by Papacostas and Xanthopoulos (1989). However, although it may be conjectured that a colliding plane wave interpretation of the complete six-parameter family of solutions exists, this has not yet been given explicitly.

In stationary axisymmetric space-times, the parameters m, a, l, α, e and g are related, respectively, to the mass, rotation, NUT parameter, acceleration and the electric and magnetic charges of the source. In the colliding plane wave context these quantify the amplitudes and relative polarisations of the initial gravitational and electromagnetic wave components.

---

[5] Chandrasekhar and Xanthopoulos (1987a) have also obtained a different type D solution for colliding electromagnetic waves that is not contained in the Plebański–Demiański family (16.18): i.e. the principal null directions of the gravitational and electromagnetic fields are not aligned.

21.6 The Bell–Szekeres (Bertotti–Robinson) solution

A purely electromagnetic plane wave with a step wavefront on the null hypersurface $u = 0$ can be represented by the metric (17.7) with $A(u) = 0$ and $B(u) = a^2\Theta(u)$, namely

$$ds^2 = -2\,du\,dr - 2a^2\Theta(u)\zeta\bar\zeta\,du^2 + 2\,d\zeta\,d\bar\zeta.$$

In this case, the only non-zero component of the curvature tensor is given by $\Phi_{22} = a^2\Theta(u)$ and the electromagnetic wave for $u > 0$ has constant amplitude a. Now consider the collision and subsequent interaction of such a wave with a similar one with constant amplitude b that is propagating in a different direction. As above, it is possible to adopt a frame of reference in which the two waves approach from opposite spatial directions, and the wavefront of the second wave may be taken to be $v = 0$, as represented in Figure 21.1. Using the transformation (17.9)–(17.10), the metrics in the regions I, II and III prior to the collision can be expressed in terms of double-null coordinates in the form (17.13), namely

I $(u < 0, v < 0)$: $ds^2 = -2\,du\,dv + dx^2 + dy^2,$

II $(u > 0, v < 0)$: $ds^2 = -2\,du\,dv + \cos^2(au)(dx^2 + dy^2),$

III $(u < 0, v > 0)$: $ds^2 = -2\,du\,dv + \cos^2(bv)(dx^2 + dy^2).$

This initial data is now well defined and it remains to determine the solution in the interaction region IV $(u > 0, v > 0)$.

The unique solution to this problem was presented by Bell and Szekeres (1974). The metric in region IV is given by

$$ds^2 = -2\,du\,dv + \cos^2(au + bv)\,dx^2 + \cos^2(au - bv)\,dy^2, \qquad (21.17)$$

which has the form (21.6), where the function $\alpha(u, v)$ is

$$\alpha = \cos(au + bv)\cos(au - bv) = 1 - \sin^2 au - \sin^2 bv.$$

This metric is in fact the conformally flat Bertotti–Robinson solution. It is obtained from the form given in (7.8) by the simple transformation

$$\tau = e(au + bv), \qquad z = e(au - bv),$$

where $e^2 = 1/2ab$. The interaction region IV in this case is locally isomorphic to the part of the Bertotti–Robinson space-time in which $y = e\phi \in (-\infty, \infty)$.

More importantly, Bell and Szekeres (1974) observed that, although the interior of region IV remains conformally flat, discontinuities necessarily occur in the derivatives of the metric functions across the initial boundaries of region IV. These manifest themselves as impulsive gravitational wave components that are given here by

$$\Psi_4 = -a \tan bv\, \delta(u)\Theta(v), \qquad \Psi_0 = -b \tan au\, \delta(v)\Theta(u).$$

It is in fact an interesting general feature of the collision of electromagnetic plane waves that impulsive gravitational waves are always generated by the collision (see Griffiths, 1991).

Since the metric representing the interaction region in this case is conformally flat, the focusing singularity that occurs at $au + bv = \pi/2$ is not a curvature singularity. It is a horizon of some type through which the space-time may be extended. On the other hand, the curvature tensor has a distributional kind of singularity (see Geroch and Traschen, 1987) at the points $u = \pi/2a$, $v = 0$ and $u = 0$, $v = \pi/2b$. These points therefore have to be removed from the space-time, and this has two further effects. Firstly, the singularities on the null hypersurfaces $u = \pi/2a$ and $v = \pi/2b$ in the prior regions II and III, respectively, acquire the character of fold singularities, as described above. And secondly, it also has the effect that any extension beyond the focusing singularity is not uniquely determined, so that this singularity is again a Cauchy horizon. The structure of this space-time is therefore exactly as illustrated in Figure 21.4. These properties have been described in detail for this case by Clarke and Hayward (1989).

21.7 Properties of other colliding plane wave space-times

In the previous sections, a number of examples of colliding plane wave space-times have been given in which the focusing singularity is a Cauchy horizon, instead of a curvature singularity. In fact, these are all of algebraic type D or, with electromagnetic waves, are conformally flat. This coincidence arises because these solutions are the simplest cases with this property. There is generally, however, no correlation between type D solutions and those with a Cauchy horizon in this context. As already mentioned, some colliding plane wave type D space-times still have a curvature singularity (e.g. the alternative degenerate Ferrari–Ibáñez solution described in Subsection 21.5.1). In addition, it must be emphasised that almost all colliding plane wave solutions with a Cauchy horizon are algebraically general. This can be seen, at least for the vacuum case with aligned linear polarisations, by including an arbitrary combination of the regular terms in the family of solutions given by Feinstein and Ibáñez (1989).

Although there exists a large class of colliding plane wave space-times with an infinite number of parameters in which the focusing singularity has the character of a Cauchy horizon through which the space-time can be extended, this is replaced in the generic situation by a scalar polynomial curvature singularity. Moreover, the exceptional cases in which Cauchy horizons occur are unstable with respect to small perturbations. This assertion is supported by arguments of Yurtsever (1987, 1989), who has considered various types of perturbations. For the vacuum case with aligned linear polarisations, the instability of these solutions with respect to bounded perturbations of the initial waves has been demonstrated rigorously by Griffiths (2005). The generic structure of these colliding plane wave space-times is therefore that illustrated in Figure 21.3.

In all the solutions described so far in this chapter, the background region into which the shock waves propagate is taken to be a flat Minkowski space. This is clearly the simplest case, and is a reasonable starting point for analysis. This assumption does, however, have a significant influence on the global properties of the interaction region. If the background is taken to be Minkowski space, any initial plane gravitational or electromagnetic waves will be non-expanding. The mutual focusing properties that occur on the interaction of such waves will, therefore, necessarily lead to the occurrence of a focusing singularity at some finite time in the future.

But, in fact, the universe we live in is observed to be expanding. If this property is reflected in the model for the background and initial data, the initial waves will be expanding. In this case, the mutual focusing properties could simply result in a reduction of their rates of expansion. On the other hand, there also exists a possibility that the expansion of the waves could be reversed and their mutual focusing could induce a contraction to a future singularity. These alternative possibilities need to be investigated.

The obvious background into which to consider the collision of shock waves is a Friedmann–Lemaître–Robertson–Walker (FLRW) model in which the matter content is "dust" or, possibly, radiation. Unfortunately, however, no exact solution is known that represents even a single shock wave propagating into such a background. Gravitational waves in these backgrounds have only been considered using approximation schemes, and these will not indicate the presence of the kind of global (nonlinear) properties that have been shown above for the simpler flat case.

Some solutions, however, are known for the case in which the source of the FLRW model is a "stiff" perfect fluid. For such a source, in which the pressure is equal to the density, the speed of sound is equal to the speed of light. In this case, the characteristics of the equations for the fluid are identical to the null characteristics of the gravitational waves, and this leads to the simplifications that are needed to obtain exact solutions. In fact, three distinct types of background can be considered. These correspond to the three FLRW models with constant zero, negative or positive spatial curvature: i.e. the cases $k = 0, -1, +1$, respectively (see Chapter 6). Because these background space-times are expanding, in all cases there exists a past (spacelike) big bang type of curvature singularity.

Exact solutions for single shock (linearly polarised) gravitational waves with arbitrary profiles and which propagate into any of these three backgrounds have been given by Griffiths (1993a), Bičák and Griffiths (1994) and Feinstein and Griffiths (1994), respectively (see also Bičák and Griffiths, 1996). In each case, the wave is back-scattered as it propagates. Its amplitude decreases both due to the expansion of the space-time, and also because it is being partially reflected. Although the region in front of the wave is conformally flat, the region behind the wavefront is algebraically general.

Exact solutions for the head-on collision of two such shock waves (with aligned polarisations) in each of these backgrounds have also been obtained. (See Griffiths, 1993b, Bičák and Griffiths, 1994, and Feinstein and Griffiths, 1994, respectively.) These show that the global structure of the interaction region is qualitatively the same as that of the corresponding region of the background FLRW space-time. The effect of the collision and interaction of waves in a closed background is at most to reduce the proper time to the final big crunch type of singularity. On the other hand, even for arbitrarily strong waves colliding in an open or flat background, their mutual interaction is not sufficiently strong to reverse the expansion and introduce a future singularity. Such waves may have large amplitudes if they collide near the initial singularity, but they are then expanding rapidly and the interaction only reduces this rate of expansion. Alternatively, if the waves are expanding less when they collide, this can only be after a longer time since the initial big bang and, by this time, their amplitudes will have been reduced sufficiently through back-scattering for the focusing effect to have little power. It may thus be concluded that the collision and subsequent interaction of gravitational waves cannot introduce a singularity into an initially expanding (stiff fluid) background. These results have been extended to include the propagation and interaction of waves in a vacuum anisotropic Kasner background (Alekseev and Griffiths, 1995) with equivalent results.

Besides considering more physically significant (expanding) backgrounds, it should be noted that gravitational waves with realistic sources are themselves expanding. It is therefore important to consider the collision and interaction of waves with expanding wavefronts. Very little progress has been made in finding exact solutions for such situations. However, one family of exact solutions for shock gravitational waves with cylindrical, spherical and toroidal wavefronts in a Minkowski background has been obtained by Alekseev and Griffiths (1996). These waves back-scatter as they propagate so that, behind the wavefront, they are algebraically general. Although they can be used to provide an exact solution for two colliding waves with exact cylindrical wavefronts with parallel axes in a flat background in which no singularities occur in the interaction region, they have unphysical sources and do not represent physically realistic situations.

22

A final miscellany

The authors are well aware that this book has only described a very limited number of the known exact solutions of Einstein's equations. We have concentrated on those we believe to be the most basic or that have particularly significant interpretations. However, many other interesting solutions have been analysed and have their own extensive published literatures. Indeed, a number of these are highly relevant – either because they represent some easily comprehended idealised physical situation, or because they demonstrate some important property. In this concluding chapter, we will therefore comment on a number of further space-times that fall into either of these categories.

After commenting on some aspects of the Bianchi and Kantowski–Sachs cosmological models, that have (at least) three spatial isometries corresponding to homogeneity, we will briefly review the main properties of some other space-times that possess two commuting Killing vectors which have not been described in previous chapters. In general, such space-times include those with cylindrical symmetry, stationary and axial symmetry, boost-rotation or boost-translation symmetry, as well as the interaction region of colliding plane waves, and the so-called Gowdy (or vacuum G_2) cosmologies.[1] Remarkably, for space-times of all these types, which possess 2-surfaces that are orthogonal to the group orbits, the resulting vacuum or electrovacuum field equations happen to be integrable. In fact, the field equations in all these cases can be written as a combination of the Ernst equations (Ernst, 1968a,b) in their elliptic or hyperbolic forms and subsidiary quadratures for

---

[1] In fact, the most important stationary, axially symmetric solutions have been reviewed in Chapter 13, while solutions with boost-rotation symmetry have been partly covered in Chapters 14 and 15, and colliding plane wave space-times have been discussed in Chapter 21. Some interesting families of solutions of the remaining types will be described in Sections 22.3 and 22.4.

the remaining metric function.[2] For the main Ernst equations, a large number of solution-generating techniques are available. In fact, many families of explicit solutions of these equations, some with an infinite number of parameters, are now known. These are widely described in the literature: see for example Hoenselaers and Dietz (1984), Griffiths (1991), Belinski and Verdaguer (2001), and Stephani *et al.* (2003). We have therefore restricted our comments, as in previous chapters, mainly to the most significant solutions of these types.

Up to this point, we have concentrated on vacuum space-times and those that include a cosmological constant, an electromagnetic or a pure radiation field. Of space-times with a perfect fluid source, we have only described the Einstein static universe in Section 3.2, the Friedmann–Lemaître–Robertson–Walker metrics in Chapter 6, interior Schwarzschild solutions in Section 8.5 and van Stockum's solutions for a rigidly rotating cylinder of dust in Section 13.1.2. However, a number of other families of perfect fluid space-times are particularly important. We will therefore conclude this book with some brief sections describing the most significant of these, namely the Gödel universe, the Lemaître–Tolman models and the Szekeres solutions.

22.1 Homogeneous Bianchi models

The idealising assumption that the universe we live in is homogeneous and isotropic led to the introduction of the Friedmann–Lemaître–Robertson–Walker cosmological models, as described in Chapter 6. However, as the real universe is not exactly homogeneous and isotropic, it is necessary to consider less restrictive classes of space-times. Relaxing the conditions of isotropy has led to the study of spatially homogeneous cosmologies. These are known as Bianchi models because they admit a three-dimensional group of isometries which act simply transitively, and can be classified according to the so-called Bianchi type.

It is possible to take the surfaces of symmetry Σ to be given by $t = $ const., and to choose a frame of vectors e_a with components $e_a{}^i$ dependent only on spatial variables. The line element for such space-times is then given by

$$ds^2 = -dt^2 + g_{ab}(t)(e^a{}_i\,dx^i)(e^b{}_j\,dx^j),$$

where $e^a{}_i$ is the matrix inverse of $e_a{}^i$. The evolution of the universe is then

---

[2] For the vacuum case, Ernst's equation in coordinate-invariant notation is given here in the alternative forms (13.11) or (13.14): for the electrovacuum case, the first form is generalised to that given in (13.26). For the hyperbolic case, these equations apply either to the Ernst potential as described in Section 13.2.1, or to metric functions as described below in Section 22.3.

represented by the time dependence of the six independent frame components g_{ab} of the metric.

The basis vectors e_a of the Bianchi models satisfy $[e_a, e_b] = C^c{}_{ab} e_c$, in which $C^c{}_{ab}$ are the structure constants of the relevant symmetry group. For the Bianchi models, this is a three-dimensional group, and it is possible to classify this according to the scheme given by Bianchi in 1897. Nine canonical types can be identified, and are labelled Bianchi I to Bianchi IX. Two of these, Bianchi VI and Bianchi VII, are one-parameter families of distinct group structures. Complex transformations formally relate type VIII to type IX, and type VI to VII. For technical details of these space-times from different points of view, see Ellis and MacCallum (1969), MacCallum (1973, 1994), Ryan and Shepley (1975), Rosquist and Jantzen (1988) and Stephani *et al.* (2003). A historical account of the development of the Bianchi classification in this context has been given by Krasiński *et al.* (2003). For a recent review, see Ellis (2006).

These space-times may be vacuum or non-vacuum but, since their main application is as cosmological models, a perfect fluid source is often assumed to be present.[3] Most solutions in this class have curvature singularities (see Collins and Ellis, 1979). These are usually of a spacelike character, corresponding to a big bang or big crunch, and coordinates are usually chosen so that the singularity occurs at $t = 0$.

The simplest solution of this type is that of Kasner (1921), which is a vacuum Bianchi I space-time. (See Harvey, 1990, for a list of rediscoveries of this solution.) It is normally expressed in terms of the line element

$$ds^2 = -dt^2 + t^{2p_1}dx^2 + t^{2p_2}dy^2 + t^{2p_3}dz^2, \qquad (22.1)$$

where p_1, p_2 and p_3 are constants that must satisfy the two constraints $p_1 + p_2 + p_3 = 1$, and $p_1^2 + p_2^2 + p_3^2 = 1$. Thus there remains just a single free parameter. In terms of one parameter a, the metric may also be expressed in the alternative form

$$ds^2 = \tilde{t}^{(a^2-1)/2}(-d\tilde{t}^2 + dz^2) + \tilde{t}^{1+a}\,dx^2 + \tilde{t}^{1-a}\,dy^2. \qquad (22.2)$$

This family of solutions contains the special cases of Minkowski space, which occurs when $(p_1, p_2, p_3) = (1, 0, 0)$ or $a = \pm 1$ as in (3.15), and the type D (plane symmetric) *AIII*-metric (9.9) in which $(p_1, p_2, p_3) = (\frac{2}{3}, \frac{2}{3}, -\frac{1}{3})$ or $a = 0$. Apart from these special cases, these space-times are algebraically general. The family may generally be interpreted as representing vacuum homogeneous but anisotropic cosmological models in which the expansion

---

[3] Taub (1951) has shown that solutions of Einstein's vacuum field equations exist for all nine Bianchi types.

(or contraction) in each of the coordinate directions may differ. Of course, the order of the parameters, as of the coordinates, is unimportant.

The Bianchi models are particularly suitable for analysing the asymptotic character of cosmological-type singularities. This was a controversial topic in the 1950s and 1960s, but a thorough understanding subsequently emerged. It is now generally accepted that, under particular but physically plausible assumptions, Einstein's equations imply that cosmological space-times are bounded by a (past or future) curvature singularity. Moreover, asymptotic behaviour near the singularity has the oscillatory character of the Bianchi IX vacuum. These oscillations can be described as sequences of distinct approximate Kasner epochs, in which the space-time has different rates of expansion or contraction in different directions. This oscillatory behaviour of the Bianchi IX vacuum was described by Misner (1969) as a "Mixmaster model". (See Misner, 1994 for a more recent review.)

Near a cosmological singularity, spatially-homogeneous but anisotropic vacuum space-times exhibit either Kasner-like behaviour, or that of the Bianchi IX models with their oscillations between distinct Kasner epochs. The time spent in each epoch decreases exponentially in time as the singularity is approached, and the sequence of particular epochs evolves according to an iterative map described by Bogoyavlenskiĭ (1985). Moreover, the oscillations appear, according to some criteria, to be chaotic. For detailed discussions of this subtle point, see e.g. Hobill *et al.* (1991), Hobill, Burd and Coley (1994) and Cornish and Levin (1997a,b).

22.2 Kantowski–Sachs space-times

Spatially homogeneous cosmological models, which are invariant under an r-parameter group of isometries whose orbits are spacelike hypersurfaces Σ, must have $r \geq 3$. If $r > 3$, then either $r = 6$ or $r = 4$ and, in *all but one* case, the complete isometry group admits a three-parameter subgroup which acts simply transitively on the three-dimensional homogeneous hypersurfaces Σ. The corresponding space-times are thus the Bianchi models described in the previous section, but with additional symmetries. In particular, if $r = 6$, the space-times are not only spatially homogeneous but are also spatially isotropic (that is, the 3-spaces Σ are maximally symmetric and have constant spatial curvature) so that they belong to the Friedmann–Lemaître–Robertson–Walker class described in Chapter 6 (see Figure 6.1). If $r = 4$, the space-times are locally rotationally symmetric. For more details see Stewart and Ellis (1968), Ellis and MacCallum (1969), MacCallum (1973), Ryan and Shepley (1975), Collins (1977) or Stephani *et al.* (2003).

The only exceptional case, with $r = 4$, for which a three-parameter subgroup acts on *two-dimensional* surfaces of constant positive curvature (without acting simply transitively on 3-spaces Σ) has been analysed independently by Kompaneets and Chernov (1964) and Kantowski and Sachs (1966). These are now known as *Kantowski–Sachs space-times* and can be written in the spherically symmetric form

$$\mathrm{d}s^2 = -\mathrm{d}t^2 + X^2(t)\,\mathrm{d}\chi^2 + Y^2(t)\,(\mathrm{d}\theta^2 + \sin^2\theta\,\mathrm{d}\phi^2), \qquad (22.3)$$

where $X(t)$ and $Y(t)$ are distinct scale functions. In general, these anisotropic cosmological space-times are of algebraic type D with principal null directions in the t, χ plane.

Many explicit solutions of Einstein's field equations for the above metric and various matter content are known. In particular, a specific vacuum solution of the form (22.3) without a cosmological constant is equivalent to the Schwarzschild solution inside the event horizon (with $r = Y$ being the timelike coordinate). Vacuum solutions with $\Lambda > 0$ include the de Sitter universe, see metric (4.23), or the Nariai universe, see metric (7.16).

Space-times (22.3) with dust were initially considered by Kantowski and Sachs (1966). Kantowski (1966) and Kompaneets and Chernov (1964) also found other perfect fluid solutions, in particular for radiation ($\gamma = \frac{4}{3}$) and a stiff fluid ($\gamma = 2$). It is possible to include an additional electromagnetic field and a cosmological constant, see Thorne (1967), Ellis (1967), Stewart and Ellis (1968), Vajk and Eltgroth (1970), Lorenz (1983), Weber (1984), Grøn and Eriksen (1987), Gair (2002) and elsewhere. For reviews, lists of references and some other details, including the description of singularities and the global behaviour of the Kantowski–Sachs cosmological models, see e.g. Collins (1977), MacCallum (1973), Ryan and Shepley (1975), Vargas Moniz (1993), Krasiński (1997) or Stephani *et al.* (2003).

22.3 Cylindrical gravitational waves

Cylindrically symmetric space-times possess two spacelike Killing vectors ∂_ϕ and ∂_z, which correspond, respectively, to a rotational symmetry about an infinite axis and a translational symmetry along it. The basic definition of a cylindrically symmetric space-time has been clarified by Carot, Senovilla and Vera (1999) in the general case in which the isometry group is not necessarily orthogonally transitive. Important static and stationary space-times with this symmetry have already been discussed in Sections 10.2 and 13.1.

The presence of the translational Killing field ∂_z implies that these space-times cannot generally be asymptotically flat. Consequently, quantities that are normally determined asymptotically, such as mass or energy, cannot be defined here. However, it is still possible to identify certain quantities that are conserved *per unit length* along the axis of symmetry.

One such conserved quantity is the *C*-energy, or "cylindrical energy", that was introduced by Thorne (1965a) (see also Chandrasekhar, 1986). This exploits properties of the local field equations without any direct reference to asymptotics. Its meaning is therefore not clear a priori. Ashtekar, Bičák and Schmidt (1997a,b) have investigated such quantities relative to the asymptotic structure of the associated 2+1-dimensional space-time. In particular, they have shown that the *C*-energy does not represent the Hamiltonian or the physical energy (per unit length) but, rather, that the physical Hamiltonian is a non-polynomial function of the *C*-energy. They have also introduced an analogue of the Bondi energy-momentum.

In the case in which there exists a family of 2-surfaces that are orthogonal to the group orbits, the line element of cylindrically symmetric space-times can be expressed in the form

$$ds^2 = e^{-2\psi}[e^{2\gamma}(-dt^2 + d\rho^2) + W^2 d\phi^2] + e^{2\psi}(dz + A\,d\phi)^2, \qquad (22.4)$$

where W, ψ, A and γ are functions of t and ρ only, and it is assumed that ρ is a measure of distance from an axis.

The vacuum field equations require that W satisfies the wave equation $W_{,\rho\rho} - W_{,tt} = 0$. If the gradient of W is spacelike, it is usually convenient to use a coordinate freedom to put $W = \rho$. They also require that the complex function $E = e^{-2\psi} + iA$ satisfies the Ernst equation

$$(E + \bar{E})\,\nabla^2 E = 2\,(\nabla E)^2, \qquad (22.5)$$

where $\nabla^2 E = (g^{\mu\nu} E_{,\nu})_{;\mu}$ and $(\nabla E)^2 = g^{\mu\nu} E_{,\mu} E_{,\nu}$.[4] This equation is known to be integrable. For any solution of this, the remaining metric function γ can then be determined (in principle) by quadratures.

For space-times with the metric (22.4) to admit an interpretation in terms of cylindrical symmetry, the metric should be regular on the axis. For this, it is necessary that $W^2 e^{-2\psi} + A^2 e^{2\psi} = O(\rho^2)$ as $\rho \to 0$. Solutions which fail to satisfy this condition have some curvature singularity on the axis. Such solutions may be acceptable as external solutions only on the understanding that they should be matched to some source along or around the axis. Away

---

[4] Since E here is a function of t and ρ, the Ernst equation (22.5) is hyperbolic. However, there is also a more subtle difference between this and the elliptic form of the Ernst equation (13.11) which arises in the context of stationary axisymmetric space-times. In that context, the imaginary part of the Ernst potential $\mathcal{E}$ is a "potential" for the non-diagonal metric function, while in this hyperbolic case it is the corresponding metric function A itself.

from the axis, as $\rho \to \infty$, solutions should normally become flat asymptotically. However, since they are invariant in the ∂_z direction, cylindrically symmetric solutions fail to be asymptotically simple.

If the two Killing vectors are hypersurface orthogonal then $A = 0$.[5] Alternatively, it is also possible to set $A = 0$ if the space-time admits the reflection symmetry $\phi \to -\phi$, $z \to -z$. In these cases, and when the gradient of W is spacelike, the line element simplifies to the form

$$\mathrm{d}s^2 = \mathrm{e}^{2(\gamma-\psi)}(-\mathrm{d}t^2 + \mathrm{d}\rho^2) + \rho^2\mathrm{e}^{-2\psi}\mathrm{d}\phi^2 + \mathrm{e}^{2\psi}\mathrm{d}z^2, \tag{22.6}$$

and the vacuum field equations are

$$\psi_{,\rho\rho} + \rho^{-1}\psi_{,\rho} - \psi_{,tt} = 0,$$

$$\gamma_{,\rho} = \rho\left(\psi_{,\rho}^2 + \psi_{,t}^2\right), \qquad \gamma_{,t} = 2\rho\,\psi_{,\rho}\,\psi_{,t}. \tag{22.7}$$

The first of these equations is identical to that for a polarised cylindrical wave propagating with the speed of light in Euclidean space. Moreover, it is linear, so that solutions can be superposed. For any wave-like solution of this equation, an expression for γ can be obtained in principle from the remaining equations using simple quadratures.

Although originally discovered by Beck (1925), the space-times (22.6) are widely known as *Einstein–Rosen waves* following the work of Einstein and Rosen (1937) and later Rosen (1954). They provided the first clear demonstration (apart from the weak-field limit) that the general theory of relativity predicts the existence of exact gravitational waves.

Early work on this topic was particularly related to the question of whether or not gravitational waves transport energy. Since these space-times are not asymptotically flat, the simplest suggested expression for energy is the C-energy that was mentioned above. For the metric (22.6), the local C-energy is defined as the quantity $\frac{1}{4}\gamma$.

One family of solutions that is of particular significance is the *pulse-wave solution* of Rosen (1954).[6] This is given by

$$\psi = \int_0^{t-\rho} \frac{F(\sigma)\mathrm{d}\sigma}{\sqrt{(t-\sigma)^2 - \rho^2}},$$

where the function $F(\sigma)$ is an arbitrary bounded function which vanishes outside a finite range $0 < \sigma < T$, for some constant T. This describes a

---

[5] For further clarification of these properties, see MacCallum (1998), MacCallum and Santos (1998) and Castejon-Amendo and MacCallum (1990).

[6] A different representation of this solution has been given by Alekseev and Griffiths (1997). This describes a general family of gravitational wave solutions with distinct wavefronts that can also be used in other contexts. In addition, this particular representation enables an expression for the remaining metric function γ to be obtained in an explicit form.

pulse of gravitational radiation which has a distinct wavefront at $t = \rho$ that can be made arbitrarily smooth by taking a sufficient number of derivatives of $F(\sigma)$ to be zero on $\sigma = 0$. However, as is usual for cylindrical waves, it retains a tail.

By superposing a Rosen pulse solution for ψ in (22.7) with the solution for an infinite static cylinder with finite density and pressures in the form $\psi = a \log \rho + b$, Marder (1958) was able to demonstrate that cylindrical gravitational waves propagate energy. Specifically, he showed that the inertial mass per unit proper length of the cylinder decreases by about a quarter of the change in γ after the emission of a Rosen pulse. This is consistent with Thorne's concept of C-energy, and supports the argument that gravitational waves carry away energy. (See also Bonnor, Griffiths and MacCallum, 1994.)

Another solution that is of particular interest is the cylindrical gravitational pulse solution of Weber and Wheeler (1957) and Bonnor (1957a). The corresponding expression for ψ can be expressed in the form

$$\psi = 2c \int_0^\infty e^{-ak} \cos(kt) \, J_0(k\rho) \, dk,$$

where a and c are constants with a determining the width of the pulse. This can be written (see Ashtekar, Bičák and Schmidt, 1997b) as

$$\psi = \sqrt{2} \, c \left(\frac{[(a^2 + \rho^2 - t^2)^2 + 4a^2t^2]^{1/2} + a^2 + \rho^2 - t^2}{(a^2 + \rho^2 - t^2)^2 + 4a^2t^2} \right)^{1/2}.$$

In this case, the expression for γ is given by

$$\gamma = \frac{c^2}{2a^2} \left(1 - \frac{2a^2\rho^2[(a^2 + \rho^2 - t^2)^2 - 4a^2t^2]}{[(a^2 + \rho^2 - t^2)^2 + 4a^2t^2]^2} + \frac{\rho^2 - a^2 - t^2}{[(a^2 + \rho^2 - t^2)^2 + 4a^2t^2]^{1/2}} \right).$$

This solution describes a cylindrical pulse of gravitational radiation which is initially incoming and then reflects symmetrically off the axis. Its main significance arises from the fact that it is everywhere regular. It is a global solution which becomes flat as $\rho \to \infty$, and is singularity-free on the axis. The character of the solution, at least for the metric function ψ, is illustrated in Figure 22.1. Of particular interest is the fact that the pulse appears to be reflected before it meets the axis. Or, relative to the axis, the wave appears to experience a phase shift. The significance of this feature, which was initially noted by Weber and Wheeler (1957), has been discussed further by Griffiths and Miccichè (1997).

When using the non-diagonal metric (22.4) with $A \neq 0$, gravitational waves have two distinct modes of polarisation, and the interaction of such waves may induce a nonlinear Faraday-type rotation between them. For details of this, see Piran, Safier and Stark (1985) and Tomimatsu (1989).

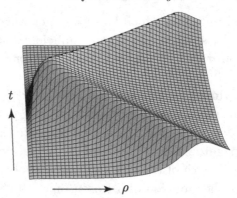

Fig. 22.1 This plot of $\sqrt{\rho}\,\psi(\rho,t)$ illustrates the character of the Weber–Wheeler–Bonnor pulse, particularly the fact that the wave appears to be reflected before it meets the axis. The scaling $\sqrt{\rho}$ has been introduced to partly compensate for the apparent decay of ψ away from the axis on which there is no singularity.

Further aspects of cylindrical gravitational waves can be found in Stachel (1966), Fustero and Verdaguer (1986), Chandrasekhar and Ferrari (1987), Ashtekar, Bičák and Schmidt (1997a,b), Mashhoon, McClune and Quevedo (2000) and references cited therein.

22.4 Vacuum G_2 cosmologies

It has already been emphasised above that, with two isometries, Einstein's vacuum field equations are integrable. Cases in which both Killing vectors are spacelike have been used above to describe cylindrical gravitational waves, as well as the interaction region of colliding plane waves. However, solutions which are locally the same can also be used to represent the simplest class of inhomogeneous vacuum cosmological models. In these, the universe is assumed to contain gravitational waves propagating in opposite spatial directions. Such space-times, which are normally algebraically general, are usually referred to as G_2 cosmologies.

In general, the (unpolarised) line element can be expressed in either of the forms (21.3) or (22.4), which can here be rewritten as

$$ds^2 = f\left(-dt^2 + dz^2\right) + \alpha\left(\chi\,dy^2 + \chi^{-1}(dx + \omega\,dy)^2\right), \qquad (22.8)$$

where α, χ, ω and f are functions of t and z only. For vacuum or electrovacuum solutions, the function α satisfies the wave equation $\alpha_{,tt} - \alpha_{,zz} = 0$, and this may have a solution whose gradient is timelike, spacelike (as used for cylindrically symmetric space-times in Section 22.3) or null. For cosmological models, it is appropriate to take α to be timelike, and it is possible to make a transformation to set $\alpha = t$. These space-times are then singular at $t = 0$, which corresponds to a big bang type of singularity. In this context, it is the complex combination of metric functions $E = \chi + i\omega$ which satisfies the Ernst equation (22.5). With any solution of this, the remaining metric function f can be determined (in principle) by quadratures.

A particularly important simplification occurs in the above procedure when $\omega = 0$. Not only is the metric then diagonal and the two Killing vectors orthogonal, but the main field equation (for $\log \chi$) becomes linear (equivalent to the first equation in (22.7)), thus enabling various particular solutions to be superposed. Space-times of this class are then said to be *polarised*. As a simple example, they include the Kasner solution (22.2).

Generally, these space-times may be considered to start at an inhomogeneous big bang type of singularity at $t = 0$. Gravitational waves emerge from this and disperse in two different directions. The time-reversed versions of these, in which interacting gravitational waves focus at a future singularity, may also be considered.

Following Gowdy (1971), it is alternatively possible to take the solution of $\alpha_{,tt} - \alpha_{,zz} = 0$ given by

$$\alpha(t, z) = \sin t \sin z.$$

Solutions of this type have both initial and final singularities at $t = 0, \pi$. It is then possible to construct space-times that are periodic in z. By identifying appropriate points, space-times with toroidal and spherical topologies may be constructed; in which cases the relevant regularity conditions also have to be satisfied. Specifically, compact spatial sections may have the topology of a 3-torus $S^1 \times S^1 \times S^1$, a 3-handle $S^1 \times S^2$ or a 3-sphere S^3. Space-times of these types are often referred to as *Gowdy cosmologies*. For a discussion of their possible topologies, see Gowdy (1974), Hanquin and Demaret (1983) and Tanimoto (1998).

Following the introduction of these space-times by Gowdy (1971), their general properties and further exact solutions have been analysed from a variety of points of view by many people including Gowdy (1974, 1975), Centrella and Matzner (1979), Moncrief (1981), Carmeli, Charach and Malin (1981), Adams *et al.* (1982), Carr and Verdaguer (1983), Ibañez and Verdaguer (1983), Feinstein and Charach (1986), Feinstein (1987), Chruściel (1990) and Kichenassamy and Rendall (1998). In particular, the occurrence of strong cosmic censorship in these space-times has been clarified by Chruściel, Isenberg and Moncrief (1990) and Grubišić and Moncrief (1993).

The well-known singularity theorems predict that cosmological models generally possess singularities. However, these indicate nothing about their structure. The character of the singularity in the Gowdy cosmologies has therefore been extensively analysed. Detailed studies have generally found that, near the singularity, time derivatives dominate over spatial derivatives in the field equations. Such behaviour is referred to as *asymptotically velocity term dominated*. As the singularity is approached, these space-times asymptotically approach a Kasner model locally. This behaviour is very different from the oscillatory character of the Mixmaster Bianchi IX models. It was first discussed in the context of dust-filled models by Eardley, Liang and Sachs (1972) and for vacuum Gowdy cosmologies by Isenberg and Moncrief (1990) and Berger and Moncrief (1993).

Another interesting feature that has recently been discovered in unpolarised Gowdy cosmologies is the possible appearance of spikes in the curvature near the singularity. These were initially found in numerical studies, but have since been studied analytically by Rendall and Weaver (2001), Garfinkle (2004) and Ringström (2004a,b).

22.5 Majumdar–Papapetrou solutions

As in Newtonian theory, it is possible in general relativity to achieve a static configuration with two sources if the gravitational attraction between them is exactly balanced by their electrostatic repulsion. For this to occur, each source must have a charge of the same sign, and this must have exactly the same magnitude as its mass. In general relativity, exact space-times describing such a situation are obtained by a superposition of two extreme Reissner–Nordström solutions. In fact, it is possible to obtain an exact solution for a random distribution of any number of such sources. This was first given by Majumdar (1947) and Papapetrou (1947), and interpreted by Hartle and Hawking (1972).

The exterior region of the extreme Reissner–Nordström space-time can be described using Weyl coordinates by the metric (10.19). In fact, it has already been pointed out in Section 10.8 that an arbitrary number of such sources can be superimposed, with each source being located on a common axis. Such a solution is given in (10.21), which has been further analysed by Azuma and Koikawa (1994). For a set of n randomly distributed sources of this type, that are not necessarily on the axis, the metric can be written in the Cartesian-like form

$$\mathrm{d}s^2 = -\left(1 + \sum_i \frac{m_i}{R_i}\right)^{-2}\mathrm{d}t^2 + \left(1 + \sum_i \frac{m_i}{R_i}\right)^2\left(\mathrm{d}x^2 + \mathrm{d}y^2 + \mathrm{d}z^2\right), \quad (22.9)$$

with $\Phi = (1 + \sum \frac{m_i}{R_i})^{-1}$, where $R_i{}^2 = (x - x_i)^2 + (y - y_i)^2 + (z - z_i)^2$ for

$i = 1, \ldots n$, and each source, which has mass m_i and charge $e_i = m_i$ (or $e_i = -m_i$), appears to be located at the "points" $x = x_i$, $y = y_i$, $z = z_i$.

As emphasised by Hartle and Hawking (1972), the metric (22.9) only represents the space-time outside the horizons of each of the sources. In fact, the degenerate horizons of each of the extreme Reissner–Nordström black holes corresponds to the apparent "points" $R_i = 0$. However, the space-time can be extended analytically through each of these horizons, and the metric for each extension is similar to (22.9) but with corresponding term in the sum occurring with a negative sign. Each of these solutions for the extended regions contains a curvature singularity where the corresponding sum vanishes. The qualitative structure of each of these sources is as described in Section 10.8.

It can be seen from the conformal diagram in Figure 9.10, that each interior region can itself be extended to further exterior regions. And it is possible, but not necessary, that the extended region through one pair of horizons may be identified with the same extended region through a pair of horizons for a different source. Thus the complete analytically extended space-time contains an infinite number of distinct regions that are not necessarily causally related. Multiply connected space-times may also occur.

The motion of test particles in this space-time can be chaotic. This was discovered and studied for two sources by Contopoulos (1990, 1991), with further analysis by Yurtsever (1995) and, for null geodesics, by Alonso, Ruiz and Sánchez-Hernández (2008). The case with multiple sources has been studied by Dettmann, Frankel and Cornish (1994).

For a discussion of the uniqueness of these solutions see Chruściel and Nadirashvili (1995) and Heusler (1997).

Similar solutions with a positive cosmological constant are also known. These represent a superposition of n "lukewarm" black holes for which $e_i = m_i$. In these, the gravitational and electrostatic forces between the sources remain balanced, but the distances between the sources vary in accordance with the cosmological expansion. The metric for this was given by Kastor and Traschen (1993, 1996) in the form

$$ds^2 = -\Omega^{-2}\, dt^2 + \Omega^2\, e^{2\sqrt{\frac{1}{3}\Lambda}\,t} \left(dx^2 + dy^2 + dz^2 \right), \qquad (22.10)$$

in which

$$\Omega = 1 + \sum_i \frac{m_i}{R_i}\, e^{-\sqrt{\frac{1}{3}\Lambda}\,t},$$

where $R_i{}^2 = (x - x_i)^2 + (y - y_i)^2 + (z - z_i)^2$ for $i = 1, \ldots n$, and $\Phi = \Omega^{-1}$. Further details have been given by Brill *et al.* (1994).

22.6 Gödel's rotating universe

The perfect-fluid solution obtained by Gödel (1949) can be expressed in comoving coordinates by the line element

$$ds^2 = - \left(dt + e^{2\omega x} \, dy \right)^2 + dx^2 + \tfrac{1}{2} e^{4\omega x} \, dy^2 + dz^2, \qquad (22.11)$$

where ω is an arbitrary constant which vanishes in the Minkowski limit. The fluid four-velocity is $u = \partial_t$, and the constant density ρ, pressure p, and cosmological constant Λ are constrained by

$$8\pi\rho + \Lambda = 8\pi p - \Lambda = 2\omega^2.$$

This implies that $4\pi(\rho - p) = -\Lambda$. Thus, when $\Lambda = 0$, the fluid has a stiff equation of state in which the pressure is equal to the density. For the traditional pressure-free dust case, the cosmological constant must be negative. Indeed, for any realistic fluid source, it is necessary that $\Lambda < 0$ (which is opposite to the condition that occurs in the Einstein static universe). The Weyl tensor is of type D and there are no curvature singularities.

This solution is characterised by the fact that the fluid has no expansion, acceleration or shear. The only non-zero kinematic quantity is the *rotation* (vorticity), which has constant magnitude ω in the sections spanned by the x and y coordinates. The Gödel solution is therefore not a realistic model of the universe in which we live. Nevertheless, it is an important pedagogical example that has some very interesting properties.

The space-time with metric (22.11) is clearly stationary. It is also spatially homogeneous. In fact, it has a five-dimensional group of isometries that is multiply transitive on space-time. It is therefore rotationally symmetric (Ellis, 1967) about any point. Moreover, Ozsváth (1965) and Farnsworth and Kerr (1966) have proved that it is the only homogeneous perfect-fluid solution with this number of symmetries. Within these symmetries, it admits three-dimensional groups of Bianchi types I, III and VIII on T^3 and of type III on S^3.

This model may be considered as a homogeneous space-time with a fluid source that is rigidly rotating. In the Newtonian analogue of such a situation, relative to any point, there will exist a cylinder of finite radius around the axis of rotation on which the particles of the fluid would move at the speed of light. For all points outside this cylinder, the particles of the classical fluid will be moving faster than light. Of course, such a situation is not possible in a relativistic theory. However, this feature is reflected here in the occurrence of *closed timelike curves* as will be shown below. Several people have followed Gödel (1949) in investigating the physical properties of a Newtonian analogue of this space-time. (See, for example, the review of Ozsváth and Schücking, 2001.)

In fact, the metric (22.11) can be rewritten in the form

$$ds^2 = -d\bar{t}^2 + dr^2 + dz^2 - \frac{2\sqrt{2}}{\omega}\sinh^2\omega r\, d\phi\, d\bar{t} + \frac{1}{\omega^2}(\sinh^2\omega r - \sinh^4\omega r)d\phi^2,$$
$$(22.12)$$

using the transformation

$$e^{2\omega x} = \cosh 2\omega r + \sinh 2\omega r \cos\phi,$$
$$\sqrt{2}\,\omega\, y\, e^{2\omega x} = \sinh 2\omega r \sin\phi,$$
$$\frac{\omega t}{\sqrt{2}} = \frac{\omega \bar{t}}{\sqrt{2}} - \frac{\phi}{2} + \arctan\left(e^{-2\omega r}\tan\frac{\phi}{2}\right),$$

where $|\frac{1}{\sqrt{2}}\omega(t - \bar{t})| < \frac{\pi}{2}$, $r \in [0, \infty)$ and $\phi \in [0, 2\pi)$.[7] These coordinates are still comoving with $\boldsymbol{u} = \partial_{\bar{t}}$, and the rotational symmetry of the metric (22.12) about an axis at $r = 0$ can be seen clearly. The spatial homogeneity of the space-time then implies that it must be rotationally symmetric about every point.

It can also be seen from the metric (22.12) that curves on which $\bar{t}$, r and z are constant are either spacelike or timelike according to whether $\sinh^2 \omega r$ is, respectively, less than or greater than 1. They are null when $\omega r = \log(1 + \sqrt{2})$. Moreover, since $\phi = 2\pi$ is identified with $\phi = 0$, these are closed timelike curves whenever r is larger than this value. However, they are not geodesics. In addition, the closed null line is not a geodesic but is an envelope of a family of null geodesics, as will be shown below and illustrated in Figure 22.3.

In view again of its spatial homogeneity, it follows that closed timelike curves must pass through every point of the space-time. Thus, although the worldlines of the particles of the fluid are all of infinite extent and are not closed, each point on a worldline of matter could theoretically communicate with its past. This violation of the concept of causality is one of the most notable features of this model. Moreover, it has been shown (Rosa and Letelier, 2007) that such closed timelike curves are stable with respect to linear perturbations, and therefore have a kind of structural stability.

The Gödel solution is globally regular and geodesically complete. However, since closed timelike curves pass through every point, there exist no embedded three-dimensional surfaces without boundary that are spacelike

---

[7] This transformation was obtained by Gödel (1949) using the group structure of the space-time, and has frequently been quoted. A simpler derivation of this has been given by Ozsváth and Schücking (2001).

everywhere. A global Cauchy hypersurface therefore does not exist in this space-time.

These and other properties of the Gödel universe have been described by Kundt (1956), Hawking and Ellis (1973), Ryan and Shepley (1975), Pfarr (1981), Novello, Svaiter and Guimarães (1993) and Ellis (2000). For example, using the metric (22.12), it can be seen that local light cones originating at any point open out and tip over as r increases. This is illustrated on a section through the space-time in Figure 22.2. (For further clarification of this, see Ozsváth and Schücking, 2003.)

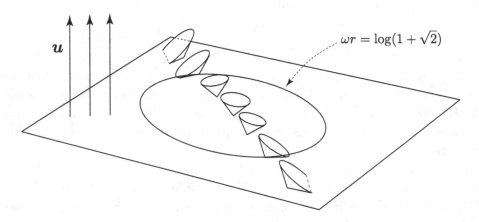

Fig. 22.2 A section through the Gödel space-time is shown using the metric (22.12) with $\bar{t}$ constant, z suppressed and the remaining coordinates r and ϕ treated as plane polar coordinates. This surface, which is everywhere orthogonal to the fluid flow $\boldsymbol{u}$, becomes locally timelike when $\omega r > \log(1 + \sqrt{2})$. Light cones originating on this surface open out and tip over as r increases. The circle on which $\omega r = \log(1 + \sqrt{2})$ is null. Circles with larger values of r are closed timelike curves, but not geodesics.

Any observer who is comoving with the fluid and looking out at neighbouring particles will actually see them at earlier times. But, since the fluid is rotating, null geodesics from these particles will appear to have rotated. In the same way, the light cone from any point on the axis $r = 0$, and with z constant, expands out in spirals to a maximum radius where the null geodesics form a circular cusp at the closed null line with $\omega r = \log(1 + \sqrt{2})$. There it experiences self-interactions and subsequently reconverges to another point on the axis. Following Hawking and Ellis (1973), this is illustrated schematically in Figure 22.3. Exact expressions for the null geodesics have been given by Kundt (1956) and Ozsváth and Schücking (2003). Interestingly, they have shown that, when projected onto the r, ϕ plane, these null geodesics appear as a family of circles which pass through the origin

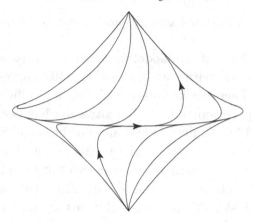

Fig. 22.3 Null geodesics from any point appear to rotate relative to a comoving observer. They spiral out to a maximum radius where they form a circular cusp with a closed null line, and then spiral back in to refocus at a future point on the worldline of the same particle of the fluid.

and touch the closed null circle. This light cone structure has been further investigated by Dautcourt and Abdel-Megied (2006).

The structure that is illustrated in Figure 22.3 has some important consequences. It immediately implies that an observer at any point can see no further than some finite distance in a direction perpendicular to the axis of rotation. (There is, however, no restriction to seeing even an infinite distance in the direction along the axis of rotation.) Specifically, the surface $wr = \log(1 + \sqrt{2})$ thus forms an event horizon relative to an observer at $r = 0$. The region beyond this, which (relative to that observer) contains closed timelike curves, always remains hidden. In addition, apart from any obstructions due to the fluid itself, any observer could actually see himself or herself at a certain finite time in the past. It is just as though each observer were living on the axis of a very large cylindrical mirror.

Rooman and Spindel (1998) have shown that the Gödel space-time, with the z-dimension suppressed, may be regarded as a deformed section of an anti-de Sitter space-time. By constructing the embedding of a generalised family of metrics in a 4+3-dimensional flat space, they have illustrated their causal structure and the unavoidability of closed timelike curves in the Gödel universe.

The stability of the Gödel universe with respect to various perturbations has been thoroughly analysed by Barrow and Tsagas (2004). They have found that the stability of the model depends primarily on the appearance of gradients in the rotational energy, and secondarily on the equation of state of the fluid.

The causal pathology of the Gödel universe is a large-scale phenomenon as, relative to any point (with $r = 0$), closed timelike curves only occur when $\omega r > \log(1 + \sqrt{2})$. There are no causal problems if a sufficiently small region of this space-time is considered. Such a situation has been investigated by Bonnor, Santos and MacCallum (1998) and Griffiths and Santos (2010) who have considered a cylinder whose interior is part of the Gödel space-time with metric (22.12), and have matched this to an exterior cylindrically symmetric vacuum solution of Santos (1993), thus constructing a globally regular space-time (see also Krasiński, 1975). These solutions are therefore similar to the van Stockum (1937) solutions that were described in Section 13.1.2, but with a non-zero cosmological constant. In particular, they found that, even when closed timelike curves are not present within the interior region, they may, or may not, appear in the exterior region depending on the choice of the arbitrary parameters in the exterior region. A number of particular cases were explicitly presented.

22.7 Spherically symmetric Lemaître–Tolman solutions

A physical situation that it is important to describe is the collapse of a cloud of particles to form a more condensed object that may subsequently develop into a star. For the simplest model of the early stages of such a process, it is appropriate to assume spherical symmetry and the presence of idealised matter in the form of dust. Such assumptions are also applicable to certain models for an inhomogeneous universe.

Exact solutions of Einstein's equations representing such a situation were investigated, initially by Lemaître (1933), Tolman (1934a) and Bondi (1947), and subsequently by many others such as Bonnor (1956, 1974) and Tomita (1992).[8] This family of type D space-times is now known as the Lemaître–Tolman solutions or sometimes as Tolman–Bondi solutions.[9]

For this brief review, it is appropriate to describe these solutions in terms of the synchronous coordinates adopted by Bondi (1947), as modified by Krasiński (1997). With these, the metric is expressed in the form

$$ds^2 = -dt^2 + \frac{R'^2}{1 + 2E}\, dr^2 + R^2(d\theta^2 + \sin^2\theta\, d\phi^2), \qquad (22.13)$$

---

[8] In the original papers of Lemaître (1933) and Tolman (1934a) a cosmological constant was also included.

[9] An equivalent family of solutions with plane symmetry also exists. See, for example, Tomita (1975) and Tomimura (1978). For the general solution covering cases with surfaces of positive, zero and negative spatial curvature, see Stephani *et al.* (2003).

where $R = R(t, r) > 0$, such that

$$\dot{R}^2 = 2E + \frac{2M}{R} + \frac{\Lambda}{3}R^2, \tag{22.14}$$

$M(r)$ and $E(r)$ are arbitrary functions, and the dot and prime denote derivatives with respect to t and r, respectively. The function $R(t, r)$, which is clearly an areal radius for the spherical surfaces on which t and r are constant, can also be interpreted here as the luminosity distance of an observer at (t, r, θ, ϕ) from the centre of symmetry. The density of the dust is then given by

$$\rho = \frac{M'}{4\pi R^2 R'},$$

and the coordinates are comoving, so that $u = \partial_t$. The fluid velocity for these solutions generally has non-zero shear – a feature that is radically different from the FLRW space-times.

In the case of a zero cosmological constant, the equation (22.14) can be completely integrated, but the character of the solutions depends on whether the function $E(r)$ is negative, zero or positive. The solutions are then described, respectively, as *bounded, marginally bounded* or *unbounded*. In each case, the solution can be written in the form

$$t - a(r) = -\frac{R^{3/2}G(-ER/M)}{\sqrt{2M}},$$

where $a(r)$ is an arbitrary function, and

$$G(y) = \begin{cases} \dfrac{\arcsin\sqrt{y}}{y^{3/2}} - \dfrac{\sqrt{1-y}}{y} & \text{for} \quad 0 < y \leq 1, \\[2ex] \dfrac{2}{3} & \text{for} \quad y = 0, \\[2ex] \dfrac{-\text{arcsinh}\sqrt{-y}}{(-y)^{3/2}} - \dfrac{\sqrt{1-y}}{y} & \text{for} \quad -\infty < y < 0. \end{cases}$$

In can be seen that $G(y)$ is a real, positive and smooth function which is bounded, monotonically increasing and strictly convex, with argument $y = -ER/M$, which has the range $-\infty < y \leq 1$.

This complete family of space-times is now characterised by three arbitrary functions of r, namely $M(r)$, $E(r)$ and $a(r)$. In particular, $M(r)$ can be physically interpreted as the total invariant mass within a shell of radius r, and $E(r)$ the total energy per unit mass of particles of the fluid. In Tolman's (1934a) alternative approach, these arbitrary functions can be determined by expressions for R, $\dot{R}$ and $\ddot{R}$ at $t = 0$, which can themselves be specified by the initial distribution and radial velocity of the matter.

For the special case in which $R(t, r) = R(t)\,r$, the metric (22.13) takes the form of the FLRW metric (6.4) with $2E(r) = -k\,r^2$ and $2M(r) = c\,r^3$. In this case (22.14) reduces to the Friedmann equation (6.15) for dust ($\gamma = 1$) and $\Lambda = 0$. The Lemaître–Tolman solutions thus include and generalise the FLRW space-times described in Chapter 6.[10] They are generally distinguished from them by the fact that they have fewer symmetries and the fluid four-velocity has non-zero shear.

In fact, the space-times given by (22.13) and (22.14) cover a wide range of situations, some of which represent particularly interesting phenomena. For example, if M' varies slower than r^2, the energy density becomes unbounded as $r \to 0$. In particular, Tolman (1934a) has used these models (with the inclusion of a cosmological constant) to demonstrate the instability of the Einstein static universe. He has shown that, in regions where the density starts to increase, the development of singularities is inevitable. Generally, the singularity that is formed in this way is a curvature singularity when $M'(0)$ is non-zero. Moreover, it can be shown to be naked under a wide range of conditions on M'.

In view of the spherical symmetry of these solutions, the fluid source can be considered to be composed of an infinite family of thin spherical shells. It can then be seen that the dust particles on different spherical shells will intersect if $R'(r, t) = 0$, as this would imply that the distance between the shells has become zero. This singularity in the density which occurs when $R' = 0$ (with $R \neq 0$) is thus referred to as a *shell-crossing singularity*.

For example, consider the simple, marginally bound class of solutions with

$$M(r) = \tfrac{2}{9}\,r^3, \qquad E(r) = 0,$$

and $a(r)$ is arbitrary. This is the "zero energy" case in which the dust particles originate from an exact state of rest at infinity. In this case, $R(t, r) = r(a - t)^{2/3}$, and the density is given by

$$\rho = \frac{1}{6\pi(a - t)(a + \tfrac{2}{3}ra' - t)}.$$

This space-time is clearly singular both at $t = a$ and at $t = a + \tfrac{2}{3}ra'$, but, as shown by Szekeres and Iyer (1993), these singularities have a different character. When $t = a$, the area of the spheres of symmetry vanishes and it is a crunch singularity similar to that which occurs in a black hole. On the

---

[10] For an analysis of the dynamics of these models (with a possible positive cosmological constant) and their relation to those of the FLRW space-times, see Wainwright and Andrews (2009).

other hand, at $t = a + \frac{2}{3}ra'$, the area radius stays finite and the singularity is not strongly censored. It is a *locally naked* shell-crossing singularity which can be interpreted in terms of the crossing of comoving shells.

It follows that solutions of this class could be considered as formal counter-examples to the cosmic censorship conjecture. However, since the matter field in this case is unphysical (shell-crossing singularities can occur for pressure-free fluids in the absence of gravity), these solutions imply nothing about naked gravitational singularities.[11]

The occurrence of naked shell-crossing singularities in Lemaître–Tolman solutions was first pointed out by Yodzis, Seifert and Müller zum Hagen (1973). Unnikrishnan (1994) has shown that a non-smooth initial density profile $M'(0)$ at the origin can give rise to a gravitationally strong naked singularity, but has argued that these are not physically acceptable counter-examples to the cosmic censorship conjecture. In particular, he has shown that the singularities when naked are weak, and when strong are strongly censored. The weakness of the shell-crossing singularities has been conclusively demonstrated by Nolan (1999).

Shell-focusing singularities also occur in these space-times. These were initially discovered by Eardley and Smarr (1979) using a numerical simulation of marginally bounded dust collapse. Christodoulou (1984), Newman (1986) and Grillo (1991) have subsequently analysed them in more detail and showed that, for an open subset of initial density distributions, these space-times violate the cosmic censorship conjecture.

The possible existence of vacuum regions (voids) within the Lemaître–Tolman space-times was discussed by Bonnor and Chamorro (1990) and Bonnor (2000). A more realistic attempt to use this family of solutions to model the kind of voids that are observed in the universe has been developed by Bolejko, Krasiński and Hellaby (2005).

The relation between the singularities that are formed in a collapsing dust cloud and the initial data specified at some definite time has been investigated by Dwivedi and Joshi (1997), Krasiński and Hellaby (2001, 2004) and many others. Indeed, because of the physical simplicity of these solutions, their mathematical tractability, and the range of interesting theoretical structures that arise in these space-times, they have attracted a very extensive literature. The review of Krasiński (1997) is particularly useful, as is the recent monograph of Bolejko *et al.* (2010).

In view of the recent observation that the rate of expansion of the universe is accelerating, and the inability of the standard FLRW model to de-

---

[11] For a discussion of the possibility of naked singularities in a fluid with a more realistic equation of state, see e.g. Rendall (1992).

scribe this without introducing the concept of dark energy, many people have debated whether or not the inhomogeneous Lemaître–Tolman space-times could represent the observed universe using only standard matter. In particular, Mustapha, Hellaby and Ellis (1997) have shown that the free functions in these models can accommodate arbitrary initial data for observed mass and luminosity distance functions. Such possibilities have been investigated among others by Célérier (2000), Iguchi, Nakamura and Nakao (2002), Alnes, Amarzguioui and Grøn (2006), Garfinkle (2006), Enqvist (2008) and Bolejko (2008). However, it was pointed out by Vanderveld, Flanagan and Wasserman (2006) that many of the models considered have a weak singularity at the location of the observer. In the absence of this singularity, the deceleration parameter is positive at the centre, in agreement with general theorems. However, other singularities tend to arise when attempting to match luminosity distance data. The inverse problem of trying to determine the particular Lemaître–Tolman model from the observed luminosity distance data has been investigated by Tanimoto and Nambu (2007). They have shown that, if a regularity condition is imposed at the centre, the solution is indistinguishable from a FLRW model up to second order in z, in agreement with a general result of Partovi and Mashhoon (1984).

22.8 Szekeres space-times without symmetry

A more general family of type D dust solutions was found by Szekeres (1975a) by assuming a line element of the form

$$\mathrm{d}s^2 = -\mathrm{d}t^2 + X^2\,\mathrm{d}z^2 + Y^2(\mathrm{d}x^2 + \mathrm{d}y^2), \qquad (22.15)$$

where X and Y may be functions of all four coordinates, which are assumed to be comoving with the fluid, so that $\boldsymbol{u} = \partial_t$. Two classes of solutions were identified according to whether or not Y depends on z. These solutions may include a non-zero cosmological constant but were initially obtained, and are summarised here, for the case with $\Lambda = 0$.

The class I solutions in which $Y_{,z} \neq 0$ generalise the Lemaître–Tolman solutions. They are given by the metric (22.15) with

$$Y = \frac{R(t,z)}{P(x,y,z)}, \qquad X = \frac{P\,Y'}{h(z)},$$

where

$$P = a(z)\big(x^2 + y^2\big) + 2b_1(z)\,x + 2b_2(z)\,y + c(z),$$

and $R(t, z)$ satisfies the dynamical equation

$$\dot{R}^2 = \frac{2M(z)}{R} - k(z),$$

where $M(z)$ is arbitrary,

$$k(z) = 4\left(ac + b_1^2 + b_2^2\right) - h^2,$$

and the dot and prime denote derivatives with respect to t and z, respectively. This solution is therefore characterised by six arbitrary functions of z, namely a, b_1, b_2, c, h and M. However, the coordinate freedom $z \to f(z)$ can be used to put $h(z)$ to any convenient form or to set $h^2 + k = \epsilon$, where $\epsilon = +1, 0, -1$. The density of the dust is then given by

$$\rho = \frac{P\,M' - 3M\,P'}{4\pi P^2\,R^2\,Y'}.$$

This can always be made positive by suitable choices of $M(z)$.

The class II solutions in which $Y_{,z} = 0$ are usually considered as generalisations of the Kantowski–Sachs and Friedmann–Lemaître–Robertson–Walker (FLRW) models. In this case, the metric can be expressed in the form given by Bonnor and Tomimura (1976), namely

$$\mathrm{d}s^2 = -\mathrm{d}t^2 + (AR + T)^2\,\mathrm{d}z^2 + R^2(\mathrm{d}x^2 + \Sigma^2\,\mathrm{d}y^2), \qquad (22.16)$$

where

$$A = A(x, y, z), \qquad R = R(t), \qquad T = T(t, z), \qquad \Sigma = \Sigma(x).$$

The 2-spaces spanned by x and y have constant curvature $k = +1, 0, -1$, so that $\Sigma = \sin x, x, \sinh x$, respectively (see Appendix A). Also, $R(t)$ satisfies the equation

$$\dot{R}^2 = \frac{2M}{R} - k,$$

where M is an arbitrary constant.

The source of both of these classes is a pressure-free perfect fluid that is non-rotating and non-accelerating, but has non-zero expansion and shear. The dust flows along a congruence of geodesics that are orthogonal to the hypersurfaces on which $t = \text{const}$. Moreover, as shown by Berger, Eardley and Olsen (1977), these hypersurfaces are conformally flat, which indicates that these space-times are nonradiative in the sense of York (1972) – a feature that could be supported by the fact that they are of algebraic type D, and which will be further commented on below.

It may be noted that the curvature of the surfaces spanned by x and y is a global constant only for the class II solutions. For the class I solutions, the curvature of these surfaces is determined by $h^2(z) + k(z)$, and this can vary over any section on which t is constant, as can also the curvature of the FLRW limit. This has the important consequence that observers in different locations in these space-times could construct locally approximate FLRW models with different curvatures (see Krasiński, 1997).

Another important characteristic of these solutions is that they generally have no Killing vectors (Bonnor, Sulaiman and Tomimura, 1977). Nevertheless, they still have very special geometrical properties. The Szekeres solutions are those for which both the second fundamental form of the spacelike hypersurfaces $t = $ const., and their Ricci tensor possess two equal eigenvectors (Collins, 1979). A covariant characterisation and classification of these solutions has been given by Collins and Szafron (1979) and Szafron and Collins (1979).

The evolution of some class I solutions has been described by Szekeres (1975b), while that of the general class II solutions has been described by Bonnor and Tomimura (1976), who have identified a number of significant limiting cases. Goode and Wainwright (1982) have noticed that the time evolution along a particular fluid worldline is the same in both classes and is governed by precisely the same functions that govern the evolution of density perturbations in FLRW dust models. They have then presented a new formulation of these solutions which unifies the two classes and reflects this observation. It was subsequently shown by Hellaby (1996) that the class II Szekeres solutions are in fact a regular limit of the class I solutions.

A particular case in which $h^2 + k = +1$ has been considered further by Szekeres (1975b). For this case, the constant-curvature surfaces on which t and z are constant are spheres even though the solution does not exhibit full spherical symmetry. Such solutions can thus be used to describe a quasi-spherical collapse of a distribution of dust. Szekeres has shown that these space-times can manifest the same kind of shell-crossing singularities that occur in the spherically symmetric case of the Lemaître–Tolman solutions. He has also argued that such singularities are globally or locally naked for certain periods as the shell collapses (see also Joshi and Królak, 1996). However, rigorous confirmation of this requires careful analysis, as has been demonstrated by Nolan and Debnath (2007).

It has also been shown by Bonnor (1976a,b) that, for some spherical surfaces, an interior Szekeres space-time can be matched to an exterior Schwarzschild solution even though the interior has no symmetry. This leads to the interesting observation that, in this asymmetric collapse, the relative motion of the dust particles does not generate gravitational waves. The conclusion is further supported by the calculations of Covarrubias (1980) who has shown that, in the linear approximation, the third derivatives of the quadrupole moments vanish for a matter distribution contained within a comoving surface. Hence the standard formula indicates zero energy loss due to gravitational radiation.

The Newtonian analogues of these solutions have been investigated by Lawitzky (1980), who has specifically shown that they represent a sequence of dust shells of irregular shape that have the same net effect as a single central source.

The general character of the collapse in these models has been discussed in detail by Barrow and Silk (1981). In particular, they have shown that the spheres on which t and z are constant expand and then collapse just as in the FLRW models. However, along the z-direction, the space first collapses and then expands to infinity as the final singularity is approached. In general, the maximum expansion of the spheres does not occur simultaneously with the minimum in the z-direction.

The properties of these solutions in the limit as the four-velocity of the dust becomes null has been investigated by Gleiser (1984), and also by Hellaby (1996). They have shown that the class I Szekeres metrics can, by a suitable limiting process, be made to yield a metric which describes a space-time containing null dust and depends on four arbitrary functions of a timelike coordinate. In fact, this is a Robinson–Trautman type D metric with generally no Killing vectors.

The cases in which $h^2 + k$ is zero or negative have been investigated by Hellaby and Krasiński (2008) and Krasiński (2008). These are quasi-planar or quasi-pseudospherical models which can represent a snake-like variable density void or overdense region in a more gently varying inhomogeneous background.

Further generalisations of these solutions in which the fluid has a non-zero pressure have been given by Szafron (1977) and thoroughly reviewed by Krasiński (1997).

Properties of the class I solutions and their possible cosmological applications have been described in a recent monograph of Bolejko *et al.* (2010).

Appendix A

2-spaces of constant curvature

For a two-dimensional Riemannian space, $ds^2 = h_{ij}\, dx^i dx^j$, where $i, j = 1, 2$, the curvature tensor has just one independent component which can be expressed in terms of its Gaussian curvature K as $\mathcal{R}_{ijkl} = K(h_{ik}\, h_{jl} - h_{il}\, h_{jk})$. For a 2-space of *constant* curvature, $K = \text{const.}$, there exists a coordinate system in which the metric takes the conformally flat form

$$ds^2 = \frac{dx^2 + dy^2}{\left(1 + \frac{1}{4}K(x^2 + y^2)\right)^2}. \qquad (A.1)$$

If the intrinsic or Gaussian curvature is expressed as

$$K = \frac{\epsilon}{a^2}, \qquad \text{where} \quad \epsilon = +1, 0, -1,$$

and the dimensionless complex coordinate $\zeta = (x + \mathrm{i}\, y)/(\sqrt{2}\, a)$ is introduced, the metric (A.1) can be written as

$$ds^2 = a^2 \frac{2\, d\zeta\, d\bar\zeta}{(1 + \frac{1}{2}\epsilon\, \zeta\bar\zeta)^2}. \qquad (A.2)$$

The constant a determines the magnitude of the curvature, while ϵ (which is the sign of K) is a discrete parameter that labels three distinct cases of 2-spaces with positive, zero and negative constant curvatures, respectively.

A.1 2-spaces of constant positive curvature

For the case with *positive curvature*, in which $\epsilon = +1$, it is convenient to put $\zeta = \sqrt{2} \tan(\theta/2)e^{\mathrm{i}\varphi}$, where $\varphi \in [0, 2\pi C)$ with $\varphi = 2\pi C$ identified with $\varphi = 0$, and $\theta \in [0, \pi]$ which is an azimuthal coordinate. With $a = 1$, the metric (A.2) then takes the form

$$ds^2 = d\theta^2 + \sin^2\theta\, d\varphi^2. \qquad (A.3)$$

462

Alternatively, putting $\varphi = C\phi$, where $\phi \in [0, 2\pi)$, the line element becomes $\mathrm{d}s^2 = \mathrm{d}\theta^2 + C^2 \sin^2\theta \, \mathrm{d}\phi^2$. When $C = 1$, this 2-space is the familiar (unit) sphere S^2 in which ζ is a stereographic coordinate. Otherwise, the spherical-type surface has a conical singularity at the poles with deficit (or excess) angle $2\pi\delta$, given by $\delta = 1 - C$.

These surfaces can be visualised as axially symmetric embeddings in a three-dimensional Euclidean space

$$\mathrm{d}s^2 = \mathrm{d}x_1{}^2 + \mathrm{d}x_2{}^2 + \mathrm{d}x_3{}^2,$$

using the parametrisation

$$x_1 = C \sin\theta \cos\phi,$$
$$x_2 = C \sin\theta \sin\phi,$$
$$x_3 = \int \sqrt{1 - C^2 \cos^2\theta} \, \mathrm{d}\theta = E(\tfrac{\pi}{2} - \theta, C),$$

where $E(\phi, C)$ is an elliptic integral of the second kind. For different values of C, these surfaces are illustrated in Figure A.1. For the case in which $C > 1$, such an embedding is only possible for values of θ that are restricted to the range $\arccos(1/C) \le \theta \le \pi - \arccos(1/C)$.

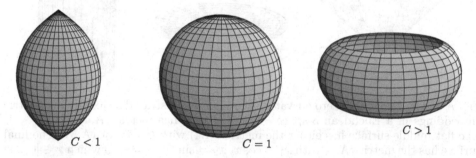

Fig. A.1 2-spaces of constant positive curvature illustrated as axially symmetric embeddings in a Euclidean 3-space for various values of C. The absence of regions near the poles for the case $C > 1$ is a feature of the embedding, not of the 2-space.

A.2 2-spaces of zero curvature

For the case of *zero curvature*, in which $\epsilon = 0$, it is convenient to consider $\zeta = \frac{1}{\sqrt{2}} \rho \, e^{i\varphi}$, with $\rho \in [0, \infty)$ and $\varphi \in [0, 2\pi C)$, and $\varphi = 2\pi C$ identified with $\varphi = 0$. With $a = 1$, the metric (A.2) then becomes

$$\mathrm{d}s^2 = \mathrm{d}\rho^2 + \rho^2 \mathrm{d}\varphi^2. \tag{A.4}$$

When $C = 1$, this 2-space is a plane spanned by polar coordinates. Otherwise, it possesses a conical singularity at $\rho = 0$ with corresponding deficit angle $2\pi\delta = 2\pi(1 - C)$. If $C > 1$, this represents an excess angle. Alternatively, by putting $\varphi = C\,\phi$, the metric can be written as $\mathrm{d}s^2 = \mathrm{d}\rho^2 + C^2\rho^2\mathrm{d}\phi^2$ as described in Section 3.4.

For the case with $C = 1$, however, it is also possible to introduce Cartesian coordinates by putting $\zeta = \frac{1}{\sqrt{2}}(x + iy)$, giving

$$\mathrm{d}s^2 = \mathrm{d}x^2 + \mathrm{d}y^2. \tag{A.5}$$

Moreover, by making the x or y coordinates periodic, the 2-space may alternatively be represented as the surface of a cylinder (or even a toroid). Although the metric for these cases is locally equivalent, their global representations differ significantly.

Some of these possibilities, which can be illustrated as axially symmetric embeddings in a Euclidean 3-space, are shown in Figure A.2. However, the case with metric (A.4) and $C > 1$, and the toroid, cannot be represented in this way.

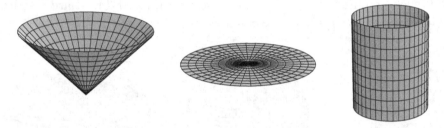

Fig. A.2 The 2-spaces of zero curvature that can be illustrated as axially symmetric embeddings in a Euclidean 3-space. The first case has metric (A.4) with $C < 1$. The flat middle surface has either the metric (A.4) with $C = 1$, or (A.5). The final surface has the metric (A.5) with $x_1 = \cos x$, $x_2 = \sin x$, $x_3 = y$, in which $x \in [0, 2\pi)$ is periodic.

A.3   2-spaces of constant negative curvature

In the remaining case, for which $\epsilon = -1$ and the constant curvature of the 2-space is *negative*, a coordinate singularity clearly occurs in the metric (A.2) when $|\zeta| = \sqrt{2}$. This can be seen to correspond to conformal infinity, and the distinct coordinate ranges satisfying $|\zeta| < \sqrt{2}$ and $|\zeta| > \sqrt{2}$ are in fact completely equivalent. Indeed, for this case, the metric (A.2) is invariant under the transformation $\zeta \to \zeta' = 2/\bar{\zeta}$, and it is sufficient to consider only the circular region for which $|\zeta| < \sqrt{2}$.

It is convenient first to put

$$\zeta = \sqrt{2}\,\tanh(R/2)e^{i\varphi}, \tag{A.6}$$

with which the metric (A.2) becomes

$$\mathrm{d}s^2 = \mathrm{d}R^2 + \sinh^2 R\,\mathrm{d}\varphi^2. \tag{A.7}$$

In these coordinates, such a space is often described as being *pseudospherical*. In this case, the singularity at $R = 0$ corresponds to a pole and it is appropriate to assume that φ is periodic with $R \in [0, \infty)$ and $\varphi \in [0, 2\pi C)$. Again, putting $\varphi = C\phi$, where $\phi \in [0, 2\pi)$, the metric (A.7) takes the form $\mathrm{d}s^2 = \mathrm{d}R^2 + C^2 \sinh^2 R\,\mathrm{d}\phi^2$.

For the case in which $C < 1$, part of this surface can be represented as an axially symmetric embedding in a three-dimensional Euclidean space using the parametrisation

$$x_1 = C \sinh R \cos \phi,$$
$$x_2 = C \sinh R \sin \phi,$$
$$x_3 = \int \sqrt{1 - C^2 \cosh^2 R}\,\mathrm{d}R = i\sqrt{1 - C^2}\,E\!\left(iR, \frac{iC}{\sqrt{1-C^2}}\right),$$

where $E(\phi, k)$ is an elliptic integral of the second kind. This surface, which is defined over the restricted range $0 < R \le \operatorname{arccosh}(1/C)$, is illustrated by the left surface in Figure A.3. In fact, this is the only surface of constant negative curvature that contains a conical singularity at $R = 0$. The above upper limit for R is a restriction of the embedding, not of the geometry.

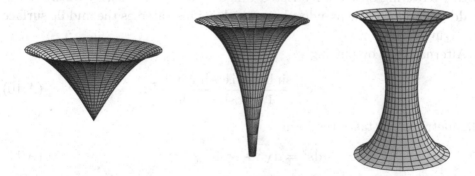

Fig. A.3 Parts of the 2-spaces of constant negative curvature represented by the metrics (A.7), (A.9) and (A.11), respectively, can, for appropriate coordinate ranges, be illustrated as axially symmetric embeddings in a Euclidean 3-space.

In fact, some alternative representations of 2-surfaces of constant negative curvature can also be obtained. For example, by putting

$$\zeta = \sqrt{2}\,\frac{\sinh\psi - \tfrac{1}{2}e^{\psi}\varphi^2 + i\,e^{\psi}\varphi}{\cosh\psi + \tfrac{1}{2}e^{\psi}\varphi^2 + 1},\tag{A.8}$$

the metric (A.2) becomes

$$ds^2 = d\psi^2 + e^{2\psi}\,d\varphi^2.\tag{A.9}$$

(These are referred to as *horocyclic coordinates*.) Moreover, it is sometimes appropriate to put $\psi = -\log z$, so that the line element can be expressed as

$$ds^2 = z^{-2}\left(dz^2 + d\varphi^2\right).$$

For $z > 0$, these are the coordinates of the Lobatchevski plane.

In the above forms, the natural range of the coordinates is that in which both $\psi, \varphi \in (-\infty, \infty)$. However, it is also possible to consider an axially symmetric-type surface for which $\varphi \in [0, 2\pi C)$, with $\varphi = 2\pi C$ identified with $\varphi = 0$, and where C is related to the curvature. It is also sometimes convenient to put $\varphi = C\phi$, where $\phi \in [0, 2\pi)$, so that the metric (A.9) takes the form $ds^2 = d\psi^2 + C^2 e^{2\psi}\,d\phi^2$. For this case, an axially symmetric embedding in a three-dimensional Euclidean space is obtained by putting

$$x_1 = Ce^{\psi}\cos\phi,$$
$$x_2 = Ce^{\psi}\sin\phi,$$
$$x_3 = \int \sqrt{1 - C^2 e^{2\psi}}\,d\psi = \sqrt{1 - C^2 e^{2\psi}} - \operatorname{arctanh}\sqrt{1 - C^2 e^{2\psi}},$$

with ψ covering only the restricted range $-\infty < \psi < -\log C$. Such an embedding produces the pseudosphere which is illustrated as the middle surface in Figure A.3.

Alternatively, by putting

$$\zeta = \sqrt{2}\,\frac{\sinh\chi + i\cosh\chi\sinh\varphi}{1 + \cosh\chi\cosh\varphi},\tag{A.10}$$

the metric (A.2) takes the form

$$ds^2 = d\chi^2 + \cosh^2\chi\,d\varphi^2,\tag{A.11}$$

in which the natural range of the coordinates is $\chi, \varphi \in (-\infty, \infty)$. (These are sometimes referred to as *equidistant* or *exocyclic* coordinates.)

An axially symmetric-type surface with metric (A.11) can again be considered for which $\varphi \in (-\pi C, \pi C]$ with $\varphi = -\pi C$ identified with $\varphi = \pi C$. For this case, with $\varphi = C\phi$, the metric becomes $ds^2 = d\chi^2 + C^2\cosh^2\chi\,d\phi^2$.

The final axially symmetric embedding in a three-dimensional Euclidean space illustrated in Figure A.3 is then obtained by putting

$$x_1 = C \cosh \chi \cos \phi,$$
$$x_2 = C \cosh \chi \sin \phi,$$
$$x_3 = \pm \int \sqrt{1 - C^2 \sinh^2 \chi} \, \mathrm{d}\chi = -\mathrm{i} \, E(\mathrm{i}\chi, \mathrm{i}C),$$

with $\phi \in (-\pi, \pi]$ and $-\mathrm{arcsinh}(1/C) \le \chi \le \mathrm{arcsinh}(1/C)]$, which cover very restricted parts of the natural coordinate range.

For the 2-spaces of constant negative curvature with metrics (A.7), (A.9) and (A.11), the coordinates $\{R, \varphi\}$, $\{\psi, \varphi\}$ and $\{\chi, \varphi\}$ are defined by the expressions for ζ given by (A.6), (A.8) and (A.10). The regions of the ζ-plane that are covered by the full ranges of the coordinates R, ψ and χ, but with the different φ coordinates restricted to $\varphi \in [-C\pi, C\pi)$ with $C < 1$, are illustrated in Figure A.4. It may also be noted that the axially symmetric embeddings that are illustrated in Figure A.3 also have limited ranges of the coordinates R, ψ and χ, respectively, that are related to the embeddings.

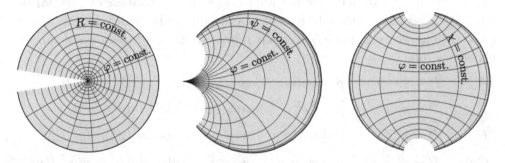

Fig. A.4 Regions of the ζ-plane that are covered by the coordinates $\{R, \varphi\}$, $\{\psi, \varphi\}$ and $\{\chi, \varphi\}$ of the metrics (A.7), (A.9) and (A.11) for 2-spaces of constant negative curvature, as given by (A.6), (A.8) and (A.10), respectively. The regions indicated have infinite ranges of the coordinates R, ψ and χ, but the different φ coordinates satisfy $\varphi \in [-\pi C, \pi C)$ with $C < 1$ here. The outer circle on which $|\zeta| = \sqrt{2}$ corresponds to infinity. The complete circular region, corresponding to the entire Lobatchevski plane, is spanned in the first case when $C = 1$ and in the other two cases when $C \to \infty$.

A.4 Other representations

It is important to note that the entire family of 2-spaces of constant curvature given by (A.1) or (A.2) can be parametrised in a number of alternative ways. In particular, they can be expressed in the form

$$ds^2 = \frac{1}{P}\,dp^2 + P\,d\sigma^2\,, \qquad \text{where} \qquad P(p) = \epsilon_0 + \epsilon_1 p - \epsilon\,p^2. \qquad (\text{A.12})$$

Here, the Gaussian curvature is again ϵ, *independent of the parameters* ϵ_0 *and* ϵ_1. These additional parameters may take any values subject to the condition that, with a possible restriction in the range of p, the signature of the metric must remain positive. Using a shift and rescaling of the coordinate p, it can be seen that there are *six distinct canonical forms*, namely:

$$
\begin{array}{llllll}
A: & \epsilon = +1 & \epsilon_1 = 0 & \epsilon_0 = +1 & P = 1 - p^2 & |p| < 1 \\[4pt]
E: & \epsilon = 0 & \epsilon_1 = +1 & \epsilon_0 = 0 & P = p & p > 0 \\[4pt]
F: & \epsilon = 0 & \epsilon_1 = 0 & \epsilon_0 = +1 & P = 1 & p \in (-\infty, \infty) \\[4pt]
C: & \epsilon = -1 & \epsilon_1 = 0 & \epsilon_0 = -1 & P = p^2 - 1 & |p| > 1 \\[4pt]
D: & \epsilon = -1 & \epsilon_1 = 0 & \epsilon_0 = 0 & P = p^2 & p > 0 \\[4pt]
B: & \epsilon = -1 & \epsilon_1 = 0 & \epsilon_0 = +1 & P = p^2 + 1 & p > 0.
\end{array}
$$

(The identifying letters on the left are those given by Kinnersley (1969a) to label his class II vacuum solutions of algebraic type D, which contain these constant-curvature 2-spaces.) Explicit representations of these cases can be chosen as follows:

$$
\begin{array}{lllll}
A: & P = 1 - p^2 & p = \cos\theta & \sigma = \varphi & ds^2 = d\theta^2 + \sin^2\theta\,d\varphi^2 \\[4pt]
E: & P = p & p = \tfrac{1}{4}\rho^2 & \sigma = 2\varphi & ds^2 = d\rho^2 + \rho^2 d\varphi^2 \\[4pt]
F: & P = 1 & p = x & \sigma = y & ds^2 = dx^2 + dy^2 \\[4pt]
C: & P = p^2 - 1 & p = \cosh R & \sigma = \varphi & ds^2 = dR^2 + \sinh^2 R\,d\varphi^2 \\[4pt]
D: & P = p^2 & p = e^{\psi} & \sigma = \varphi & ds^2 = d\psi^2 + e^{2\psi}\,d\varphi^2 \\[4pt]
B: & P = p^2 + 1 & p = \sinh\chi & \sigma = \varphi & ds^2 = d\chi^2 + \cosh^2\chi\,d\varphi^2.
\end{array}
$$

These clearly cover the six metric forms (A.3), (A.4), (A.5), (A.7), (A.9) and (A.11), respectively. However, it should be noted that the coordinates φ in these metrics are not the same, as is also evident for surfaces of negative curvature from Figure A.4.

For positive curvature $\epsilon = +1$, there is only one possibility, namely case A. Here, the 2-space is (part of) a sphere, and the stereographic coordinate ζ is given in terms of p and σ by $\zeta = \sqrt{2}\sqrt{\frac{1-p}{1+p}}\,e^{i\sigma}$, where $p \in [-1, 1]$ and $\sigma \in [0, 2\pi C)$.

When $\epsilon = 0$, there are two completely equivalent cases $E: P = p$ and

F: $P = 1$. These clearly correspond, respectively, to the familiar polar and Cartesian coordinates as used in (A.4) and (A.5). Their different expressions for ζ are given by

$$E: \qquad \zeta = \sqrt{2p}\,e^{i\sigma/2} \qquad\qquad p \in [0, \infty) \qquad\qquad \sigma \in [0, 4\pi C)$$

$$F: \qquad \zeta = \tfrac{1}{\sqrt{2}}(p + i\sigma) \qquad\quad p \in (-\infty, \infty) \quad \sigma \in (-\infty, \infty).$$

When $\epsilon = -1$, there are three equivalent cases C, D and B. These may be considered as different *local* coordinate representations of the same 2-space, with coordinates related through their different expressions for ζ, namely

$$C: \qquad \zeta = \sqrt{2}\,\sqrt{\frac{p-1}{p+1}}\,e^{i\sigma} \qquad\qquad p \in [1, \infty) \qquad\quad \sigma \in [0, 2\pi C)$$

$$D: \qquad \zeta = \sqrt{2}\,\frac{p^2(1+i\sigma)^2 - 1}{(p+1)^2 + p^2\sigma^2} \qquad\quad p \in [0, \infty) \qquad\quad \sigma \in (-\infty, \infty)$$

$$B: \qquad \zeta = \sqrt{2}\,\frac{p + i\sqrt{p^2+1}\,\sinh\sigma}{1 + \sqrt{p^2+1}\,\cosh\sigma} \qquad p \in (-\infty, \infty) \quad \sigma \in (-\infty, \infty),$$

which correspond to the parametrisations (A.6), (A.8), (A.10), respectively. These span the required circular region of the ζ-plane in different ways, as illustrated in Figure A.4. Moreover, these surfaces have different topological properties that are partly illustrated in their different embeddings as given in Figure A.3.

It may finally be concluded that, although the metric (A.12) appears to have six distinct canonical forms A–F, there are in fact *locally* only three distinct types of 2-spaces of constant curvature that may be classified according to the sign of the Gaussian curvature (i.e. $\epsilon = +1, 0, -1$). However, when a constant-curvature 2-space appears as one *component* in the metric of a four-dimensional space-time, all six distinct canonical forms may be required in some cases. In particular, if the remaining metric components depend on p in different ways, it will not normally be possible as above to identify the cases with the same values of ϵ. For example, the algebraic type D space-times of class II, as classified by Kinnersley (1969a), have *all six* distinct forms that he has identified. On the other hand, in the particular case of the NUT metric, the forms of the additional metric components are such that a further coordinate transformation can be made to identify the apparently distinct cases with the same values of ϵ. Thus the NUT metric includes just the *three* cases that were originally identified by Newman, Tamburino and Unti (1963) (see Section 12.3).

Appendix B
3-spaces of constant curvature

A Riemannian 3-space $ds^2 = h_{ij}\, dx^i dx^j$, where $i, j = 1, 2, 3$, is said to be of constant curvature if all components of its curvature tensor at any point are determined by the same single quantity K, namely

$$\mathcal{R}_{ijkl} = K\,(\,h_{ik}\,h_{jl} - h_{il}\,h_{jk}), \qquad K = \text{const.} \tag{B.1}$$

Since all points and all directions are, in the above sense, equivalent, such a space is homogeneous and isotropic – it admits 6 isometries and is therefore maximally symmetric. From (B.1) it also follows that the Ricci tensor is $\mathcal{R}_{ij} = 2K\,h_{ij}$ and the Ricci scalar is $\mathcal{R} = 6K$, so that the space of constant curvature is an Einstein space.

There exists a suitable coordinate system in which the metric of such a space takes the form

$$ds^2 = \frac{dx^2 + dy^2 + dz^2}{\left(1 + \frac{1}{4}K(x^2 + y^2 + z^2)\right)^2}. \tag{B.2}$$

This shows explicitly that the constant-curvature 3-space is conformally flat. It is also convenient to express the curvature K as

$$K = \frac{k}{R^2}, \qquad \text{where} \quad k = +1, 0, -1.$$

The dimensionless parameter $k = \text{sign}\,K$ thus distinguishes three possible cases of positive, zero or negative curvature. By rescaling the coordinates, the metric (B.2) can easily be rewritten as

$$ds^2 = R^2 \frac{dx^2 + dy^2 + dz^2}{\left(1 + \frac{1}{4}k(x^2 + y^2 + z^2)\right)^2}. \tag{B.3}$$

In this form, the coordinates x, y, z are dimensionless, and the magnitude of the curvature is determined by R, which has the proper dimension of length.

Introducing spherical polar coordinates $x = \tilde{r} \sin \theta \cos \phi$, $y = \tilde{r} \sin \theta \sin \phi$, $z = \tilde{r} \cos \theta$, the metric (B.3) immediately becomes

$$ds^2 = R^2 \left(1 + \tfrac{1}{4} k \, \tilde{r}^2\right)^{-2} \left(d\tilde{r}^2 + \tilde{r}^2 (d\theta^2 + \sin^2 \theta \, d\phi^2)\right). \qquad \text{(B.4)}$$

The further transformation $r = \tilde{r}/(1 + \tfrac{1}{4} k \, \tilde{r}^2)$ gives another standard form of the metric of the 3-spaces of constant curvature:

$$ds^2 = R^2 \left(\frac{dr^2}{1 - k \, r^2} + r^2 (d\theta^2 + \sin^2 \theta \, d\phi^2)\right). \qquad \text{(B.5)}$$

According to k, it is now possible to distinguish three possibilities:

$k = 0$: This case corresponds to a *flat space* which is naturally covered by the Cartesian coordinates (B.3) or by spherical coordinates (B.5). Renaming the dimensionless parameter r by χ, the latter becomes

$$ds^2 = R^2 \left(d\chi^2 + \chi^2 (d\theta^2 + \sin^2 \theta \, d\phi^2)\right). \qquad \text{(B.6)}$$

$k = +1$: This represents a 3-space of constant positive curvature, which is a *3-sphere S^3*. By introducing $r = \sin \chi$, the metric (B.5) takes the form

$$ds^2 = R^2 \left(d\chi^2 + \sin^2 \chi \, (d\theta^2 + \sin^2 \theta \, d\phi^2)\right). \qquad \text{(B.7)}$$

$k = -1$: This is a Lobatchevski space of constant negative curvature, or a *hyperbolic 3-space H^3*. With $r = \sinh \chi$, the metric (B.5) reads

$$ds^2 = R^2 \left(d\chi^2 + \sinh^2 \chi \, (d\theta^2 + \sin^2 \theta \, d\phi^2)\right). \qquad \text{(B.8)}$$

The spherically symmetric forms (B.6), (B.7), (B.8) of the above constant-curvature 3-spaces look remarkably similar and simple. The metrics differ only in the factors χ, $\sin \chi$ and $\sinh \chi$, respectively. However, it should be emphasised that these are different metrics of three different spaces. Moreover, while $\theta \in [0, \pi]$, $\phi \in [0, 2\pi)$ in all three cases,[1] the ranges of the coordinates χ are *not* the same: $\chi \in [0, \infty)$ for $k = 0$ and $k = -1$, whereas for $k = +1$ it is necessary to consider only a restricted range $\chi \in [0, \pi]$ since the 3-sphere is compact.

More details about the cases $k = +1$ and $k = -1$ will be given in the following two sections.

---

[1] Unlike Appendix A, conical-type singularities are not considered here.

B.1 The 3-sphere

As described above, the 3-sphere S^3 is a three-dimensional closed space of constant positive curvature $k = +1$ which has no boundary. It has no privileged points or axes, and is invariant with respect to arbitrary rotations.

A 3-sphere of radius R can most easily be understood as the three-dimensional hypersurface

$$x_1{}^2 + x_2{}^2 + x_3{}^2 + x_4{}^2 = R^2, \tag{B.9}$$

in the four-dimensional Euclidean space

$$ds^2 = dx_1{}^2 + dx_2{}^2 + dx_3{}^2 + dx_4{}^2.$$

This can be naturally represented by the parameters χ, θ, ϕ defined as

$$x_1 = R \cos \chi,$$
$$x_2 = R \sin \chi \cos \theta,$$
$$x_3 = R \sin \chi \sin \theta \cos \phi,$$
$$x_4 = R \sin \chi \sin \theta \sin \phi.$$

In order to cover all points of the 3-sphere with both positive and negative values of the coordinates x_i, it is necessary that $\chi \in [0, \pi]$, $\theta \in [0, \pi]$, $\phi \in [0, 2\pi)$. In these coordinates the metric of S^3 takes the form

$$ds^2 = R^2 \left(d\chi^2 + \sin^2 \chi \, (d\theta^2 + \sin^2 \theta \, d\phi^2) \right), \tag{B.10}$$

which is exactly the canonical form (B.7).

Sections of the 3-sphere on which χ is constant are clearly 2-spheres of radius $R \sin \chi$ covered by the standard spherical polar coordinates θ and ϕ. As χ increases from 0 to π, the radius of these 2-spheres increases from zero to R and then decreases back to zero. The 3-sphere can thus be visualised as the combination of the sequence of these 2-spheres for the complete range of $\chi \in [0, \pi]$.

It is helpful at this point to consider the analogous simpler representation of a 2-sphere. This can be visualised as a combination of circles (lines of latitude) whose radius increases from zero at the North pole to a maximum R at the equator, and then decreases back to zero at the South pole. Each hemisphere can be represented by a plane map that is created by projecting it perpendicularly onto the equatorial plane. Such projections combine the family of circles (lines of latitude) on each hemisphere to form a disc. In this representation, the complete 2-sphere is thus pictured as *two* separate discs that are each bounded by the equator $\chi = \pi/2$. Corresponding points on each of these boundaries have to be identified.

In a similar way, a 3-sphere can be represented as two separate families of 2-spheres for $\chi \in [0, \pi/2]$ and $\chi \in [\pi/2, \pi]$. These can be combined to form *two solid spheres* for which points on the outer boundaries $\chi = \pi/2$ are identified.

Taking now $r = \sin \chi$ as a new coordinate, the metric of the 3-sphere may be rewritten as

$$ds^2 = R^2 \left(\frac{dr^2}{1 - r^2} + r^2(d\theta^2 + \sin^2 \theta \, d\phi^2) \right), \tag{B.11}$$

where $r \in [0, 1]$, which is the $k = +1$ subcase of (B.5). It is important to emphasise in this case that the complete 3-sphere requires *two coverings* of the metric in this form, corresponding to the two ranges $\chi \in [0, \pi/2]$ and $\chi \in [\pi/2, \pi]$. These correspond to each of the two solid spheres in the representation described above.

There is, however, an alternative representation of a 3-sphere which pictures it as the *interior* of a *single* sphere. In this case, the sections on which χ is a constant are represented as 2-spheres of radius χ (rather than $R \sin \chi$). This family of 2-spheres together make up the interior of a 2-sphere of radius π. According to this mapping, the section of S^3 on which $\theta = \pi/2$ is a spherical surface S^2 which maps to a disc, as shown in Figure B.1.

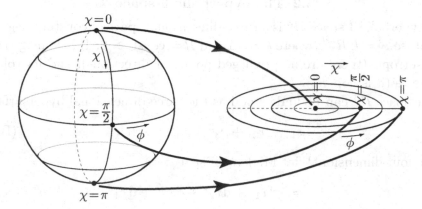

Fig. B.1 The schematic representation of sections of a 3-sphere on which $\theta = \pi/2$. The circles of constant latitude (χ is a constant) are transformed to circles on a disc, but their radii are not conserved.

The figure that is constructed in this way is a schematic representation of the 3-sphere. The sections on which ϕ is constant are hemispheres of radius R spanned by $\chi \in [0, \pi]$ and $\theta \in [0, \pi]$. More significantly, sections on which χ is constant are 2-spheres with physical radius $R \sin \chi$. However, these are represented as 2-spheres of radius χ. As the representational radius

χ increases from 0 to π, the physical radius increases to a maximum (at $\chi = \pi/2$) and then decreases back to zero. The mathematical representation of the 3-sphere that is constructed in this way is often more useful in practice as it avoids the double covering of the space (B.11), distinguishing χ from $\pi - \chi$. However, it is less physically intuitive as the outer surface $\chi = \pi$ in this case corresponds to a single point, namely the South pole.

Another alternative parametrisation of the 3-sphere (B.9) is given by

$$
\begin{aligned}
x_1 &= R \cos \tfrac{1}{2}\theta \, \cos \tfrac{1}{2}(\phi + \psi), \\
x_2 &= R \cos \tfrac{1}{2}\theta \, \sin \tfrac{1}{2}(\phi + \psi), \\
x_3 &= R \sin \tfrac{1}{2}\theta \, \cos \tfrac{1}{2}(\phi - \psi), \\
x_4 &= R \sin \tfrac{1}{2}\theta \, \sin \tfrac{1}{2}(\phi - \psi),
\end{aligned}
\tag{B.12}
$$

in which the parameters have the ranges $\theta \in [0, \pi]$, $\phi \in [0, 2\pi]$, $\psi \in [0, 2\pi)$. In this case, the metric of S^3 is expressed in the form

$$
ds^2 = \tfrac{1}{4}R^2 \left((d\psi + \cos\theta \, d\phi)^2 + d\theta^2 + \sin^2\theta \, d\phi^2 \right).
\tag{B.13}
$$

B.2 The hyperbolic 3-space

The hyperbolic 3-space H^3 is a three-dimensional space of constant negative curvature $K = k\,R^{-2}$, where $k = -1$ and $R = \text{const.} > 0$. It is homogeneous and isotropic (there are no privileged points or directions), and its volume is infinite (it is "open").

The space H^3 can be understood as the three-dimensional hypersurface

$$
-x_1{}^2 + x_2{}^2 + x_3{}^2 + x_4{}^2 = -R^2,
\tag{B.14}
$$

in the four-dimensional flat Minkowski space[2]

$$
ds^2 = -dx_1{}^2 + dx_2{}^2 + dx_3{}^2 + dx_4{}^2.
$$

The most natural parameters χ, θ, ϕ on this hypersurface are

$$
\begin{aligned}
x_1 &= R \cosh\chi, \\
x_2 &= R \sinh\chi \cos\theta, \\
x_3 &= R \sinh\chi \sin\theta \cos\phi, \\
x_4 &= R \sinh\chi \sin\theta \sin\phi,
\end{aligned}
\tag{B.15}
$$

---

[2] It is not possible to embed H^3 into a four-dimensional Euclidean space.

with the ranges $\chi \in [0, \infty)$, $\theta \in [0, \pi]$, $\phi \in [0, 2\pi)$. In these coordinates the metric of H^3 takes the form

$$ds^2 = R^2 \left(d\chi^2 + \sinh^2 \chi \, (d\theta^2 + \sin^2 \theta \, d\phi^2) \right), \qquad (B.16)$$

which is exactly the canonical form (B.8). Introducing then the coordinate $r = \sinh \chi$, where $r \in [0, \infty)$, this metric becomes

$$ds^2 = R^2 \left(\frac{dr^2}{1 + r^2} + r^2 (d\theta^2 + \sin^2 \theta \, d\phi^2) \right), \qquad (B.17)$$

which is the $k = -1$ subcase of (B.5).

Obviously, 2-spaces of constant χ (or, equivalently, r) on H^3 are 2-spheres of radius $R \sinh \chi = R r$ and surface area $4\pi R^2 \sinh^2 \chi = 4\pi R^2 r^2$, and θ and ϕ are the standard spherical polar coordinates on these spheres. As the parameter χ (or r) increases from 0 to ∞, both the radius and the area of these successive 2-spheres monotonically increase from zero to infinity. Of course, due to the homogeneity of H^3, the "origin" where $\chi = 0$ can be chosen arbitrarily.

The space H^3 can be visualised by plotting the surface given by expression (B.14) with $x_1 > 0$, as in Figure B.2. (An alternative sheet of (B.14) with $x_1 < 0$ would be parametrised by (B.15) with $x_1 = -R \cosh \chi$.)

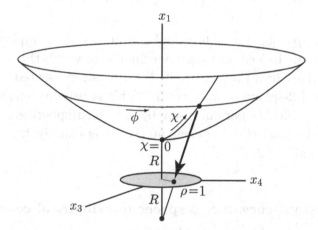

Fig. B.2 Visualisation of the equatorial section $\theta = \pi/2$ of a hyperbolic 3-space H^3. All points on such a hyperboloid can also be represented by a stereographic-like projection onto a horizontal disk of radius R in the plane $x_1 = 0$. The polar coordinates ρ, ϕ, where $R\rho$ is the radial distance of the projected point from the origin of the shaded disk, are coordinates in which the hyperbolic space takes the conformally flat Poincaré form.

As in the previously discussed case of a 3-sphere S^3, it is possible to obtain a useful representation of the hyperbolic 3-space H^3 by considering its one-to-one projection onto the *interior* of a *single solid sphere* in flat space. In fact, due to the non-compact (open) character of this space such a construction is simpler than in the case of S^3.

Restricting to the equatorial section $\theta = \pi/2$ through H^3, as in Figure B.2, each point on the hyperboloid can be uniquely mapped by a stereographic-like projection onto the horizontal plane $x_1 = 0$. Such a mapping is obtained by taking the intersection of the plane with the straight line connecting a given point on H^3 with the reference point at $x_1 = -R$, $x_3 = 0 = x_4$. All points on the hyperboloid are thus mapped onto an interior of a circle of radius R, called the Poincaré disk. Introducing usual polar coordinates, where $R\rho$ is the radial distance from the centre of the disk, and ϕ is the angle, the mapping identifies $\chi = 0$ with $\rho = 0$ and $\chi = \infty$ with $\rho = 1$. Straightforward trigonometry leads to the explicit relations

$$\rho = \tanh \frac{\chi}{2} = \frac{r}{1 + \sqrt{1 + r^2}}.$$

Taking ρ as a new dimensionless coordinate, the metric of the hyperbolic 3-space H^3 can then be written as

$$ds^2 = \frac{4R^2}{(1-\rho^2)^2}\Big(d\rho^2 + \rho^2(d\theta^2 + \sin^2\theta\, d\phi^2)\Big). \qquad (B.18)$$

With the range $\rho \in [0, 1)$, $\theta \in [0, \pi]$ and $\phi \in [0, 2\pi)$, the complete H^3 is thus naturally covered. Indeed, in these coordinates, in which the metric is manifestly conformally flat, the hyperbolic 3-space is represented as an interior of a single solid 2-sphere of unit radius. This is usually referred to as the Poincaré ball, which is parametrised by standard spherical polar coordinates ρ, θ, ϕ. In fact, the line element (B.18) is exactly the metric (B.4) with $k = -1$ and $\tilde{r} = 2\rho$.

B.3 Constant-curvature 3-spaces in cylindrical coordinates

So far in this appendix spherical-like coordinates have been used. However, it is sometimes convenient to express these spaces in cylindrical-like coordinates. By putting

$$r\sin\theta = \hat{\rho}, \qquad \frac{r\cos\theta}{\sqrt{1 - kr^2}} = \begin{cases} \tan\hat{z} & \text{for} \quad k = +1, \\ \hat{z} & \text{for} \quad k = 0, \\ \tanh\hat{z} & \text{for} \quad k = -1, \end{cases}$$

the above metric (B.5) can be expressed in the manifestly cylindrical form

$$ds^2 = R^2 \left(\frac{d\hat{\rho}^2}{1 - k\hat{\rho}^2} + (1 - k\hat{\rho}^2)d\hat{z}^2 + \hat{\rho}^2 d\phi^2 \right). \qquad \text{(B.19)}$$

The three distinct cases must now be considered separately according to corresponding values of k:

$k = 0$: This is simply *flat space* in cylindrical coordinates, with $\hat{\rho} \in [0, \infty)$, $\hat{z} \in (-\infty, \infty)$ and $\phi \in [0, 2\pi)$. It does not need to be elaborated here.

$k = +1$: This case is a *3-sphere* S^3 of constant positive curvature. As described in Section B.1, the complete 3-sphere is naturally represented in terms of the parameter $\chi \in [0, \pi]$, and this requires two copies of the metric (B.11) with $r \in [0, 1]$. In fact, the complete space S^3 is given by (B.19) with $k = +1$ for $\hat{\rho} \in [0, 1]$ and $\phi \in [0, 2\pi)$, and the two ranges $\hat{z} \in [-\pi/2, \pi/2]$ and $\hat{z} \in [\pi/2, 3\pi/2]$ give the required two copies of (B.11).

Clearly, the 3-sphere is a closed space, and this is obtained by identifying $\hat{z} = 3\pi/2$ with $\hat{z} = -\pi/2$, thus producing a periodic coordinate. Putting

$$\hat{z} = \psi - \frac{\pi}{2},$$

the metric becomes

$$ds^2 = R^2 \left(\frac{d\hat{\rho}^2}{1 - \hat{\rho}^2} + (1 - \hat{\rho}^2)d\psi^2 + \hat{\rho}^2 d\phi^2 \right), \qquad \text{(B.20)}$$

in which $\hat{\rho} \in [0, 1]$ and $\psi, \phi \in [0, 2\pi)$.

In fact, the surfaces on which $\hat{\rho}$ is a constant are a family of tori. This can most easily be seen by representing the space as a three-dimensional surface (B.9) in a four-dimensional Euclidean space. A natural parametrization of the metric (B.20) is given by

$$
\begin{aligned}
x_1 &= R\sqrt{1 - \hat{\rho}^2}\, \cos\psi, \\
x_2 &= R\sqrt{1 - \hat{\rho}^2}\, \sin\psi, \\
x_3 &= R\hat{\rho}\, \cos\phi, \\
x_4 &= R\hat{\rho}\, \sin\phi.
\end{aligned}
$$

Consequently

$$x_1{}^2 + x_2{}^2 = R^2(1 - \hat{\rho}^2), \qquad x_3{}^2 + x_4{}^2 = R^2\hat{\rho}^2,$$

$$\frac{x_2}{x_1} = \tan\psi, \qquad \frac{x_4}{x_3} = \tan\phi.$$

Thus any surface on which $\hat{\rho} = \text{const.}$ is clearly a torus $S^1 \times S^1$, spanned by ψ and ϕ, which are both periodic.

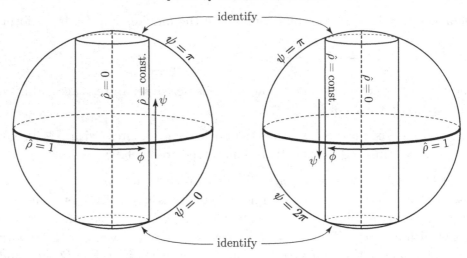

Fig. B.3 A complete 3-sphere can be represented as two solid spheres. These may be regarded as two copies of the unit sphere in which the dimensionless coordinates r, θ, ϕ of the metric (B.11) are treated as spherical polar coordinates in a flat space. Points on the two upper and two lower hemispherical boundaries are identified. Surfaces on which $\hat{\rho}$ is a constant are thus seen to be explicitly toroidal.

In the cylindrical coordinates of (B.20), the complete 3-sphere can be represented as two solid spheres with $\psi \in [0, \pi]$ and $\psi \in [\pi, 2\pi]$, as illustrated in Figure B.3. This construction has been described by Misner (1963). (For further details, see Appendix B of Gowdy (1974) and the appendix of Bičák and Griffiths (1996).) The surfaces $\hat{\rho} = $ const. are cylinders around the axis at $\hat{\rho} = 0$, while the surfaces $\psi = $ const. are rotational ellipsoids with semi-major axes $(1, 1, |\cos\psi|)$. By identifying corresponding points on the two hemispherical surfaces on which $\psi = \pi$ and points on the two hemispherical surfaces on which $\psi = 0$ and $\psi = 2\pi$, the surfaces on which $\hat{\rho}$ is a constant are seen to be tori. In fact, each of these surfaces divides the complete 3-sphere into two solid tori. One is centred on the axis $\hat{\rho} = 0$, which is aligned with the ψ-direction and with rotational coordinate ϕ. The other is complementarily centred on an axis at $\hat{\rho} = 1$, which is aligned with the ϕ-direction and has rotational coordinate ψ.

Also, putting $\hat{\rho} = \sin(\rho/R)$, the metric (B.20) becomes

$$ds^2 = d\rho^2 + R^2 \cos^2\left(\frac{\rho}{R}\right) d\psi^2 + R^2 \sin^2\left(\frac{\rho}{R}\right) d\phi^2, \qquad (B.21)$$

where $\rho/R \in [0, \pi/2]$ and $\psi, \phi \in [0, 2\pi)$. In this form, the coordinate ρ is the proper radial distance from the axis at $\rho = 0$. This metric clearly possesses toroidal symmetry, which is the natural equivalent of cylindrical symmetry in a closed space.

$k = -1$: This *hyperbolic 3-space* H^3 is a Lobatchevski space of constant negative curvature. Putting

$$\hat{\rho} = \sinh(\rho/R), \qquad \hat{z} = z/R,$$

the metric (B.19) takes the form

$$ds^2 = d\rho^2 + \cosh^2(\rho/R)\,dz^2 + R^2 \sinh^2(\rho/R)\,d\phi^2, \qquad \text{(B.22)}$$

where $\rho \in [0,\infty)$, $z \in (-\infty,\infty)$ and $\phi \in [0,2\pi)$. In this case, ρ is the proper distance from a single infinite axis at $\rho = 0$, and the surfaces on which $\rho = $ const. are cylindrical.

The space H^3, with metrics (B.19) with $k = -1$ or (B.22), can be represented as a three-dimensional hypersurface (B.14) in a flat Minkowski space using the parametrizations

$$
\begin{aligned}
x_1 &= R\sqrt{1+\hat{\rho}^2}\,\cosh\hat{z} &&= R\cosh(\rho/R)\cosh(z/R),\\
x_2 &= R\sqrt{1+\hat{\rho}^2}\,\sinh\hat{z} &&= R\cosh(\rho/R)\sinh(z/R),\\
x_3 &= R\hat{\rho}\cos\phi &&= R\sinh(\rho/R)\cos\phi,\\
x_4 &= R\hat{\rho}\sin\phi &&= R\sinh(\rho/R)\sin\phi.
\end{aligned}
$$

Consequently

$$-x_1{}^2 + x_2{}^2 = -R^2(1+\hat{\rho}^2), \qquad x_3{}^2 + x_4{}^2 = R^2\hat{\rho}^2,$$

$$\frac{x_2}{x_1} = \tanh\hat{z}, \qquad \frac{x_4}{x_3} = \tan\phi.$$

The surfaces $\hat{\rho} = $ const. have the geometry $H^1 \times S^1$, and the axis $\hat{\rho} = 0$ is the hyperbola $x_1^2 - x_2^2 = R^2$, $x_3 = 0 = x_4$, parametrized by $x_1 = R\cosh\hat{z}$, $x_2 = R\sinh\hat{z}$.

References

Adams, P. J., Hellings, R. W., Zimmerman, R. L., Farhoosh, H., Levine, D. I. and Zeldich, S. (1982). Inhomogeneous cosmology: gravitational radiation in Bianchi backgrounds, *Astrophys. J.*, **253**, 1–18. §22.4

Aichelburg, P. C. (1971). Remark on the superposition principle for gravitational waves, *Acta Phys. Austriaca*, **34**, 279–284. §17.3

Aichelburg, P. C. (1991). Lightlike contractions – singular limits of spacetimes, in *Relativity and gravitation: classical and quantum*, eds. D'Olivo *et al.*, (World Scientific, Singapore), 198–203. §20.2

Aichelburg, P. C. and Balasin, H. (1998). ADM and Bondi four-momenta for the ultrarelativistic Schwarzschild black hole, *Class. Quantum Grav.*, **15**, 3841–3844. §20.2

Aichelburg, P. C. and Balasin, H. (2000). Generalized asymptotic structure of the ultrarelativistic Schwarzschild black hole, *Class. Quantum Grav.*, **17**, 3645–3662. §20.2

Aichelburg, P. C. and Sexl, R. U. (1971). On the gravitational field of a massless particle, *Gen. Rel. Grav.*, **2**, 303–312. §20

Alekseev, G. A. (1985). The method of the inverse scattering problem and the singular integral equations for interacting massless fields, *Sov. Phys. Dokl.*, **30**, 565–568. §21.2

Alekseev, G. A. (1988). Exact solutions in the general theory of relativity, *Proc. Steklov Inst. Maths.*, **3**, 215–262. §21.2

Alekseev, G. A. (2001). New integral equation form of integrable reductions of Einstein equations, *Teor. Mat. Fiz.*, **129**, 184–206; *Theor. Math. Phys.*, **129**, 1466–1483, (2001). §21.2

Alekseev, G. A. and Belinski, V. A. (2007). Equilibrium configurations of two charged masses in general relativity, *Phys. Rev. D*, **76**, 021501. §10.8

Alekseev, G. A. and Griffiths, J. B. (1995). Propagation and interaction of gravitational waves in some expanding backgrounds, *Phys. Rev. D*, **52**, 4497–4502. §21.7

Alekseev, G. A. and Griffiths, J. B. (1996). Exact solutions for gravitational waves with cylindrical, spherical and toroidal wavefronts, *Class. Quantum Grav.*, **13**, 2191–2209. §21.7

Alekseev, G. A. and Griffiths, J. B. (1997), Gravitational waves with distinct wavefronts, *Class. Quantum Grav.*, **14**, 2869–2880. §22.3

Alekseev, G. A. and Griffiths, J. B. (2001). Solving the characteristic initial value problem for colliding plane gravitational and electromagnetic waves, *Phys. Rev. Lett.*, **87**, 221101. §21.2

Alekseev, G. A. and Griffiths, J. B. (2004), Collision of plane gravitational and electromagnetic waves in a Minkowski background: solution of the characteristic initial value problem, *Class. Quantum Grav.*, **21**, 5623–5654. §21.2

Alekseev, G. A. and Khlebnikov, V. I. (1978). Newman–Penrose formalism and its applications in general relativity, *Sov. J. Part. & Nucl.*, **9**, 421–451. §2.1.3

Aliev, A. N. and Nutku, Y. (2001). Impulsive spherical gravitational waves, *Class. Quantum Grav.*, **18**, 891–906. §20.4

Alnes, H., Amarzguioui, M. and Grøn, Ø. (2006). Inhomogeneous alternative to dark energy? *Phys. Rev. D*, **73**, 083519. §22.7

Alonso, D., Ruiz, A. and Sánchez-Hernández, M. (2008). Escape of photons from two fixed extreme Reissner–Nordström black holes, *Phys. Rev. D*, **78**, 104024. §22.5

Åminneborg, S., Bengtsson, I., Holst, S. and Peldán, P. (1996). Making anti-de Sitter black holes, *Class. Quantum Grav.*, **13**, 2707–2714. §5.1

Anderson, M. R. (2003). *The mathematical theory of cosmic strings*, (Institute of Physics Publishing, Bristol). §3.4

Arianrhod, R., Fletcher, S. and McIntosh, C. B. G. (1991). Principal null directions of the Curzon metric, *Class. Quantum Grav.*, **8**, 1519–1528. §10.5

Arnowitt, R., Deser, S. and Misner, C. W. (1962). The dynamics of general relativity, in *Gravitation: an introduction to current research*, ed. L. Witten, (Wiley, New York), 227–265. §8.1

Ashtekar, A. (1980). Asymptotic structure of the gravitational field at spatial infinity, in *General relativity and gravitation, one hundred years after the birth of Albert Einstein*, vol. 2, ed. A. Held, (Plenum, New York), 37–69. §8.4

Ashtekar, A., Bičák, J. and Schmidt, B. G. (1997a). Asymptotic structure of symmetry-reduced general relativity, *Phys. Rev. D*, **55**, 669–686. §22.3

Ashtekar, A., Bičák, J. and Schmidt, B. G. (1997b). Behaviour of Einstein–Rosen waves at null infinity, *Phys. Rev. D*, **55**, 687–694. §22.3

Ashtekar, A. and Dray, T. (1981). On the existence of solutions to Einstein's equation with non-zero Bondi news, *Commun. Math. Phys.*, **79**, 581–589. §14

Ashtekar, A. and Hansen, R. O. (1978). A unified treatment of null and spatial infinity in general relativity. I. Universal structure, asymptotic symmetries, and conserved quantities at spatial infinity, *J. Math. Phys.*, **19**, 1542–1566. §8.4

Ashtekar, A. and Schmidt, B. G. (1980). Null infinities and Killing fields, *J. Math. Phys.*, **21**, 862–867. §15.1

Ashtekar, A. and Xanthopoulos, B. C. (1978). Isometries compatible with asymptotic flatness at null infinity: A complete description, *J. Math. Phys.*, **19**, 2216–2222. §15.1

Avis, S. J., Isham, C. J. and Storey, D. (1978). Quantum field theory in anti-de Sitter space-time, *Phys. Rev. D*, **18**, 3565–3576. §§5.2, 5.4

Azuma, T. and Koikawa, T. (1994). Equilibrium condition in the axisymmetric *N*-Reissner–Nordström solution, *Prog. Theor. Phys.*, **92**, 1095–1104. §§10.8.1, 22.5

Bach, R. and Weyl, H. (1922). Neue Lösungen der Einsteinschen Gravitationsgleichungen. B. Explizite Aufstellung statischer axialsymmetrischer Felder. Miteinem Zusattz über das statische Zweikörperproblem von H. Weyl, *Math. Zeits*, **13**, 134–145. §§10.4, 10.7

Balasin, H. and Nachbagauer, H. (1995). The ultrarelativistic Kerr geometry and its energy-momentum tensor, *Class. Quantum Grav.*, **12**, 707–713. §20.2

Balasin, H. and Nachbagauer, H. (1996). Boosting the Kerr geometry in an arbitrary direction, *Class. Quantum Grav.*, **13**, 731–737. §20.2

Balbinot, R., Bergamini, R. and Comastri, A. (1988). Solution of the Einstein–Straus problem with a Λ term, *Phys. Rev. D*, **38**, 2415–2418. §8.6

Baldwin, O. R. and Jeffery, G. B. (1926). The relativity theory of plane waves, *Proc. Roy. Soc. A*, **111**, 95–104. §§7.1, 17.5

Bañados, M. (1998). Constant curvature black holes, *Phys. Rev. D*, **57**, 1068–1072. §9.6

Bardeen, J. M. (1968). General relativistic collapse of charged dust, *Bull. Am. Phys. Soc.*, **13**, 41. §9.2

Bardeen, J. M. (1973). Timelike and null geodesics in the Kerr metric, in *Black holes – les astres occlus*, eds. C. DeWitt and B. S. DeWitt, (Gordon and Breach, New York), 215–239. §11.1.5

Barnes, A. (2001). On the symmetries of the Edgar–Ludwig metric, *Class. Quantum Grav.*, **18**, 5287–5291. §§18.2, 18.3.3

Barrabès, C. (1989). Singular hypersurfaces in general relativity: a unified approach, *Class. Quantum Grav.*, **6**, 581–588. §§2.1.5, 20.6

Barrabès, C. and Hogan, P. A. (2003a). *Singular null hypersurfaces in general relativity*, (World Scientific). §20

Barrabès, C. and Hogan, P. A. (2003b). Light-like boost of the Kerr gravitational field, *Phys. Rev. D*, **67**, 084028. §20.2

Barrabès, C. and Hogan, P. A. (2007). Inhomogeneous high frequency expansion-free gravitational waves, *Phys. Rev. D*, **75**, 124012. §18.4.4

Barrabès, C. and Israel, W. (1991). Thin shells in general relativity and cosmology: the lightlike limit, *Phys. Rev. D*, **43**, 1129–1142. §§2.1.5, 13.2.2, 20.6

Barrow, J. D. and Götz, G. (1989). Newtonian no-hair theorems, *Class. Quantum Grav.*, **6**, 1253–1265. §19.4

Barrow, J. D. and Silk, J. (1981). The growth of anisotropic structures in a Friedmann universe, *Astrophys. J.*, **250**, 432–449. §22.8

Barrow, J. D. and Tsagas, C. G. (2004). Dynamics and stability of the Gödel universe, *Class. Quantum Grav.*, **21**, 1773–1790. §22.6

Bażański, S. L. and Ferrari, V. (1986). Analytic extension of the Schwarzschild–de Sitter metric, *Nuovo Cimento B*, **91**, 126–142. §9.4.1

Beciu, M. I. (1984). Evaporating black hole in Vaidya metric, *Phys. Lett. A*, **100**, 77–79. §9.5.1

Beck, G. (1925). Zur Theorie binärer Gravitationsfelder, *Z. Phys.*, **33**, 713. §22.3

Bedran, M. L., Calvão, M. O., Paiva, F. M. and Damião Soares, I. (1997). Taub's plane-symmetric vacuum spacetime reexamined, *Phys. Rev. D*, **55**, 3431–3439. §9.1.2

Bekenstein, J. D. (1971). Hydrostatic equilibrium and gravitational collapse of relativistic charged fluid balls, *Phys. Rev. D*, **4**, 2185–2190. §9.2

Belinski, V. and Verdaguer, E. (2001). *Gravitational solitons*, (Cambridge University Press). §§13.2.2, 22

Bell, P. and Szekeres, P. (1974). Interacting electromagnetic shock waves in general relativity, *Gen. Rel. Grav.*, **5**, 275–286. §21.6

Berger, B. K., Eardley, D. M. and Olsen, D. W. (1977). Note on the spacetimes of Szekeres, *Phys. Rev. D*, **16**, 3086–3089. §22.8

Berger, B. K. and Moncrief, V. (1993). Numerical investigation of cosmological singularities, *Phys. Rev. D*, **48**, 4676–4687. §22.4

Bertotti, B. (1959). Uniform electromagnetic field in general relativity, *Phys. Rev.*, **116**, 1331–1333. §7.1

Bičák, J. (1968). Gravitational radiation from uniformly accelerated particles in general relativity, *Proc. Roy. Soc. A*, **302**, 201–224. §§14, 15.2, 16.6

Bičák, J. (1971). Selected topics in the problem of energy and radiation, in *Relativity and gravitation*, eds. C. G. Kuper and A. Peres, (Gordon and Breach), 47–67. §15.2

Bičák, J. (1985). On exact radiative solutions representing finite sources, in *Galaxies, axisymmetric systems and relativity*, ed. M. A. H. MacCallum, (Cambridge University Press), 99–114. §§14, 15.2

Bičák, J. (1987). Radiative properties of space-times with the axial and boost symmetries, in *Gravitation and geometry*, eds. W. Rindler and A. Trautman, (Bibliopolis, Naples), 55–69. §15.2

Bičák, J. (1989). Exact radiative space-times, in *The fifth Marcel Grossmann meeting*, eds. D. G. Blair and M. J. Buckingham, (World Scientific), 309–341. §§1, 19.4

Bičák, J. (1990). Is there a news function for an infinite cosmic string? *Astron. Nachr.*, **311**, 189–192. §§15.2.5, 20.4

Bičák, J. (1997). Radiative spacetimes: exact approaches, in *Relativistic gravitation and gravitational radiation*, eds. J.-A. Marck and J.-P. Lasota, (Cambridge University Press), 67–87. §1

Bičák, J. (2000a). Selected solutions of Einstein's field equations: their role in general relativity and astrophysics, in *Einstein's field equations and their physical implications*, Lecture notes in physics, **540**, (Springer), 1–126. §§1, 4, 9.2.2, 12.1.3

Bičák, J. (2000b). Exact radiative space-times: some recent developments, *Ann. Physik*, **9**, 207–216. §1

Bičák, J. (2006). Einstein equations: exact solutions, in *Encyclopedia of mathematical physics*, eds. J.-P. Francoise, G. L. Naber and S. T. Tsou, (Elsevier, Oxford), volume 2, 165–173. §1

Bičák, J. and Griffiths, J. B. (1994). Scattering and collision of gravitational waves in Friedmann–Robertson–Walker open universes, *Phys. Rev. D*, **49**, 900–906. §21.7

Bičák, J. and Griffiths, J. B., (1996). Gravitational waves propagating into Friedmann–Robertson–Walker universes, *Ann. Physics*, **252**, 180–210. §§6.3.3, 21.7, B.3, B.3

Bičák, J. and Hoenselaers, C. (1985). Two equal Kerr–Newman sources in stationary equilibrium, *Phys. Rev. D*, **31**, 2476–2479. §13.3.2

Bičák, J., Hoenselaers, C. and Schmidt, B. G. (1983a). The solutions of the Einstein equations for uniformly accelerated particles without nodal singularities. I. Freely falling particles in external fields, *Proc. Roy. Soc. A*, **390**, 397–409. §15

Bičák, J., Hoenselaers, C. and Schmidt, B. G. (1983b). The solutions of the Einstein equations for uniformly accelerated particles without nodal singularities. II. Self-accelerating particles, *Proc. Roy. Soc. A*, **390**, 411–419. §15

Bičák, J. and Janiš, V. (1985). Magnetic fluxes across black holes, *Mon. Not. R. Astron. Soc.*, **212**, 899–915. §9.3

Bičák, J. and Kofroň, D. (2010). Rotating charged black holes accelerated by an electric field, *Phys. Rev. D*, **82**, 024006. §14.3

Bičák, J. and Krtouš. P. (2003). Radiative fields in spacetimes with Minkowski and de Sitter asymptotics, in *Recent developments in gravity*, eds. K. D. Kokkotas and N. Stergioulas, (World Scientific), 3–25. §1

Bičák, J. and Krtouš. P. (2005). Fields of accelerated sources: Born in de Sitter, *J. Math. Phys.*, **46**, 102504. §§4, 14.4

Bičák, J. and Ledvinka, T., (1993). Relativistic disks as sources of the Kerr metric, *Phys. Rev. Lett.*, **71**, 1669–1672. §§11.1, 13.2.2

Bičák, J., Ledvinka, T., Schmidt, B. G. and Žofka, M. (2004). Static fluid cylinders and their fields: global solutions, *Class. Quantum Grav.*, **21**, 1583–1608. §10.2

Bičák, J., Lynden-Bell, D. and Katz, J. (1993). Relativistic discs as sources of static vacuum spacetimes, *Phys. Rev. D*, **47**, 4334–4343. §10.7

Bičák, J., Lynden-Bell, D. and Pichon, C. (1993). Relativistic discs and flat galaxy models, *Mon. Not. R. Astron. Soc.*, **265**, 126–144. §10.7

Bičák, J. and Perjés, Z. (1987). Asymptotic behaviour of Robinson–Trautman pure radiation solutions, *Class. Quantum Grav.*, **4**, 595–597. §19.5.3

Bičák, J. and Podolský, J. (1995). Cosmic no-hair conjecture and black-hole formation: An exact model with gravitational radiation, *Phys. Rev. D*, **52**, 887–895. §§9.4, 19.4

Bičák, J. and Podolský, J. (1997). Global structure of Robinson–Trautman radiative space-times with cosmological constant, *Phys. Rev. D*, **55**, 1985–1993. §§9.4, 19.4

Bičák, J. and Podolský, J. (1999a). Gravitational waves in vacuum spacetimes with cosmological constant. I. Classification and geometrical properties of non-twisting type N solutions, *J. Math. Phys.*, **40**, 4495–4505. §§18.2, 18.3.2, 19.2

Bičák, J. and Podolský, J. (1999b). Gravitational waves in vacuum spacetimes with cosmological constant. II. Deviation of geodesics and interpretation of non-twisting type N solutions, *J. Math. Phys.*, **40**, 4506–4517. §§2.1.3, 18.2, 18.3.2, 19.2

Bičák, J. and Pravda, V. (1998). Curvature invariants in type-N spacetimes, *Class. Quantum Grav.*, **15**, 1539–1555. §§18.3.2, 18.3.3, 19.1

Bičák, J. and Pravda, V. (1999). Spinning C metric: Radiative spacetime with accelerating, rotating black holes, *Phys. Rev. D*, **60**, 044004. §§14.3, 16.3.2

Bičák, J. and Pravdová, A. (1998). Symmetries of asymptotically flat electrovacuum space-times and radiation, *J. Math. Phys.*, **39**, 6011–6039. §§15.1, 16.6

Bičák, J. and Schmidt, B. (1984). Isometries compatible with gravitational radiation, *J. Math. Phys.*, **25**, 600–606. §15.1

Bičák, J. and Schmidt, B. (1989a). On the asymptotic structure of axisymmetric radiative spacetimes, *Class. Quantum Grav.*, **6**, 1547–1559. §20.4

Bičák, J. and Schmidt, B. (1989b). Asymptotically flat radiative space-times with boost-rotation symmetry, *Phys. Rev. D*, **40**, 1827–1853. §§14.1, 15.1, 15.2, 15.3.3

Bičák, J. and Žofka, M. (2002). Notes on static cylindrical shells, *Class. Quantum Grav.*, **19**, 3653–3664. §10.2

Birell, N. D. and Davies, P. C. W. (1982), *Quantum fields in curved spaces*, (Cambridge University Press). §4.5

Birkhoff, G. D. (1923). *Relativity and modern physics*, (Harvard University Press). §8.1

Bogoyavlenskiĭ, O. I. (1985). *Methods in the qualitative theory of dynamical systems in astrophysics and gas dynamics*, (Springer). §22.1

Bolejko, K. (2008). Supernovae Ia observations in the Lemaitre–Tolman model, *PMC Physics A*, **2**:1. §22.7

Bolejko, K., Krasiński, A. and Hellaby, C. (2005). Formation of voids in the universe within the Lemaître–Tolman model, *Mon. Not. R. Astron. Soc.*, **362**, 213–228. §22.7

Bolejko, K., Krasiński, A., Hellaby, C. and Célérier, M-N. (2010). *Structures in the universe by exact methods*, (Cambridge University Press). §§22.7, 22.8

Bondi, H. (1947). Spherically symmetrical models in general relativity, *Mon. Not. R. Astron. Soc.*, **107**, 410–425. §22.7

Bondi, H. (1957a). Negative mass in general relativity, *Rev. Mod. Phys.*, **29**, 423–428. §§3.5, 15

Bondi, H. (1957b). Plane gravitational waves in general relativity, *Nature*, **179**, 1072–1073. §17.5

Bondi, H. (1961). *Cosmology*, (Cambridge University Press). §6.3.3

Bondi, H. and Pirani, F. A. E. (1989). Gravitational waves in general relativity. XIII. Caustic property of plane waves, *Proc. Roy. Soc. A*, **421**, 395–410. §17.5.2

Bondi, H., Pirani, F. A. E. and Robinson, I. (1959). Gravitational waves in general relativity. III. Exact plane waves, *Proc. Roy. Soc. A*, **251**, 519–533. §§17.2, 17.5

Bonnor, W. B. (1954). Static magnetic fields in general relativity, *Proc. Phys. Soc. A*, **67**, 225–232. §§7.3, 10.8

Bonnor, W. B. (1956). The Formation of the nebulae, *Z. Astrophysik*, **39**, 143–159; reprinted in *Gen. Rel. Grav.*, **30**, 1113–1132, (1998). §22.7

Bonnor, W. B. (1957a). Non-singular fields in general relativity, *J. Math. Mech.*, **6**, 203–214. §22.3

Bonnor, W. B. (1957b). Les ondes gravitationnelles en Relativité générale, *Ann. Inst. Henri Poincaré*, **15**, 146–157. §17.5

Bonnor, W. B. (1966a). An exact solution of the Einstein–Maxwell field equations referring to a magnetic dipole, *Z. Phys.*, **190**, 444–445. §10.8.1

Bonnor, W. B. (1966b). An exact solution for uniformly accelerated particles, *Wissenschaft. Z. der Friedrich-Schiller-Universität Jena*, **15**, 71–79. §15.2

Bonnor, W. B. (1969a). A new interpretation of the NUT metric in general relativity, *Proc. Camb. Phil. Soc.*, **66**, 145–151. §§12.1.3, 16.3

Bonnor, W. B. (1969b). The gravitational field of light, *Comm. Math. Phys.*, **13**, 163–174. §§17.3, 20.2

Bonnor, W. B. (1970a). Charge moving with the speed of light in Einstein-Maxwell theory, *Int. J. Theoret. Phys.*, **3**, 57–65. §17.3

Bonnor, W. B. (1970b). Spinning null fluid in general relativity, *Int. J. Theoret. Phys.*, **3**, 257–266. §18.5

Bonnor, W. B. (1974). Evolution of inhomogeneous cosmological models, *Mon. Not. R. Astron. Soc.*, **167**, 55–61. §22.7

Bonnor, W. B. (1976a). Do freely falling bodies radiate? *Nature*, **263**, 301. §22.8

Bonnor, W. B. (1976b). Non-radiative solutions of Einstein's equations for dust, *Commun. Math. Phys.*, **51**, 191–199. §22.8

Bonnor, W. B. (1977). A rotating dust cloud in general relativity, *J. Phys. A*, **10**, 1673–1677. §13.1.2

Bonnor, W. B. (1979a). Solution of Einstein's equations for a line mass of perfect fluid, *J. Phys. A*, **12**, 847–851. §10.2

Bonnor, W. B. (1979b). A source for Petrov's homogeneous vacuum space-time, *Phys. Lett. A*, **75**, 25–26. §13.1.2

Bonnor, W. B. (1980). The rigidly rotating relativistic dust cylinder, *J. Phys. A*, **13**, 2121–2132. §§13.1.1, 13.1.2

Bonnor, W. B. (1981). The equilibrium of two charged masses in general relativity, *Phys. Lett. A*, **83**, 414–416. §10.8.1

Bonnor, W. B. (1982). Globally regular solutions of Einstein's equations, *Gen. Rel. Grav.*, **14**, 807–821. §§1, 10.5

Bonnor, W. B. (1983). The sources of the *C*-metric, *Gen. Rel. Grav.*, **15**, 535–551. §14

Bonnor, W. B. (1988). An exact solution of Einstein's equations for two particles falling freely in an external gravitational field, *Gen. Rel. Grav.*, **20**, 607–622. §14.2.1

Bonnor, W. B. (1990a). The *C*-metric in Bondi's coordinates, *Class. Quantum Grav.*, **7**, L229–L230. §14.1.7

Bonnor, W. B. (1990b). The physical interpretation of a certain static vacuum space-time, *Wiss. Z. Friedrich-Schiller-Univ. Jena Nat. wiss. Reihe*, **39**, 25–29. §9.1.2

Bonnor, W. B. (1992). Physical interpretation of vacuum solutions of Einstein's equations. Part I. Time-independent solutions, *Gen. Rel. Grav.*, **24**, 551–574. §§1, 12.1.3, 13.1.2

Bonnor, W. B. (1993). The equilibrium of a charged test particle in the field of a spherical charged mass in general relativity, *Class. Quantum Grav.*, **10**, 2077–2082. §10.8.1

Bonnor, W. B. (1994). The photon rocket, *Class. Quantum Grav.*, **11**, 2007–2012. §19.5.2

Bonnor, W. B. (1996). Another photon rocket, *Class. Quantum Grav.*, **13**, 277–282. §19.5.4

Bonnor, W. B. (1999). The static cylinder in general relativity, in *On Einstein's path: essays in honour of Englebert Schucking*, ed. A. Harvey, (Springer), 113–119. §10.2

Bonnor, W. B. (2000). A generalization of the Einstein–Straus vacuole, *Class. Quantum Grav.*, **17**, 2739–2748. §22.7

Bonnor, W. B. (2001). The interactions between two classical spinning particles, *Class. Quantum Grav.*, **18**, 1381–1388. §§3.4.1, 12.1.4

Bonnor, W. B. (2002). Classical gravitational spin-spin interaction, *Class. Quantum Grav.*, **19**, 143–147. §13.2.2

Bonnor, W. B. (2005). A rotating dust cloud in general relativity II, *Gen. Rel. Grav.*, **37**, 2245–2250. §13.1.2

Bonnor, W. B. (2008). A second anti-de Sitter universe, *Class. Quantum Grav.*, **25**, 225005. §10.2

Bonnor, W. B. and Chamorro, A. (1990). Models of voids in the expanding universe, *Astrophys. J.*, **361**, 21–26. §22.7

Bonnor, W. B. and Davidson, W. (1992). Interpreting the Levi-Civita vacuum metric, *Class. Quantum Grav.*, **9**, 2065–2068. §10.2

Bonnor, W. B., Griffiths, J. B. and MacCallum, M. A. H. (1994). Physical interpretation of vacuum solutions of Einstein's equations. Part II. Time-dependent solutions, *Gen. Rel. Grav.*, **26**, 687–729. §§1, 15.2, 22.3

Bonnor, W. B. and Martins, M. A. P. (1991). The interpretation of some static vacuum metrics, *Class. Quantum Grav.*, **8**, 727–738. §10.1.1

Bonnor, W. B. and Sackfield, A. (1968). The interpretation of some spheroidal metrics, *Commun. Math. Phys.*, **8**, 338–344. §10.4

Bonnor, W. B., Santos, N. O. and MacCallum, M. A. H. (1998). An exterior for the Gödel space-time, *Class. Quantum Grav.*, **15**, 357–366. §§13.1.3, 22.6

Bonnor, W. B. and Steadman, B. R. (2004). The double-Kerr solution, *Class. Quantum Grav.*, **21**, 2723–2732. §13.2.2

Bonnor, W. B. and Steadman, B. R. (2005). Exact solutions of the Einstein-Maxwell equations with closed timelike curves, *Gen. Rel. Grav.*, **37**, 1833–1844. §13.3.2

Bonnor, W. B., Sulaiman, A. H. and Tomimura, N. (1977). Szekeres's space-times have no Killing vectors, *Gen. Rel. Grav.*, **8**, 549–559. §22.8

Bonnor, W. B. and Swaminarayan, N. S. (1964). An exact solution for uniformly accelerated particles in general relativity, *Z. Phys.*, **177**, 240–256. §15

Bonnor, W. B. and Tomimura, N. (1976). Evolution of Szekeres's cosmological models, *Mon. Not. R. Astron. Soc.*, **175**, 85–93. §22.8

Bonnor, W. B. and Vickers, P. A. (1981). Junction conditions in general relativity, *Gen. Rel. Grav.*, **13**, 29–36. §§2.1.5, 20.6

Bonnor, W. B. and Ward, J. P. (1973). The field of charged, spinning, magnetic particles, *Commun. Math. Phys.*, **28**, 323–330. §13.3.2

Bonometto, S., Gorini, V. and Moschella, U. eds. (2002). *Modern cosmology*, (Institute of Physics Publishing). §6.5

Booth, I. S. and Mann, R. B. (1998). Complex instantons and charged rotating black hole pair creation, *Phys. Rev. Lett.*, **81**, 5052–5055. §14.4

Booth, I. S. and Mann, R. B. (1999). Cosmological pair production of charged and rotating black holes, *Nucl. Phys. B*, **539**, 267–306. §14.4

Bose, S. K. and Estaban, E. (1981). A null tetrad analysis of the Ernst metric, *J. Math. Phys.*, **22**, 3006–3009. §9.3

Boucher, W., and Gibbons, G. W. (1984). Uniqueness theorem for anti-de Sitter spacetime, *Phys. Rev. D*, **30**, 2447–2451. §5.4

Boulware, D. G. (1973). Naked singularities, thin shells, and the Reissner-Nordström metric, *Phys. Rev. D*, **8**, 2363–2368. §9.2

Bousso, R. and Hawking, S. W. (1996). Pair creation of black holes during inflation, *Phys. Rev. D*, **54**, 6312–6322. §7.2.1

Boyer, R. H. and Lindquist, R. W. (1967). Maximal analytic extension of the Kerr metric, *J. Math. Phys.*, **8**, 265–281. §11.1

Boyer, R. H. and Price, T. G. (1965). An interpretation of the Kerr metric in general relativity, *Proc. Camb. Phil. Soc.*, **61**, 531–534. §11.1.5

Bradley, M., Fodor, G., Gergely, L. Á., Marklund, M. and Perjés, Z. (1999). Rotating perfect fluid sources of the NUT metric, *Class. Quantum Grav.*, **16**, 1667–1675. §§12.1, 12.2

Brady, P. R. and Poisson, E. (1992). Cauchy horizon instability for Reissner-Nordström black holes in de Sitter space, *Class. Quantum Grav.*, **9**, 121–125. §9.6

Bratek, Ł., Jałocha, J. and Kutschera, M. (2007). Van Stockum–Bonnor spacetimes of rigidly rotating dust, *Phys. Rev. D*, **75**, 107502. §13.1.2

Brdička, M. (1951). On gravitational waves, *Proc. Roy. Irish Acad. A*, **54**, 137–142. §§7.1, 17.5

Bretón, N., Manko, V. S. and Aguilar Sánchez, J. A. (1998). On the equilibrium of charged masses in general relativity: the electrostatic case, *Class. Quantum Grav.*, **15**, 3071–3083. §10.8.1

Bretón, N., Manko, V. S. and Aguilar Sánchez, J. (1999). On the equilibrium of charged masses in general relativity: II. The stationary electrovacuum case, *Class. Quantum Grav.*, **16**, 3725–3734. §13.3.2

Brill, D. R. (1964). Electromagnetic fields in a homogeneous, nonisotropic universe, *Phys. Rev. B*, **133**, 845–848. §12.4

Brill, D. R. and Hayward, S. A. (1994). Global structure of a black hole cosmos and its extremes, *Class. Quantum Grav.*, **11**, 359–370. §9.6

Brill, D. R., Horowitz, G. T., Kastor, D. and Traschen, J. (1994). Testing cosmic censorship with black hole collisions, *Phys. Rev. D*, **49**, 840–852. §§9.6, 22.5

Brill, D. R., Louko, J. and Peldán, P. (1997). Thermodynamics of (3+1)-dimensional black holes with toroidal or higher genus horizons, *Phys. Rev. D*, **56**, 3600–3610. §9.6

Brinkmann, H. W. (1925). Einstein spaces which are mapped conformally on each other, *Math. Ann.*, **94**, 119–145. §17.1

Buchdahl, H. A. (1971). Conformal flatness of the Schwarzschild interior solution, *Amer. J. Phys.*, **39**, 158–162. §8.5

Cahen, M. and Leroy, J. (1966). Exact solutions of Einstein–Maxwell equations, *J. Math. Mech.*, **16**. 501–508. §7.1

Cai, R.-G. and Zhang, Y.-Z. (1996). Black plane solutions in four-dimensional spacetimes, *Phys. Rev. D*, **54**, 4891–4898. §9.6

Caldarelli, M. M., Dias, Ó. J. C., Monteiro, R. and Santos, J. E. (2011). Black funnels and droplets in thermal equilibrium, *arXiv:1102.4337*. §16.2

Carmeli, M., Charach, Ch. and Malin, S. (1981). Survey of cosmological models with gravitational, scalar and electromagnetic waves, *Phys. Rep.*, **76**, 79–156. §22.4

Carminati, J. and McLenaghan, R. G. (1991). Algebraic invariants of the Riemann tensor in a four-dimensional Lorentzian space, *J. Math. Phys.*, **32**, 3135–3140. §2.1.2

Carot, J., Senovilla, J. M. M. and Vera, R. (1999). On the definition of cylindrical symmetry, *Class. Quantum Grav.*, **16**, 3025–3034. §22.3

Carr, B. J. and Verdaguer, E. (1983). Soliton solutions and cosmological gravitational waves, *Phys. Rev. D*, **28**, 2995–3006. §22.4

Carroll, S. M., Press, W. H. and Turner, E. L. (1992). The cosmological constant, *Annu. Rev. Astron. Astrophys.*, **30**, 499–542. §6.5

Carter, B. (1966a). The complete analytic extension of the Reissner-Nordström metric in the special case $e^2 = m^2$, *Phys. Lett.*, **21**, 423–424. §9.2.3

Carter, B. (1966b). Complete analytic extension of the symmetry axis of Kerr's solution of Einstein's equations, *Phys. Rev.*, **141**, 1242–1247. §11.1

Carter, B. (1968a). Global structure of the Kerr family of gravitational fields, *Phys. Rev.*, **174**, 1559–1571. §§11.1, 11.2

Carter, B. (1968b). Hamilton–Jacobi and Schrödinger separable solutions of Einstein's equations, *Commun. Math. Phys.*, **10**, 280–310. §§11.1.5, 16, 18.6

Carter, B. (1971a). Causal structure in space-time, *Gen. Rel. Grav.*, **1**, 349–391. §2.3.3

Carter, B. (1971b). Axisymmetric black hole has only two degrees of freedom, *Phys. Rev. Lett.*, **26**, 331–333. §11

Carter, B. (1973). Black hole equilibrium states, in *Black holes – les astres occlus*, eds. C. DeWitt and B. S. DeWitt, (Gordon and Breach), 75–214. §9.2.3

Carter, B. (1987). Mathematical foundations of the theory of relativistic stellar and black hole configurations, in *Gravitation in astrophysics*, eds. B. Carter and J. B. Hartle, (Plenum Press), 63–122. §§9.2, 11.2

Castejon-Amendo, J. and MacCallum, M. A. H. (1990). On finding hypersurface-orthogonal Killing fields, *Gen. Rel. Grav.*, **22**, 393–415. §22.3

Célérier, M-N. (2000). Do we really see a cosmological constant in the supernovae data? *Astron. Astrophys.*, **353**, 63–71. §22.7

Centrella, J. and Matzner, R. A. (1979). Plane-symmetric cosmologies, *Astrophys. J.*, **230**, 311–324. §22.4

Chambers, C. M. (1997). The Cauchy horizon in black hole-de Sitter spacetimes, in *Internal structure of black holes and spacetime singularities*, eds L. M. Burko and A. Ori, (Ann. Israel Phys. Soc., No 13, IOP Publishing). §9.6

Chandrasekhar, S. (1983). *The mathematical theory of black holes*, (Oxford University Press). §11.1, 13.2.2

Chandrasekhar, S. (1986). Cylindrical waves in general relativity, *Proc. Roy. Soc. A*, **408**, 209–232. §22.3

Chandrasekhar, S. and Ferrari, V. (1984). On the Nutku–Halil solution for colliding impulsive gravitational waves, *Proc. Roy. Soc. A*, **396**, 55–74. Errata, *Proc. Roy. Soc. A*, **398**, 429, (1985). §21.3

Chandrasekhar, S. and Ferrari, V. (1987). On the dispersion of cylindrical impulsive gravitational waves, *Proc. Roy. Soc. A*, **412**, 75–91. §22.3

Chandrasekhar, S. and Hartle, J. B. (1982). On crossing the Cauchy horizon of a Reissner–Nordström black hole, *Proc. Roy. Soc. A*, **384**, 301–315. §9.2.2

Chandrasekhar, S. and Xanthopoulos, B. C. (1985). On colliding waves in the Einstein–Maxwell theory, *Proc. Roy. Soc. A*, **398**, 223–259. §21.3

Chandrasekhar, S. and Xanthopoulos, B. C. (1986). A new type of singularity created by colliding gravitational waves, *Proc. Roy. Soc. A*, **408**, 175–208. §§2.1.2, 21.5.2

Chandrasekhar, S. and Xanthopoulos, B. C. (1987a). On colliding waves that develop time-like singularities: a new class of solutions of the Einstein–Maxwell equations, *Proc. Roy. Soc. A*, **410**, 311–336. §21.5.4

Chandrasekhar, S. and Xanthopoulos, B. C. (1987b). The effect of sources on horizons that develop when plane gravitational waves collide, *Proc. Roy. Soc. A*, **414**, 1–30. §21.5.4

Chandrasekhar, S. and Xanthopoulos, B. C. (1989). Two black holes attached to strings, *Proc. Roy. Soc. A*, **423**, 387–400. §10.8.1

Chazy, J. (1924). Sur la champ de gravitation de deux masses fixes dans la théory de la relativité, *Bull. Soc. Math. France*, **52**, 17. §§10.1.2, 10.5, 10.6

Chng, B., Mann, R. and Stelea, C. (2006). Accelerating Taub–NUT and Eguchi–Hanson solitons in four dimensions, *Phys. Rev. D*, **74**, 084031. §16.3

Christodoulou, D. (1984). Violation of cosmic censorship in the gravitational collapse of a dust cloud, *Commun. Math. Phys.*, **93**, 171–195. §22.7

Chruściel, P. T. (1990). On space-times with $U(1) \times U(1)$ symmetric compact Cauchy surfaces, *Ann. Physics*, **202**, 100–150. §22.4

Chruściel, P. T. (1991). Semi-global existence and convergence of solutions of the Robinson–Trautman (2-dimensional Calabi) equation, *Commun. Math. Phys.*, **137**, 289–313. §19.4

Chruściel, P. T. (1992). On the global structure of Robinson–Trautman spacetimes, *Proc. Roy. Soc. A*, **436**, 299–316. §19.4

Chruściel, P. T. (1994). "No hair" theorems – folklore, conjectures, results, *Contemp. Math.*, **170**, 23–49. §§9.2, 11.2

Chruściel, P. T. (1996). Uniqueness of stationary, electro-vacuum black holes revisited, *Helvet. Phys. Acta*, **69**, 529–543. §§11.2, 19.4

Chruściel, P. T., Isenberg, J., and Moncrief, V. (1990). Strong cosmic censorship in polarised Gowdy spacetimes, *Class. Quantum Grav.*, **7**, 1671–1680. §22.4

Chruściel, P. T. and Nadirashvili, P. T. (1995). All electrovacuum Majumdar–Papapetrou spacetimes with non-singular black holes, *Class. Quantum Grav.*, **12**, L17–L23. §22.5

Chruściel, P. T., Reall, H. S. and Tod, P. (2006). On Israel–Wilson–Perjés black holes, *Class. Quantum Grav.*, **23**, 2519–2540. §13.3.2

Chruściel, P. T. and Singleton, D. B. (1992). Non-smoothness of event horizons of Robinson–Trautman black holes, *Commun. Math. Phys.*, **147**, 137–162. §19.4

Chruściel, P. T. and Tod, P. (2007). The classification of static electro-vacuum space-times containing an asymptotically flat spacelike hypersurface with compact interior, *Commun. Math. Phys.*, **271**, 577–589. §10.8.1

Clarke, C. J. S. (1993). *The analysis of space-time singularities*, (Cambridge University Press). §2.3.1

Clarke, C. J. S. (1994). A review of cosmic censorship, *Class. Quantum Grav.*, **11**, 1375–1386. §8.2

Clarke, C. J. S. and Dray, T. (1987). Junction conditions for null hypersurfaces, *Class. Quantum Grav.*, **4**, 265–275. §§2.1.5, 20.6

Clarke, C. J. S. and Hayward, S. A. (1989). The global structure of the Bell–Szekeres solution, *Class. Quantum Grav.*, **6**, 615–622. §21.6

Cohen, J. M. (1968). Angular momentum and the Kerr metric, *J. Math. Phys.*, **9**, 905–906. §11.1.5

Cohen, J. M. and Gautreau, R. (1979). Naked singularities, event horizons, and charged particles, *Phys. Rev. D*, **19**, 2273–2279. §9.2

Coley, A., Hervik, S., Papadopoulos, G. and Pelavas, N. (2009). Kundt spacetimes, *Class. Quantum Grav.*, **26**, 105016. §18.1

Coley, A., Hervik, S. and Pelavas, N. (2009). Spacetimes characterized by their scalar curvature invariants, *Class. Quantum Grav.*, **26**, 025013. §18.1

Collins, C. B. (1977). Global structure of the "Kantowski–Sachs" cosmological models, *J. Math. Phys.*, **18**, 2116–2124. §22.2

Collins, C. B. (1979). Intrinsic symmetries in general relativity, *Gen. Rel. Grav.*, **10**, 925–929. §22.8

Collins, C. B. and Ellis, G. F. R. (1979). Singularities in Bianchi cosmologies, *Phys. Rep.*, **56**, 65–106. §22.1

Collins, C. B. and Szafron, D. A. (1979). A new approach to inhomogeneous cosmologies: Intrinsic symmetries. I, *J. Math. Phys.*, **20**, 2347–2353. §22.8

Collinson, C. D. (1976). The uniqueness of the Schwarzschild interior metric, *Gen. Rel. Grav.*, **7**, 419–422. §8.5

Contopoulos, G. (1990). Periodic orbits and chaos around two black holes, *Proc. Roy. Soc. A*, **431**, 183–202. §22.5

Contopoulos, G. (1991). Periodic orbits and chaos around two fixed black holes. II, *Proc. Roy. Soc. A*, **435**, 551–562. §22.5

Cooperstock, F. I. and de la Cruz, V. (1979). Static and stationary solutions of the Einstein–Maxwell equations, *Gen. Rel. Grav.*, **10**, 681–697. §10.8.1

Cooperstock, F. I. and Junevicus, G. J. (1974). Singularities in Weyl gravitational fields, *Int. J. Theoret. Phys.*, **9**, 59–64. §10.5

Cornish, F. H. J. (2000). Robinson–Trautman radiating metrics with zero news and photon rockets, *Class. Quantum Grav.*, **17**, 3945–3950. §§19.5.2, 19.5.4

Cornish, F. H. J. and Uttley, W. J. (1995a). The interpretation of the C metric. The vacuum case, *Gen. Rel. Grav.*, **27**, 439–454. §14

Cornish. F. H. J. and Uttley, W. J. (1995b). The interpretation of the C metric – the charged case when $e^2 \le m^2$, *Gen. Rel. Grav.*, **27**, 735–749. §14.2

Cornish, N. J. and Levin, J. J. (1997a). The mixmaster universe is chaotic, *Phys. Rev. Lett.*, **78**, 998–1001. §22.1

Cornish, N. J. and Levin, J. J. (1997b). The mixmaster universe: A chaotic Farey tale, *Phys. Rev. D*, **55**, 7489–7510. §22.1

Covarrubias, G. M. (1980). Gravitational radiation in Szekeres's quasi-spherical space-times, *J. Phys. A: Math. Gen.*, **13**, 3023–3028. §22.8

Cruz, N., Olivares, M. and Villanueva, J. R. (2005). The geodesic structure of the Schwarzschild anti-de Sitter black hole, *Class. Quantum Grav.*, **22**, 1167–1190. §9.4.3

Curzon, H. E. J. (1924). Cylindrical solutions of Einstein's gravitation equations, *Proc. London Math. Soc*, **23**, 477–480. §§10.1.2, 10.5, 10.6

D'Eath, P. D. (1978). High-speed black-hole encounters and gravitational radiation, *Phys. Rev. D*, **18**, 990–1019. §20.2

da Silva, M. F. A., Herrera, L., Paiva, F. M. and Santos, N. O. (1995a). On the parameters of the Lewis metric for the Lewis class, *Class. Quantum Grav.*, **12**, 111–118. §13.1.1

da Silva, M. F. A., Herrera, L., Paiva, F. M. and Santos, N. O. (1995b). The parameters of the Lewis metric for the Weyl class, *Gen. Rel. Grav.*, **27**, 859–871. §§13.1.1, 13.1.3

da Silva, M. F. A., Herrera, L., Paiva, F. M. and Santos, N. O. (1995c). The Levi–Civita space-time, *J. Math. Phys.*, **36**, 3625–3631. §10.2

da Silva, M. F. A., Herrera, L., Santos, N. O. and Wang, A. Z. (2002). Rotating cylindrical shell source for Lewis spacetime, *Class. Quantum Grav.*, **19**, 3809–3819. §13.1.3

da Silva, M. F. A., Wang, A. Z., Paiva, F. M. and Santos, N. O. (2000). Levi-Civita solutions with a cosmological constant, *Phys. Rev. D*, **61**, 044003. §10.2

Dadhich, N., Hoenselaers, C. and Vishveshwara, C. V. (1979). Trajectories of charged particles in the static Ernst space-time, *J. Phys. A*, **12**, 215–221. §9.3

Dain, S., Moreschi, O. M. and Gleiser, R. J. (1996). Photon rockets and the Robinson–Trautman geometries, *Class. Quantum Grav.*, **13**, 1155–1160. §§19.1, 19.5.2

Damour, T. (1995). Photon rockets and gravitational radiation, *Class. Quantum Grav.*, **12**, 725–737. §19.5.2

Darmois, G. (1927). *Mémorial des sciences mathématiques*, (Gauthier-Villars, Paris). §§2.1.5, 10.4, 20

Das, A. (1962). A class of exact solutions of certain classical field equations in general relativity, *Proc. Roy. Soc. A*, **267**, 1–10. §10.8

Dautcourt, G. and Abdel-Megied, M. (2006). Revisiting the light cone of the Gödel universe, *Class. Quantum Grav.*, **23**, 1269–1288. §22.6

Davidson, W. (2000). A cylindrically symmetric stationary solution of Einstein's equations describing a perfect fluid of finite radius, *Class. Quantum Grav.*, **17**, 2499–2507. §13.1.3

Davidson, W. (2001). The Lewis metric as a vacuum exterior for a rotating perfect-fluid cylinder, *Class. Quantum Grav.*, **18**, 3721–3733. §13.1.3

Davies, H. and Caplan, T. A. (1971), The space-time metric inside a rotating cylinder, *Proc. Camb. Phil. Soc.*, **69**, 325–327. §13.1.1

de Felice, F. and Bradley, M. (1988). Rotational anisotropy and repulsive effects in the Kerr metric, *Class. Quantum Grav.*, **5**, 1577–1585. §11.1.5

de Felice, F. and Maeda, K. (1982). Topology of collapse in conformal diagrams, *Prog. Theoret. Phys.*, **68**, 1967–1978. §9.2.2

De Groote, L., Van den Bergh, N. and Wylleman, L. (2010). Petrov type D pure radiation fields of Kundt's class, *J. Math. Phys.*, **51**, 102501. §18.2

de la Cruz, V. and Israel, W. (1967). Gravitational bounce, *Nuovo Cimento A*, **51**, 744–760. §9.2

de Sitter, W. (1917a). Over de relativiteit der traagheid: Beschouingen naar aanleiding van Einstein's hypothese, *Koninklijke Akademie van Wetenschappen te Amsterdam*, **25**, 1268–1276; *Proc. Akad. Amsterdam*, **19**, 1217–1225, (1918). §§4.1, 6.1, 6.3

de Sitter, W. (1917b). Over de Kromming der ruimte, *Koninklijke Akademie van Wetenschappen te Amsterdam*, **26**, 222–236; *Proc. Akad. Amsterdam*, **20**, 229–243, (1918). §§4.1, 6.1, 6.3

Debever, R. (1969). Sur les espaces de Brandon Carter, *Bull. Cl. Sci. Acad. R. Belg.*, **55**, 8–16. §16

Debever, R. (1971). On type D expanding solutions of Einstein–Maxwell equations, *Bull. Soc. Math. Belg.*, **23**, 360–376. §16

Debever, R. and Kamran, N. (1980). Coordonnées symétriques et coordonnées isotropes des solutions de type D des équations d'Einstein–Maxwell avec constante cosmologique, *Bull. Cl. Sci. Acad. Roy. Belg.*, **66**, 585–599. §16

Debever, R., Kamran, N. and McLenaghan, R. G. (1983). A single expression for the general solution of Einstein's vacuum and electrovac field equations with cosmological constant for Petrov type D admitting a non-singular aligned Maxwell field, *Phys. Lett. A*, **93**, 399–402. §16.1

Debever, R., Kamran, N. and McLenaghan, R. G. (1984). Exhaustive integration and a single expression for the general solution of the type D vacuum and electrovac field equations with cosmological constant for a nonsingular aligned Maxwell field, *J. Math. Phys.*, **25**, 1955–1972. §16

Delgaty, M. S. R. and Lake, K. (1998). Physical acceptability of isolated, static, spherically symmetric, perfect fluid solutions of Einstein's equations, *Comput. Phys. Commun.*, **115**, 395–415. §8.5

Demiański, M. and Newman, E. T. (1966). A combined Kerr–NUT solution of the Einstein field equations, *Bull. Acad. Polon. Sci. Math. Astron. Phys.*, **14**, 653. §12.1.3

Deser, S., Jackiw, R. and 't Hooft, G. (1984). Three-dimensional Einstein gravity: Dynamics of flat space, *Ann. Physics*, **152**, 220–235. §3.4.1

Dettmann, C. P., Frankel, N. E. and Cornish, N. J. (1994). Fractal basins and chaotic trajectories in multi-black-hole spacetimes, *Phys. Rev. D*, **50**, R618–R621. §22.5

Dias, Ó. J. C. (2004). Pair creation of anti-de Sitter black holes on a cosmic string background, *Phys. Rev. D*, **70**, 024007. §14.4

Dias, Ó. J. C. and Lemos, J. P. S. (2003a). Pair of accelerated black holes in an anti-de Sitter background: the AdS C metric, *Phys. Rev. D*, **67**, 064001. §14.4

Dias, Ó. J. C. and Lemos, J. P. S. (2003b). Pair of accelerated black holes in a de Sitter background: the dS C metric, *Phys. Rev. D*, **67**, 084018. §14.4

Dias, Ó. J. C. and Lemos, J. P. S. (2004). Pair creation of de Sitter black holes on a cosmic string background, *Phys. Rev. D*, **69**, 084006. §14.4

Dietz, W. and Hoenselaers, C. (1982). Stationary system of two masses kept apart by their gravitational spin-spin interaction, *Phys. Rev. Lett.*, **48**, 778–780. §13.2.2

Dietz, W. and Hoenselaers, C. (1985). Two mass solutions of Einstein's vacuum equations: the double Kerr solution, *Ann. Physik*, **165** (497), 319–383. §13.2.2

DiNunno, B. S. and Matzner, R. A. (2010). The volume inside a black hole, *Gen. Rel. Grav.*, **42**, 63–76. §8.2

Dowker, F., Gauntlett, J. P., Giddings, S. B. and Horowitz, G. T. (1994). Pair creation of extremal black holes and Kaluza-Klein monopoles, *Phys. Rev. D*, **50**, 2662–2679. §14.5

Dowker, H. F. and Thambyahpillai, S. N. (2003). Many accelerating black holes, *Class. Quantum Grav.*, **20**, 127–135. §14.5

Dowker, J. S. (1974). The NUT solution as a gravitational dyon, *Gen. Rel. Grav.*, **5**, 603–613. §§12.1.3, 12.2

Dray, T. (1982). On the asymptotic flatness of the *C* metrics at spatial infinity, *Gen. Rel. Grav.*, **14**, 109–112. §14

Dray, T. and 't Hooft, G. (1985). The gravitational shock wave of a massless particle, *Nucl. Phys. B*, **253**, 173–188. §20.6

Dray, T. and Walker, M. (1980). On the regularity of Ernst's generalized C-metric, *Lett. Math. Phys.*, **4**, 15–18. §14.2.1

Dwivedi, I. H. and Joshi, P. S. (1989). On the nature of naked singularities in Vaidya spacetimes, *Class. Quantum Grav.*, **6**, 1599–1606. §9.5.2

Dwivedi, I. H. and Joshi, P. S. (1991). On the nature of naked singularities in Vaidya spacetimes: II, *Class. Quantum Grav.*, **8**, 1339–1348. §9.5.2

Dwivedi, I. H. and Joshi, P. S. (1997). Initial data and the final fate of inhomogeneous dust collapse, *Class. Quantum Grav.*, **14**, 1223–1236. §22.7

Dyer, C. C. (1976). The gravitational perturbation of the cosmic background radiation by density concentrations, *Mon. Not. R. Astron. Soc.*, **175**, 429–447. §8.6

Eardley, D. M., Horowitz, G. T., Kastor, D. A. and Traschen, J. (1995). Breaking cosmic strings without monopoles, *Phys. Rev. Lett.*, **75**, 3390–3393. §14.5

Eardley, D., Liang, E. and Sachs, R. (1972). Velocity-dominated singularities in irrotational dust cosmologies, *J. Math. Phys.*, **13**, 99–107. §22.4

Eardley, D. M. and Smarr, L. (1979). Time functions in numerical relativity: Marginally bound dust collapse, *Phys. Rev. D*, **19**, 2239–2259. §22.7

Economou, J. E. (1976). Approximate form of the Tomimatsu–Sato $\delta = 2$ solution near the poles $x = 1$, $y = \pm 1$, *J. Math. Phys.*, **17**, 1095-1098. §13.2.2

Eddington, A. S. (1924). A comparison of Whitehead's and Einstein's formulas, *Nature*, **113**, 192. §8.2

Edgar, S. B. and Ludwig, G. (1997a). All conformally flat pure radiation metrics, *Class. Quantum Grav.*, **14**, L65–L68. §§18.2, 18.3.3

Edgar, S. B. and Ludwig, G. (1997b). Integration in the GHP formalism III: finding conformally flat radiation metrics as an example of an 'optimal solution', *Gen. Rel. Grav.*, **29**, 1309–1328. §§18.2, 18.3.3

Edwards, D. (1972). Exact expressions for the properties of the zero-pressure Friedmann models, *Mon. Not. R. Astron. Soc.*, **159**, 51–66. §§6.3, 6.4.2

Ehlers, J. and Kundt, W. (1962). Exact solutions of the gravitational field equations, in *Gravitation: an introduction to current research*, ed. L. Witten, (Wiley), 49–101. §§1, 9.1, 14, 16.4.1, 17.1, 18.1, 18.6

Einstein, A. (1917). Kosmologische Betrachtungen zur allgemeinen Relativitätstheorie, *Sitz. Preuss. Akad. Wiss. Berlin*, 142–152. §§3.2, 6

Einstein, A. and de Sitter, W. (1932). On the relation between the expansion and the mean density of the universe, *Proc. Nat. Acad. Sci. USA*, **18**, 213–214. §6.3.2

Einstein, A. and Rosen, N. (1936). Two-body problem in general relativity, *Phys. Rev.*, **49**, 404–405. §10.6

Einstein, A. and Rosen, N. (1937). On gravitational waves, *J. Franklin Inst.*, **223**, 43–54. §22.3

Einstein, A. and Straus, E. G. (1945). The influence of the expansion of space on the gravitation fields surrounding the individual stars, *Rev. Mod. Phys.*, **17**, 120–124; **18**, 148–149, (1946). §8.6

Ellis, G. F. R. (1967). Dynamics of pressure-free matter in general relativity, *J. Math. Phys.*, **8**, 1171–1194. §§22.2, 22.6

Ellis, G. F. R. (2000). Editor's note on reprints of Gödel's papers, *Gen. Rel. Grav.*, **32**, 1399–1408. §22.6

Ellis, G. F. R. (2006). The Bianchi models: Then and now, *Gen. Rel. Grav.*, **38**, 1003–1015. §22.1

Ellis, G. F. R. and MacCallum, M. A. H. (1969). A class of homogeneous cosmological models, *Commum. Math. Phys.*, **12**, 108–141. §§22.1, 22.2

Ellis, G. F. R. and Schmidt, B. G. (1977). Singular space-times, *Gen. Rel. Grav.*, **8**, 915–953. §2.3.1

Emparan, R. (1995). Pair creation of black holes joined by cosmic strings, *Phys. Rev. Lett.*, **75**, 3386–3389. §14.5

Emparan, R. (2000). Black diholes, *Phys. Rev. D*, **61**, 104009. §10.8.1

Emparan, R., Horowitz, G. T. and Myers, R. C. (2000a). Exact description of black holes on branes, *J. High Energy Phys.*, no. 1, paper 7. §14.4

Emparan, R., Horowitz, G. T. and Myers, R. C. (2000b). Exact description of black holes on branes. II. Comparison with BTZ black holes and black strings, *J. High Energy Phys.*, no. 1, paper 21. §14.4

Emparan, R. and Teo, E. (2001). Macroscopic and microscopic description of black diholes, *Nucl. Phys. B*, **610**, 190–214. §10.8.1

Enqvist, K. (2008). Lemaitre–Tolman–Bondi model and accelerating expansion, *Gen. Rel. Grav.*, **40**, 451–466. §22.7

Erez, G. and Rosen, N. (1959). The gravitational field of a particle possessing a multipole moment, *Bull. Res. Counc. of Israel*, **8F**, 47–50. §10.1.2

Eriksen, E. and Grøn, O. (1995). The de Sitter universe models, *Int. J. Mod. Phys. D*, **4**, 115–159. §4

Ernst, F. J. (1968a). New formulation of the axially symmetric gravitational wave problem, *Phys. Rev.*, **167**, 1175–1178. §§13, 21.2, 22

Ernst, F. J. (1968b). New formulation of the axially symmetric gravitational wave problem II, *Phys. Rev.*, **168**, 1415–1417. §§13, 21.2, 22

Ernst, F. J. (1976a). Removal of the nodal singularity of the C-metric, *J. Math. Phys.*, **17**, 515–516. §§9.3, 14.2.1

Ernst, F. J. (1976b). New representation of the Tomimatsu–Sato solution, *J. Math. Phys.*, **17**, 1091–1094. §13.2.2

Ernst, F. J. (1978). Generalized C-metric, *J. Math. Phys.*, **19**, 1986–1987. §14.2.1

Ernst, F. J., Manko, V. S. and Ruiz, E. (2006). Equatorial symmetry/antisymmetry of stationary axisymmetric electrovac spacetimes, *Class. Quantum Grav.*, **23**, 4945–4952. §13.3.2

Ernst, F. J. and Wild, W. J. (1976). Kerr black holes in a magnetic universe, *J. Math. Phys.*, **17**, 182–184. §9.3

Esposito, F. P. and Witten, L. (1975). On a static axisymmetric solution of the Einstein equations, *Phys. Lett. B*, **58**, 357–360. §10.4

Evans, A. B. (1977). Static fluid cylinders in general relativity, *J. Phys. A: Math. Gen.*, **10**, 1303–1311. §10.2

Farhoosh, H. and Zimmerman, R. L. (1979). Stationary charged C-metric, *J. Math. Phys.*, **20**, 2272–2279. §16.6

Farhoosh, H. and Zimmerman, R. L. (1980a). Killing horizons and dragging of the inertial frame about a uniformly accelerating particle, *Phys. Rev. D*, **21**, 317–327. §14

Farhoosh, H. and Zimmerman, R. L. (1980b). Surfaces of infinite red-shift around a uniformly accelerating and rotating particle, *Phys. Rev. D*, **21**, 2064–2074. §§14.3, 16.3.2

Farhoosh, H. and Zimmerman, R. L. (1980c). Killing horizons around a uniformly accelerating and rotating particle, *Phys. Rev. D*, **22**, 797–801. §§14.3, 16.3.2

Farnsworth, D. L. and Kerr, R. P. (1966). Homogeneous dust-filled cosmological solutions, *J. Math. Phys.*, **7**, 1625–1632. §22.6

Fayos, F., Martín-Prats, M. M. and Senovilla, J. M. M. (1995). On the extension of Vaidya and Vaidya–Reissner–Nordström spacetimes, *Class. Quantum Grav.*, **12**, 2565–2576. §9.5

Fayos, F. and Torres, R. (2008). A class of interiors for Vaidya's radiating metric: singularity-free gravitational collapse, *Class. Quantum Grav.*, **25**, 175009. §9.5.1

Feinstein, A. (1987). Effective energy of gravitational solitons, *Phys. Rev. D*, **35**, 3263–3265. §22.4

Feinstein, A. and Charach, Ch. (1986). Gravitational solitons and spatial flatness, *Class. Quantum Grav.*, **3**, L5–L8. §22.4

Feinstein, A. and Griffiths, J. B. (1994). Colliding gravitational waves in a closed Friedmann–Robertson–Walker background, *Class. Quantum Grav.*, **11**, L109–L113. §21.7

Feinstein, A. and Ibáñez, J. (1989). Curvature singularity free solutions for colliding plane gravitational waves with broken $(u-v)$ symmetry, *Phys. Rev. D*, **39**, 470–473. §21.7

Felten, J. E. and Isaacman, R. (1986). Scale factors $R(t)$ and critical values of the cosmological constant Λ in Friedmann universes, *Rev. Mod. Phys.*, **58**, 689–698. §6.5

Ferrari, V. and Ibáñez, J. (1987). On the collision of gravitational plane waves: a class of soliton solutions, *Gen. Rel. Grav.*, **19**, 405–425. §§21.4, 21.5

Ferrari, V. and Ibáñez, J. (1988). Type D solutions describing the collision of plane-fronted gravitational waves, *Proc. Roy. Soc. A*, **417**, 417–431. §§21.4, 21.5

Ferrari, V. and Pendenza, P. (1990). Boosting the Kerr metric, *Gen. Rel. Grav.*, **22**, 1105–1117. §20.2

Finkelstein, D. (1958). Past-future asymmetry of the gravitational field of a point particle, *Phys. Rev.*, **110**, 965–967. §8.2

Foster, J. and Newman, E. T. (1967). Note on the Robinson–Trautman solutions, *J. Math. Phys.*, **8**, 189–194. §§19.2, 19.4

Frauendiener, J. (2004). Conformal infinity, *Living Revs. in Rel.*, lrr-2004-1. §2.3.3

Frauendiener, J. and Klein, C. (2001). Exact relativistic treatment of stationary counterrotating dust disks: Physical properties, *Phys. Rev. D*, **63**, 084025. §13.2.2

Frehland, E. (1971). The general stationary gravitational vacuum field of cylindrical symmetry, *Commun. Math. Phys.*, **23**, 127–131. §13.1.1

Friedmann, A. (1922). Über die Krümmung des Raumes, *Z. Physik*, **10**, 377–386. §§6.1, 6.3.2

Friedmann, A. (1924). Über die Möglichkeit einer Welt mit konstanter negativer Krümmung des Raumes, *Z. Physik*, **21**, 326–332. §§6.1, 6.3.2

Friedrich, H. (1995). Einstein equations and conformal structure: Existence of Anti-de Sitter type space-times, *J. Geom. Phys.*, **17**, 125–184. §§5.2, 5.4

Frittelli, S., and Moreschi, O. M. (1992). Study of the Robinson–Trautman metrics in the asymptotic future, *Gen. Rel. Grav.*, **24**, 575–597. §19.4

Frolov, V. P. (1974a). The motion of charged radiating shells in the general theory of relativity and Friedman states, *Zh. Eksp. Teor. Fiz.*, **66**, 813–825; *Sov. Phys. JETP*, **39**, 393–398. §9.6

Frolov, V. P. (1974b). Kerr and Newman–Unti–Tamburino type solutions of Einstein's equations with cosmological term, *Teor. Mat. Fiz.*, **21**, 213–223; *Theoret. & Math. Phys.*, **21**, 1088–1096, (1974). §16.3.1

Frolov, V. P. (1979). The Newman–Penrose method in the theory of general relativity, in *Problems in the general theory of relativity and theory of group representations*, ed. N. G. Basov, (Plenum). §2.1.3

Frolov, V. P. and Fursaev, D. V. (2005). Gravitational field of a spinning radiation beam pulse in higher dimensions, *Phys. Rev. D*, **71**, 104034. §18.5

Frolov, V. P., Israel, W. and Zelnikov, A. (2005). Gravitational field of relativistic gyratons, *Phys. Rev. D*, **72**, 084031. §18.5

Frolov, V. P. and Novikov, I. D. (1998). *Black hole physics: Basic concepts and new developments*, (Kluwer, Dortrecht). §11.1

Frolov, V. P. and Zelnikov, A. (2005). Relativistic gyratons in asymptotically AdS spacetime, *Phys. Rev. D*, **72**, 104005. §18.5

Frolov, V. P. and Zelnikov, A. (2006). Gravitational field of charged gyratons, *Class. Quantum Grav.*, **23**, 2119–2128. §18.5

Fustero, X. and Verdaguer, E. (1986). Einstein–Rosen metrics generated by the inverse scattering transform, *Gen. Rel. Grav.*, **18**, 1141–1158. §22.3

Gair, J. R. (2002). Kantowski–Sachs universes with counter-rotating dust, *Class. Quantum Grav.*, **19**, 6345–6357. §22.2

Gal'tsov, D. V. and Masar, E. (1989). Geodesics in spacetimes containing cosmic strings, *Class. Quantum Grav.*, **6**, 1313–1341. §3.4

García D., A. (1984). Electrovac type D solutions with cosmological constant, *J. Math. Phys.*, **25**, 1951–1954. §16

García D., A. and Alvarez, C. M. (1984). Shear-free special electrovac type-II solutions with cosmological constant, *Nuovo Cimento B*, **79**, 266–270. §18.7

García D., A. and Macias, A. (1998). Black holes as exact solutions of the Einstein–Maxwell equations of Petrov type D, in *Black holes: Theory and observation*, Lecture notes in Physics **514**, eds. F. W. Hehl, C. Kiefer and R. J. K. Metzler, (Springer), 203–223. §16.1

García Díaz, A. and Plebański, J. F. (1981). All non-twisting N's with cosmological constant, *J. Math. Phys.*, **22**, 2655–2658. §§18.2, 18.3.2, 19.2

García Díaz, A. and Plebański, J. F. (1982). Solutions of type D possessing a group with null orbits as contractions of the seven-parameter solution, *J. Math. Phys.*, **23**, 1463–1465. §16.1

García Díaz, A. and Salazar, I. H. (1983). All null orbit type d electrovac solutions with cosmological constant, *J. Math. Phys.*, **24**, 2498–2503. §16.1

García-Parrado, A. and Senovilla, J. M. M. (2003). Causal relationship: a new tool for the causal characterization of Lorentzian manifolds, *Class. Quantum Grav.*, **20**, 625–664. §6.4.2

García-Parrado, A. and Senovilla, J. M. M. (2005). Causal structures and causal boundaries, *Class. Quantum Grav.*, **22**, R1–R84. §§2.3.1, 2.3.3

Garfinkle, D. (2004). The fine structure of Gowdy spacetimes, *Class. Quantum Grav.*, **21**, S219–231. §22.4

Garfinkle, D. (2006). Inhomogeneous spacetimes as a dark energy model, *Class. Quantum Grav.*, **23**, 4811–4818. §22.7

Garfinkle, D. and Melvin, M. A. (1992). Traveling waves on a magnetic universe, *Phys. Rev. D*, **45**, 1188–1191. §7.3

Gautreau, R. (1983), Geodesic coordinates in the de Sitter universe, *Phys. Rev. D*, **27**, 764–778. §4.5

Gautreau, R. and Anderson, J. L. (1967). Directional singularities in Weyl gravitational fields, *Phys. Lett. A*, **25**, 291–292. §§10.4, 10.5

Gautreau, R. and Hoffman, R. B. (1969). Exact solutions of the Einstein vacuum field equations in Weyl coordinates, *Nuovo Cimento B*, **61**, 411–424. §§10.1.1, 10.2

Gautreau, R. and Hoffman, R. B. (1972). Generating potential for the NUT metric in general relativity, *Phys. Lett. A*, **39**, 75. §12.1.1

Gautreau, R., Hoffman, R. B. and Armenti, A. (1972). Static multiparticle systems in general relativity, *Nuovo Cimento B*, **7**, 71–98. §10.8

Geroch, R. (1968). What is a singularity in general relativity? *Ann. Physics*, **48**, 526–540. §2.3.1

Geroch, R, and Hartle, J. B. (1982). Distorted black holes, *J. Math. Phys.*, **23**, 680–692. §14.1.3

Geroch, R. and Traschen, J. (1987). Strings and other distributional sources in general relativity, *Phys. Rev. D*, **36**, 1017–1031. §21.6

Geyer, K. H. (1980). Geomtrie de Raum-Zeit der Massbestimmung von Kottler, Weyl und Trefftz, *Astron. Nachr.*, **301**, 135–149. §9.4

Gibbons, G. W. and Hawking, S. W. (1977). Cosmological event horizons, thermodynamics, and particle creation, *Phys. Rev. D*, **15**, 2738–2751. §§9.4.1, 11.3, 19.4

Gibbons, G. W. and Herdeiro, C. A. R. (2001). The Melvin universe in Born–Infeld theory and other theories of nonlinear electrodynamics, *Class. Quantum Grav.*, **18**, 1677–1690. §7.3

Gibbons, G. W. and Russell-Clark, R. A. (1973). Note on the Sato–Tomimatsu solution of Einstein's equations, *Phys. Rev. Lett.*, **30**, 398–399. §13.2.2

Ginsparg, P. and Perry, M. J. (1983). Semiclassical perdurance of de Sitter space, *Nucl. Phys. B*, **222**, 245–268. §7.2.1

Girotto, F. and Saa, A. (2004). Semianalytical approach for the Vaidya metric in double-null coordinates, *Phys. Rev. D*, **70**, 084014. §9.5.2

Gleiser, R. J. (1984). A relation between the Szekeres quasispherical gravitational collapse solution and the Robinson–Trautman metrics, *Gen. Rel. Grav.*, **16**, 1039–1043. §22.8

Gleiser, R. J. and Dotti, G. (2006). Instability of the negative mass Schwarzschild naked singularity, *Class. Quantum Grav.*, **23**, 5063–5077. §8.4

Gleiser, R. and Pullin, J. (1989). Are cosmic strings stable topological defects? *Class. Quantum Grav.*, **6**, L141–L144. §20.4

Gödel, K. (1949). An example of a new type of cosmological solutions of Einstein's

498   *References*

field equations of gravitation, *Rev. Mod. Phys.*, **21**, 447–450. [*Gen. Rel. Grav*, **32**, 1409–1417, (2000).] §§13.1.3, 22.6

Goldberg, J. N. and Sachs, R. K. (1962). A theorem on Petrov types, *Acta Phys. Polon. Suppl.*, **22**, 13–23. §2.1.3

González, G. A. and Letelier, P. S. (2000). Rotating relativistic thin disks, *Phys. Rev. D*, **62**, 064025, 1–8. §12.1.3

Goode, S. W. and Wainwright, J. (1982). Singularities and evolution of the Szekeres cosmological models, *Phys. Rev. D*, **26**, 3315–3326. §22.8

Gott, J. R. (1974). Tachyon singularity: A spacelike counterpart of the Schwarzschild black hole, *Nuovo Cimento B*, **22**, 49–69. §§9.1.1, 16.4.1

Gowdy, R. H. (1971). Gravitational waves in closed universes, *Phys. Rev. Lett.*, **27**, 826–829. §22.4

Gowdy, R. H. (1974). Vacuum spacetimes with two-parameter spacelike isometry groups and compact invariant hypersurfaces: Topologies and boundary conditions, *Ann. Physics*, **83**, 203–241. §§22.4, B.3

Gowdy, R. H. (1975). Closed gravitational-wave universes: Analytic solutions with two-parameter symmetry, *J. Math. Phys.*, **16**, 224–226. §22.4

Graves, J. C. and Brill, D. R. (1960). Oscillatory character of Reissner–Nordström metric for an ideal charged wormhole, *Phys. Rev.*, **120**, 1507–1513. §9.2.2

Griffiths, J. B. (1972). Some physical properties of neutrino-gravitational fields, *Int. J. Theoret. Phys.*, **5**, 141–150. §18.5

Griffiths, J. B. (1974). Neutrino radiation in spherically-symmetric gravitational fields III, *Gen. Rel. Grav.*, **5**, 453–458. §9.5

Griffiths, J. B. (1991), *Colliding plane waves in general relativity*, (Oxford University Press). §§21, 22

Griffiths, J. B. (1993a). On the propagation of a gravitational wave in a stiff perfect fluid, *Class. Quantum Grav.*, **10**, 975–983. §21.7

Griffiths, J. B. (1993b). The collision of gravitational waves in a stiff perfect fluid, *J. Math. Phys.*, **34**, 4064–4069. §21.7

Griffiths, J. B. (2005). The stability of Killing–Cauchy horizons in colliding plane wave space-times, *Gen. Rel. Grav.*, **37**, 1119–1128. §21.7

Griffiths, J. B. and Docherty, P. (2002). A disintegrating cosmic string, *Class. Quantum Grav.*, **19**, L109–112. §19.2.3

Griffiths, J. B., Docherty, P. and Podolský, J. (2004). Generalized Kundt waves and their physical interpretation, *Class. Quantum Grav.*, **21**, 207–222. §18

Griffiths, J. B. and Halburd, R. G. (2007). The *C*-metric as a colliding plane wave space-time, *Class. Quantum Grav.*, **27**, 1049–1054. §21.5.4

Griffiths, J. B., Krtouš, P. and Podolský, J. (2006). Interpreting the *C*-metric, *Class. Quantum Grav.*, **23**, 6745–6766. §14

Griffiths, J. B. and Miccichè, S. (1997). The Weber–Wheeler–Bonnor pulse and phase shifts in gravitational soliton interactions, *Phys. Lett. A*, **233**, 37–42. §22.3

Griffiths, J. B. and Newing, R. A. (1974). Neutrino radiation in spherically-symmetric gravitational fields II, *Gen. Rel. Grav.*, **5**, 345–352. §9.5

Griffiths, J. B. and Podolský, J. (1997). Null multipole particles as sources of *pp*-waves, *Phys. Lett. A*, **236**, 8–10. §20.2

Griffiths, J. B. and Podolský, J. (1998). Interpreting a conformally flat pure radiation space-time, *Class. Quantum Grav.*, **15**, 3863–3871. §§18.2, 18.4.1

Griffiths, J. B. and Podolský, J. (2005). Accelerating and rotating black holes, *Class. Quantum Grav.*, **22**, 3467–3479. §§14.3, 16.1, 16.3

Griffiths, J. B. and Podolský, J. (2006a). Global aspects of accelerating and rotating black hole space-times, *Class. Quantum Grav.*, **23**, 555–568. §14.3

Griffiths, J. B. and Podolský, J. (2006b). A new look at the Plebański–Demiański family of solutions, *Int. J. Mod. Phys. D*, **15**, 335–369. §§16, 18.6

Griffiths, J. B. and Podolský, J. (2010). The Linet–Tian solution with a positive cosmological constant in four and higher dimensions, *Phys. Rev. D*, **81**, 064015. §10.2

Griffiths, J. B., Podolský, J. and Docherty, P. (2002). An interpretation of Robinson–Trautman type N solutions, *Class. Quantum Grav.*, **19**, 4649–4662. §19.2

Griffiths, J. B. and Santano-Roco, M. (2002). The characteristic initial value problem for colliding plane waves: The linear case, *Class. Quantum Grav.*, **19**, 4273–4286. §21.2

Griffiths, J. B. and Santos, N. O. (2010). A rotating cylinder in an asymptotically locally anti-de Sitter background. *Class. Quantum Grav.*, **27**, 125004. §§13.1.3, 22.6

Grillo, G. (1991). On a class of naked strong-curvature singularities, *Class. Quantum Grav.*, **8**, 739–749. §22.7

Grøn, Ø. and Eriksen, E. (1987). A dust-filled Kantowski–Sachs universe with $\Lambda > 0$, *Phys. Lett. A*, **121**, 217–220. §22.2

Grøn, Ø. and Hervik, S. (2007). *Einstein's general theory of relativity*, (Springer). §§4.5, 6.5

Grubišić, B. and Moncrief, V. (1993). Asymptotic behaviour of the $T^3 \times R$ Gowdy space-times, *Phys. Rev. D*, **47**, 2371–2382. §22.4

Gürses, M. (1977). Conformal uniqueness of the Schwarzschild interior metric, *Lett. Nuovo Cimento*, **18**, 327–328. §8.5

Hackmann, E. and Lämmerzahl, C. (2008). Geodesic equation in Schwarzschild–(anti-)de Sitter space-times: Analytical solutions and applications, *Phys. Rev. D*, **78**, 024035. §9.4

Haggag, S. and Desokey, F. (1996). Perfect-fluid sources for the Levi-Civita metric, *Class. Quantum Grav.*, **13**, 3221–3228. §10.2

Hájíček, P. (1970). Extension of the Taub and NUT spaces and extensions of their tangent bundles, *Commun. Math. Phys.*, **17**, 109. §12.2.1

Halilsoy, M. (1993). Interpolation of the Schwarzschild and Bertotti–Robinson solutions, *Gen. Rel. Grav.*, **25**, 275–280. §16.5

Hall, G. S. (2004). *Symmetries and curvature structure in general relativity*, (World Scientific). §2.1.4

Hanquin, J. L. and Demaret, J. (1983). Gowdy $S^1 \otimes S^2$ and S^3 inhomogeneous cosmological models, *J. Phys. A: Math. Gen.*, **16**, L5–L10. §22.4

Harrison, E. R. (1967). Classification of uniform cosmological models, *Mon. Not. R. Astron. Soc.*, **137**, 69–79. §§6.2, 6.3.2, 6.3.3

Hartle, J. B. and Hawking, S. W. (1972). Solutions of the Einstein–Maxwell equations with many black holes, *Commun. Math. Phys.*, **26**, 87–101. §§10.8, 13.3.2, 22.5

Harvey, A. (1990). Will the real Kasner metric please stand up, *Gen. Rel. Grav.*, **22**, 1433–1445. §22.1

Hauser, I. and Ernst, F. J. (1980). A homogeneous Hilbert problem for the Kinnersley–Chitre transformations, *J. Math. Phys.*, **21**, 1126–1140. §13.2.2

Hauser, I. and Ernst, F. J. (1989a). Initial value problem for colliding gravitational plane waves I, *J. Math. Phys.*, **30**, 872–887. §21.2

Hauser, I. and Ernst, F. J. (1989b). Initial value problem for colliding gravitational plane waves II, *J. Math. Phys.*, **30**, 2322–2336. §21.2

Hauser, I. and Ernst, F. J. (1990). Initial value problem for colliding gravitational plane waves III, *J. Math. Phys.*, **31**, 871–881. §21.2

Hauser, I. and Ernst, F. J. (1991). Initial value problem for colliding gravitational plane waves IV, *J. Math. Phys.*, **32**, 198–209. §21.2

Hauser, I. and Ernst, F. J. (2001). Proof of a generalized Geroch conjecture for the hyperbolic Ernst equation, *Gen. Rel. Grav.*, **33**, 195–293. §21.2

Havrdová, L. and Krtouš P. (2007). Melvin universe as a limit of the C-metric, *Gen. Rel. Grav.*, **39**, 291–296. §7.3

Hawking, S. W. (1983). The boundary conditions for gauged supergravity, *Phys. Lett. B*, **126**, 175–177. §5.2

Hawking, S. W. and Ellis, G. F. R. (1973). *The large scale structure of space-time*, (Cambridge University Press). §§1, 2.3, 4, 5.2, 5.4, 6.5, 9.5, 11.1, 22.6

Hawking, S. W. and Ross, S. F. (1995). Pair production of black holes on cosmic strings, *Phys. Rev. Lett.*, **75**, 3382–3385. §14.4

Hawking, S. W. and Ross, S. F. (1997). Loss of quantum coherence through scattering off virtual black holes, *Phys. Rev. D*, **56**, 6403–6415. §14.2

Hayward, S. A. (1989). Colliding waves and black holes, *Class. Quantum Grav.*, **6**, 1021–1032. §21.4

Hayward, S. A., Shirumizo, T. and Nakao, K. (1994). A cosmological constant limits the size of black holes, *Phys. Rev. D*, **49**, 5080–5095. §9.4.2

Heckmann, O. (1932). Die Ausdehnung der Welt in ihrer Abhängigkeit von der Zeit, *Nachr. Ges. Wiss. Göttingen*, 97–106. §6.3.3

Hellaby, C. (1996). The null and KS limits of the Szekeres model, *Class. Quantum Grav.*, **13**, 2537–2546. §22.8

Hellaby, C. and Krasiński, A. (2008). Physical and geometrical interpretation of the $\epsilon \leq 0$ Szekeres models, *Phys. Rev. D*, **77**, 023529. §22.8

Hernandez, W. C. (1967). Static, axially symmetric, interior solution in general relativity, *Phys. Rev.*, **153**, 1359–1363. §10.5

Herrera, L., Paiva, F. M. and Santos, N. O. (1999). The Levi-Civita space-time as a limiting case of the γ space-time, *J. Math. Phys.*, **40**, 4064–4071. §10.4

Herrera, L., Santos, N. O., Teixeira, A. F. F. and Wang, A. Z. (2001). On the interpretation of the Levi-Civita spacetime for $0 \leq \sigma < \infty$, *Class. Quantum Grav.*, **18**, 3847–3855. §10.2

Heusler, M. (1997). On the uniqueness of the Papapetrou–Majumdar metric, *Class. Quantum Grav.*, **14**, L129–L134. §22.5

Heusler, M. (1998). Stationary black holes: uniqueness and beyond, *Living Revs. in Rel.*, Irr-1998-6. §11.2

Hindmarsh, M. B. and Kibble, T. W. B. (1995). Cosmic strings, *Rep. Prog. Phys.*, **58**, 477–562. §3.4

Hiscock, W. A. (1981a). Models of evaporating black holes I, *Phys. Rev. D*, **23**, 2813–2822. §9.5.1

Hiscock, W. A. (1981b). Models of evaporating black holes II. Effects of the outgoing created radiation, *Phys. Rev. D*, **23**, 2823–2827. §9.5.1

Hiscock, W. A. (1981c). On the topology of charged spherical collapse, *J. Math. Phys.*, **22**, 215–218. §9.2.2

Hiscock, W. A. (1981d). On black holes in magnetic universes, *J. Math. Phys.*, **22**, 1828–1833. §9.3

Hobill, D., Bernstein, D., Welge, M. and Simkins, D. (1991). The Mixmaster cosmology as a dynamical system, *Class. Quantum Grav.*, **8**, 1155–1171. §22.1

Hobill, D., Burd, A. and Coley, A. A. eds. (1994). *Deterministic chaos in general relativity*, *NATO Adv. Sci. Inst. Ser. B Phys.*, **332**, (Plenum). §22.1

Hoenselaers, C. (1978). Directional singularities reexamined, *Progr. Theoret. Phys.*, **59**, 1170–1172. §10.4

Hoenselaers, C. (1990). Axisymmetric stationary vacuum solutions near a ring singularity, *Class. Quantum Grav.*, **7**, 581–584. §10.7

Hoenselaers, C. (1995). The Weyl solution for a ring in a homogeneous field, *Class. Quantum Grav.*, **12**, 141–148. §§10.6, 10.7

Hoenselaers, C. and Dietz, W. eds. (1984). *Solutions of Einstein's equations: techniques and results*, Lecture Notes in Physics No. 205. (Springer). §22

Hoenselaers, C. and Ernst, F. J. (1983). Remarks on the Tomimatsu–Sato metrics, *J. Math. Phys.*, **24**, 1817–1820. §13.2.2

Hoenselaers, C. and Ernst, F. J. (1990). Matching pp-waves to the Kerr metric, *J. Math. Phys.*, **31**, 144–146. §21.5.2

Hoenselaers, C. and Perjés, Z. (1993). Remarks on the Robinson–Trautman solutions, *Class. Quantum Grav.*, **10**, 375–383. §§19.1, 19.4

Hogan, P. A. (1992). A spherical gravitational wave in the de Sitter universe, *Phys. Lett. A*, **171**, 21–22. §§20.4, 20.5

Hogan, P. A. (1993). A spherical impulse gravity wave, *Phys. Rev. Lett.*, **70**, 117–118. §20.4

Hogan, P. A. (1994). Lorentz group and spherical impulsive gravity waves, *Phys. Rev. D*, **49**, 6521–6525. §20.4

Hogan, P. A. (1995). Imploding-exploding gravitational waves, *Lett. Math. Phys.*, **35**, 277–280. §20.4

Holst, S. and Peldán, P. (1997). Black holes and causal structure in anti-de Sitter isometric spacetimes, *Class. Quantum Grav.*, **14**, 3433–3452. §9.6

Hong, K. and Teo, E. (2003). A new form of the C-metric, *Class. Quantum Grav.*, **20**, 3269–3277. §14.1

Hong, K. and Teo, E. (2005). A new form of the rotating C-metric, *Class. Quantum Grav.*, **22**, 109–117. §§14.3, 16.3.2

Horský, J. and Novotný, J. (1982). On the analogy between the plane and the spherical solutions of Einstein equations in the vacuum, *Czech. J. Phys. B*, **32**, 1321–1324. §9.1.2

Horowitz, G. T. and Itzhaki, N. (1999). Black holes, shock waves, and causality in the AdS/CFT correspondence, *J. High Energy Phys.*, **2**, 154. §20.3

Hotta, M. and Tanaka, T. (1993). Shock-wave geometry with non-vanishing cosmological constant, *Class. Quantum Grav.*, **10**, 307–314. §20.3

Huang, C-G. and Liang, C-B. (1995). A torus-like black hole, *Phys. Lett. A*, **201**, 27–32. §9.6

Hubeny, V. E., Marolf, D. and Rangamani, M. (2010). Black funnels and droplets from the AdS C-metrics, *Class. Quantum Grav.*, **27**, 025001. §16.2

Ibañez, J. and Verdaguer, E. (1983). Soliton collision in general relativity, *Phys. Rev. Lett.*, **51**, 1313–1315. §22.4

Iguchi, H., Nakamura, T. and Nakao, K. I. (2002). Is dark energy the only solution to the apparent acceleration of the present universe? *Prog. Theor. Phys.*, **108**, 809–818. §22.7

Isenberg, J. and Moncrief, V. (1990). Asymptotic behavior of the gravitational field and the nature of singularities in Gowdy spacetimes, *Ann. Physics*, **199**, 84–122. §22.4

Ishikawa, K. and Miyashita, T. (1982). Classification of Petrov type D empty Einstein spaces with diverging null geodesic congruences, *Prog. Theor. Phys.*, **67**, 828–843. §16

Ishikawa, K. and Miyashita, T. (1983). Three classes of space-time with a uniformly accelerating and nonrotating gravitational source, *Gen. Rel. Grav.*, **15**, 1009–1026. §§14, 16.6, 19.3

Islam, J. N. (1985). *Rotating fields in general relativity*, (Cambridge University Press). §13.2.1, 13.2.2

Islam, J. N. (2002). *An introduction to mathematical cosmology, 2nd edition*, (Cambridge University Press). §6.5

Israel, W. (1958). Discontinuities in spherically symmetric gravitational fields and shells of radiation, *Proc. Roy. Soc. A*, **248**, 404–414. §9.5

Israel, W. (1966a). Singular hypersurfaces and thin shells in general relativity, *Nuovo Cimento B*, **44**, 1–14. §§2.1.5, 13.2.2, 20.6

Israel, W. (1966b). New interpretation of the extended Schwarzschild manifold, *Phys. Rev.*, **143**, 1016–1021. §9.2.2

Israel, W. (1967). Gravitational collapse of a radiating star, *Phys. Lett. A*, **24**, 184–186. §9.5

Israel, W. (1977). Line sources in general relativity, *Phys. Rev. D*, **15**, 935–941. §§3.4, 10.2, 10.6

Israel, W. and Khan, K. A. (1964). Collinear particles and Bondi dipoles in general relativity, *Nuovo Cimento*, **33**, 331–344. §§10.7, 15

Israel, W. and Wilson, G. A. (1972). A class of stationary electromagnetic vacuum fields, *J. Math. Phys.*, **13**, 865–867. §13.3.2

Ivanov, B. V. (2002). On rigidly rotating perfect fluid cylinders, *Class. Quantum Grav.*, **19**, 3851–3861. §13.1.3

Ivanov, B. V. (2005). No news for Kerr–Schild fields, *Phys. Rev. D*, **71**, 044012. §19.5.4

Jacob, U. and Piran, T. (2006). Embedding the Reissner–Nordström spacetime in Euclidean and Minkowski spaces, *Class. Quantum Grav.*, **23**, 4035–4045. §9.2.2

Jaklitsch, M. J., Hellaby, C. and Matravers, D. R. (1989). Particle motion in the spherically symmetric vacuum solution with positive cosmological constant, *Gen. Rel. Grav.*, **21**, 941–951. §9.4.1

Jensen, B. and Kučera, J. (1994). A reinterpretation of the Taub singularity, *Phys. Lett. A*, **195**, 111–115. §10.2

Jensen, B. and Soleng, H. H. (1992). General-relativistic model of a spinning cosmic string, *Phys. Rev. D*, **45**, 3528–3533. §3.4.1

Johansen, N. V. and Ravndal, F. (2006). On the discovery of Birkhoff's theorem, *Gen. Rel. Grav.*, **38**, 537–540. §8.1

Jordan, P., Ehlers, J. and Sachs, R. (1961). Beiträge zur Theorie des reinen Gravitationsstrahlung, *Akad. Wiss. Lit. Mainz, Abhandl. Math.-Nat. Kl*, **1**, 1–62. §2.1.3

Joshi, P. S. and Królak, A. (1996). Naked strong curvature singularities in Szekeres spacetimes, *Class. Quantum Grav.*, **13**, 3069–3074. §22.8

Kagramanova, V. G. and Ahmedov, B. J. (2006). On properties of vacuum axial symmetric spacetime of gravitomagnetic monopole in cylindrical coordinates, *Gen. Rel. Grav.*, **38**, 823–835. §12.1.1

Kaiser, N. (1982). Background radiation fields as a probe of the large-scale matter distribution in the universe, *Mon. Not. R. Astron. Soc.*, **198**, 1033–1052. §8.6

Kantowski, R. (1966). *Some relativistic cosmological models*, Ph.D. thesis, University of Texas at Austin. §22.2

Kantowski, R. (1969). Corrections in the luminosity-redshift relations of the homogeneous Friedmann models, *Astrophys. J.*, **155**, 89–103. §8.6

Kantowski, R. and Sachs, R. K. (1966). Some spatially homogeneous anisotropic relativistic cosmological models, *J. Math. Phys.*, **7**, 443–446. §22.2

Kar, S. C. (1926). Das Gravitationsfeld einer geladenen Edene, *Phys. Zeitschr.*, **27**, 208. §9.6

Karas, V. and Vokrouhlický, D. (1992). Chaotic motion of test particles in the Ernst space-time, *Gen. Rel. Grav.*, **24**, 729–743. §9.3

Kasner, E. (1921). Geometrical theorems on Einstein's cosmological equations, *Amer. J. Math.*, **43**, 217–221. §22.1

Kasner, E. (1925). An algebraic solution of the Einstein equations, *Trans. Am. Math. Soc.*, **27**, 101–105. §7.2.1

Kastor, D. and Traschen, J. (1993). Cosmological multi-black-hole solution, *Phys. Rev. D*, **47**, 5370–5375. §§9.6, 22.5

Kastor, D. and Traschen, J. (1996). Particle production and positive energy theorems for charged black holes in DeSitter, *Class. Quantum Grav.*, **13**, 2753–2761. §22.5

Katz, A. (1968). Derivation of Newton's law of gravitation from general relativity, *J. Math. Phys.*, **9**, 983–985. §10.6

Kermack, W. O. and McCrea W. H. (1933). On Milne's theory of world structure, *Mon. Not. R. Astron. Soc.*, **93**, 519–529. §6.3.1

Kerr, R. P. (1963). Gravitational field of a spinning mass as an example of algebraically special metrics, *Phys. Rev. Lett.*, **11**, 237–238. §11

Kerr, R. P. (2008). Discovering the Kerr and Kerr–Schild metrics, in *The Kerr spacetime: rotating black holes in general relativity*, eds. D. L. Wiltshire, M. Visser and S. M. Scott, (Cambridge University Press). §11.1

Khan, K. A. and Penrose, R. (1971). Scattering of two impulsive gravitational plane waves, *Nature*, **229**, 185–186. §21.3

Khlebnikov, V. I. (1978). Gravitational field of radiating non-twisting charged systems, I, *Teor. Mat. Fiz.*, **35**, 296–311; *Theoret. & Math. Phys.*, **35**, 470–479, (1978). §19.5.4

Khlebnikov, V. I. (1986). Gravitational radiation in electromagnetic universes, *Class. Quantum Grav.*, **3**, 169–173. §18.7

Khlebnikov, V. I. and Shelkovenko, A. É. (1976). On exact solutions of the Einstein–Maxwell equations in the Newman–Penrose formalism. II. Conformally plane metrics, *Sov. Phys. J.*, **19**, 960–962. §7.1

Kichenassamy, S. and Rendall, A. D. (1998). Analytic description of singularities in Gowdy spacetimes, *Class. Quantum Grav.*, **15**, 1339–1355. §22.4

Kinnersley, W. (1969a). Type D vacuum metrics, *J. Math. Phys.*, **10**, 1195–1203. §§16, 18.2, 18.6, A.4

Kinnersley, W. (1969b). Field of an arbitrarily accelerating point mass, *Phys. Rev.*, **186**, 1335–1336. §19.5.2

Kinnersley, W. (1975). Recent progress in exact solutions, in *General relativity and gravitation*, Proc. GR7, eds. G. Shaviv and J. Rosen, (Wiley), 109–135. §16.4

Kinnersley, W. and Kelley, E. F. (1974). Limits of the Tomimatsu–Sato gravitational field, *J. Math. Phys.*, **15**, 2121–2126. §13.2.2

Kinnersley, W. and Walker, M. (1970). Uniformly accelerating charged mass in general relativity, *Phys. Rev. D*, **2**, 1359–1370. §§14, 19.3

Klein, C. (2001). Exact relativistic treatment of stationary counterrotating dust disks: Boundary value problems and solutions, *Phys. Rev. D*, **63**, 064033. §13.2.2

Klösch, T. and Strobl, T. (1996). Explicit global coordinates for Schwarzschild and Reissner–Nordström solutions, *Class. Quantum Grav.*, **13**, 1191–1199. §9.2.2

Kodama, H. and Hikida, W. (2003). Global structure of the Zipoy–Voorhees–Weyl spacetime and the $\delta = 2$ Tomimatsu–Sato spacetime, *Class. Quantum Grav.*, **20**, 5121–5140. §10.4

Kolb, E. W. and Turner, M. S. (1990). *The early universe*, (Addison–Wesley, Redwood City). §6.5

Komar, A. (1959). Covariant conservation laws in general relativity, *Phys. Rev.*, **113**, 934–936. §8.1

Kompaneets, A. S. and Chernov, A. S. (1964). Solution of the gravitation equations for a homogeneous anisotropic model, *Zh. Eksp. Teor. Fiz.*, **47**, 1939–1944. English translation: *Sov. Phys. JETP*, **20**, 1303–1306 (1965). §22.2

Konkowski, D. A., Helliwell, T. M. and Shepley, L. C. (1985). Cosmologies with quasiregular singularities. I. Spacetimes and test waves, *Phys. Rev. D*, **31**, 1178–1194. §12.2.1

Kottler, F. (1918). Über die physikalischen Grundlagen der Einsteinschen Gravitationstheorie, *Ann. Physik*, **56** (361), 401–462. §9.4

Koutras, A. and McIntosh, C. (1996). A metric with no symmetries or invariants, *Class. Quantum Grav.*, **13**, L47–L49. §§18.2, 18.3.3

Kowalczyński, J. K. and Plebański, J. F. (1977). Metric and Ricci tensors for a certain class of space-times of type D, *Int. J. Theoret. Phys.*, **16**, 371–388; erratum **17**, 388 (1978). §16.1

Kozameh, C., Kreiss, H.-O. and Reula, O. (2008). On the well posedness of Robinson–Trautman–Maxwell solutions, *Class. Quant. Grav.*, **25**, 025004. §19.6

Kozameh, C., Newman, E. T. and Silva-Ortigoza, G. (2006). On the physical meaning of the Robinson–Trautman–Maxwell fields, *Class. Quant. Grav.*, **23**, 6599–6620. §19.6

Kramer, D. (1984). Kerr solution endowed with magnetic dipole moment, *Class. Quantum Grav.*, **1**, L45–L50. §13.3.2

Kramer, D. (1986). Two Kerr–NUT constituents in equilibrium, *Gen. Rel. Grav.*, **18**, 497–509. §13.2.2

Kramer, D. (1988). The nonlinear evolution equations for Robinson–Trautman space-times, in *Relativity today*, ed. Z. Perjés, (World Scientific), 71–83. §19.1

Kramer, D. and Neugebauer, G. (1980). The superposition of two Kerr solutions, *Phys. Lett. A*, **75**, 259–261. §13.2.2

Kramer, D., Stephani, H., MacCallum, M. and Herlt, E. (1980). *Exact solutions of Einstein's field equations, 1st edition*, (Cambridge University Press), (see Stephani *et al.*, 2003). §1

Kraniotis. G. V. (2004). Precise relativistic orbits in Kerr and Kerr–(anti) de Sitter spacetimes, *Class. Quantum Grav.*, **21**, 4743–4769. §11.1.5

Kraniotis. G. V. (2005). Frame dragging and bending of light in Kerr and Kerr–(anti) de Sitter spacetimes, *Class. Quantum Grav.*, **22**, 4391–4424. §11.1.5

Kraniotis. G. V. (2007). Periapsis and gravitomagnetic precessions of stellar orbits in Kerr and Kerr–de Sitter black hole spacetimes, *Class. Quantum Grav.*, **24**, 1775–1808. §11.1.5

Krasiński, A. (1975). Some solutions of the Einstein field equations for a rotating perfect fluid, *J. Math. Phys.*, **16**, 125–131. §§13.1.3, 22.6

Krasiński, A. (1997). *Inhomogeneous cosmological models*, (Cambridge University Press). §§1, 6.5, 8.6, 22.2, 22.7, 22.8

Krasiński, A. (2008). Geometry and topology of the quasi-plane Szekeres model, *Phys. Rev. D*, **78**, 064038. §22.8

Krasiński, A., Behr, C. G., Schücking, E., Estabrook, F. B., Wahlquist, H. D., Ellis, G. F. R., Jantzen, R. and Kundt, W. (2003). The Bianchi classification in the Schücking–Behr approach, *Gen. Rel. Grav.*, **35**, 475–489. §22.1

Krasiński, A. and Hellaby, C. (2001). Structure formation in the Lemaître–Tolman model, *Phys. Rev. D*, **65**, 023501. §22.7

Krasiński, A. and Hellaby, C. (2004). More examples of structure formation in the Lemaître–Tolman model, *Phys. Rev. D*, **69**, 023502. §22.7

Krtouš, P. (2005). Accelerated black holes in an anti-de Sitter universe, *Phys. Rev. D*, **72**, 124019. §§5.1, 5.4, 14.4

Krtouš, P. and Podolský, J. (2003). Radiation from accelerating black holes in a de Sitter universe, *Phys. Rev. D*, **68**, 024005. §§14.4, 19.3

Kruskal, M. D. (1960). Maximal extension of Schwarzschild metric, *Phys. Rev.*, **119**, 1743–1745. §8.3

Kuchowicz, B. (1971). General relativistic fluid spheres. IV. Differential equations for non-charged spheres of perfect fluid, *Acta Phys. Polon. B*, **2**, 657–667. §8.5

Kundt, W. (1956). Trägheitsbahnen in einem von Gödel angegebenen kosmologischen Modell, *Z. Physik*, **145**, 611–620. §22.6

Kundt, W. (1961). The plane-fronted gravitational waves, *Z. Physik*, **163**, 77–86. §§17, 18

Kundt, W. (1962). Exact solutions of the field equations: twist-free pure radiation fields, *Proc. Roy. Soc. A*, **270**, 328–334. §18

Kundt, W. and Thompson, A. (1962). Le tenseur de Weyl et une congruence associée de géodésiques isotropes sans distortion, *C.R. Acad. Sci.*, **254**, 4257–4259. §2.1.3

Kundt, W. and Trümper, M. (1962). Beiträge zur Theorie der Gravitations-Strahlungsfelder, *Akad. Wiss. Lit. Mainz, Abhandl. Math. Nat. Kl.* no. 12. §§2.1.3, 18.1

Kunzinger, M. and Steinbauer, R. (1999a). A rigorous solution concept for geodesic and geodesic deviation equations in impulsive gravitational waves, *J. Math. Phys.*, **40**, 1479–1489. §20.2

Kunzinger, M. and Steinbauer, R. (1999b). A note on the Penrose junction conditions, *Class. Quantum Grav.*, **16**, 1255–1264. §20.2

Kuroda, Y. (1984). Vaidya spacetime as an evaporating black hole, *Progr. Theoret. Phys.*, **71**, 1422–1425. §9.5.1

Lake, K. (2006). Maximally extended, explicit and regular coverings of the Schwarzschild–de Sitter vacua in arbitrary dimension, *Class. Quantum Grav.*, **23**, 5883–5895. §9.2.2

Lake, K. and Roeder, R. C. (1977). Effects of a nonvanishing cosmological constant on the spherically symmetric vacuum manifold, *Phys. Rev. D*, **15**, 3513–3519. §9.4

Lanczos, C. (1922). Bemerkung zur de Sitterschen Welt, *Phys. Zeitschr.*, **23**, 539–543. §§4.1, 6.3

Lanczos, C. (1924). Über eine stationäre kosmologie im sinne der Einsteinischen Gravitationstheories, *Z. Physik.*, **21**, 73. §13.1.1

Landau, L. D. and Lifshitz, E. M. (2000). *The classical theory of fields, 4th revised English edition*, (Butterworth–Heinemann, Oxford). §6.5

Lathrop, J. D. and Orsene, M. S. (1980). Dust cylinders in static space-times, *J. Math. Phys.*, **21**, 152–153. §10.2

Laue, H. and Weiss, M. (1977). Maximally extended Reissner–Nordström manifold with cosmological constant, *Phys. Rev. D*, **16**, 3376–3379. §§9.4.1, 9.6

Lawitzky, G. (1980). Newtonian analogs of Szekeres' space-times, *Gen. Rel. Grav.*, **12**, 903–912. §22.8

Lazkoz, R. and Valiente-Kroon, J. A. (2000). Boost-rotation symmetric type D radiative metrics in Bondi coordinates, *Phys. Rev. D*, **62**, 084033. §16.6

Ledvinka, T. (1998). *Thin disks as sources of stationary axisymmetric electro-vacuum spacetimes*, Ph.D. thesis, Charles University in Prague. §13.2.2

Lemaître, G. (1925). Note on de Sitter's universe, *J. Math. and Physics (M.I.T.)*, **4**, 189–192. §6.3

Lemaître, G. (1927). Un Univers homogène de masse constante et de rayon croissant rendant compte de la vitesse radiale des nébuleuses extragalactiques, *Ann. Soc. Sci. Bruxelles A*, **47**, 19–59. §6.1
English translation: *Mon. Not. R. Astron. Soc.*, **91**, 483–490, (1931).

Lemaître, G. (1933). L'Univers en expansion, *Ann. Soc. Sci. Bruxelles A*, **53**, 51–85. §§6.3.3, 22.7
English translation with notes: *Gen. Rel. Grav.*, **29**, 637–680, (1997).

Lemos, J. P. S. (1988). Remarkable properties of the limiting counter rotating disc, *Mon. Not. R. Astron. Soc.*, **230**, 451–456. §10.7

Lemos, J. P. S. (1995a). Two-dimensional black holes and planar general relativity, *Class. Quantum Grav.*, **12**, 1081–1086. §9.6

Lemos, J. P. S. (1995b). Three dimensional black holes and cylindrical general relativity, *Phys. Lett. B*, **353**, 46–51. §9.6

Lemos, J. P. S. and Letelier, P. S. (1994). Exact general relativistic thin disks around black holes, *Phys. Rev. D*, **49**, 5135–5143. §10.7

Lemos, J. P. S. and Zanchin, V. T. (1996). Rotating charged black strings and three-dimensional black holes, *Phys. Rev. D*, **54**, 3840–3853. §9.6

Leroy, J. (1978). Sur une classe d'espaces-temps solutions des equations d'Einstein–Maxwell, *Bull. Acad. Roy. Belg. Cl. Sci.*, **64**, 130. §16.1

Letelier, P. S. and Oliveira, S. R. (1998). Superposition of Weyl solutions: the equilibrium forces, *Class. Quantum Grav.*, **15**, 421–433. §10.7

Letelier, P. S. and Oliveira, S. R. (2001). Uniformly accelerated black holes, *Phys. Rev. D*, **64**, 064005. §§14.3, 16.3.2

Levi-Civita, T. (1917). Realtà fisica di alcuni spazi normali del Bianchi, *Rend. Accad. Lincei*, **26**, 519–531. §7.1

Levi-Civita, T. (1918). ds^2 einsteiniani in campi newtoniani. I, *Rend. Accad. Lincei*, **27**, 343. §10, 14

Levi-Civita, T. (1919). ds^2 einsteiniani in campi newtoniani. IX: L'analogo del potenziale logaritmico, *Rend. Accad. Lincei*, **28**, 101–109. §§7.3, 10.2

Levy, H. and Robinson, W. J. (1963). The rotating body problem, *Proc. Camb. Phil. Soc.*, **60**, 279–285. §13

Lewandowski, J. (1992). Reduced holonomy group and Einstein equations with a cosmological contant, *Class. Quantum Grav.*, **9**, L147–L151. §18.7

Lewis, T. (1932). Some special solutions of the equations of axially symmetric gravitational fields, *Proc. Roy. Soc. A*, **136**, 176–192. §13

Lichnerowicz, A. (1955). *Théories relativistes de la gravitation et de l'électro-magnétisme*, (Masson, Paris). §§2.1.5, 20

Lindquist, R. W., Schwartz, R. A. and Misner, C. W. (1965). Vaidya's radiating Schwarzschild metric, *Phys. Rev. B*, **137**, 1364–1368. §9.5

Linet, B. (1986). The static, cylindrically symmetric strings in general relativity with cosmological constant, *J. Math. Phys.*, **27**, 1817–1818. §10.2

Lorenz, D. (1983). Exact Bianchi–Kantowski–Sachs solutions of Einstein's field equations, *J. Phys. A*, **16**, 575–584. §22.2

Louko, J. (1987). Fate of singularities in Bianchi type-III quantum cosmology, *Phys. Rev. D*, **35**, 3760–3767. §9.1.1

Loustó, C. O. and Sánchez, N. (1989). Gravitational shock waves of ultra-high energetic particles on curved space-times, *Phys. Lett. B*, **220**, 55–60. §20.6

Loustó, C. O. and Sánchez, N. (1992). The ultra-high limit of the boosted Kerr–Newman geometry and the scattering of spin-$\frac{1}{2}$ particles, *Nucl. Phys. B*, **383**, 377–394. §20.2

Loustó, C. O. and Sánchez, N. (1995). Scattering at the Planck scale, in *Second Paris cosmology colloquium*, eds. H. J. de Vega and N. Sánchez, (World Scientific), 339–370. §20.2

Lukács, B., Perjés, Z., Porter, J., and Sebestyén, Á. (1984). Lyapunov functional approach to radiative metrics, *Gen. Rel. Grav.*, **16**, 691–701. §19.4

Lynden-Bell, D. and Nouri-Zonoz, M. (1998). Classical monopoles: Newton, NUT space, gravitomagnetic lensing, and atomic spectra, *Rev. Mod. Phys.*, **70**, 427–445. §12.1.3

Lynden-Bell, D. and Pineault, S. (1978). Relativistic disks – I. Counter-rotating disks, *Mon. Not. R. Astron. Soc.*, **185**, 679–694. §10.7

MacCallum, M. A. H. (1973). Cosmological models from a geometrical point of view, in *Cargése lectures in physics*, vol. 6, ed. E. Schatzman, (Gordon and Breach), 61–174. §§22.1, 22.2

MacCallum, M. A. H. (1994). Relativistic cosmologies, in *Deterministic chaos in general relativity, NATO Adv. Sci. Inst. Ser. B Phys.*, **332**, eds. D. Hobill *et al.*, (Plenum), 179–201. §22.1

MacCallum, M. A. H. (1998). Hypersurface-orthogonal generators of an orthogo-nally transitive G_2I, topological identifications, and axially and cylindrically symmetric spacetimes, *Gen. Rel. Grav.*, **30**, 131–150. §22.3

MacCallum, M. A. H. (2006). On singularities, horizons, invariants, and the results of Antoci, Liebscher and Mihich (GRG 38, 15 (2006) and earlier), *Gen. Rel. Grav.*, **38**, 1887–1899. §2.3.1

MacCallum, M. A. H. and Santos, N. O. (1998). Stationary and static cylindrically symmetric Einstein spaces of the Lewis form, *Class. Quantum Grav.*, **15**, 1627–1636. §§13.1.1, 13.1.3, 22.3

Maeda, K. (1989). Inflation and cosmic no hair conjecture, in *The fifth Marcel Grossmann meeting*, eds. D. G. Blair and M. J. Buckingham, (World Scientific), 145–155. §19.4

Majumdar, S. D. (1947). A class of exact solutions of Einstein's field equations, *Phys. Rev.*, **72**, 390–398. §22.5

Manko, V. S. (1993). New generalization of the Kerr metric referring to a magne-tized spinning mass, *Class. Quantum Grav.*, **10**, L239–L242. §13.3.2

Manko, V. S. (2007). Double-Reissner–Nordström solution and the interaction force between two spherical charged masses in general relativity, *Phys. Rev. D*, **76**, 124032. §10.8

Manko, V. S., Martín, J. and Ruiz, E. (1994). Metric of two arbitrary Kerr–Newman sources located on the symmetry axis, *J. Math. Phys.*, **35**, 6644–6657. §13.3.2

Manko, V. S. and Ruiz, E. (2001). Exact solution of the double-Kerr equilibrium problem, *Class. Quantum Grav.*, **18**, L11–L15. §13.2.2

Manko, V. S. and Sibgatullin, N. R. (1992a). Exact solution of the Einstein-Maxwell equations for the exterior gravitational field of a magnetized rotating mass, *Phys. Rev. D*, **46**, R4122–R4124. §13.3.2

Manko, V. S. and Sibgatullin, N. R. (1992b). Kerr metric endowed with magnetic dipole moment, *Class. Quantum Grav.*, **9**, L87–L92. §13.3.2

Manko, V. S. and Sibgatullin, N. R. (1993). Kerr–Newman metric endowed with magnetic dipole moment, *J. Math. Phys.*, **34**, 170–177. §13.3.2

Mann, R. B. (1997). Pair production of topological anti-de Sitter black holes, *Class. Quantum Grav.*, **14**, L109–L114. §§9.6, 14.4, 19.3

Mann, R. B. (1998). Charged topological black hole pair creation, *Nucl. Phys. B*, **516**, 357–381. §19.3

Mann, R. B. and Ross, S. F. (1995). Cosmological production of charged black hole pairs, *Phys. Rev. D*, **52**, 2254–2265. §14.4

Marder, L. (1958). Gravitational waves in general relativity I. Cylindrical waves, *Proc. Roy. Soc. A*, **244**, 524–537. §§10.2, 22.3

Marek, J. J. J. (1967). Static axisymmetric interior solution in general relativity, *Phys. Rev.*, **163**, 1373–1376. §10.5

Mariot, L. (1954). Le champ électromagnétique singulier, *C. R. Acad. Sci.*, **238**, 2055–2056. §2.2.1

Mars, M. (2001). On the uniqueness of the Einstein–Straus model, *Class. Quantum Grav.*, **18**, 3645–3663. §8.6

Martins, M. A. P. (1996). The sources of the A and B degenerate static vacuum fields, *Gen. Rel. Grav.*, **28**, 1309–1320. §§9.1.1, 16.4.1

Mashhoon, B., McClune, J. C. and Quevedo, H. (2000). On a class of rotating gravitational waves, *Class. Quantun Grav.*, **17**, 533–549. §22.3

Matzner, R. A. and Tipler, F. J. (1984). Metaphysics of colliding self-gravitating plane waves, *Phys. Rev. D*, **29**, 1575–1583. §21.3

Matzner, R. A., Zamorano, N. and Sandberg, V. D. (1979). Instability of the Cauchy horizon of Reissner–Nordström black holes, *Phys. Rev. D*, **19**, 2821–2826. §9.2.2

Mazur, P. O. (1986). Spinning cosmic strings and quantization of energy, *Phys. Rev. Lett.*, **57**, 929–932. §3.4.1

McLenaghan, R. G., Tariq, N. and Tupper, B. O. J. (1975). Conformally flat solutions of the Einstein–Maxwell equations for null electromagnetic fields, *J. Math. Phys.*, **16**, 829–831. §§7.1, 18.3.3

Meinel, R. (2004). Quasistationary collapse to the extreme Kerr black hole, *Ann. Physik*, **13**, 600–603. §11.1.5

Mellor, F. and Moss, I. (1989). Black holes and gravitational instantons, *Class. Quantum Grav.*, **6**, 1379–1385. §9.6

Mellor, F. and Moss, I. (1990). Stability of black holes in de Sitter space, *Phys. Rev. D*, **41**, 403–409. §9.6

Mellor, F. and Moss, I. (1992). A reassessment of the stability of the Cauchy horizon in de Sitter space, *Class. Quantum Grav.*, **9**, L43–L46. §9.6

Melvin, M. A. (1964). Pure magnetic and electric geons, *Phys. Lett.*, **8**, 65–68. §7.3

Melvin, M. A. (1965). Dynamics of cylindrical electromagnetic universe, *Phys. Rev.*, **139**, B225–B243. §7.3

Melvin, M. A. and Wallingford J. S. (1966). Orbits in a magnetic universe, *J. Math. Phys.*, **7**, 333–340. §7.3

Mena, F. C., Natário, J. and Tod, P. (2008). Avoiding closed timelike curves with a collapsing rotating null dust shell, *Class. Quantum Grav.*, **25**, 045016. §3.4.1

Miller, J. G., Kruskal, M. D. and Godfrey, B. B. (1971). Taub–NUT (Newman, Unti, Tamburino) metric and incompatible extensions, *Phys. Rev. D*, **4**, 2945–2948. §§12.1.5, 12.2

Milne, E. A. (1933). World-structure and the expansion of the universe, *Zeitschrift für Astrophysik*, **6**, 1–95. §6.3.1

Milson, R. and Pelavas, N. (2008). The type N Karlhede bound is sharp, *Class. Quantum Grav.*, **25**, 012001. §18.3.2

Misner, C. W. (1963). The flatter regions of Newman, Unti and Tamburino's generalized Schwarzschild space, *J. Math. Phys.*, **4**, 924–937. §§12, B.3

Misner, C. W. (1967). Taub–NUT space as a counter-example to almost anything, in *Relativity theory and astrophysics I. Relativity and cosmology*, ed. J. Ehlers, Amer. Math. Soc. Lectures in Applied Mathematics, **8**, (AMS, New York), 160–169. §12

Misner, C. W. (1969). Quantum Cosmology. I, *Phys. Rev.*, **186**, 1319–1327. §22.1

Misner, C. W. (1994). The mixmaster cosmological metrics, in *Deterministic chaos in general relativity*, NATO Adv. Sci. Inst. Ser. B Phys., **332**, eds. D. Hobill *et al.*, (Plenum), 317–328. §22.1

Misner, C. W. and Taub, A. H. (1968). A singularity-free empty universe, *Zh. Eks. Teor. Fiz.*, **55**, 233–255; *Sov. Phys. JETP*, **28**, 122–133, (1969). §12.2

Misner, C. W., Thorne, K. S. and Wheeler, J. A. (1973). *Gravitation*, (W. H. Freeman and Co., San Francisco). §§6.5, 8.5, 11.1, 17.2

Møller, C. (1972), *The theory of relativity*, 2nd ed. (Oxford University Press). §4.5

Moncrief, V. (1981). Infinite-dimensional family of vacuum cosmological models with Taub-NUT (Newman-Unti-Tamburino)-type extensions, *Phys. Rev. D*, **23**, 312–315. §22.4

Morgan, T. and Morgan, L. (1969). The gravitational field of a disc, *Phys. Rev.*, **183**, 1097–1101. §10.7

Mottola, E. (1985). Particle creation in de Sitter space, *Phys. Rev. D*, **31**, 754–766. §4.5

Mustapha, M., Hellaby, C. and Ellis, G. F. R. (1997). Large scale inhomogeneity versus source evolution – Can we distinguish them observationally? *Mon. Not. R. Astron. Soc.*, **292**, 817–830. §22.7

Nagar, A. and Rezzolla, L. (2005). Gauge-invariant non-spherical metric perturbations of Schwarzschild black-hole spacetimes, *Class. Quantum Grav.*, **22**, R167–R192. §8.2

Nakao, K. (1992). The Oppenheimer–Snyder space-time with a cosmological constant, *Gen. Rel. Grav.*, **24**, 1069–1081. §9.4.1

Nariai, H. (1951). On a new cosmological solution of Einstein's field equations of gravitation, *Sci. Rep. Tôhoku Univ.*, **35**, 62–67. §7.2.1

Neugebauer, G., Kleinwächter, A. and Meinel, R. (1996). Relativistically rotating dust, *Helv. Phys. Acta*, **69**, 472–489. §13.2.2

Neugebauer, G. and Meinel, R. (1993). The Einsteinian gravitational field of the rigidly rotating disc of dust, *Astrophys. J.*, **414**, L97–L99. §13.2.2

Neugebauer, G. and Meinel, R. (1994). General relativistic gravitational field of a rigidly rotating disk of dust: Axis potential, disk metric, and surface mass density, *Phys. Rev. Lett.*, **73**, 2166–2168. §13.2.2

Neugebauer, G. and Meinel, R. (1995). General relativistic gravitational field of a rigidly rotating disk of dust: Solution in terms of ultraelliptic functions, *Phys. Rev. Lett.*, **75**, 3046–3047. §13.2.2

Newman, E. T., Couch, E., Chinnapared, K., Exton, E., Prakash, A. and Torrence, R. (1965). Metric of a rotating, charged mass, *J. Math. Phys.*, **6**, 918–919. §11.2

Newman, E. and Penrose, R. (1962). An approach to gravitational radiation by a method of spin coefficients, *J. Math. Phys.*, **3**, 566–579; **4**, 998, (1963). §§2.1.1, 2.1.3

Newman, E. T. and Tamburino, L. A. (1962). Empty space metrics containing hypersurface orthogonal geodesic rays, *J. Math. Phys.*, **3**, 902–907. §19.1

Newman, E. T., Tamburino, L. A. and Unti, T. (1963). Empty-space generalisation of the Schwarzschild metric, *J. Math. Phys.*, **4**, 915–923. §§12, A.4

Newman, E. T. and Unti, T. W. J. (1963). A class of null flat-space coordinate systems, *J. Math. Phys.*, **4**, 1467–1469. §19.5.2

Newman, R. P. A. C. (1986). Strengths of naked singularities in Tolman–Bondi spacetimes, *Class. Quantum Grav.*, **3**, 527–539. §22.7

Nolan, B. C. (1999). Strengths of singularities in spherical symmetry, *Phys. Rev. D*, **60**, 024014. §22.7

Nolan, B. C. (2001). Sectors of spherical homothetic collapse, *Class. Quantum Grav.*, **18**, 1651–1675. §9.5.2

Nolan, B. C. (2007). Odd-parity perturbations of self-similar Vaidya spacetime, *Class. Quantum Grav.*, **24**, 177–200. §9.5.2

Nolan, B. C. and Debnath, U. (2007). Is the shell-focusing singularity of Szekeres space-time visible? *Phys. Rev. D*, **76**, 104046. §22.8

Nolan, B. C. and Vera, R. (2005). Axially symmetric equilibrium regions of Friedmann–Lemaître–Robertson–Walker universes, *Class. Quantum Grav.*, **22**, 4031–4050. §8.6

Nolan, B. C. and Waters, T. J. (2005). Even perturbations of self-similar Vaidya space-time, *Phys. Rev. D*, **71**, 104030. §9.5.2

Nordström, G. (1918). On the energy of the gravitational field in Einstein's theory, *Proc. Kon. Ned. Akad. Wet.*, **20**, 1238. §9.2

Novello, M., Svaiter, N. F. and Guimarães, M. E. X. (1993). Synchronized frames for Gödel's universe, *Gen. Rel. Grav.*, **25**, 137–164. §22.6

Novikov, I. D. (1966). Change of relativistic collapse into anticollapse and kinematics of a charged sphere, *Zh. Eksp. Teor. Fiz. Pis'ma*, **3**, 223–227: *JETP Lett.*, **3**, 142–144. §9.2

Nutku, Y. (1991). Spherical shock waves in general relativity, *Phys. Rev. D*, **44**, 3164–3168. §19.2.3

Nutku, Y. and Halil, M. (1977). Colliding impulsive gravitational waves, *Phys. Rev. Lett.*, **39**, 1379–1382. §21.3

Nutku, Y. and Penrose, R. (1992). On impulsive gravitational waves, *Twistor Newsletter*, No. 34, 11 May, 9–12. §20.4

O'Brien, S. and Synge, J. L. (1952). Jump conditions at discontinuities in general relativity, *Commun. Dublin Inst. Adv. Stud. A*, no 9. §§2.1.5, 20.6, 21.1

O'Neill, B. (1995). *Geometry of Kerr black holes*, (A. K. Peters, Wellesley, Mass.). §11.1

Opher, R., Santos, N. O. and Wang, A. (1996). Geodesic motion and confinement in van Stockum space-time, *J. Math. Phys.*, **37**, 1982–1990. §13.1.2

Oppenheimer, J. R. and Snyder, H. (1939). On continued gravitational contraction, *Phys. Rev.*, **56**, 455–459. §8.5

Ortaggio, M. (2002). Impulsive waves in the Nariai universe, *Phys. Rev. D*, **65**, 084046. §§7.2.1, 20.6

Ortaggio, M. (2004). Ultrarelativistic black hole in an external electromagnetic field and gravitational waves in the Melvin universe, *Phys. Rev. D*, **69**, 064034. §7.3

Ortaggio, M and Podolský, J. (2002). Impulsive waves in electrovac direct product spacetimes with Λ, *Class. Quantum Grav.*, **19**, 5221–5227. §§7.2, 18.6, 20.6

Ozsváth, I. (1965). New homogeneous solutions of Einstein's field equations with incoherent matter obtained by a spinor technique, *J. Math. Phys.*, **6**, 590–610. §22.6

Ozsváth, I. (1987). All homogeneous solutions of Einstein's vacuum field equations with a non vanishing cosmological term, in *Gravitation and geometry*, eds. W. Rindler and A. Trautman, (Bibliopolis), 309–340. §5.4

Ozsváth, I., Robinson, I. and Rózga, K. (1985). Plane-fronted gravitational and electromagnetic waves in spaces with cosmological constant, *J. Math. Phys.*, **26**, 1755–1761. §18

Ozsváth, I. and Schücking, E. (2001). Approaches to Gödel's rotating universe, *Class. Quantum Grav.*, **18**, 2243–2252. §22.6

Ozsváth, I. and Schücking, E. (2003). Gödel's trip, *Am. J. Phys.*, **71**, 801–805. §22.6

Paolino, A. and Pizzi, M. (2008). Electric force lines of the double Reissner–Nordstrom exact solution, *Int. J. Mod. Phys. D*, **17**, 1159–1177. §10.8.1

Papacostas, T. and Xanthopoulos, B. C. (1989). Collisions of gravitational and electromagnetic waves that do not develop curvature singularities, *J. Math. Phys.*, **30**, 97–103. §21.5.4

Papadopoulos, D., Stewart, B. and Witten, L. (1981). Some properties of a particular static, axially symmetric space-time, *Phys. Rev. D*, **24**, 320–326. §10.4

Papapetrou, A. (1947). A static solution of the equations of the gravitational field for an arbitrary charge distribution, *Proc. Roy. Irish Acad.*, **51**, 191–205. §22.5

Papapetrou, A. (1966). Champs gravitationnels stationnaires à symétrie axiale, *Ann. Inst. H. Poincaré A*, **4**, 83–105. §13

Papapetrou, A. (1985). Formation of a singularity and causality, in *A random walk in relativity and cosmology*, eds. N. Dadhich, J. Krishna Rao, J. V. Narlikar and C. V. Vishveshwara, (Wiley Eastern Ltd.), 184–191. §9.5.2

Partovi, M. H. and Mashhoon, B. (1984). Toward verification of large-scale homogeneity in cosmology, *Astrophys. J.*, **276**, 4–12. §22.7

Peacock, J. A. (1999). *Cosmological physics*, (Cambridge University Press). §6.5

Peebles, P. J. E. (1993). *Principles of physical cosmology*, (Princeton University Press). §6.5

Pelavas, N., Neary, N. and Lake, K. (2001). Properties of the instantaneous ergo surface of a Kerr black hole, *Class. Quantum Grav.*, **18**, 1319–1331. §11.1.3

Penrose, R. (1963). Asymptotic properties of fields and space-times, *Phys. Rev. Lett.*, **10**, 66–68. §§3.3, 8.4

Penrose, R. (1964). Conformal treatment of infinity, in *Relativity, groups and topology, Les Houches 1963*, eds. C. DeWitt and B. DeWitt, (Gordon and Breach), 563–584. §§2.3, 3.2, 3.3, 6.4, 6.5

Penrose, R. (1965a). A remarkable property of plane waves in general relativity, *Rev. Mod. Phys.*, **37**, 215–220. §17.5.2

Penrose, R. (1965b). Zero rest-mass fields including gravitation: asymptotic behaviour, *Proc. Roy. Soc. A*, **284**, 159–203. §§2.3.3, 8.4

Penrose, R. (1966). General-relativistic energy flux and elementary optics, in *Perspectives in geometry and relativity*, ed. B. Hoffmann, (Indiana University Press), 259–274. §17.2

Penrose, R. (1968a). Structure of space-time, in *Batelle rencontres* (1967 lectures in mathematics and physics), eds. C. M. DeWitt and J. A. Wheeler, (Benjamin, New York), 121–235. §§2.3, 3.3, 4.5, 6.4, 6.5, 20

Penrose, R. (1968b). Twistor quantisation and curved space-time, *Int. J. Theoret. Phys.*, **1**, 61–99. §§20.1, 20.2

Penrose, R. (1969). Gravitational collapse: The role of general relativity, *Rivista del Nuovo Cimento, numero speziale*, **1**, 252–276; Reprinted in *Gen. Rel. Grav.*, **34**, 1141–1165, (2002). §8.2

Penrose, R. (1972). The geometry of impulsive gravitational waves, in *General relativity: papers in honour of J. L. Synge*, ed. L. O'Raifeartaigh, (Oxford University Press), 101–115. §§2.1.5, 20

Penrose, R. and Rindler, W. (1984). *Spinors and space-time, volume 1: Two-spinor calculus and relativistic fields*, (Cambridge University Press). §20.4

Penrose, R. and Rindler, W. (1986). *Spinors and space-time, volume 2: Spinor and twistor methods in space-time geometry*, (Cambridge University Press). §§2.1.2, 2.1.3, 2.3, 6.4

Peres, A. (1959). Some gravitational waves, *Phys. Rev. Lett.*, **3**, 571–572. §17.1

Peres, A. (1960). Null electromagnetic fields in general relativity theory, *Phys. Rev.*, **118**, 1105–1110. §17.3

Peres, A. (1970). Gravitational field of a tachyon, *Phys. Lett. A*, **31**, 361–362. §9.1.1

Perjés, Z. (1971). Solutions of the coupled Einstein–Maxwell equations representing the fields of spinning sources, *Phys. Rev. Lett.*, **27**, 1668–1670. §13.3.2

Perry, G. P. and Cooperstock, F. I. (1997). Electrostatic equilibrium of two spherical charged masses in general relativity, *Class. Quantum Grav.*, **14**, 1329–1345. §10.8.1

Petrov, A. Z. (1962). Gravitational field geometry as the geometry of automorphisms, in *Recent developments in general relativity*, (Pergamon Press, Oxford), 379–386. §13.1.2

Pfarr, J. (1981). Time travel in Gödel's space, *Gen. Rel. Grav.*, **13**, 1073–1091. §22.6

Philbin, T. G. (1996). Perfect-fluid cylinders and walls - sources for the Levi-Civita spacetime, *Class. Quantum Grav.*, **13**, 1217–1232. §10.2

Pichon, C. and Lynden-Bell, D. (1996). New sources for Kerr and other metrics: rotating relativistic disks with pressure support, *Mon. Not. R. Astron. Soc.*, **280**, 1007–1026. §11.1.5

Piran, T., Safier, P. N. and Stark, R. F. (1985). General numerical solution of cylindrical gravitational waves, *Phys. Rev. D*, **32**, 3101–3107. §22.3

Pirani, F. A. E. (1957). Invariant formulation of gravitational radiation theory, *Phys. Rev.*, **105**, 1089–1099. §17.2

Pirani, F. A. E. (1965). Introduction to gravitational radiation theory, in *Lectures on general relativity*, 249–373. Brandeis Summer Institute of Theoretical

Physics, **1**, 346, (Prentice Hall, Englewood Cliffs, New Jersey). §§2.1.3, 2.1.5, 20.6

Plebański, J. F. (1975). A class of solutions of the Einstein–Maxwell equations, *Ann. Physics*, **90**, 196–255. §§16.4, 16.5, 18.6

Plebański, J. F. (1979). The nondiverging and nontwisting type D electrovac solutions with λ, *J. Math. Phys.*, **20**, 1946–1962. §§7.2.2, 7.3

Plebański, J. and Demiański, M. (1976). Rotating, charged and uniformly accelerating mass in general relativity, *Ann. Physics*, **98**, 98–127. §§14.3, 14.4, 16, 18.2, 18.6

Plebański, J. F. and Hacyan, S. (1979). Some exceptional electrovac type D metrics with cosmological constant, *J. Math. Phys.*, **20**, 1004–1010. §7.2

Plebański, J. and Krasiński, A. (2006). *An introduction to general relativity and cosmology*, (Cambridge University Press). §11.1.5

Podolský, J. (1998a). Interpretation of the Siklos solutions as exact gravitational waves in the anti-de Sitter universe, *Class. Quantum Grav.*, **15**, 719–733. §§18.2, 18.3.2

Podolský, J. (1998b). Nonexpanding impulsive gravitational waves, *Class. Quantum Grav.*, **15**, 3229–3239. §§20.2, 20.3

Podolský, J. (1999). The structure of the extreme Schwarzschild-de Sitter spacetime, *Gen. Rel. Grav.*, **31**, 1703–1725. §9.4.2

Podolský, J. (2001). Exact non-singular waves in the anti-de Sitter universe, *Gen. Rel. Grav.*, **33**, 1093–1113. §18.4.4

Podolský, J. (2002a). Accelerating black holes in anti-de Sitter universe, *Czech. J. Phys.*, **52**, 1–10. §§5.1, 14.4

Podolský, J. (2002b). Exact impulsive gravitational waves in spacetimes of constant curvature, in *Gravitation: following the Prague inspiration*, eds. O. Semerák, J. Podolský and M. Žofka, (World Scientific), 205–246. §20

Podolský, J. (2008). Photon rockets in the (anti-)de Sitter universe, *Phys. Rev. D*, **78**, 044029. §19.5.4

Podolský, J. and Beláň, M. (2004). Geodesic motion in Kundt spacetimes and the character of the envelope singularity, *Class. Quantum Grav.*, **21**, 2811–2829. §18.4.1

Podolský, J. and Griffiths, J. B. (1997). Impulsive gravitational waves generated by null particles in de Sitter and anti-de Sitter backgrounds, *Phys. Rev. D*, **56**, 4756–4767. §§4.5, 20.3

Podolský, J. and Griffiths, J. B. (1998a). Impulsive waves in de Sitter and anti-de Sitter space-times generated by null particles with an arbitrary multipole structure, *Class. Quantum Grav.*, **15**, 453–463. §20.3

Podolský, J. and Griffiths, J. B. (1998b). Boosted static multipole particles as sources of impulsive gravitational waves, *Phys. Rev. D*, **58**, 124024, 1–5. §20.2

Podolský, J. and Griffiths, J. B. (1999a). Expanding impulsive gravitational waves, *Class. Quantum Grav.*, **16**, 2937–2946. §§20.4, 20.5

Podolský, J. and Griffiths, J. B. (1999b). Nonexpanding impulsive gravitational waves with an arbitrary cosmological constant, *Phys. Lett. A*, **261**, 1–4. §20.3

Podolský, J. and Griffiths, J. B. (2000). The collision and snapping of cosmic strings generating spherical impulsive gravitational waves, *Class. Quantum Grav.*, **17**, 1401–1413. §20.4

Podolský, J. and Griffiths, J. B. (2001a). Null limits of generalised Bonnor–Swaminarayan solutions, *Gen. Rel. Grav.*, **33**, 37–57. §§15.3.3, 20.4

Podolský, J. and Griffiths, J. B. (2001b). Null limits of the C-metric, *Gen. Rel. Grav.*, **33**, 59–64. §20.4

Podolský, J. and Griffiths, J. B. (2001c). Uniformly accelerating black holes in a de Sitter universe, *Phys. Rev. D*, **63**, 024006. §§4.3, 14.4, 16.1

Podolský, J. and Griffiths, J. B. (2004). A snapping cosmic string in a de Sitter or anti-de Sitter universe, *Class. Quantum Grav.*, **21**, 2537–2547. §20.5

Podolský, J. and Griffiths, J. B. (2006). Accelerating Kerr–Newman black holes in (anti-) de Sitter space-time, *Phys. Rev. D*, **73**, 044018, 1–5. §§14.4, 16.3.2

Podolský, J. and Kofroň, D. (2007). Chaotic motion in Kundt spacetimes, *Class. Quantum Grav.*, **24**, 3413–3424. §18.3.2

Podolský, J. and Ortaggio, M. (2003). Explicit Kundt type II and N solutions as gravitational waves in various type D and O universes, *Class. Quantum Grav.*, **20**, 1685–1701. §§7.1, 18.2, 18.6, 18.7

Podolský, J., Ortaggio, M. and Krtouš, P. (2003). Radiation from accelerated black holes in an anti-de Sitter universe, *Phys. Rev. D*, **68**, 124004. §§5.4, 14.4

Podolský, J. and Svítek, O. (2005). Radiative spacetimes approaching the Vaidya metric, *Phys. Rev. D*, **71**, 124001. §19.5.3

Podolský, J. and Veselý, K. (1998a). New examples of sandwich gravitational waves and their impulsive limit, *Czech. J. Phys.*, **48**, 871–878. §20.2

Podolský, J. and Veselý, K. (1998b). Continuous coordinates for all impulsive pp-waves, *Phys. Lett. A*, **241**, 145–147. §20.2

Podolský, J. and Veselý, K. (1998c). Chaos in pp-wave spacetimes, *Phys. Rev. D*, **58**, 081501. §17.6

Podolský, J. and Veselý, K. (1998d). Chaotic motion in pp-wave spacetimes, *Class. Quantum Grav.*, **15**, 3505–3521. §17.6

Podolský, J. and Veselý, K. (1999). Smearing of chaos in sandwich pp-waves, *Class. Quantum Grav.*, **16**, 3599–3618. §17.6

Poisson, E. and Israel, W. (1990). Internal structure of black holes, *Phys. Rev. D*, **41**, 1796–1809. §9.2.2

Polarski, D. (1989). The scalar wave equation on static de Sitter space and anti-de Sitter spaces, *Class. Quantum Grav.*, **6**, 893–900. §5.4

Pravda, V. (1999). Curvature invariants in type III spacetimes, *Class. Quantum Grav.*, **16**, 3321–3326. §19.1

Pravda, V. and Pravdová, A. (2000). Boost-rotation symmetric spacetimes – review, *Czech. J. Phys.*, **50**, 333–440. §14

Pravda, V. and Pravdová, A. (2002). On the spinning C-metric, in *Gravitation: following the Prague inspiration*, eds. O. Semerák, J. Podolský and M. Žofka, (World Scientific), 247–262. §§14.3, 15.1, 16.3.2

Pravda, V., Pravdová, A., Coley, A. and Milson, R. (2002). All spacetimes with vanishing curvature invariants, *Class. Quantum Grav.*, **19**, 6213-6236. §§18.3.2, 18.3.3

Pravda, V. and Zaslavskii, O. B. (2005). Curvature tensors on distorted Killing horizons and their algebraic classification, *Class. Quantum Grav.*, **22**, 5053-5071. §9.3

Pravdová, A. and Pravda, V. (2002). Boost-rotation symmetric vacuum spacetimes with spinning sources, *J. Math. Phys.*, **43**, 1536–1546. §15.1

Pretorius, F. and Israel, W. (1998). Quasi-spherical light cones of the Kerr geometry, *Class. Quantum Grav.*, **15**, 2289–2301. §11.1.5

Price, R. H. (1972). Nonspherical perturbations of relativistic gravitational collapse. I. Scalar and gravitational perturbations, *Phys. Rev. D*, **5**, 2419–2438. §8.2

Punsly, B. (1985). A physical interpretation of the Kerr solution, *J. Math. Phys.*, **26**, 1728–1739. §11.1.2

Punsly, B. (1990). The Kerr space-time near the ring singularity, *Gen. Rel. Grav.*, **22**, 1169–1206. §11.1.2

Quevedo, H. (1990). Multipole moments in general relativity – static and stationary vacuum solutions, *Fortschr. Phys.*, **38**, 733–840. §10.1.2

Ray, J. R. and Wei, M. S. (1977). A solution-generating theorem with applications in general relativity, *Nuovo Cimento B*, **42**, 151–164. §16.5

Raychaudhuri, A. K. (1953). Reine Strahlungsfelder mit Zentralsymmetrie in der Allgemeinen Relativitätstheorie, *Z. Phys.*, **135**, 225–231. §9.5

Raychaudhuri, A. K. and Maiti, S. R. (1979). Conformal flatness and the Schwarzschild interior solution, *J. Math. Phys.*, **20**, 245–246. §8.5

Rees, M. J. and Sciama, D. W. (1968). Large-scale density inhomogeneities in the Universe, *Nature*, **217**, 511–516. §8.6

Regge, T. and Wheeler, J. A. (1957). Stability of a Schwarzschild singularity, *Phys. Rev.*, **108**, 1063–1069. §8.2

Reissner, H. (1916). Über die Eigengravitation des elektrischen Feldes nach der Einsteinschen Theorie, *Ann. Physik*, **50** (355), 106–120. §9.2

Rendall, A. D. (1988). Existence and asymptotic properties of global solutions of the Robinson–Trautman equation, *Class. Quantum Grav.*, **5**, 1339–1347. §19.4

Rendall, A. D. (1992). Cosmic censorship and the Vlasov equation, *Class. Quantum Grav.*, **9**, L99–L104. §22.7

Rendall, A. D. and Weaver, M. (2001). Manufacture of Gowdy spacetimes with spikes, *Class. Quantum Grav.*, **18**, 2959–2975. §22.4

Rindler, W. (1956). Visual horizons in world-models, *Mon. Not. R. Astron. Soc.*, **116**, 662–677; *Gen. Rel. Grav.*, **34**, 133–153, (2002). §6.5

Rindler, W. (1966). Kruskal space and the uniformly accelerated frame, *Am. J. Phys.*, **34**, 1174–1178. §3.5

Ringström, H. (2004a). On Gowdy vacuum spacetimes, *Math. Proc. Camb. Phil. Soc.*, **136**, 485–512. §22.4

Ringström, H. (2004b). Asymptotic expansions close to the singularity in Gowdy spacetimes, *Class. Quant. Grav.*, **21**, S305–S322. §22.4

Robertson, H. P. (1929). On the foundations of relativistic cosmology, *Proc. Nat. Acad. Sci. USA*, **15**, 822–829. §6.1

Robertson, H. P. (1933). Relativistic cosmology, *Rev. Mod. Phys.*, **5**, 62–90. §6

Robertson, H. P. and Noonan T. W. (1968). *Relativity and cosmology*, (Saunders, Philadelphia). §6.3.1

Robinson, I. (1959). A solution of the Maxwell–Einstein equations, *Bull. Acad. Polon. Sci. Ser. Sci. Math. Astron. Phys.*, **7**, 351–352. §7.1

Robinson, I. (1961). Null electromagnetic fields, *J. Math. Phys.*, **2**, 290–291. §2.2.1

Robinson, I. and Trautman, A. (1960). Spherical gravitational waves, *Phys. Rev. Lett.*, **4**, 431–432. §19

Robinson, I. and Trautman, A. (1962). Some spherical gravitational waves in general relativity, *Proc. Roy. Soc. A*, **265**, 463–473. §19

Robson, E. H. (1973). Null hypersurfaces in general relativity theory, *Ann. Inst. Henri Poincaré*, **18**, 77–88. §§2.1.5, 20.6

Romans, L. J. (1992). Supersymmetric, cold and lukewarm black holes in cosmological Einstein–Maxwell theory, *Nucl. Phys. B*, **383**, 395–415. §9.6

Rooman, M. and Spindel, Ph. (1998). Gödel metric as a squashed anti-de Sitter geometry, *Class. Quant. Grav.*, **15**, 3241–3249. §22.6

Rosa, V. and Letelier, P. S. (2007). Stability of closed timelike curves in the Gödel universe, *Gen. Rel. Grav.*, **39**, 1419–1435. §22.6

Rosen, N. (1937). Plane polarised waves in the general theory of relativity, *Phys. Z. Sowjet.*, **12**, 366–372. §17.5

Rosen, N. (1954). Some cylindrical gravitational waves, *Bull. Res. Council Israel.*, **3**, 328–332. §22.3

Rosquist, K. and Jantzen, R. T. (1988). Unified regularization of Bianchi cosmology, *Phys. Rep.*, **166**, 89–124. §22.1

Ruback, P. (1988). A new uniqueness theorem for charged black holes, *Class. Quantum Grav.*, **5**, L155–L159. §9.2

Ruban, V. A. (1972). Nonsingular Taub–Newman–Unti–Tamburino metrics, *Dokl. Acad. Nauk CCCP*, **204**, 1085–1089: *Sov. Phys. Dokl.*, **17**, 568–571. §12.4

Ryan, M. P. and Shepley, L. C. (1975). *Homogeneous relativistic cosmologies*, (Princeton University Press). §§1, 12.2, 22.1, 22.2, 22.6

Sachs, R. (1961). Gravitational waves in general relativity. VI. The outgoing radiation condition, *Proc. Roy. Soc. A*, **264**, 309–338. §2.1.3

Sachs, R. K. (1962). Gravitational waves in general relativity. VIII. Waves in asymptotically flat space-time, *Proc. Roy. Soc. A*, **270**, 103–126. §2.1.3

Sackfield, A. (1971). Physical interpretation of the N.U.T. metric, *Proc. Camb. Phil. Soc.*, **70**, 89–94. §12.1.3

Salazar, H. I., García, A. and Plebański, J. F. (1983). Symmetries of the nontwisting type-N solutions with cosmological constant, *J. Math. Phys.*, **24**, 2191–2196. §18.3.2

Santos, N. O. (1993). Solution of the vacuum Einstein equations with non-zero cosmological constant for a stationary cylindrically symmetric spacetime, *Class. Quantum Grav.*, **10**, 2401–2406. §§13.1.3, 22.6

Schmidt, B. G. (1988). Existence of solutions of the Robinson–Trautman equation and spatial infinity, *Gen. Rel. Grav.*, **20**, 65–70. §19.4

Schmidt, H-J. (1993). On the de Sitter space-time – the geometric foundation of inflationary cosmology, *Fortschr. Phys.*, **41**, 179–199. §§4.4, 4.5

Schneider, P., Ehlers, J. and Falco, E. E. (1999). *Gravitational lenses*, (Springer). §8.1

Schneider, P., Kochanek, C. and Wambsgass, J. (2006). *Gravitational lensing: strong, weak and micro*, (Springer). §8.1

Schrödinger, E. (1956). *Expanding universes*, (Cambridge University Press). §4.5

Schücking, E. (1954). Das Schwarzschildsche Linienelement und die Expansion des Weltalls, *Z. Phys.*, **137**, 595–603. §8.6

Schwarzschild, K. (1916a). Über das Gravitationsfeld eines Massenpunktes nach der Einsteinschen Theorie, *Sitz. Preuss. Akad. Wiss. Berlin*, 189–196. §8.1 English translation: *Gen. Rel. Grav.*, **35**, 951–959, (2003).

Schwarzschild, K. (1916b). Über das Gravitationsfeldeiner Kugal aus inkompressibler Flüssigkeit nach der Einsteinschen Theorie, *Sitz. Preuss. Akad. Wiss. Berlin*, 424–434. §8.5

Scott, S. M. (1989). A survey of the Weyl metrics, *Proc. Centre Math. Anal. Austral. Nat. Univ.*, **19**, 175–195. §10.7

Scott, S. M. and Szekeres, P. (1986a). The Curzon singularity I: spatial sections, *Gen. Rel. Grav.*, **18**, 557–570. §10.5

Scott, S. M. and Szekeres, P. (1986b). The Curzon singularity II: global picture, *Gen. Rel. Grav.*, **18**, 571–583. §10.5

Semerák, O. (2002). Towards gravitating discs around stationary black holes, in *Gravitation: following the Prague inspiration*, eds. O. Semerák, J. Podolský and M. Žofka, (World Scientific), 111–160. §10.7

Semerák, O., Zellerin, T. and Žáček, M. (1999). The structure of superposed Weyl fields, *Mon. Not. R. Astron. Soc.*, **308**, 691–704. §10.7

Senovilla, J. M. M. (1998). Singularity theorems and their consequences, *Gen. Rel. Grav.*, **30**, 701–848. §6.4.2

Sfetsos, K. (1995). On gravitational shock waves in curved spacetimes, *Nucl. Phys. B*, **436**, 721–745. §§20.3, 20.6

Sharp, N. A. (1979). Geodesics in black hole space-times, *Gen. Rel. Grav.*, **10**, 659–670. §11.1.5

Siklos, S. T. C. (1976). Two completely singularity-free NUT space-times, *Phys. Lett. A*, **59**, 173. §12.3

Siklos, S. T. C. (1985). Lobatchevski plane gravitational waves, in *Galaxies, axisymmetric systems and relativity*, ed. M. A. H. MacCallum, (Cambridge University Press), 247–274. §§5.4, 18.2, 18.3

Silberstein, L. (1936). Two-centers solution of the gravitational field equations, and the need for a reformed theory of matter, *Phys. Rev.*, **49**, 268–270. §10.6

Simpson, M. and Penrose, R. (1973). Internal instability in a Reissner–Nordström black hole, *Int. J. Theoret. Phys.*, **7**, 183–197. §9.2.2

Singh, K. P. and Roy, S. R. (1966). Electromagnetic behviour in space-times conformal to some well-known empty space-times, *Proc. Nat. Inst. Sci. India*, **A32**, 223. §7.1

Singleton, D. (1990). On global existence and convergence of vacuum Robinson–Trautman solutions, *Class. Quant. Grav.*, **7**, 1333–1343. §19.4

Skea, J. E. F. (1997). The invariant classification of conformally flat pure radiation spacetimes, *Class. Quantum Grav.*, **14**, 2393–2404. §§18.2, 18.3.3

Sládek, P. and Finley III, J. D. (2010). Asymptotic properties of the C-metric, *Class. Quantum Grav.*, **27**, 205020. §14

Smarr, L. (1973). Surface geometry of charged rotating black holes, *Phys. Rev. D*, **7**, 289–295. §11.1

Smith, W. L. and Mann, R. B. (1997). Formation of topological black holes from gravitational collapse, *Phys. Rev. D*, **56**, 4942–4947. §9.6

Sod-Hoffs, J. and Rodchenko, E. D. (2007). On the properties of the Ernst–Manko–Ruiz equatorially antisymmetric solutions, *Class, Quantum Grav.*, **24**, 4617–4629. §13.3.2

Som, M. M., Teixeira, A. F. F. and Wolk, I. (1976). Stationary vacuum fields with cylindrical symmetry, *Gen. Rel. Grav.*, **7**, 263–267. §13.1.1

Stachel, J. J. (1966). Cylindrical gravitational news, *J. Math. Phys.*, **7**, 1321–1331. §22.3

Stachel, J. J. (1968). Structure of the Curzon metric, *Phys. Lett. A*, **27**, 60–61. §10.5

Stachel, J. (1982). Globally stationary but locally static space-times: A gravitational analog of the Aharonov-Bohm effect, *Phys. Rev. D*, **26**, 1281–1290. §13.1.1

Starobinskii, A. A. (1983). Isotropization of arbitrary cosmological expansion given an effective cosmological constant, *JETP Lett.*, **37**, 66–69. §19.4

Steadman, B. R. (1998). Confinement of null geodesics in the van Stockum exterior, *Class. Quantum Grav.*, **15**, 1357–1365. §13.1.2

Steadman, B. R. (2003). Causality violation on van Stockum geodesics, *Gen. Rel. Grav.*, **35**, 1721–1726. §13.1.2

Steinbauer, R. (1998). Geodesics and geodesic deviation for impulsive gravitational waves, *J. Math. Phys.*, **39**, 2201–2212. §20.2

Steinbauer, R. and Vickers, J. A. (2006). The use of generalised functions and distributions in general relativity, *Class. Quantum Grav.*, **23**, R91–R114. §3.4

Steinmüller, B., King, A. R. and Lasota, J. P. (1975). Radiating bodies and naked singularities, *Phys. Lett. A*, **51**, 191–192. §9.5.2

Stellmacher, K. (1938). Ausbreitungsgesetze für charakteristische Singularitäten der Gravitationsgleichungen, *Math. Annalen.*, **115**, 740–783. §§2.1.5, 20.6

Stephani, H. (1967). Konform flache Gravitationsfelder, *Commun. Math. Phys.*, **5**, 337–342. §7.1

Stephani, H., Kramer, D., MacCallum, M., Hoenselaers, C. and Herlt, E. (2003). *Exact solutions of Einstein's field equations, 2nd edition*, (Cambridge University Press). §§1, 2, 6.2, 7.1, 7.3, 8.5, 12.1.1, 13.2.2, 16.1, 17.1, 18, 19, 21, 22

Stewart, B. W., Papadopoulos, D., Witten, L., Berezdivin, R. and Herrera, L. (1982). An interior solution for the gamma metric, *Gen. Rel. Grav.*, **14**, 97–103. §10.4

Stewart, J. M. and Ellis, G. F. R. (1968). Solutions of Einstein's equations for a fluid which exhibit local rotational symmetry, *J. Math. Phys.*, **9**, 1072–1082. §22.2

Stuchlík, Z. and Hledík, S. (1999). Photon capture cones and embedding diagrams of the Ernst spacetime, *Class. Quantum Grav.*, **16**, 1377–1387. §9.3

Sussman, R. A. (1985). Conformal structure of a Schwarzschild black hole immersed in a Friedman universe, *Gen. Rel. Grav.*, **17**, 251–291. §8.6

Synge, J. L. (1957). A model in general relativity for an instantaneous transformation of a massive particle into radiation, *Proc. Roy. Irish Acad. A*, **59**, 1–13. §20.6

Szafron, D. A. (1977). Inhomogeneous cosmologies: New exact solutions and their evolution, *J. Math. Phys.*, **18**, 1673–1677. §22.8

Szafron, D. A. and Collins, C. B. (1979). A new approach to inhomogeneous cosmologies: Intrinsic symmetries. II. Conformally flat slices and an invariant classification, *J. Math. Phys.*, **20**, 2354–2361. §22.8

Szekeres, G. (1960). On the singularities of a Riemannian manifold, *Publ. Mat. Debrecen*, **7**, 285–301. §8.3

Szekeres, P. (1965). The gravitational compass, *J. Math. Phys.*, **6**, 1387–1391. §§2.1.3, 17.2

Szekeres, P. (1968). Multipole particles in equilibrium in general relativity, *Phys. Rev.*, **176**, 1446–1450. §10.7

Szekeres, P. (1972). Colliding plane gravitational waves, *J. Math. Phys.*, **13**, 286–294. §21.2

Szekeres, P. (1975a). A class of inhomogeneous cosmological models, *Commun. Math. Phys.*, **41**, 55–64. §22.8

Szekeres, P. (1975b). Quasi-spherical gravitational collapse, *Phys. Rev. D*, **12**, 2941–2948. §22.8

Szekeres, P. and Iyer, V. (1993). Spherically symmetric singularities and strong cosmic censorship, *Phys. Rev. D*, **47**, 4362–4371. §22.7

Szekeres, P. and Morgan, F. H. (1973). Extensions of the Curzon metric, *Commun. Math. Phys.*, **32**, 313–318. §10.5

Takeno, H. (1961a). On the space-time of Peres, *Tensor, N S*, **11**, 99–109. §17.1

Takeno, H. (1961b). The mathematical theory of plane gravitational waves in general relativity, *Sci. Rep. Res. Inst. Theor. Phys. Hiroshima Univ.*, **no. 1**. §17.1

Tanimoto, M. (1998). New varieties of Gowdy space-times, *J. Math. Phys.*, **39**, 4891–4898. §22.4

Tanimoto, M. and Nambu, Y. (2007). Luminosity distance–redshift relation for the LTB solution near the centre, *Class. Quantum Grav.*, **24**, 3843–3857. §22.7

Tariq, N. and Tupper, B. O. J. (1974). The uniqueness of the Bertotti–Robinson electromagnetic universe, *J. Math. Phys.*, **15**, 2232–2235. §7.1

Taub, A. H. (1951). Empty space-times admitting a three-parameter group of motions, *Ann. Math.*, **53**, 472–490. [*Gen. Rel. Grav.*, **36**, 2689–2719, (2004).] §§9.1.2, 10.2, 10.3, 12, 22.1

Taub, A. H. (1980). Space-times with distribution-valued curvature tensors, *J. Math. Phys.*, **21**, 1423–1431. §§2.1.5, 20.6

Tauber, G. E. (1967). Expanding universe in conformally flat coordinates, *J. Math. Phys.*, **8**, 118–123. §6.3.2

Taylor, J. P. W. (2005). Unravelling directional singularities, *Class. Quantum Grav.*, **22**, 4961–4971. §§10.5, 10.6

Thorne, K. S. (1965a). Energy of infinitely long, cylindrically symmetric systems in general relativity, *Phys. Rev. B*, **138**, 251–266. §22.3

Thorne, K. S. (1965b). Absolute stability of Melvin's magnetic universe, *Phys. Rev. B*, **139**, 244–254. §7.3

Thorne, K. S. (1967). Primordial element formation, primordial magnetic fields, and the isotropy of the universe, *Astrophys. J.*, **148**, 51–68. §22.2

Thorne, K. S. (1975). A toroidal solution of the vacuum Einstein field equations, *J. Math. Phys.*, **16**, 1860–1865. §10.7

Tian, Q. (1986). Cosmic strings with cosmological constant, *Phys. Rev. D*, **33**, 3549–3555. §10.2

Tipler, F. J. (1974). Rotating cylinders and the possibility of global causality violation, *Phys. Rev. D*, **9**, 2203–2206. §§13.1.1, 13.1.2

Tipler, F. J. (1980). Singularities from colliding plane gravitational waves, *Phys. Rev. D*, **22**, 2929–2932. §21.4

Tipler, F. J. (1986). Penrose diagrams for the Einstein, Eddington–Lemaitre, Eddington–Lemaitre–Bondi, and anti-de Sitter universes, *J. Math. Phys.*, **27**, 559–561. §5.2

Tipler, F. J., Clarke, C. S. J. and Ellis, G. F. R. (1980). Singularities and horizons – A review article, in *General relativity and gravitation, one hundred years after the birth of Albert Einstein*, vol. 2, ed. A. Held, (Plenum), 97–206. §2.3.1

Tod, K. P. (1989). Analogues of the past horizon in the Robinson–Trautman metrics, *Class. Quantum Grav.*, **6**, 1159–1163. §19.4

Tolman, R. C. (1931). On the theoretical requirements for a periodic behaviour of the universe, *Phys. Rev.*, **38**, 1758–1771. §6.3.2

Tolman, R. C. (1934a). Effect of inhomogeneity on cosmological models, *Proc. Nat. Acad. Sci. US*, **20**, 169–176. §22.7

Tolman, R. C. (1934b). *Relativity, thermodynamics and cosmology*, (Oxford University Press). §6.3.2

Tomimatsu, A. (1989). The gravitational Faraday rotation for cylindrical gravitational solitons, *Gen. Rel. Grav.*, **21**, 613–621. §22.3

Tomimatsu, A. and Sato, H. (1972). New exact solution for the gravitational field of a spinning mass, *Phys. Rev. Lett.*, **29**, 1344–1345. §§10.4, 13.2.2

Tomimatsu, A. and Sato, H. (1973). New series of exact solutions for gravitational fields of spinning masses, *Prog. Theor. Phys.*, **50**, 95–110. §§10.4, 13.2.2

Tomimura, N. (1978). Evolution of inhomogeneous plane-symmetric cosmological models, *Nuovo Cimento B*, **44**, 372–380. §22.7

Tomita, K. (1975). Evolution of irregularities in a chaotic early universe, *Prog. Theoret. Phys.*, **54**, 730–739. §22.7

Tomita, K. (1992). Density perturbations driven by the irregular spatial curvature in inhomogeneous cosmological models, *Astrophys. J.*, **394**, 401–408. §22.7

Trefftz, E. (1922). Das statische Gravitationsfeld zweier Massenpunkte in der Einsteinschen Theorie, *Mathem. Ann.*, **86**, 317–326. §9.4

Unnikrishnan, C. S. (1994). Naked singularities in spherically symmetric gravitational collapse: a critique, *Gen. Rel. Grav.*, **26**, 655–662. §22.7

Urbantke, H. (1979). On rotating plane-fronted waves and their Poincaré-invariant differential geometry, *J. Math. Phys.*, **20**, 1851–1860. §18.4

Vaidya, P. C. (1943). The external field of a radiating star in general relativity, *Current Sci. (India)*, **12**, 183. §9.5

Vaidya, P. C. (1951a). The gravitational field of a radiating star, *Proc. Indian Acad. Sci.*, **33**, 264–276. §9.5

Vaidya, P. C. (1951b). Nonstatic solutions of Einstein's field equations for spheres of fluids radiating energy, *Phys. Rev.*, **83**, 10–17. §9.5

Vaidya, P. C. (1953). "Newtonian" time in general relativity, *Nature*, **171**, 260–261. §9.5

Vaidya, P. C. (1955). The general relativity field of a radiating star, *Bull. Calcutta Math. Soc.*, **47**, 77–80. §9.5

Vajk, J. P. (1969). Exact Robertson–Walker cosmological solutions containing relativistic fluids, *J. Math. Phys.*, **10**, 1145–1151. §6.3.2

Vajk, J. P. and Eltgroth, P. G. (1970). Spatially homogeneous anisotropic cosmological models containing relativistic fluid and magnetic field, *J. Math. Phys.*, **11**, 2212–2222. §22.2

Van den Bergh, N., Gunzig, E. and Nardone, P. (1990). Exact radiative solutions for conformally invariant fields on a conformally flat background, *Class. Quantum Grav.*, **7**, L175–L179. §§18.2, 18.3.3

Van den Bergh, N. and Wils, P. (1985). The rotation axis for stationary and axisymmetric space-times, *Class. Quantum Grav.*, **2**, 229–240. §10.1

van Stockum, W. J. (1937). The gravitational field of a distribution of particles rotating about an axis of symmetry, *Proc. Roy. Soc. Edinburgh A*, **57**, 135–154. §§13.1.2, 22.6

Vanderveld, R. A., Flanagan, É. É. and Wasserman, I. (2006). Mimicking dark energy with Lemaître-Tolman-Bondi models: Weak central singularities and critical points, *Phys. Rev. D*, **74**, 023506. §22.7

Vandyck, M. A. J. (1987). On the time evolution of some Robinson–Trautman solutions: II, *Class. Quantum Grav.*, **4**, 759–768. §19.4

Vanzo, L. (1997). Black holes with unusual topology, *Phys. Rev. D*, **56**, 6475–6483. §9.6

Vargas Moniz, P. (1993). Kantowski–Sachs universes and the cosmic no-hair conjecture, *Phys. Rev. D*, **47**, 4315–4321. §22.2

Vilenkin, A. (1985). Cosmic strings and domain walls, *Phys. Rep.*, **121**, 263–315. §3.4

Vilenkin, A. and Shellard, P. (1994). *Cosmic strings and other topological defects*, (Cambridge University Press). §3.4

Vishveshwara, C. V. (1970). Stability of the Schwarzschild metric, *Phys. Rev. D*, **1**, 2870–2879. §8.2

Vishveshwara, C. V. and Winicour, J. (1977). Relativistically rotating dust cylinders, *J. Math. Phys.*, **18**, 1280–1284. §13.1.3

Visser, M. (2008). The Kerr spacetime: A brief introduction, in *The Kerr spacetime: rotating black holes in general relativity*, eds. D. L. Wiltshire, M. Visser and S. M. Scott, (Cambridge University Press). §11.1

von der Gönna, U. and Kramer, D. (1998). Pure and gravitational radiation, *Class. Quantum Grav.*, **15**, 215–223. §§19.5.2, 19.5.4

Voorhees, B. H. (1970). Static axially symmetric gravitational fields, *Phys. Rev. D*, **2**, 2119–2122. §10.4

Wainwright, J. and Andrews, S. (2009). The dynamics of Lemaître–Tolman cosmologies, *Class. Quantum Grav.*, **26**, 085017. §22.7

Wald, R. M. (1979). Note on the stability of the Schwarzschild metric, *J. Math. Phys.*, **20**, 1056–1058. §8.2

Wald, R. M. (1983). Asymptotic behavior of homogeneous cosmological models in the presence of a positive cosmological constant, *Phys. Rev. D*, **28**, 2118–2120. §19.4

Wald, R. M. (1984). *General relativity*, (University of Chicago Press). §§2.3.2, 6.5, 8.4, 9.5

Walker, A. G. (1935). On Riemannian spaces with spherical symmetry about a line, and the conditions for isotropy in general relativity, *Q. J. Math.*, **6**, 81–93. §6.1

Walker, M. (1970). Block diagrams and the extension of timelike two-surfaces, *J. Math. Phys.*, **11**, 2280–2286. §9.2.2

Ward, J. P. (1973). Equilibrium of charged, spinning, magnetic particles, *Commun. Math. Phys.*, **34**, 123–130. §13.3.2

Waugh, B. and Lake, K. (1986). Double-null coordinates for the Vaidya metric, *Phys. Rev. D*, **34**, 2978–2984. §9.5.2

Weber, E. (1984). Kantowski–Sachs cosmological models approaching isotropy, *J. Math. Phys.*, **25**, 3279–3285. §22.2

Weber, J. and Wheeler, J. A. (1957). Reality of the cylindrical gravitational waves of Einstein and Rosen, *Rev. Mod. Phys.*, **29**, 509–515. §22.3

Weinberg, S. (1972). *Gravitation and cosmology: Principles and applications of the general theory of relativity*, (Wiley). §§4.5, 6.5

Weinberg, S. (2008). *Cosmology*, (Oxford University Press). §6.5

Weir, G. J. and Kerr, R. P. (1977). Diverging type-D metrics, *Proc. Roy. Soc. A*, **355**, 31–52. §16

Weyl, H. (1917). Zur Gravitationstheorie, *Ann. Physik*, **54** (359), 117–145. §§9.2, 10, 10.8

Weyl, H. (1919a). Bemerkung über die axisymmetrischen Lösungen der Einsteinschen gravitationsgleichungen, *Ann. Physik*, **59** (364), 185–188. §14

Weyl, H. (1919b). Über die statischen, kugelsymmetrischen Lösungen von Einsteins „kosmologischen" Gravitationsgleichungen, *Phys. Z.*, **20**, 31–34. §9.4

Wheeler, J. A. (1980). The beam and stay of the Taub universe, in *Essays in general relativity*, ed. F. J. Tipler, (Academic Press, New York). §12.1.2

Wild, W. J. and Kerns, R. M. (1980). Surface geometry of a black hole in a magnetic field, *Phys. Rev. D*, **21**, 332–335. §9.3

Wild, W. J., Kerns, R. M. and Drish, W. F. (1981). Stretching a black hole, *Phys. Rev. D*, **23**, 829–831. §9.3

Wilkins, D. C. (1972). Bound geodesics in the Kerr metric, *Phys. Rev. D*, **5**, 814–822. §11.1.5

Will, C. M. (1993). *Theory and experiment in gravitational physics*, (Cambridge University Press), 2nd edition. §8.1

Will, C. M. (2006). The confrontation between general relativity and experiment, *Living Revs. in Rel.*, lrr-2006-3. §8.1

Wils, P. (1989). Homogeneous and conformally Ricci flat pure radiation fields, *Class. Quantum Grav.*, **6**, 1243–1251. §§18.2, 18.3.3

Wils, P. and Van den Bergh, N. (1990). Petrov type D pure radiation fields of Kundt's class, *Class. Quantum Grav.*, **7**, 577–580. §§18.2, 18.6

Wiltshire, D. L., Visser, M. and Scott, S. M. eds. (2008). *The Kerr spacetime: rotating black holes in general relativity*, (Cambridge University Press). §11.1

Yamazaki, M. (1977). On the Kerr–Tomimatsu–Sato family of spinning mass solutions, *J. Math. Phys.*, **18**, 2502–2508. §13.2.2

Yamazaki, M. (1983). Stationary line of N Kerr masses kept apart by gravitational spin-spin interaction, *Phys. Rev. Lett.*, **50**, 1027–1030. §13.2.2

Yodzis, P., Seifert, H.-J. and Müller zum Hagen, H. (1973). On the occurrence of naked singularities in general relativity, *Commun. Math. Phys.*, **34**, 135–148. §22.7

York, J. W. (1972). Role of conformal three-geometry in the dynamics of gravitation, *Phys. Rev. Lett.*, **28**, 1082–1085. §22.8

Yurtsever, U. (1987). Instability of Killing–Cauchy horizons in plane-symmetric spacetimes, *Phys. Rev. D*, **36**, 1662–1672. §21.7

Yurtsever, U. (1988c). Structure of the singularities produced by colliding plane waves, *Phys. Rev. D*, **38**, 1706–1730. §21.2

Yurtsever, U. (1989). Singularities and horizons in the collision of gravitational waves, *Phys. Rev. D*, **40**, 329–359. §21.7

Yurtsever, U. (1995). Geometry of chaos in the two-center problem in general relativity, *Phys. Rev. D*, **52**, 3176–3183. §22.5

Zakharov, V. D. (1973). *Gravitational waves in Einstein's theory of gravitation*, (Halsted Press, New York). §17.1

Zakhary, E. and McIntosh, C. B. G. (1997). A complete set of Riemann invariants, *Gen. Rel. Grav.*, **29**, 539–581. §2.1.2

Zerilli, F. J. (1970a). Effective potential for even-parity Regge-Wheeler gravitational perturbation equations, *Phys. Rev. Lett.*, **24**, 737–738. §8.2

Zerilli, F. J. (1970b). Gravitational field of a particle falling in a Schwarzschild geometry analyzed in tensor harmonics, *Phys. Rev. D*, **2**, 2141–2160. §8.2

Zhao, Z., Yang, C-Q. and Ren, Q. (1994). Hawking effect in Vaidya–de Sitter space-time, *Gen. Rel. Grav.*, **26**, 1055–1065. §9.5.1

Zimmerman, R. L. and Shahir, B. Y. (1989). Geodesics for the NUT metric and gravitational monopoles, *Gen. Rel. Grav.*, **21**, 821–848. §12.1.3

Zipoy, D. M. (1966). Topology of some spheroidal metrics, *J. Math. Phys.*, **7**, 1137–1143. §10.4

Žofka, M. and Bičák, J. (2008). Cylindrical spacetimes with a cosmological constant and their sources, *Class. Quantum Grav.*, **25**, 015011. §10.2

Index

Printed in the United States
By Bookmasters